Advances in Carbohydrate Chemistry and Biochemistry

Editor

R. STUART TIPSON

Associate Editor

DEREK HORTON

Board of Advisors

W. W. PIGMAN   S. ROSEMAN   WILLIAM J. WHELAN   ROY L. WHISTLER

Board of Advisors for the British Commonwealth

A. B. FOSTER   SIR EDMUND HIRST   J. K. N. JONES   MAURICE STACEY

Volume 26

ACADEMIC PRESS   New York and London   1971

ACADEMIC PRESS, INC.
111 Fifth Avenue, New York, New York 10003

United Kingdom Edition published by
ACADEMIC PRESS, INC. (LONDON) LTD.
24/28 Oval Road, London NW1 7DD

LIBRARY OF CONGRESS CATALOG CARD NUMBER: 45-11351

PRINTED IN THE UNITED STATES OF AMERICA

CONTENTS

Melville Lawrence Wolfrom (1900–1969)

DEREK HORTON

Conformational Analysis of Sugars and Their Derivatives

PHILIPPE L. DURETTE AND DEREK HORTON

Cyclic Acyloxonium Ions in Carbohydrate Chemistry

HANS PAULSEN

Cyclic Acetals of Ketoses

ROBERT F. BRADY, JR.

Tables of the Properties of Deoxy Sugars and Their Simple Derivatives

ROGER F. BUTTERWORTH AND STEPHEN HANESSIAN

Morphology and Biogenesis of Cellulose and Plant Cell-walls

F. SHAFIZADEH AND G. D. MCGINNIS

Biosynthesis of Saccharides from Glycopyranosyl Esters of Nucleoside Pyrophosphates ("Sugar Nucleotides")

H. NIKAIDO AND W. Z. HASSID

LIST OF CONTRIBUTORS

Numbers in parentheses indicate the pages on which the authors' contributions begin.

ROBERT F. BRADY, JR., *Institute for Materials Research, National Bureau of Standards, Washington, D.C. 20234* (197)

ROGER F. BUTTERWORTH, *Department of Chemistry, University of Montreal, Montreal 101, Quebec, Canada* (279)

PHILIPPE L. DURETTE, *Department of Chemistry, The Ohio State University, Columbus, Ohio 43210* (49)

STEPHEN HANESSIAN, *Department of Chemistry, University of Montreal, Montreal 101, Quebec, Canada* (279)

W. Z. HASSID, *Department of Bacteriology and Immunology and Department of Biochemistry, University of California, Berkeley, California 94720* (351)

DEREK HORTON, *Department of Chemistry, The Ohio State University, Columbus, Ohio 43210* (1, 49)

G. D. McGINNIS, *Wood Chemistry Laboratory, School of Forestry and Department of Chemistry, University of Montana, Missoula, Montana 59801* (297)

H. NIKAIDO, *Department of Bacteriology and Immunology and Department of Biochemistry, University of California, Berkeley, California 94720* (351)

HANS PAULSEN, *Institut für Organische Chemie und Biochemie, Universität Hamburg, Bundesrepublik Deutschland* (127)

F. SHAFIZADEH, *Wood Chemistry Laboratory, School of Forestry and Department of Chemistry, University of Montana, Missoula, Montana 59801* (297)

PREFACE

In this volume, the twenty-sixth of *Advances in Carbohydrate Chemistry and Biochemistry*, is presented an article on conformational analysis of sugars and their derivatives by Durette and Horton (Columbus), who have brought into focus many of the recent advances in this field. In a discussion of cyclic acyloxonium ions in carbohydrate chemistry, Paulsen (Hamburg) provides many striking illustrations of stable ions long considered only as transient reaction-intermediates. Their transformations furnish fascinating examples of the way in which the sugars can be used in furthering our basic knowledge of the stereochemical control of organic reactions. Brady (Washington) offers an up-to-date account of cyclic acetals of ketoses that complements earlier articles on cyclic acetals of hexitols, pentitols, and tetritols (Barker and Bourne, Vol. 7) and of the aldoses and aldosides (de Belder, Vol. 20). Butterworth and Hanessian (Montreal) have compiled useful tables of the properties of deoxy sugars and their simple derivatives; these tables are intended for use in conjunction with Hanessian's earlier chapter on deoxy sugars in Vol. 21. The biological elaboration of complex saccharides receives in-depth treatment in two articles in the present volume. Shafizadeh and McGinnis (Missoula) have contributed an authoritative treatment of the morphology and biogenesis of cellulose and plant cell-walls, and a "companion piece" by Nikaido and Hassid (Berkeley) is a masterly account of the biosynthesis of sacharides from glycopyranosyl esters of nucleoside pyrophosphates. An obituary of Melville L. Wolfrom has been written by Horton, a colleague of Wolfrom's at The Ohio State University. Because of the unusual significance of Wolfrom's contributions to carbohydrate chemistry, a list of his publications has been provided that will, we trust, prove useful to our readers.

The Subject Index was prepared by Dr. L. T. Capell.

Kensington, Maryland
Columbus, Ohio
October, 1971

R. STUART TIPSON
DEREK HORTON

Advances in
Carbohydrate Chemistry and Biochemistry

Volume 26

MELVILLE LAWRENCE WOLFROM

1900–1969

Melville Wolfrom was born on April 2, 1900, in Bellevue, Ohio, the youngest in a family of nine children. His father Friedrich (Frederick) Wolfrom was fifty-two at the time, and, in turn, his paternal grandfather Johann Lorenz Wolfrum had been sixty-five years of age when Friedrich was born. Johann Lorenz Wolfrum (1783–1856) was a native of the Sudeten German border town of Asch in Bohemia (now Czechoslovakia), where he worked as a master weaver, and the son Johannes of his first marriage founded a textile-manufacturing concern in Asch. Johann's second wife, Elisabeth Raab (1812–1904), bore him a daughter Ernestina in 1845 and a son Friedrich in 1848; both children were born in Asch. The family in Asch were German-speaking and of the Lutheran faith.

In 1854, Johann Lorenz Wolfrum, his wife, and their two children emigrated to America. The sailing-ship voyage from Hamburg to New York took fifty-four days. The family settled in a log cabin near Weaver's Corners, Sherman Township, Huron County, Ohio, where Lorenz attempted to make a living as a weaver. He died in 1856, leaving his widow and children practically destitute.

Friedrich Wolfrum attended the county schools, and worked from an early age to help support the family. For many years he was in the dry-goods business, in a store in nearby Bellevue, Ohio. Later, he worked as secretary-treasurer of a local telephone company, organized by his brother-in-law, Frank A. Knapp, that was the forerunner of the Ohio Northern Telephone Company, at present the largest independent telephone company in the State of Ohio. Friedrich's skill as a bookkeeper was an important factor in the success of the company. He was one of the principal members of the Bellevue Lutheran Church. A staunch Democrat, he held several local offices, including membership of the town Board of Education. In 1878, he married Maria Louisa Sutter, the eldest daughter of the Rev. John Jakob Sutter, a Lutheran minister who, at the age of thirteen, had accompanied his parents in emigrating to America from their native Switzerland. The Rev. Sutter, who was originally apprenticed in New York City as a shoemaker, had entered the new Lutheran seminary in Columbus, Ohio, and, after ordination, was called to a charge at Sugar Grove, near Lancaster, Ohio, where he served until accepting a charge near Bellevue, Ohio, in 1874. The Rev. J. J. Sutter was the first of four successive generations of Lutheran ministers in the Sutter family. The minister's daughter, Melville Wolfrom's mother, grew up in a bilingual family where German was the preferred language at home.

Some time before his marriage, Friedrich Wolfrum had anglicized the spelling of his name to Frederick Wolfrom. The Bellevue community at the turn of the century was principally a farmers' trading center and a railroad town, and its five thousand inhabitants were a mixture of New England stock, the German immigrants of the mid-nineteenth century, and later groups of Irish immigrants.

Melville was the last to be born in a family of five brothers and four sisters, and was the only one of the immediate family who was not taught German in the home. His father died when the youngest son was only seven years old, although Melville's mother lived to the age of ninety-three. From an early age, the boy was instilled with the need for self-reliance, to depend entirely on his own efforts, and not to expect any support from the family. When still quite young, he went out to earn money from odd jobs, especially during the summers. His mother had a great respect for music and good books, and stimulated, in her youngest son especially, an interest in serious reading. Melville received a very strict and orthodox Lutheran teaching, and, although in later years he did not adhere to this strict religious background, he consistently advocated some type of formal religious training for his children during their formative years.

In his early 'teens, Melville became deeply involved in a small manufacturing business maintained in their home by his brother Ralph. The three oldest brothers in the family had bought the patent on a type of horse harness-snap that was used successfully by several fire departments. After school each day, on Saturdays and holidays, and throughout the summer vacations, Melville worked on the production of these harness snaps, and received a remuneration of ten to fifteen cents an hour for his labors. During this time, he was always trying out new ideas and experiments to improve the devices, and, as a result of these experiences, determined to become a college graduate engineer with a view to a career in manufacturing experimentation. His second eldest brother, Carl, was an engineer; he graduated from the University of Michigan with the aid of funds loaned by the electric power company where he worked. This brother eventually became manager of the Salt Lake City district of the Utah Power and Light Company.

While in high school, Melville joined the town YMCA and took time off one summer to attend a camp; this was the beginning of an enduring interest in camping and the outdoors. He also took time from the shop during his senior year to play tackle on the high-school football team, but found little natural aptitude or enthusiasm for such participation, although in later years he found much enjoyment in watching The Ohio State University football team in action.

Like all of the members of the family who attended the Bellevue schools, Melville was a good student, and he graduated second in his high-school class of 1917. Stimulating teachers helped him develop an early interest in nature study, in fine art, in mathematics, and in German. His first encounter with science was in high school, where he learned physical geography and botany from a Mr. S. A. Kurtz, with whom he worked personally in conducting experimental observations. Later, he was much influenced by Mr. W. A. Hammond, with whom he studied chemistry and, later, physics. The chemistry text used was written by McPherson and Henderson of The Ohio State University, and Mr. Hammond had completed an M. Sc. degree in analytical chemistry at that university under Professor C. W. Foulk. Hammond's influence was sufficient to make Melville decide to become a chemist, or, more specifically, a chemical engineer; the more "practical" aspect of the applied field was appealing as a result of his earlier experience in the workshop.

Upon graduation from high school, there was no way in which Melville could enter college on his own resources, and, with no family help forthcoming, he decided to work for a year to save some money. He obtained a position in the nearby town of Fremont, in the laboratories of the National Carbon Company, a firm manufacturing wet and dry batteries. There, in the works laboratory, he tested the quality of the daily production. Within six months he was placed, at the age of seventeen, in charge of the laboratory, with about six persons under him. Here he conducted his first research project, an evaluation of the physical properties of carbon dry-cell electrodes as a function of the conditions used in baking the electrodes. During the winter, he took an evening course in qualitative, inorganic analysis, given at the plant. Early the following summer, he resigned his post to go to Cleveland, with the idea of entering Western Reserve University in the autumn and earning his board by waiting at table in a boarding house in the university district. He worked during the summer at the boarding house and also at a variety of jobs in factories and laboratories in Cleveland that were busy at that time with war production. That autumn, the Government established the Students Army Training Corps, and Melville entered the naval unit at Western Reserve. The prescribed course of study included physics, but no chemistry. The courses were uninspiring, much disorganization resulted from the influenza epidemic, and he disliked the barracks life and the snobbish fraternity system of the school. The armistice came in November,

1918, and after being mustered out at Christmas, he was disgusted and returned home to Bellevue.

Encouraged by his eldest brother Elmer, Melville went to New York City to assist Elmer in his work as the eastern advertising representative for a trade paper. The work involved travelling to outlying parts of the territory, but Melville found selling not to his liking, although he valued the experience. In the autumn of 1919, he entered Washington Square College of New York University to embark on an arts course. The college unit was new and was not operating well; no chemistry was offered; the whole tenor of the post-war period was unrest and discontent, and Melville heartily disliked the entire situation. Again, he gave up his studies and returned to Bellevue, where he felt regarded as a disgrace, a misfit, and a real problem. Still interested in chemical engineering, he had been consistently discouraged from entering this field by his eldest brother, although one of his temporary employers in New York had influenced him strongly with the advice that persons should bend all their efforts in following their true interests, no matter wheresoever they might lead.

During the spring and summer of 1920, he worked in Bellevue as a laborer in the engine roundhouse of the Nickel Plate Railroad, as a factory hand at the Ohio Cultivator Company, and as a bookkeeper for an unsuccessful business venture of his brother Ralph's. This brother was a competent accountant, and Melville, adept at such work, received good training; the meticulous records that he kept throughout his career show the influence of the several members of his immediate family who at one time or another worked as accountants or bookkeepers. Ralph had, in fact, succeeded his father, Frederick Wolfrom, in the telephone company position; later in life, he joined with his brother Elmer in the small-town newspaper-publishing business, where they were very successful. Melville enjoyed the experience of work in the roundhouse, and developed a great respect for skilled and unskilled labor alike, but, at the Cultivator Company, he had to fit in with the spirit of the factory hands and was forced to loaf along with them; this he disliked.

Finally, in the autumn of 1920, he entered The Ohio State University in Columbus, and embarked on a course in chemical engineering, an endeavor which at last held his attention and interest, and which he enjoyed greatly. For his board, he worked at boarding houses, restaurants, and cafeterias, as waiter, counter man, and dishwasher; he preferred the last kind of work. He also spent part of the time as a store clerk, but, in general, he always preferred working with material things rather than with people; this preference was evident throughout his life, although he had an unexpectedly perceptive insight into the character of those people he got to know.

The first encounter that the young Wolfrom had with the chemistry of carbohydrates came at the end of his sophomore year, when Professor C. W. Foulk recommended him for a post of student research assistant to Professor William Lloyd Evans of the Department of Chemistry. The stipend was $250 per year, and Wolfrom put in all of the extra time that he had on the work. During his junior year, he carried out quantitative oxidations of maltose with permanganate at various temperatures and concentrations of alkali. In his senior year, he attempted unsuccessfully to synthesize amino acid esters of glycerol. None of this work was published, but it was a good introduction to chemical research. Professor Evans, a student of J. U. Nef's, was very research-minded and inspirational.

During every summer of his college career, Wolfrom worked at Gypsum, Ohio, with his high-school chemistry teacher, W. A. Hammond, who was plant chemist for that installation of the United States Gypsum Company. Hammond, who had served in the army, was an analytical chemist during World War I, had entered industry after the war, and subsequently went into business for himself. Later, Hammond completed his Ph. D. degree in chemical engineering at The Ohio State University with a thesis on the essentially non-existent vapor pressure of the soluble anhydrite form of calcium sulfate. He then took out a patent on this substance, and made a fortune by manufacturing and marketing the material as the desiccant Drierite. Wolfrom's services with Hammond consisted of a combination of analytical and development work. He carried out a very precise determination of the effect of the concentration of hydrochloric acid on the precipitation of barium sulfate in the presence of calcium ions. The work yielded data worthy of publication, but none of it was submitted.

The well-to-do uncle, Frank A. Knapp, impressed by his nephew's progress, offered to loan any amount of money needed for Melville to complete his college work, but Melville's mother sternly forbade him to take advantage of this offer, as it did not fit into the scheme of Spartan training that his mother wished to give him.

Both Hammond and Evans wanted Wolfrom to drop engineering and do graduate work in chemistry. The engineering schedule in the senior year was such that there would have been no time to continue the work with Professor Evans, but, with the connivance of William E. Henderson, then Dean of the College of Arts, Wolfrom transferred from engineering to arts, and thus found time to continue his work with Professor Evans. He received the A. B. degree (cum laude) in 1924. During his stay at Ohio State, he had been exceptionally stimulated by Professor Evans; he also held in high regard William McPherson for his development of organic chemistry through the historical approach, C. W. Foulk for his teaching of analytical chemistry and his precision in the use of English, and E. Mack, Jr., from whom he learned physical chemistry. Other courses that were strongly to his liking included crystallography and mineralogy, zoology, and engineering drawing. The influence of Professor Evans and the other inspirational teachers of his undergraduate days was to endure throughout his career; the broad interdisciplinary approach that he took to research, and the insistence upon careful observation, clear expression, and historical accuracy, can all be traced to the early roots of his undergraduate training.

Following graduation from The Ohio State University, Wolfrom moved, at Professor Evans' suggestion, to Northwestern University in Chicago, to carry out his graduate work under Professor W. Lee Lewis, a Nef student. Just at that time, Lewis had resigned the Chairmanship of the Department of Chemistry to accept an industrial position, but his successor, Frank C. Whitmore, had persuaded Lewis to continue as a part-time Research Professor. About once a month, Lewis would spend Saturday afternoon with his students, and, between times, they would solve their own problems; Professor Whitmore saw to it that the students were working, but paid no attention to the results. Wolfrom appreciated this system as good training in independence of action, and it worked out well, as the problems assigned by Lewis were well designed and fruitful. Whitmore was a rank disbeliever in the lecture system in advanced teaching—Wolfrom had only one course in advanced organic chemistry and two seminars; for the rest, he read books. He began research immediately, and worked on it night and day, completing the M. Sc. degree in 1925, and the Ph. D. in 1927. His problem was to provide experimental evidence for the enediol theory advanced by Wohl and Neuberg to explain the Lobry de Bruyn–Alberda van Ekenstein interconversion of sugars in alkaline media. He observed that 2,3,4,6-tetra-O-methyl-D-glucose could be equilibrated with the D-*manno* epimer in aqueous alkali, and that there was no loss of the 2-O-methyl group and no formation of keto sugars. The result pointed to an enediol intermediate common to the two methylated sugars, and showed that the mechanism of enol formation was not one of selective hydration and dehydration, as had been suggested by Nef, but rather was consistent with a simple keto–enol tautomerism. This work, published with Lewis in 1928, was the first in what was to become Wolfrom's remarkably prolific output of research papers on the sugars, extending over more than four decades and numbering more than five hundred individual reports. He had a phenomenal memory for detail from his early work; forty years after his paper with Lewis was published, he could still describe it in exact detail without preparation, even to remembering the values of some of the physical constants.

At Northwestern, Wolfrom held an unusual teaching post provided by the association of fire-insurance underwriters, who sponsored a technical course to selected scholarship holders at the university. The program included a course in chemistry, and Wolfrom taught this course in the laboratories of the nearby College of Dentistry, holding the rank of assistant instructor and receiving the rather good annual stipend, for the time, of $1,800.

In 1926, he was married to Agnes Louise Thompson, of Auburn, Indiana. She had been trained at Depauw University in Greencastle, Indiana, as a public-school music teacher, and later did advanced work at Northwestern University, where she met her future husband. Throughout their married life, she continued to be involved in music teaching and in musical activities in the community, and was ever a sympathetic and stimulating helpmeet to her husband.

After graduating from Northwestern, Wolfrom received a National Research Council Fellowship that enabled him to undertake a period of postdoctoral study with some of the leading investigators in his field of interest. First of all, he went to study with Claude S. Hudson, then at the National Bureau of Standards in Washington, D. C., and the undisputed leader on the American scene in research on the carbohydrates. Hudson, as a student of Van't Hoff's had a strong background in physical chemistry, and his individualistic philosophy of research impressed Wolfrom greatly: his continuity of purpose, his exacting standards in experimental work, and his conservatism in theorizing until a thorough basis of facts had been recorded. All of these attributes, together with Hudson's concise and lucid style of writing, were to serve as models to Wolfrom throughout his career. It is doubtful that two such strong personalities could have long coexisted in the same institution, but Wolfrom ever after regarded Hudson as an inspiring teacher and colleague, to whom he owed a great deal; years later, they were to be closely associated in editorial and nomenclatural work, and they much enjoyed each other's company.

After a few months in Washington, Wolfrom moved in September, 1927, to New York City in order to work in the laboratory of P. A. Levene at the Rockefeller Institute for Medical Research. Levene, the outstanding master on the North American continent in the young discipline of biochemistry, was an ardent genius with a remarkable capacity for hard work, an urbane and cosmopolitan personality, and a warm interest in all of those who worked with him. In his contact with Levene, Wolfrom was able to assimilate at first hand some of the valuable aspects of the European traditions in science that Levene was able to convey to his coworkers at the Rockefeller Institute, and he was simultaneously exposed to the enormous challenge to the structural chemist offered by the seemingly hopeless slimes and mucins that were components of animal tissues. Levene had realized that more needed to be known about the structures of the simple sugars, especially the linkage positions in the disaccharides and the ring size in cyclic monosaccharide derivatives, before he could ever hope to achieve his goal of structural elucidation in the nucleic acids. Wolfrom worked with him on these aspects, and, within a few months, a paper resulted on the Wohl degradation of cellobiose and its use in determining interglycosidic linkage-position. In quick succession thereafter were published two more papers, on the acetylated methyl D-lyxosides and evidence for their ring forms. Levene did not then encourage his coworkers to become deeply involved in developing the ideas and the planning of research, and Wolfrom returned to The Ohio State University in the summer of 1928 to finish up his two-year Fellowship; Professor Evans was now prepared to let Wolfrom work more or less independently on a problem of his own choosing, and Wolfrom decided to study the synthesis of stable derivatives of the acyclic forms of the sugars, since such acyclic intermediates had been so often proposed as transient species in reactions of the sugars. By removing the thioacetal groups from the pentaacetate of D-glucose diethyl dithioacetal, he was able to obtain and characterize the acetate of the free aldehyde form of D-glucose, and this work was submitted for publication in March, 1929; similar work in the D-galactose series followed later.

In the autumn of 1929, Wolfrom was appointed Instructor in Chemistry at The Ohio State University, and one year later was raised to the rank of Assistant Professor. He remained on the faculty of the Department of Chemistry at Ohio State for the whole of his career, becoming Associate Professor in 1936, and Professor in 1940. In 1939, he was awarded a Fellowship by the John Simon Guggenheim Memorial Foundation, and, in February of that year, he travelled to Switzerland, to work in the laboratory of Professor P. Karrer of the University of Zürich, but he returned to the United States at the outbreak of hostilities in Western Europe.

With the aid of the successive generations of students who came to carry out their graduate research under his direction, Professor Wolfrom was able to launch a wide-ranging program of research, with problems of structure and reactivity in the carbohydrate field constituting the principal theme. The procedures used for obtaining acetylated *aldehydo*-D-glucose were systematically extended through the sugar series, and new types of aldose derivatives containing substituents on the hydrated carbonyl group were obtained; these showed the predictable behavior in being isolable in two isomeric forms, epimeric at C-1. The now well-established fact that acyclic structures can exist as reactive sugar-intermediates, sometimes having considerable stability, rests largely on his pioneering work. His first Ph. D. student, Alva Thompson, showed that the acetylated oxime of D-glucose undergoes conversion from a cyclic to an acyclic form during the Wohl degradation, and for this work, Thompson received the Ph. D. degree in 1931. During much of his professional career, until his death in 1962, Dr. Thompson remained associated with Professor Wolfrom at Ohio State.

Although organic chemical aspects of the carbohydrates formed the central theme of Wolfrom's work, he never accepted the limitations of being classified strictly as an organic chemist, neither in his selection of research problems nor in the tools that he used for their solution. The scope of interests of his early teachers, especially Hudson and Levene, encouraged him to look broadly at science as a whole for general concepts and their implications, from basic considerations of molecular architecture to the understanding of life processes. He also had a strong feeling for the symbiotic relationship between "pure" science and technology; he was scornful of those who expressed disdain for applied science, and he strove to bridge the gap between basic and mission-oriented research. He welcomed opportunities to make fundamental contributions relevant to problems in such applied areas as food technology, textile and paper technology, and pharmacology; as a result, he was frequently called upon as a consultant to industrial laboratories and to the Government. Although interested in the practical consequences of research, Wolfrom never allowed himself to be diverted from a scholarly approach to the solution of a problem because of any considerations of expediency; whatever the problem was, it had to be reduced to a set of meaningful experiments that might provide definitive answers that could eventually be presented in the form of a published report. He was not a dreamer, and he did not care to speculate openly or make assumptions not based on hard facts. Lengthy debate and hypothetical ideas in chemistry held little interest for him. He respected people who could get things done. In developing a research problem, he would read deeply on the subject and listen to independent viewpoints, especially from persons in other fields, but, once having been satisfied that the research could and should be done, he addressed himself to dividing up the problem into unit steps, each of which could be carefully and completely worked out. He did not care to build up an elaborate edifice upon tenuous threads with the idea of subsequently trying to consolidate it, and he had little respect for those who performed superficially impressive research by such means. He saw the structure of solid research as being built up by the careful putting together of many small "building stones," dramatic breakthroughs in research being seldom the result of work by any one individual, and the new developments being only as good as the quality of the parts from which they were assembled.

To the new graduate student, Professor Wolfrom often appeared rather formidable and awesome, even though he was physically only of medium height and build. He suffered fools not at all, and expected of his colleagues the standards of work that he set for himself. It was often difficult for lesser people to live up to his standards. His own experimental research was always done with precision, and he was proud to show that the samples he had prepared himself in the 'thirties were undecomposed several decades later and that their purity was unimpeachable, even by the chromatographic techniques later developed. He expected that all melting points and optical rotations recorded by his students should be as authoritative as his own experimental values. If ever any of the published values were challenged, he would leave no stone unturned, no matter how much work would be involved, until the question was resolved and the result put into print. The unfortunate researcher found wanting in the accuracy of his data could expect at the least a harrowing ordeal in being verbally torn to shreds by a grim and uncompromising taskmaster, and he might possibly face permanent loss of confidence in any of his previous or subsequent work. Wolfrom was very quick to penetrate half-truths and evasions of any sort, and could rapidly judge a person's character. He expected his students to develop the kind of self-reliance that had been expected of *him* when he worked with W. Lee Lewis. He never undertook to "spoon-feed" these workers in developing their research, although he willingly helped the student who had reached the limit of his own resources and ingenuity in solving a problem. Frequently, the more irascible he seemed to the student, the more he was taking a close personal interest in the development of that individual; but he demanded that each coworker expend his own efforts to the limit of his ability. Wolfrom did not believe in an elaborate system of graduate course-work in chemistry, but expected his students to be well read in the literature of science, and encouraged each of them to build a personal library. He would give the new student a research problem, acquaint him with the appropriate techniques and the background literature, and then turn him loose to see (in his words) "if he can swim." During much of Wolfrom's career at Ohio State, he had the collaboration of one or more long-term research associates, notably Drs. Alva Thompson and Wendell Binkley, to whom the new researcher could turn for help and advice; in addition, the other graduate students and visiting postdoctoral researchers in the laboratory helped in building a group spirit. For some of the students, the first real contact beyond the often gruff and austere outward personality of Wolfrom came when they presented to him their first report, or draft of a thesis. At that point, Professor Wolfrom would go through the report line by line with the student. Precision of experimental detail and historical accuracy in citing earlier published work would always be the first items for consideration. His questions and comments would be incisive and frequently devastating; the author of a sloppily prepared report might find himself quickly ushered out with a few terse comments. He had a remarkable eye for errors and inconsistencies in a document that might have seemed polished to perfection by its author. Every fault had to be corrected. There are probably very few scientists who have published as extensively as Wolfrom and at the same time been so close to the experimental details of the published work.

Not every student or colleague who came into contact with Wolfrom could accept his uncompromising standards. Professor Wolfrom chose to expend his energies in those areas where the problem could be clearly defined, and, once he had decided what was the right course, he held steadfast to that position, regardless of any outside pressures, and would not back down. He tended to avoid becoming involved in situations where negotiations and compromises had to be made, or where the issues could not be stated in precise terms and a firm position not taken.

A hard taskmaster, Wolfrom earned the greater respect of many of his students *after* they had completed their work with him. Despite his rather retiring and diffident attitude toward groups of people not known to him, and despite the apparent gruffness and terseness that so often characterized his day-to-day contacts with his coworkers, he actually took a deep interest in the welfare of every colleague and student who had a genuine interest in, and aptitude for, science. He went to considerable lengths to help each of his students become established in a suitable post after graduation, and kept in touch with a surprisingly large proportion of them long after their graduation. He had a deep insight into human personality, and found it intriguing to delve into the background and motivations of each of the persons with whom he worked. This interest is reflected in the number of biographical memoirs that he chose to write, especially of his early mentors; these were done with characteristic thoroughness and show his perceptive qualities in understanding human nature.

Although he never regarded lightly any of the work he undertook,

Wolfrom had a very strong sense of humor, not always recognized by those who did not know him well. He had an endless store of anecdotes concerning the personalities of science, based on his own contacts with other scientists and on his wide reading of the history of science; the humor of his dry remarks would once in a while be betrayed by a fleeting smile. This side of his personality was most in evidence when he was with small groups of people he knew well, and with small classes of advanced students who were perceptive enough to appreciate the subtleties of his comments.

His approach to teaching was always based on a solid, historical foundation that traced the development of science through the major milestones of factual knowledge, rather than through rationalizations and correlations that involved extrapolation of existing information. Inevitably, his lectures would involve lengthy discourses on the major personalities of science; his course on the carbohydrates was a particularly fascinating, personalized account of the subject for the scholarly student, but a disconcerting one for the student who expected a routine, "lecture notes" type of course. For much of his career, Wolfrom taught large classes of undergraduate students the principles of general chemistry and elementary organic chemistry, in addition to courses for graduate students, although it is probably true to say that, in the large classes, only the more strongly motivated students appreciated his subtle comments and expended the individual effort expected by Wolfrom.

Professor Wolfrom's major educational contribution was at the graduate level, where he supervised almost a hundred Ph. D. students and numerous M. S. candidates. With these students, he was able to pursue research on several broad fronts in the field of the carbohydrates. A list of Wolfrom's published articles and the participating coworkers is given at the end of this memoir. In the early days, most of the research students were employed as part-time teaching assistants in chemistry at Ohio State. Later, and especially after World War II, outside funding through grants and contracts from Government and industry became available, and Wolfrom was able to expand his research program further. The research group was enriched by a regular succession of postdoctoral associates who came from other institutions for one or two years of experience in Professor Wolfrom's laboratory. The group became very cosmopolitan, always containing members from countries in Europe and Asia, and Wolfrom particularly appreciated the new ideas and techniques brought in by these colleagues who had received their doctoral training in other laboratories. On occasion, the research seminars would be conducted in the

German language, to the dismay of those graduate students who questioned the value of the language requirement in the doctoral program.

Wolfrom's early theme of research on the acyclic forms of the sugars continued throughout his career, and, in fact, one of his posthumous articles is a book chapter on the subject. Extending the route developed for *aldehydo*-D-glucose pentaacetate, he devised general methods for obtaining crystalline acetates of those sugars in which the carbonyl group, aldehydic or ketonic, was present in the free form, uncombined with any hydroxyl group of the sugar chain, and the general chemistry of the hydrated carbonyl group was explored. A synthesis of higher-carbon ketoses by the action of diazomethane on acetylated aldonyl chlorides was established that led to the preparation of acyclic *keto*-acetates which, on deacetylation, gave ketoses of higher carbon content. It was shown that the *keto* acetates could be used for the synthesis of branched-chain structures. In cooperation with T. M. Lowry of Cambridge University, Wolfrom conducted pioneer work on the optical rotatory dispersion of the acyclic sugar acetates, and demonstrated the Cotton effects attributable to the asymmetrically perturbed absorption of the carbonyl group. It was demonstrated that many of the hydrazones and osazones of the sugars were either totally acyclic or contained such a structure as a significant tautomeric form. He recognized at an early stage the potential of nuclear magnetic resonance spectroscopy in structural chemistry, and applied it in 1962 to show that an "anhydro-phenylosazone" that had been prepared in his laboratory in 1946 possessed an unexpected, unsaturated phenylazo structure.

The chemistry of the dithioacetals of the sugars was explored in detail, and many useful synthetic transformations were demonstrated. A notable development was the reductive desulfurization of the dithioacetals to the hydrocarbon stage. This reaction was used to establish a major milestone in the chemistry of natural products, the unambiguous correlation between the configurational standards of D-glyceraldehyde for the sugars and L-serine for the amino acids; the correlation was achieved by way of the diethyl dithioacetal of 2-amino-2-deoxy-D-glucose, which was transformed into a derivative of L-serine without disturbing the configuration of the asymmetric center at carbon atom two.

Other developments in the chemistry of dithioacetals included preparation of the first dithioacetal of a ketose (D-fructose), and the establishing of the technique of mercaptolysis for the fragmentation of polysaccharides. In other hands, the technique of mercaptolysis has been applied successfully for determination of structure, notably with the seaweed polysaccharides agar and carrageenan. The acetylated dithioacetals were shown to be useful characterizing derivatives for the sugars, and subsequent workers have utilized these derivatives extensively for determination of the gross structures of sugars by mass spectrometry. Dithioacetal derivatives were also significant in Wolfrom's work on the structure of the antitubercular antibiotic streptomycin, playing a role in the elucidation of structure of the streptose component. The configuration of the streptidine entity was established by its synthesis from 2-amino-2-deoxy-D-glucose, and further contributions were made on the structure and configuration of the entire streptomycin molecule.

Methods were developed for the synthesis of amino sugars by displacement of sulfonyloxy groups by nitrogen nucleophiles, and were applied especially for the synthesis of 2-amino-2-deoxypentoses, until the complete series of eight stereoisomers had been elaborated. Procedures for protecting the amino group were established that led to the successful synthesis of nucleosides containing 2-amino-2-deoxy sugars in the furanosyl form. This work formed part of an extensive program of synthesis of nucleoside analogs having structural variation in the carbohydrate moiety, as potential anticancer agents.

Professor Wolfrom was ever concerned with planning research in a logical, orderly way, and he undertook to fill in some of the gaps left by Emil Fischer in the systematic elaboration of the simple sugars. These included the crystalline forms of racemic glucose, racemic glucitol, D-glucose dimethyl acetal, L-fructose, racemic talitol, L-talitol, and xylitol. For key crystalline compounds, he made an especial point of recording the data from an X-ray powder diagram as an unequivocal fingerprint of the compound in that particular crystalline modification; he had little faith in syrups unless a suitable crystalline derivative could be prepared. He would accept chromatographic evidence as a tool for monitoring reactions, but only as a preliminary guide to characterization by a definitive method, preferably on a crystalline basis.

Wolfrom undertook a number of investigations having implications in technology. The formation of color in sugar solutions as present in food products and during sugar refining was investigated. Products of the dehydrating reactions favored by acidity were identified, and the mechanism of the non-enzymic browning (Maillard) reaction between sugars and amino acids was examined; a reactive 3-deoxyhexosulose intermediate was established for the latter reaction, a finding that provided the basis for extensive work in other laboratories. It was demonstrated that the action of alkali on reducing sugars leads to stepwise enolization down the sugar chain. The composition of cane-sugar molasses was examined in detail. In a study on the alkaline elec-troreduction of D-glucose, it was shown that carbonyl groups are reduced completely to the hydrocarbon stage, and that a side-product is a twelve-carbon atom derivative formed by way of an aldol reaction.

The effect of ionizing radiation on sugars was the subject of pio-neering work by Professor Wolfrom and his group; the reactions were interpreted on the basis of free-radical processes, and the oxidizing role of water was studied in this connection. Electron paramagnetic resonance studies were made on the remarkably stable, free radicals formed when sugars in the solid state are irradiated. The chemical transformations taking place during the controlled ignition of cellu-lose nitrate were investigated extensively in a project for the armed services. Other nitrated polyhydroxy compounds were also inves-tigated for potential uses as explosive polymers.

Incidental to the studies on cellulose nitrate, specifically ^{14}C-la-beled cotton celluloses were prepared biologically, and some insight into biosynthesis in the cotton boll was obtained. α-D-Glu-copyranosyl phosphate, the anomer of the Cori ester, was synthe-sized, and a synthetic route to L-iduronic acid was devised; both of these products were subsequently found by others to occur naturally.

Important contributions were made by Wolfrom and his associates in establishing new techniques for working with carbohydrates and their derivatives, notably in separation methods. Extrusive column chromatography was developed as a valuable tool for the separation of mixtures of acylated sugars, and was utilized particularly in studies on oligosaccharides, including the characterization of several polymer-homologous series of oligosaccharides. Procedures involving ion-exchange resins were introduced into the carbohydrate field, and the use of microcrystalline cellulose for thin-layer chromatography of un-substituted sugars was developed. The use of sodium borohydride for reducing the free sugars to alditols was first recorded by Wolfrom's group, as were the first examples of arsenate, benzeneboronate, and urethan derivatives of the sugars, and the use of absolute hydrogen sulfate as a solvent. Also developed were reliable analytical proce-dures for the determination of acetyl and methoxyl groups in carbohy-drates containing them.

Many years of persistent effort were spent by Professor Wolfrom in the determination of the structure of various polysaccharides. The most challenging of these was heparin, the natural blood an-ticoagulant. Methods were found for modifying the intractable "back-bone" chain of this polymer, notably by use of diborane to reduce the

uronic acid moieties, to give a derivative amenable to structural characterization by the method of fragmentation analysis. By means of
crystalline, disaccharide fragments that were unequivocally characterized, it was shown that N- and O-sulfated 2-amino-2-deoxy-D-
glucopyranose and D-glucopyranuronic acid residues, connected by
α-D-(1→4) linkages, are present in the polymer, and that L-iduronic
acid residues also occur. Other animal polysaccharides investigated
by Wolfrom were chondroitinsulfuric acid and the galactan of beef
lung.

The fine structures of starch and glycogen were extensively investigated; evidence for the branch points at carbon atom six was placed
on a crystalline, isolative basis by the method of fragmentation analysis. Incidental to this work, the nature of reversion of sugars by acids
was interpreted. Synthetic confirmation of the structure of the branch-
point disaccharide, namely, isomaltose, involved the difficult step of
introducing the α-D linkage in the interglycosidic position; this was
achieved by a modification of the Koenigs–Knorr synthesis, with the
use of a non-participating protecting group at carbon atom two in the
glycosyl halide derivative. A similar approach was subsequently used
for the synthesis of panose, a trisaccharide fragment involved at the
branch points in the polysaccharide. Structures were also established
for the mannan and arabinogalactan of the green coffee bean, and the
presence of these polysaccharides in commercial coffee extracts was
established.

In other work related to starch, detailed structural investigations
were made on the pyrodextrins, or British gums, produced commer-
cially by the heat treatment of starch. In the quest for novel derivatives of starch having potential utility in industry, various acetal and
unsaturated ether derivatives were studied, and routes were developed for the synthesis of amino derivatives of starch having the
hydroxyl group at carbon two replaced by an amino group; the latter
were used for preparing polymers having structures related to that of
heparin. In the cellulose field, comparative studies were made on
various series of cello-oligosaccharide derivatives as models for the
parent polymer; these investigations included oxidation with alkaline
hypochlorite as related to the industrial extraction and bleaching of
cellulose fibers.

Outside the field of carbohydrates, Professor Wolfrom had a long-
standing interest in the pigments occurring in the osage orange
(*Maclura pomifera* Raf.), a common hedge-tree found in Ohio. The
chemical nature of two complex phenolic pigments present in the fruit
of this plant was elucidated and a synthesis of their skeletal compo-

nents was effected. These compounds were the first for which it was established that isoprenoid units were condensed on the nucleus of a common plant-pigment, in this case, an isoflavone; other examples have since been found in plants. Three phenolic pigments containing isoprene units condensed on a xanthone nucleus were discovered in the root bark of the same plant, and it was found possible to elucidate their structures, mainly by use of spectroscopic techniques; one of the pigments was synthesized. Two of them were found to contain an isoprenoid unit in the form of a 1,1-dimethylallyl group, and they were the first examples discovered of natural phenolic compounds so constituted.

In the Department of Chemistry at Ohio State, Professor Wolfrom assumed the duties of Head of the Organic Division in 1948, the Department then being under the chairmanship of a physical chemist, Edward Mack, Jr., who, in 1941, had succeeded William Lloyd Evans as departmental chairman. Wolfrom's responsibilities included co-ordination of the courses offered and of the requirements for graduate degrees in organic chemistry. He also served for many years on the departmental Library Committee. Thorough as always in the tasks he undertook, he played an important part in developing an excellent chemistry library, both as regards the extent of coverage and the completeness of the collection of early books and periodicals. In 1960, Wolfrom was named Research Professor, and the responsibilities of the Organic Division were passed on to M. S. Newman. In his new position, Professor Wolfrom was able to concentrate more on his indi-vidual teaching effort at the graduate level, although he continued to teach courses in the chemistry of carbohydrates. His office was a very modest one, in a long corridor of small research laboratories affec-tionately known to successive generations of occupants as "Sugar Alley." He spurned the opportunity to move into more spacious and modern quarters when the new Evans Laboratory was added to the department in 1960; he felt a sentimental attachment to the antiquated laboratories whose dust was, as legend had it, rich in the seeds of a myriad crystal species. Fact or fiction, there is little doubt that the Wolfrom group had an impressive record of success in bringing recal-citrant syrups of sugars to crystallization.

Professor Wolfrom influenced the carbohydrate field in many ways far beyond even his own exceedingly prolific contributions to the liter-ature on the subject. He provided the motivation for many others to pursue work in the area. A surprising number of the persons who, at one time or another, have worked in Wolfrom's laboratory have con-tinued independent research on the carbohydrates; those in academic positions include H. El Khadem (Alexandria), A. B. Foster (London), S. Hanessian (Montreal), R. U. Lemieux (Edmonton), G. E. Mc-Casland (San Francisco), R. Montgomery (Iowa), K. Onodera (Kyoto), A. Rosenthal (Vancouver), F. Shafizadeh (Montana), the late J. C. Sowden (St. Louis), W. A. Szarek (Kingston), J. R. Vercellotti (Virginia), R. L. Whistler (Purdue), and this writer.

Professor Wolfrom worked long hours both at home and in his office. He was exceptionally well organized, and, with his keen eye for selecting objectives that were to prove fruitful, he was able to complete a formidable amount of work in a day. He hated wasting time, and handled much of his business by telephone, even with persons in the room next to his office. These calls would be brief and often blunt, as were his letters and notes; he always made his point with maximum impact in as few words as possible.

Although he derived great satisfaction from his work, chemistry was by no means Professor Wolfrom's sole preoccupation. He read widely in the classics and history, and enjoyed building a fine library in his home. Not an instrumentalist himself, he nevertheless shared his wife's love of music, and worked with her in helping the development of the Columbus Symphony Orchestra, and in other musical and cultural activities in the community. Together, they also enjoyed the theater, ballet, and the fine arts, both in the local community and during their travels. The Wolfroms frequently received groups of colleagues, students, and visiting scientists in their spacious home on the north side of Columbus. Mrs. Wolfrom, a most gracious hostess, would often entertain the guests with a musical recital at the piano, and Professor Wolfrom was happy to show guests the pleasant garden the cultivation of which was a source of great enjoyment to him. On other occasions, he would lead the members of his research group and their families on picnic trips into the surrounding Ohio countryside; strictly, these expeditions were recreational, although it must be said that, should Wolfrom spot some osage-orange trees along the way, a "work gang" might rapidly be delegated to collect samples.

Five children were born to the Wolfroms. Their firstborn, Frederick Lorenz, died shortly after birth in 1933. A daughter, Eva Magdalena, was born a year later, and twin daughters, Anne Marie and Betty Jane, were born in 1938. Their son Carl Thompson, born in 1942, received his Juris Doctor degree in 1967 from the College of Law at Ohio Northern University, and set up law practice in Fostoria, Ohio, a short distance from his father's birthplace of Bellevue.

Wolfrom felt deeply concerning the best interests of his community, the university, his country, and the world; he was never afraid to take a strong public stand and to work actively on issues on which he felt that he was right. He believed in the responsibility of the individual to work hard to the maximum of his ability, and to control his own destiny, but equally he believed that the social organization had the responsibility of permitting the exercise of individual initiative. He abhorred rigid systems of social class-structure, economic domination, or authoritarian control that caused any individual to be subjected to exploitation, discrimination, or regimentation.

His precise style of writing, and his insistence on the accurate historical record, led Wolfrom increasingly into the secondary literature in the carbohydrate field: the books and monographs that interpret and evaluate the primary research literature. His reviews of published books were succinct, to the point, and often tart. More and more, he felt the need for a periodic series of authoritative articles on various aspects of research on the carbohydrates, to be written by qualified specialists and supervised editorially through a rigorous policy, in order to ensure extremely high standards of consistency and accuracy. From this idea there developed, near the end of World War II, the *Advances in Carbohydrate Chemistry*. The policies were formulated by an Executive Committee consisting of W. L. Evans, H. O. L. Fischer, R. Maximilian Goepp, Jr., W. N. Haworth, and C. S. Hudson, together with Wolfrom and his co-editor, W. Ward Pigman. With the enthusiastic help and collaboration of publisher Kurt Jacoby and the then-fledgling Academic Press, the first issue of the series was launched in 1945. Wolfrom remained a prime mover in this annual series for the rest of his life, and was editor or co-editor of all volumes through Volume 24, except for those of 1950 and 1951. Under his guidance, the series has reflected all of the major and significant developments in carbohydrate chemistry and biochemistry, through timely contributions written by authorities in the field. His broad knowledge and critical ability, his attention to editorial detail, and his insistence on "getting it right," have given the series an excellent reputation for quality and reliability. Not even the most eminent of authors were immune from his pungent remarks if their manuscripts failed in any way to meet the standards demanded. His respected colleague Claude S. Hudson worked closely with Professor Wolfrom on the early volumes of *Advances*, and, following the death of Hudson in 1952, Wolfrom invited R. S. Tipson to join in editing the series; the Wolfrom–Tipson editorial partnership for *Advances* continued for eighteen years thereafter. With the addition of representation from the British Isles on the board of *Advances*, starting with the second volume in the series, a close link was established between British and American carbohy-

drate chemists that in subsequent years led to much fruitful cooperation, especially in the field of carbohydrate nomenclature. Professor
Wolfrom established a particularly firm friendship with Professor M.
Stacey of Birmingham University; the Wolfroms and Staceys enjoyed
exchanging visits and, over the years, a regular succession of Birmingham graduates came to Ohio State for postdoctoral work in
Wolfrom's laboratory.

In collaboration with R. L. Whistler, Wolfrom also served as co-
editor or consulting editor for the series *Methods in Carbohydrate
Chemistry.* These collections of experimental procedures in the carbohydrate field have proved an invaluable standby for research
workers. The international journal *Carbohydrate Research*, inaugu-
rated in 1965, also received strong support from Professor Wolfrom,
who served on the Editorial Advisory Board of that journal.

For a quarter of a century, Professor Wolfrom worked on the system-
atization and codification of carbohydrate nomenclature. He was a
member of an A. C. S. committee chaired by C. S. Hudson that developed, during the period 1945–1948, the rules of carbohydrate nomenclature that were approved in 1948 by the Council of the American
Chemical Society. The committee continued to consolidate and ex-
tend the rules, this time in cooperation with chemists in Great Britain,
and, in 1951, Professor Wolfrom became chairman of the committee.
The joint study of carbohydrate nomenclature by British and American chemists furnished an excellent example of effective cooperation
to improve the language of science for clear and exact reporting of scientific information. A set of rules under joint British–American spon-
sorship was published in 1953, and, as a result of continued coopera-
tive work, the rules were extended and further clarified; a revised set
of jointly approved rules was issued in 1963. In 1955, Professor
Wolfrom joined the American Chemical Society Committee on Nomenclature, Spelling, and Pronunciation, and he also served on the
U. S. National Research Council Committee on Chemical Nomenclature; through these groups, he was able to ensure that essential
links were maintained between the specialized nomenclature for the
carbohydrates and the general body of chemical nomenclature.

After publication of the 1963 Rules, Wolfrom's committee continued to work in the development of nomenclature systems for the
carbohydrates, in order to accommodate the special requirements of
newly developing research areas and to encompass areas, such as the
polysaccharides and conformational terminology, not covered in the
published rules. At the same time, he sought to develop full international acceptance of the rules by working actively with the Special

Committee on Carbohydrate Nomenclature of the Organic Commission of the International Union of Pure and Applied Chemistry. He met several times with the international committee, and laid much of the important ground-work for the set of international rules drafted by that body.

An important factor contributing to Professor Wolfrom's effectiveness in the field of nomenclature was his close liaison with the nomenclature specialists and indexing staff at Chemical Abstracts Service, in particular with L. T. Capell and K. L. Loening. The proximity of the Service to the Ohio State campus was a useful asset in this regard; indeed, for many years, the CA facility was located just a few steps from Wolfrom's office. Through these contacts, he came to appreciate the problems involved in so refining carbohydrate nomenclature as both to permit effective indexing and be compatible with techniques then being developed for automatic data-processing and information retrieval, eventual objectives being the establishment of systems that could permit the translation by purely mechanical means of a systematic name into a digital record or into a three-dimensional structure.

In 1959, Professor Wolfrom assumed the duties of Section Editor for the Carbohydrates section of *Chemical Abstracts*, a task that he undertook with characteristic thoroughness. All abstracts for that section were carefully checked to ensure that the names used conformed to the approved terminology, and the scientific content of the abstracts was also checked, usually against the original article. Not infrequently, he himself re-wrote those abstracts he found unsatisfactory. In 1964, he was made a member of the Board of Advisors for *Chemical Abstracts*.

In recognition of his outstanding services in the field of chemical documentation, Professor Wolfrom received in 1967 the Austin Patterson Award, sponsored by the Dayton Section of the American Chemical Society. At the award ceremony, many of his old friends and colleagues were present, including Dr. W. A. Hammond, his high-school chemistry teacher.

In addition to his contributions in the field of chemical documentation, Professor Wolfrom also served on numerous committees and held a number of offices in professional societies. From 1940 until 1945, he was an Official Investigator of the National Defense Research Committee. A member of the American Chemical Society, he served as Chairman of the Columbus Section and also of the Cellulose Division in 1940, and of the Division of Sugar Chemistry in 1948. In 1958, he was chairman of Symposium I of the International Union of

Biochemistry in Vienna, and, for this, he was honored by a citation from the Austrian government. He was a member of the American Society of Biological Chemists, Phi Beta Kappa, Sigma Xi, Phi Lambda Upsilon, Pi Mu Epsilon, and Alpha Chi Sigma. He was a member of the National Committee of the Phi Beta Kappa Book Award in Science from 1961 to 1963, and served as its chairman in 1963. He was a Fellow of the American Academy of Arts and Sciences, the New York Academy of Science, the Ohio Academy of Science, the American Association for the Advancement of Science, and The Chemical Society (London). In 1959, he was an invited lecturer in the Biochemistry Department at Tufts University Medical School in Boston, Massachusetts.

Professor Wolfrom received numerous other honors and recognitions for his work. In 1950, he was elected a Fellow of the U. S. National Academy of Sciences, and, in 1952, was presented the Honor Award (now the Hudson Award) of the Division of Carbohydrate Chemistry of the American Chemical Society. In 1965, he was honored by The Ohio State University by being named Regents' Professor, a title created at that time to recognize exceptional distinction in scholarly activity at the University. In 1967, he was honored by the Kansas City Section of the American Chemical Society with the Kenneth A. Spencer Award for his contributions to agricultural chemistry.

Still at the height of his effectiveness as a scientist, Professor Wolfrom was, in mid-1969, actively planning new research programs to take effect beyond the nominal age for retirement at seventy years. Tragically, an aortic aneurysm, found during a routine physical examination, ruptured several days after its discovery, just hours before a proposed surgical repair could be effected; Professor Wolfrom died in Columbus on June 20, 1969. It is particularly indicative of his methodical and organized personality that, on the day before he was to enter the hospital, he visited each of his research students to plan work for the following few weeks, answered all of the correspondence on his desk, and left instructions for handling the various items of business that were expected to arise. He was survived by his widow, four children, three sisters, and seven grandchildren.

A special issue of the journal *Carbohydrate Research*, comprised of research contributions by ex-students of Professor Wolfrom's, was published as the Wolfrom Memorial Issue in April, 1970, on the seventieth anniversary of his birth. Also dedicated to his memory was the program of papers presented in Toronto, Ontario, in May, 1970, at the joint meeting of the Carbohydrate Division of the American Chemical Society and the Canadian Institute of Chemistry. In addition, a special tribute to him was paid at the Fifth International Conference on Carbohydrates held in Paris in August, 1970.

Professor Wolfrom has left impressed in the pages of science a monolithic achievement that can serve as an inspiration and a challenge to others. Even more important, his spirit and ideas made a lasting influence on a whole generation of new scholars, so that the qualities for which he stood can continue to flourish.

DEREK HORTON

APPENDIX

The following is a chronological list of the scientific publications of Professor Wolfrom and his research colleagues. A number of posthumous papers remain to be published.

"Acrolein," H. Adkins, W. H. Hartung, F. C. Whitmore, and M. L. Wolfrom, *Org. Syn.*, **6**, 1–5 (1926).

"The Reactivity of the Methylated Sugars. II. The Action of Dilute Alkali on Tetramethyl Glucose," M. L. Wolfrom and W. Lee Lewis, *J. Amer. Chem. Soc.*, **50**, 837–854 (1928).

"Lactone Formation of Cellobionic and of Glucoarabonic Acids and Its Bearing on the Structure of Cellobiose," P. A. Levene and M. L. Wolfrom, *J. Biol. Chem.*, **77**, 671–683 (1928).

"Acetyl Monoses. IV. Two Isomeric Triacetyl Methyllyxosides," P. A. Levene and M. L. Wolfrom, *J. Biol. Chem.*, **78**, 525–533 (1928).

"Acetyl Monoses. V. The Rates of Hydrolysis of Tetraacetylmethylmannosides and of Triacetylmethyllyxosides," P. A. Levene and M. L. Wolfrom, *J. Biol. Chem.*, **79**, 471–474 (1928).

"The Acetate of the Free Aldehyde Form of Glucose," M. L. Wolfrom, *J. Amer. Chem. Soc.*, **51**, 2188–2193 (1929).

"The Fifth Penta-acetate of Galactose, Its Alcoholate and Aldehydrol," M. L. Wolfrom, *J. Amer. Chem. Soc.*, **52**, 2464–2473 (1930).

"Aldehydo-*l*-arabinose Tetra-acetate," M. L. Wolfrom and Mildred R. Newlin, *J. Amer. Chem. Soc.*, **52**, 3619–3623 (1930).

"Carbohydrates," W. Lloyd Evans and M. L. Wolfrom, in "Annual Survey of American Chemistry," Chap. XX, Vol. IV, National Research Council by the Chemical Catalog Co., 1930, pp. 244–257.

"The Constitution of Sugars," by W. N. Haworth (book review), M. L. Wolfrom, *J. Chem. Educ.*, **7**, 484 (1930).

"The Reactive Form of Glucose Oxime," M. L. Wolfrom and Alva Thompson, *J. Amer. Chem. Soc.*, **53**, 622–632 (1931).

"The Mutarotation of the Alcoholate and Aldehydrol of Aldehydo-Galactose Pentaacetate," M. L. Wolfrom, *J. Amer. Chem. Soc.*, **53**, 2275–2279 (1931).

"The Rotatory Dispersion of Several Aldehydo Sugar Acetates," M. L. Wolfrom and Wallace R. Brode, *J. Amer. Chem. Soc.*, **53**, 2279–2281 (1931).

"The Occurrence of True Hydrazone Structures in the Sugar Series," M. L. Wolfrom and Clarence C. Christman, *J. Amer. Chem. Soc.*, **53**, 3413–3419 (1931).

"Aldehydo-*d*-xylose Tetra-acetate and the Mercaptals of Xylose and Maltose," M. L. Wolfrom, Mildred R. Newlin, and Eldon E. Stahly, *J. Amer. Chem. Soc.*, **53**, 4379–4383 (1931).

"Hemi-acetals of Aldehydo-Galactose Pentaacetate and Their Optical Properties," M. L. Wolfrom and William M. Morgan, *J. Amer. Chem. Soc.*, **54**, 3390–3393 (1932).

"Ring–Chain Isomerism in the Acetates of Galactose Oxime," M. L. Wolfrom, Alva Thompson, and L. W. Georges, *J. Amer. Chem. Soc.*, **54**, 4091–4095 (1932).

"The Acetylation of Galactose Oxime," Venancio Deulofeu, M. L. Wolfrom, Pedro Cattaneo, C. C. Christman, and L. W. Georges, *J. Amer. Chem. Soc.*, **55**, 3488–3493 (1933).

"The Rotatory Dispersion of Organic Compounds. Part XXIII. Rotatory Dispersion and Circular Dichroism of Aldehydic Sugars," H. Hudson, M. L. Wolfrom, and T. M. Lowry, *J. Chem. Soc.*, 1179–1192 (1933).

"Keto-Fructose Pentaacetate," M. L. Wolfrom and Alva Thompson, *J. Amer. Chem. Soc.*, **56**, 880–882 (1934).

"The Free Aldehyde Form of Fucose Tetraacetate," M. L. Wolfrom and J. A. Orsino, *J. Amer. Chem. Soc.*, **56**, 985–987 (1934).

"Ring Opening of Galactose Acetates," Jack Compton and M. L. Wolfrom, *J. Amer. Chem. Soc.*, **56**, 1157–1162 (1934).

"A New Synthesis of Aldehydo Sugar Acetates," M. L. Wolfrom, L. W. Georges, and S. Soltzberg, *J. Amer. Chem. Soc.*, **56**, 1794–1797 (1934).

"The Structure of *d*-Glucoheptulose Hexaacetate," M. L. Wolfrom and Alva Thompson, *J. Amer. Chem. Soc.*, **56**, 1804–1806 (1934).

"The Tritylation of Sugar Mercaptals," M. L. Wolfrom, Joseph L. Quinn, and Clarence C. Christman, *J. Amer. Chem. Soc.*, **56**, 2789–2790 (1934).

"The Tritylation of Sugar Mercaptals," M. L. Wolfrom, Joseph L. Quinn, and Clarence C. Christman, *J. Amer. Chem. Soc.*, **57**, 713–717 (1935).

"The Rotatory Dispersion of Organic Compounds. Part XXV. Open-chain Derivatives of Arabinose, Fructose, and Fucose. Optical Cancellation in Penta-acetyl μ-Fructose," W. C. G. Baldwin, M. L. Wolfrom, and T. M. Lowry, *J. Chem. Soc.*, 696–704 (1935).

"The Carbohydrates," by E. F. and K. F. Armstrong (book review), M. L. Wolfrom, *J. Amer. Chem. Soc.*, **57**, 1387 (1935).

"Esters of the Aldehydrol Form of Sugars," M. L. Wolfrom, *J. Amer. Chem. Soc.*, **57**, 2498–2500 (1935).

"Ring Closure Studies in the Sugar Benzoates," M. L. Wolfrom and Clarence C. Christman, *J. Amer. Chem. Soc.*, **58**, 39–43 (1936).

"Estimation of O-Acetyl and N-Acetyl and the Structure of Osazone Acetates," M. L. Wolfrom, M. Konigsberg, and S. Soltzberg, *J. Amer. Chem. Soc.*, **58**, 490–491 (1936).

"Open Chain Derivatives of *d*-Mannose," M. L. Wolfrom and L. W. Georges, *J. Amer. Chem. Soc.*, **58**, 1781–1782 (1936).

"Semicarbazone and Oxime Acetates of Maltose and Cellobiose. *Aldehydo*-Cellobiose Octaacetate," M. L. Wolfrom and S. Soltzberg, *J. Amer. Chem. Soc.*, **58**, 1783–1785 (1936).

"A Study of Cellulose Hydrolysis by Means of Ethyl Mercaptan," M. L. Wolfrom and Louis W. Georges, *J. Amer. Chem. Soc.*, **59**, 282–286 (1937).

"The Beta to Alpha Conversion of Fully Acetylated Sugars by Alkali," M. L. Wolfrom and Donald R. Husted, *J. Amer. Chem. Soc.*, **59**, 364–365 (1937).

"2,3,6-Trimethylglucose Diethyl Mercaptal and its Use in the Preparation of 2,3,6-Trimethylglucose," M. L. Wolfrom and Louis W. Georges, *J. Amer. Chem. Soc.*, **59**, 601–603 (1937).

"Aldehydo Derivatives of Dibenzylideneglucose," M. L. Wolfrom and Leo J. Tanghe, *J. Amer. Chem. Soc.*, **59**, 1597–1602 (1937).

"The 1,5-Anhydride of 2,3,4,6-Tetramethylglucose 1,2-Enediol," M. L. Wolfrom and Donald R. Husted, *J. Amer. Chem. Soc.*, **59**, 2559–2561 (1937).

"Acetals of Galactose and of Dibenzylideneglucose," M. L. Wolfrom, Leo J. Tanghe, R. W. George, and S. W. Waisbrot, *J. Amer. Chem. Soc.*, **60**, 132–134 (1938).

"Esters of the Aldehydrol Form of Sugars. II," M. L. Wolfrom and M. Konigsberg, *J. Amer. Chem. Soc.*, **60**, 288–289 (1938).

"Crystalline Lactositol," M. L. Wolfrom, W. J. Burke, K. R. Brown, and Robert S. Rose, Jr., *J. Amer. Chem. Soc.*, **60**, 571–573 (1938).

"A Yellow Pigment from the Osage Orange (*Maclura pomifera* Raf.)," E. D. Walter, M. L. Wolfrom, and W. W. Hess, *J. Amer, Chem. Soc.*, **60**, 574–577 (1938).

"The Dimethyl Acetal of *d*-Glucose," M. L. Wolfrom and S. W. Waisbrot, *J. Amer. Chem. Soc.*, **60**, 854–855 (1938).

"Carbohydrates I," M. L. Wolfrom, in "Organic Chemistry," Vol. II. Chap. XVI, Henry Gilman, Ed., John Wiley and Sons, Inc., 1938, pp. 1399–1476.

"Studies of Cellulose Hydrolysis by Means of Ethyl Mercaptan. II," M. L. Wolfrom, Louis W. Georges, and John C. Sowden, *J. Amer. Chem. Soc.*, **60**, 1026–1031 (1938).

"Studies of Cellulose Hydrolysis by Means of Ethyl Mercaptan. III," M. L. Wolfrom and John C. Sowden, *J. Amer. Chem. Soc.*, **60**, 3009–3013 (1938).

"*Aldehydo-d*-Mannose Pentaacetate Ethyl Hemiacetal," M. L. Wolfrom, M. Konigsberg, and D. I. Weisblat, *J. Amer. Chem. Soc.*, **61**, 574–576 (1939).

"The Molecular Size of Methylated Cellulose," M. L. Wolfrom, John C. Sowden, and E. N. Lassettre, *J. Amer. Chem. Soc.*, **61**, 1072–1076 (1939).

"The Behavior of the Dimethyl Acetals of Glucose and Galactose Under Hydrolytic and Glycoside-forming Conditions," M. L. Wolfrom and S. W. Waisbrot, *J. Amer. Chem. Soc.*, **61**, 1408–1411 (1939).

"Tritylation Experiments in the Sugar Alcohol Series," M. L. Wolfrom, W. J. Burke, and S. W. Waisbrot, *J. Amer. Chem. Soc.*, **61**, 1827–1829 (1939).

"The Molecular Size of Starch by the Mercaptalation Method," M. L. Wolfrom, D. R. Myers, and E. N. Lassettre, *J. Amer. Chem. Soc.*, **61**, 2172–2175 (1939).

"Osage Orange Pigments. II. Isolation of a New Pigment, Pomiferin," M. L. Wolfrom, F. L. Benton, A. S. Gregory, W. W. Hess, J. E. Mahan, and P. W. Morgan, *J. Amer. Chem. Soc.*, **61**, 2832–2836 (1939).

"Osage Orange Pigments. III. Fractionation and Oxidation," M. L. Wolfrom and A. S. Gregory, *J. Amer. Chem. Soc.*, **62**, 651–652 (1940).

"Monothioacetals of Galactose," M. L. Wolfrom and D. I. Weisblat, *J. Amer. Chem. Soc.*, **62**, 878–880 (1940).

"The Action of Phosphorus Pentachloride upon *aldehydo*-Galactose Pentaacetate. The 1,1-Dichloride of *aldehydo*-Galactose Pentaacetate," M. L. Wolfrom and D. I. Weisblat, *J. Amer. Chem. Soc.*, **62**, 1149–1151 (1940).

"Crystalline Phenylurethans (Carbanilates) of Sugar Glycosides," M. L. Wolfrom and D. E. Pletcher, *J. Amer. Chem. Soc.*, **62**, 1151–1153 (1940).

"*aldehydo*-Maltose Octaacetate," M. L. Wolfrom and M. Konigsberg, *J. Amer. Chem. Soc.*, **62**, 1153–1154 (1940).

"Osage Orange Pigments. IV. Degree of Unsaturation and Flavone Nature," M. L. Wolfrom, P. W. Morgan, and F. L. Benton, *J. Amer. Chem. Soc.*, **62**, 1484–1489 (1940).

"Derivatives of the Aldehydrol Form of Sugars. III. Carbon One Asymmetry," M. L. Wolfrom, M. Konigsberg, and F. B. Moody, *J. Amer. Chem. Soc.*, **62**, 2343–2349 (1940).

"Melibiotol and Maltitol," M. L. Wolfrom and Thomas S. Gardner, *J. Amer. Chem. Soc.*, **62**, 2553–2555 (1940).

"*d*-Glucose S-Ethyl O-Methyl Monothioacetal," M. L. Wolfrom, D. I. Weisblat, and A. R. Hanze, *J. Amer. Chem. Soc.*, **62**, 3246–3250 (1940).

"Acyclic Derivatives of *d*-Lyxose," M. L. Wolfrom and F. B. Moody, *J. Amer. Chem. Soc.*, **62**, 3465–3466 (1940).

"The Action of Diazomethane upon Acyclic Sugar Derivatives. I," M. L. Wolfrom, D. I. Weisblat, W. H. Zohpy, and S. W. Waisbrot, *J. Amer. Chem. Soc.*, **63**, 201–203 (1941).

"Osage Orange Pigments. V. Isomerization," M. L. Wolfrom, F. L. Benton, A. S. Gregory, W. W. Hess, J. E. Mahan, and P. W. Morgan, *J. Amer. Chem. Soc.*, **63**, 422–426 (1941).

"The Action of Diazomethane upon Acyclic Sugar Derivatives," M. L. Wolfrom, D. I. Weisblat, and S. W. Waisbrot, *J. Amer. Chem. Soc.*, **63**, 632 (1941).

"The Structure of the Cori Ester," M. L. Wolfrom and D. E. Pletcher, *J. Amer. Chem. Soc.*, **63**, 1050–1053 (1941).

"Derivatives of the Aldehydrol Form of Sugars. IV," M. L. Wolfrom and Robert L. Brown, *J. Amer. Chem. Soc.*, **63**, 1246–1247 (1941).

"Osage Orange Pigments. VI. Isoflavone Nature of Osajin," M. L. Wolfrom, J. E. Mahan, P. W. Morgan, and G. F. Johnson, *J. Amer. Chem. Soc.*, **63**, 1248–1253 (1941).

"Osage Orange Pigments. VII. Isoflavone Nature of Pomiferin," M. L. Wolfrom and J. E. Mahan, *J. Amer. Chem. Soc.*, **63**, 1253–1256 (1941).

"Molecular Size of Polysaccharides by the Mercaptalation Method; Methylated Potato Starch," M. L. Wolfrom and D. R. Myers, *J. Amer. Chem. Soc.*, **63**, 1336–1339 (1941).

"Mesylated Cellulose and Derivatives," M. L. Wolfrom, John C. Sowden, and E. A. Metcalf, *J. Amer. Chem. Soc.*, **63**, 1688–1691 (1941).

"Osage Orange Pigments. VIII. Oxidation," M. L. Wolfrom and A. S. Gregory, *J. Amer. Chem. Soc.*, **63**, 3356–3358 (1941).

"The β-Form of the Cori Ester (*d*-Glucopyranose 1-Phosphate)," M. L. Wolfrom, C. S. Smith, D. E. Pletcher, and A. E. Brown, *J. Amer. Chem. Soc.*, **64**, 23–26 (1942).

"The Transformation of Tetramethylglucoseen-1,2 into 5-(Methoxymethyl)-2-furaldehyde," M. L. Wolfrom, E. G. Wallace, and E. A. Metcalf, *J. Amer. Chem. Soc.*, **64**, 265–269 (1942).

"Osage Orange Pigments. IX. Improved Separation; Establishment of the Isopropylidene Group," M. L. Wolfrom and John Mahan, *J. Amer. Chem. Soc.*, **64**, 308–311 (1942).

"Osage Orange Pigments. X. Oxidation," M. L. Wolfrom and Sam M. Moffett, *J. Amer. Chem. Soc.*, **64**, 311–315 (1942).

"Survey of the Literature of Cellulose and Allied Substances, 1938–1940. I. General Chemical Properties of Cellulose," M. L. Wolfrom and Paul W. Morgan, *Paper Trade J.*, Technical Assoc. Papers, Ser. 25, 1–6 (June 1942).

"The Action of Diazomethane upon Acyclic Sugar Derivatives. II," M. L. Wolfrom, S. W. Waisbrot, and Robert L. Brown, *J. Amer. Chem. Soc.*, **64**, 1701–1704 (1942).

"Crystalline Xylitol," M. L. Wolfrom and E. J. Kohn, *J. Amer. Chem. Soc.*, **64**, 1739 (1942).

"O-Pentaacetyl-*d*-gluconates of Polyhydric Alcohols and Cellulose," M. L. Wolfrom and P. W. Morgan, *J. Amer. Chem. Soc.*, **64**, 2026–2028 (1942).

"The Action of Diazomethane upon Acyclic Sugar Derivatives. III. A New Synthesis of Ketoses and of their Open Chain (*keto*) Acetates," M. L. Wolfrom, S. W. Waisbrot, and Robert L. Brown, *J. Amer. Chem. Soc.*, **64**, 2329–2331 (1942).

"Application of the Mercaptalation Assay to Synthetic Starch," M. L. Wolfrom, C. S. Smith, and A. E. Brown, *J. Amer. Chem. Soc.*, **65**, 255–259 (1943).

"Sorbityl Glycosides and 2,3,4,5,6-O-Pentamethyl Sorbitol," M. L. Wolfrom and

Thomas S. Gardner, *J. Amer. Chem. Soc.*, **65**, 750–752 (1943).

"Derivatives of the Aldehydrol Form of Sugars. V. Rotatory Power," M. L. Wolfrom and Robert L. Brown, *J. Amer. Chem. Soc.*, **65**, 951–953 (1943).

"Chemical Studies on Crystalline Barium Acid Heparinate," M. L. Wolfrom, D. I. Weisblat, R. J. Morris, C. D. DeWalt, J. V. Karabinos, and Jay McLean, *Science*, **97**, 450 (1943).

"Chemical Studies on Crystalline Barium Acid Heparinate," M. L. Wolfrom, D. I. Weisblat, J. V. Karabinos, W. H. McNeely, and Jay McLean, *J. Amer. Chem. Soc.*, **65**, 2077–2085 (1943).

"The Action of Diazomethane upon Acyclic Sugar Derivatives. IV. Ketose Synthesis," M. L. Wolfrom, Robert L. Brown, and Evan F. Evans, *J. Amer. Chem. Soc.*, **65**, 1021–1027 (1943).

"The Structures of Osajin and Pomiferin," M. L. Wolfrom, George F. Johnson, W. D. Harris, and B. S. Wildi, *J. Amer. Chem. Soc.*, **65**, 1434 (1943).

"The Action of Diazomethane upon Acyclic Sugar Derivatives. V. Halogen Derivatives," M. L. Wolfrom and Robert L. Brown, *J. Amer. Chem. Soc.*, **65**, 1516–1521 (1943).

"Carbohydrates. I." M. L. Wolfrom, in "Organic Chemistry," Vol. II. Chap. 20, Henry Gilman, Ed., 2nd edition, John Wiley and Sons, Inc., 1944, pp. 1532–1604.

"The Action of Diazomethane upon Acyclic Sugar Derivatives. VI. D-Sorbose," M. L. Wolfrom, S. M. Olin, and Evan F. Evans, *J. Amer. Chem. Soc.*, **66**, 204–206 (1944).

"An Acyclic Sugar Orthoacetate," M. L. Wolfrom and D. I. Weisblat, *J. Amer. Chem. Soc.*, **66**, 805–806 (1944).

"Carbonyl Reduction by Thioacetal Hydrogenolysis," M. L. Wolfrom and J. V. Karabinos, *J. Amer. Chem. Soc.*, **66**, 909–911 (1944).

"Cellulose and Cellulose Derivatives," by Emil Ott, Editor (book review), M. L. Wolfrom, *J. Chem. Educ.*, **21**, 519–520 (1944).

"The Chemistry of Cellulose," by Emil Heuser (book review), M. L. Wolfrom, *Chem. Met. Eng.*, **51**, 199 (1944).

"Ring Closure Studies in the D-Glucose Structure," M. L. Wolfrom, S. W. Waisbrot, D. I. Weisblat, and A. Thompson, *J. Amer. Chem. Soc.*, **66**, 2063–2065 (1944).

"The Reactivity of the Monothioacetals of Glucose and Galactose in Relation to Furanoside Synthesis," M. L. Wolfrom, D. I. Weisblat, and A. R. Hanze, *J. Amer. Chem. Soc.*, **66**, 2065–2068 (1944).

"The Identification of Aldose Sugars by their Mercaptal Acetates," M. L. Wolfrom and J. V. Karabinos, *J. Amer. Chem. Soc.*, **67**, 500–501 (1945).

"Separation of Sugar Acetates by Chromatography," W. H. McNeely, W. W. Binkley, and M. L. Wolfrom, *J. Amer. Chem. Soc.*, **67**, 527–529 (1945).

"Heparin—Hydrolytic Characteristics," M. L. Wolfrom and J. V. Karabinos, *J. Amer. Chem. Soc.*, **67**, 679–680 (1945).

"The Relation between the Structure of Heparin and its Anticoagulant Activity," M. L. Wolfrom and W. H. McNeely, *J. Amer. Chem. Soc.*, **67**, 748–753 (1945).

"L-Talitol," F. L. Humoller, M. L. Wolfrom, B. W. Lew, and R. Max Goepp, Jr., *J. Amer. Chem. Soc.*, **67**, 1226 (1945).

"The Species Specificity of Heparin," M. L. Wolfrom, J. V. Karabinos, C. S. Smith, P. H. Ohliger, J. Lee, and O. Keller, *J. Amer. Chem. Soc.*, **67**, 1624 (1945).

"Isolation of Constituents of Cane Juice and Blackstrap Molasses by Chromatographic Methods," W. W. Binkley, Mary Grace Blair, and M. L. Wolfrom, *J. Amer. Chem. Soc.*, **67**, 1789–1793 (1945).

"The Action of Diazomethane upon Acyclic Sugar Derivatives. VII. D-Psicose," M. L. Wolfrom, A. Thompson, and Evan F. Evans, *J. Amer. Chem. Soc.*, **67**, 1793–1797 (1945).

"Chromatography of Carbohydrates and Some Related Compounds," B. W. Lew, M. L. Wolfrom, and R. Max Goepp, Jr., *J. Amer. Chem. Soc.*, **67**, 1865 (1945).

Letter to Editor, M. L. Wolfrom, *J. Chem. Educ.*, **22**, 299 (1945).

Advances in Carbohydrate Chemistry, Vol. 1, edited by W. W. Pigman and M. L. Wolfrom, Academic Press, New York, 1945.

"Sugar Interconversion under Reducing Conditions. I," M. L. Wolfrom, M. Konigsberg, F. B. Moody, and R. Max Goepp, Jr., *J. Amer. Chem. Soc.*, **68**, 122–126 (1946).

"Osage Orange Pigments. XI. Complete Structures of Osajin and Pomiferin," M. L. Wolfrom, Walter D. Harris, George F. Johnson, J. E. Mahan, Sam M. Moffett, and Bernard Wildi, *J. Amer. Chem. Soc.*, **68**, 406–418 (1946).

"The Uronic Acid Component of Heparin," M. L. Wolfrom and F. A. H. Rice, *J. Amer. Chem. Soc.*, **68**, 532 (1946).

"Organic Preparations," by C. Weygand (book review), M. L. Wolfrom, *Science*, **103**, 408 (1946).

"Sugar Interconversion under Reducing Conditions. II," M. L. Wolfrom, F. B. Moody, M. Konigsberg, and R. Max Goepp, Jr., *J. Amer. Chem. Soc.*, **68**, 578–580 (1946).

"A New Aldehyde Synthesis," M. L. Wolfrom and J. V. Karabinos, *J. Amer. Chem. Soc.*, **68**, 724 (1946).

"Wood Chemistry," by Louis E. Wise, Editor (book review), M. L. Wolfrom, *J. Amer. Chem. Soc.*, **68**, 918 (1946).

"L-Fructose," M. L. Wolfrom and Alva Thompson, *J. Amer. Chem. Soc.*, **68**, 791–793 (1946).

"Sugar Interconversion under Reducing Conditions. III," M. L. Wolfrom, B. W. Lew, and R. Max Goepp, Jr., *J. Amer. Chem. Soc.*, **68**, 1443–1448 (1946).

"Chromatography of Sugars and Related Polyhydroxy Compounds," B. W. Lew, M. L. Wolfrom, and R. Max Goepp, Jr., *J. Amer. Chem. Soc.*, **68**, 1449–1453 (1946).

"D-Gluco-L-*tagato*-octose," M. L. Wolfrom and Alva Thompson," *J. Amer. Chem. Soc.*, **68**, 1453–1455 (1946).

"A New Aldehyde Synthesis," M. L. Wolfrom and J. V. Karabinos, *J. Amer. Chem. Soc.*, **68**, 1455–1456 (1946).

"Isolation of Aldonic Acid Lactones through their Hydrazides," Alva Thompson and M. L. Wolfrom, *J. Amer. Chem. Soc.*, **68**, 1509–1510 (1946).

"An Improved Synthesis of N-Methyl-L-glucosaminic Acid," M. L. Wolform, Alva Thompson, and I. R. Hooper, *Science*, **104**, 276–277 (1946).

"Chromatographic Isolation of Cane Juice Constituents," W. W. Binkley and M. L. Wolfrom, *J. Amer. Chem. Soc.*, **68**, 1720–1721 (1946).

"Chemical Interactions of Amino Compounds and Sugars. I," Liebe F. Cavalieri and M. L. Wolfrom, *J. Amer. Chem. Soc.*, **68**, 2022–2025 (1946).

"Acetylation of D-Mannose Phenylhydrazone," M. L. Wolfrom and Mary Grace Blair, *J. Amer. Chem. Soc.*, **68**, 2110 (1946).

"Degradative Studies on Streptomycin," I. R. Hooper, L. H. Klemm, W. J. Polglase, and M. L. Wolfrom, *J. Amer. Chem. Soc.*, **68**, 2120 (1946).

"Chromatography of Sugars and their Derivatives," L. W. Georges, R. S. Bower, and M. L. Wolfrom, *J. Amer. Chem. Soc.*, **68**, 2169–2171 (1946).

"Isosucrose Synthesis," W. W. Binkley and M. L. Wolfrom, *J. Amer. Chem. Soc.*, **68**, 2171–2172 (1946).

"Sugar Interconversion under Reducing Conditions. IV. D-L-Glucitol," M. L. Wolfrom, B. W. Lew, R. A. Hales, and R. Max Goepp, Jr., *J. Amer. Chem. Soc.*, **68**, 2342–2343 (1946).

"N-Methyl-L-glucosaminic Acid," M. L. Wolfrom, Alva Thompson, and I. R. Hooper, *J. Amer. Chem. Soc.*, **68**, 2343–2345 (1946).

"Degradative Studies on Streptomycin," R. U. Lemieux, W. J. Polglase, C. W. DeWalt, and M. L. Wolfrom, *J. Amer. Chem. Soc.*, **68**, 2747 (1946).

Advances in Carbohydrate Chemistry, Vol. 2, edited by W. W. Pigman, M. L. Wolfrom, and S. Peat, Academic Press, New York, 1946.

"The Reductive Acetolysis of Nitrate Esters," D. O. Hoffman, R. S. Bower, and M. L. Wolfrom, *J. Amer. Chem. Soc.*, **69**, 249–250 (1947).

"The Crystalline Octaacetate of 6-α-D-Glucopyranosido-β-D-glucose," L. W. Georges, I. L. Miller, and M. L. Wolfrom, *J. Amer. Chem. Soc.*, **69**, 473 (1947).

"Recovery of Sucrose from Cane Blackstrap and Beet Molasses," W. W. Binkley and M. L. Wolfrom, *J. Amer. Chem. Soc.*, **69**, 664–665 (1947).

"Determination of Methoxyl and Ethoxyl Groups in Acetals and Easily Volatile Alcohols," D. O. Hoffman and M. L. Wolfrom, *Anal. Chem.*, **19**, 225–228 (1947).

"Degradative Studies on Streptomycin. I.," I. R. Hooper, L. H. Klemm, W. J. Polglase, and M. L. Wolfrom, *J. Amer. Chem. Soc.*, **69**, 1052–1056 (1947).

"The Composition of Crystalline Secretin Picrolonate," Harry Greengard, M. L. Wolfrom, and R. K. Ness, *Fed. Proc.*, **6**, 115 (1947).

"Laboratory Disposal of Mercaptan Vapors," Hubert M. Hill and M. L. Wolfrom, *J. Amer. Chem. Soc.*, **69**, 1539 (1947).

"Dibenzylidene-xylitol and its Reaction with Tetraacetyl-D-glucosyl Bromide; Dibenzylidene-L-fucitol," M. L. Wolfrom, W. J. Burke, and E. A. Metcalf, *J. Amer. Chem. Soc.*, **69**, 1667–1668 (1947).

"The Uronic Acid Component of Mucoitinsulfuric Acid," M. L. Wolfrom and F. A. H. Rice, *J. Amer. Chem. Soc.*, **69**, 1833 (1947).

"Degradative Studies on Streptomycin," R. U. Lemieux, C. W. DeWalt, and M. L. Wolfrom, *J. Amer. Chem. Soc.*, **69**, 1838 (1947).

"Scission of Semicarbazones with Nitrous Acid," M. L. Wolfrom, *Rec. Trav. Chim. Pays-Bas*, **66**, 238 (1947).

"A Galactogen from Beef Lung," M. L. Wolfrom, D. I. Weisblat, and J. V. Karabinos, *Arch. Biochem.*, **14**, 1–6 (1947).

"Derivatives of N-Methyl-L-glucosaminic Acid; N-Methyl-L-mannosaminic Acid," M. L. Wolfrom and Alva Thompson, *J. Amer. Chem. Soc.*, **69**, 1847–1849 (1947).

"Chemical Interactions of Amino Compounds and Sugars. II. Methylation Experiments," M. L. Wolfrom, Liebe F. Cavalieri, and Doris K. Cavalieri, *J. Amer. Chem. Soc.*, **69**, 2411–2413 (1947).

"Electrophoretic Resolution of Heparin and Related Polysaccharides," M. L. Wolfrom and F. A. H. Rice, *J. Amer. Chem. Soc.*, **69**, 2918 (1947).

Letter to Editor, M. L. Wolfrom, *Chem. Eng. News*, **26**, 278 (1948).

"Chromatography of Cuban Blackstrap Molasses on Clay; Some Constituents of an Odor and Pigment Fraction," W. W. Binkley and M. L. Wolfrom, *J. Amer. Chem. Soc.*, **70**, 290–292 (1948).

"Chemical Interactions of Amino Compounds and Sugars. III. The Conversion of D-Glucose to 5-(Hydroxymethyl)-2-furaldehyde," M. L. Wolfrom, R. D. Schuetz, and Liebe F. Cavalieri, *J. Amer. Chem. Soc.*, **70**, 514–517 (1948).

"A Synthesis of Streptidine," M. L. Wolfrom and W. J. Polglase, *J. Amer. Chem. Soc.*, **70**, 1672 (1948).

"Action of Heat on D-Fructose. Isolation of Diheterolevulosan and a New Di-D-fructose Dianhydride," M. L. Wolfrom and Mary Grace Blair, *J. Amer. Chem. Soc.*, **70**, 2406–2409 (1948).

"Ribitol Pentaacetate," W. W. Binkley and M. L. Wolfrom, *J. Amer. Chem. Soc.*, **70**, 2809 (1948).

"Degradative Studies on Streptomycin," M. L. Wolfrom and W. J. Polglase, *J. Amer. Chem. Soc.*, **70**, 2835 (1948).

"The Configuration of Streptose," M. L. Wolfrom and C. W. DeWalt, *J. Amer. Chem. Soc.*, **70**, 3148 (1948).

"The Homogeneity of the Phenylosazone Prepared from D-Fructose," W. W. Binkley and M. L. Wolfrom, *J. Amer. Chem. Soc.*, **70**, 3507 (1948).

"Acetylation of D-Psicose," W. W. Binkley and M. L. Wolfrom, *J. Amer. Chem. Soc.*, **70**, 3940 (1948).

"Chromatography of Sugars and Related Substances," Wendell W. Binkley and Melville L. Wolfrom (monograph), Scientific Report Series, No. 10, Sugar Research Foundation, New York (1948).

Rudolph Maximillian Goepp, Jr., [Obituary of], M. L. Wolfrom, *Advan. Carbohyd. Chem.*, **3**, xv–xxiii (1948).

"The Chemistry of Streptomycin," R. U. Lemieux and M. L. Wolfrom, *Advan. Carbohyd. Chem.*, **3**, 337–384 (1948).

Advances in Carbohydrate Chemistry, Vol. 3, edited by W. W. Pigman, M. L. Wolfrom, and S. Peat, Academic Press, New York, 1945.

"Synthetic Methods of Organic Chemistry," by W. Theilheimer (book review), M. L. Wolfrom, *Arch. Biochem.*, **22**, 162 (1949).

"Crystalline Derivatives of Isomaltose," M. L. Wolfrom, L. W. Georges, and I. L. Miller, *J. Amer. Chem. Soc.*, **71**, 125–127 (1949).

"A Polymer-homologous Series of Sugar Acetates from the Acetolysis of Cellulose," E. E. Dickey and M. L. Wolfrom, *J. Amer. Chem. Soc.*, **71**, 825–828 (1949).

"The Chemistry of the Polysaccharides," by Robert J. McIlroy (book review), M. L. Wolfrom, *J. Polym. Sci.*, **4**, 404 (1949).

"Synthesis of Perseulose (L-Galaheptulose)," M. L. Wolfrom, J. M. Berkebile, and A. Thompson, *J. Amer. Chem. Soc.*, **71**, 2360–2362 (1949).

"Two Ketoöctoses from the D-Galaheptonic Acids," M. L. Wolfrom and Pascal W. Cooper, *J. Amer. Chem. Soc.*, **71**, 2668–2671 (1949).

"Configurational Correlation of L-(*levo*)-Glyceraldehyde with Natural (*dextro*)-Alanine by a Direct Chemical Method," M. L. Wolfrom, R. U. Lemieux, and S. M. Olin, *J. Amer. Chem. Soc.*, **71**, 2870–2873 (1949).

"Enzymic Hydrolysis of Amylopectin. Isolation of a Crystalline Trisaccharide Hendecaacetate," M. L. Wolfrom, L. W. Georges, Alva Thompson, and I. L. Miller, *J. Amer. Chem. Soc.*, **71**, 2873–2875 (1949).

"Racemic Glucose," M. L. Wolfrom and H. B. Wood, *J. Amer. Chem. Soc.*, **71**, 3175–3176 (1949).

"An Introduction to the Chemistry of Carbohydrates," by John Honeyman, (book review), M. L. Wolfrom, *J. Amer. Chem. Soc.*, **71**, 3268 (1949).

"Maltotriose and its Crystalline β-D-Hendecaacetate," J. M. Sugihara and M. L. Wolfrom, *J. Amer. Chem. Soc.*, **71**, 3357–3359 (1949).

"2-Methylcellulose," J. M. Sugihara and M. L. Wolfrom, *J. Amer. Chem. Soc.*, **71**, 3509–3510 (1949).

"Chemical Interactions of Amino Compounds and Sugars. IV. Significance of Furan Derivatives in Color Formation," M. L. Wolfrom, R. D. Schuetz, and Liebe F. Cavalieri, *J. Amer. Chem. Soc.*, **71**, 3518–3523 (1949).

Advances in Carbohydrate Chemistry, Vol. 4, edited by W. W. Pigman, M. L. Wolfrom, and S. Peat, Academic Press, New York, 1949.

"Degradation of Glycogen to Isomaltose," M. L. Wolfrom and A. N. O'Neill, *J. Amer. Chem. Soc.*, **71**, 3857 (1949).

"Configurational Correlation of (*levo*)-Glyceraldehyde with (*dextro*)-Lactic Acid by a New Chemical Method," M. L. Wolfrom, R. U. Lemieux, S. M. Olin, and D. I. Weisblat, *J. Amer. Chem. Soc.*, **71**, 4057–4059 (1949).

"D-Manno-L-*fructo*-octose," M. L. Wolfrom and Pascal W. Cooper, *J. Amer. Chem. Soc.*, **72**, 1345–1347 (1950).

"Acid Degradation of Amylopectin to Isomaltose," M. L. Wolfrom, J. T. Tyree, T. T. Galkowski, and A. N. O'Neill, *J. Amer. Chem. Soc.*, **72**, 1427 (1950).

"Chemistry of the Carbohydrates," M. L. Wolfrom, in "Agricultural Chemistry—A Reference Text," Donald E. H. Frear, Ed., Chap. I of Part I, D. Van Nostrand Co., Inc., New York, 1950, pp. 3–59.

"Chromatographic Separation of Carbohydrates," U. S. Patent 2,504,169 (April 18, 1950), Melville L. Wolfrom and Wilfred Wendell Binkley.

"A Synthesis of Streptidine," M. L. Wolfrom, S. M. Olin, and W. J. Polglase, *J. Amer. Chem. Soc.*, **72**, 1724–1729 (1950).

"The Chemistry of Penicillin," by H. T. Clarke, J. R. Johnson, and R. Robinson, Editors (book review), M. L. Wolfrom, *J. Chem. Educ.*, **27**, 352 (1950).

Carbohydrate Chemistry," M. L. Wolfrom and J. M. Sugihara, *Ann. Rev. Biochem.*, **19**, 67–88 (1950).

"Acetylation Desulfation of Carbohydrate Acid Sulfates," M. L. Wolfrom and Rex Montgomery, *J. Amer. Chem. Soc.*, **72**, 2859–2861 (1950).

"Improvements in or Relating to the Separation of Sugars and Related Substances," Cuban Patent No. 13,919 (August 10, 1950), Melville L. Wolfrom and Wilfred W. Binkley.

"The Structures of β-Diacetone-D-fructose and β-Monoacetone-D-fructose," M. L. Wolfrom, Wilbur L. Shilling, and W. W. Binkley, *J. Amer. Chem. Soc.*, **72**, 4544–4545 (1950).

"Chromatographic Fractionation of Cane Blackstrap Molasses and of Its Fermentation Residue," W. W. Binkley and M. L. Wolfrom, *J. Amer. Chem. Soc.*, **72**, 4778–4782 (1950).

"Organic Chemistry," by Paul Karrer (book review), *J. Amer. Oil Chem. Soc.*, **27**, 27–28 (1950).

"Chemical Interactions of Amino Compounds and Sugars. V. Comparative Studies with D-Xylose and 2-Furaldehyde," Tzi-Lieh Tan, M. L. Wolfrom, and A. W. Langer, Jr., *J. Amer. Chem. Soc.*, **72**, 5090–5095 (1950).

"Chromatographic Separation of Carbohydrates," U. S. Patent 2,524,414 (October 3, 1950), Melville L. Wolfrom and Baak W. Lew.

"Reduktone," by H. von Euler and H. Hasselquist (book review), M. L. Wolfrom, *Chem. Eng. News*, **28**, 4208 (1950).

"The Structure of the Caryophyllenes," M. L. Wolfrom and Abraham Mishkin, *J. Amer. Chem. Soc.*, **72**, 5350 (1950).

"The Structure of Heparin," M. L. Wolfrom, Rex Montgomery, J. V. Karabinos, and P. Rathgeb, *J. Amer. Chem. Soc.*, **72**, 5796 (1950).

"Chemistry and Industry of Starch," edited by Ralph W. Kerr (book review), M. L. Wolfrom, *J. Polym. Sci.*, **6**, 254 (1951).

"Osage Orange Pigments. XII. Synthesis of Dihydro-iso-osajin and of Dihydro-isopomiferin," M. L. Wolfrom and Bernard S. Wildi, *J. Amer. Chem. Soc.*, **73**, 235–241 (1951).

"Degradation of Glycogen to Isomaltose," M. L. Wolfrom, E. N. Lassettre, and A. N. O'Neill, *J. Amer. Chem. Soc.*, **73**, 595–599 (1951).

"L-Mannoheptulose (L-Manno-L-*tagato*-heptose)," M. L. Wolfrom and Harry B. Wood, *J. Amer. Chem. Soc.*, **73**, 730–733 (1951).

"A Simple Acetolysis of Nitrate Esters," M. L. Wolfrom, R. S. Bower, and G. G. Maher, *J. Amer. Chem. Soc.*, **73**, 874 (1951).

"Chromatography of Sugars and Their Derivatives; Aldonamides," M. L. Wolfrom, R.

S. Bower, and G. G. Maher, *J. Amer. Chem. Soc.*, **73**, 875 (1951).

"Vegetable Gums and Resins," by F. N. Howes (book review), M. L. Wolfrom, *Arch. Biochem. Biophys.*, **31**, 156–157 (1951).

"Biochemistry of Glucuronic Acid," by Neal E. Artz and Elizabeth M. Osman (book review), M. L. Wolfrom, *Chem. Eng. News*, **29**, 2324 (1951).

"Sodium Borohydride as a Reducing Agent for Sugar Lactones," M. L. Wolfrom and Harry B. Wood, *J. Amer. Chem. Soc.*, **73**, 2933–2934 (1951).

"The Gentiobiose Heptaacetates," Alva Thompson and M. L. Wolfrom, *J. Amer. Chem. Soc.*, **73**, 2966 (1951).

"A Dodecitol from the Alkaline Electroreduction of D-Glucose," M. L. Wolfrom, W. W. Binkley, C. C. Spencer, and B. W. Lew, *J. Amer. Chem. Soc.*, **73**, 3357–3358 (1951).

"Action of Heat on D-Fructose. II. Structure of Diheterolevulosan II," M. L. Wolfrom, W. W. Binkley, W. L. Shilling, and H. W. Hilton, *J. Amer. Chem. Soc.*, **73**, 3553–3557 (1951).

"Action of Heat on D-Fructose. III. Interconversion to D-Glucose," M. L. Wolfrom and Wilbur L. Shilling, *J. Amer. Chem. Soc.*, **73**, 3557–3558 (1951).

"Some Carbohydrate and Natural Product Research in the Department of Chemistry," M. L. Wolfrom, *The Graduate School Record*, The Ohio State University, **4**, No. 11, 4–6 (1951).

"4-α-Isomaltopyranosyl-D-glucose," M. L. Wolfrom, A. Thompson, and T. T. Galkowski, *J. Amer. Chem. Soc.*, **73**, 4093–4095 (1951).

"Acid Degradation of Amylopectin to Isomaltose and Maltotriose," M. L. Wolfrom, J. T. Tyree, T. T. Galkowski, and A. N. O'Neill, *J. Amer. Chem. Soc.*, **73**, 4927–4929 (1951).

"Degradation of Amylopectin to Panose," A. Thompson and M. L. Wolfrom, *J. Amer. Chem. Soc.*, **73**, 5849–5850 (1951).

"Isomaltitol," M. L. Wolfrom, A. Thompson, A. N. O'Neill, and T. T. Galkowski, *J. Amer. Chem. Soc.*, **74**, 1062–1064 (1952).

"Lactitol Dihydrate," M. L. Wolfrom, Raymond M. Hann, and C. S. Hudson, *J. Amer. Chem. Soc.*, **74**, 1105 (1952).

"The Structure of Chondrosine and of Chondroitinsulfuric Acid," M. L. Wolfrom, R. K. Madison, and M. J. Cron, *J. Amer. Chem. Soc.*, **74**, 1491–1494 (1952).

"Synthesis of Streptamine," U. S. Patent No. 2,590,831 (March 25, 1952), M. L. Wolfrom and S. M. Olin.

"Acyl Derivatives of D-Glucosaminic Acid," M. L. Wolfrom and M. J. Cron, *J. Amer. Chem. Soc.*, **74**, 1715–1716 (1952).

"Molasses: Important but Neglected Product of Sugar Cane," M. L. Wolfrom, W. W. Binkley, and L. F. Martin, *Sugar*, **47**, No. 5, 33–34 (1952).

"Synthesis of Sedoheptulose (D-Altroheptulose)," M. L. Wolfrom, J. M. Berkebile, and A. Thompson, *J. Amer. Chem. Soc.*, **74**, 2197–2198 (1952).

"A New Di-D-fructose Dianhydride," M. L. Wolfrom, H. W. Hilton, and W. W. Binkley, *J. Amer. Chem. Soc.*, **74**, 2867–2870 (1952).

"The Structure of Maltotriose," A. Thompson and M. L. Wolfrom, *J. Amer. Chem. Soc.*, **74**, 3612–3614 (1952).

"Correction Concerning Some Reported Derivatives of D-Talitol," C. S. Hudson, M. L. Wolfrom, and T. Y. Shen, *J. Amer. Chem. Soc.*, **74**, 4456 (1952).

"Molecular Structure of the Galactogen from Beef Lung," M. L. Wolfrom, Gordon Sutherland, and Max Schlamowitz, *J. Amer. Chem. Soc.*, **74**, 4883–4886 (1952).

"Effect of Moisture on the Chromatographic Properties of a Synthetic Hydrated Magnesium Silicate," M. L. Wolfrom, Alva Thompson, T. T. Galkowski, and E. J. Quinn, *Anal. Chem.*, **24**, 1670–1671 (1952).

"Some Studies on the Chromatography of Cane Juice and Blackstrap Molasses," M. L. Wolfrom, *El Crisol*, **6**, 67–69 (1952).

"The Polymer-homologous Series of Oligosaccharides from Cellulose," M. L. Wolfrom and J. C. Dacons, *J. Amer. Chem. Soc.*, **74**, 5331–5333 (1952).

"Bimolecular Dianhydrides of L-Sorbose," M. L. Wolfrom and H. W. Hilton, *J. Amer. Chem. Soc.*, **74**, 5334–5336 (1952).

"Sodium Borohydride as a Reducing Agent in the Sugar Series," M. L. Wolfrom and Kimiko Anno, *J. Amer. Chem. Soc.*, **74**, 5583–5584 (1952).

"Acetylated Thioacetals of D-Glucosamine," M. L. Wolfrom and Kimiko Anno, *J. Amer. Chem. Soc.*, **74**, 6150–6151 (1952).

"Improved Preparation of Stachyose," M. L. Wolfrom, R. C. Burrell, A. Thompson, and S. S. Furst, *J. Amer. Chem. Soc.*, **74**, 6299 (1952).

Advances in Carbohydrate Chemistry, Vol. 7, edited by C. S. Hudson, M. L. Wolfrom, S. M. Cantor, S. Peat, and M. Stacey, Academic Press, Inc., New York, 1952.

"Chemical Interactions of Amino Compounds and Sugars. VI. The Repeating Unit in Browning Polymers," M. L. Wolfrom, R. C. Schlicht, A. W. Langer, Jr., and C. S. Rooney, *J. Amer. Chem. Soc.*, **75**, 1013 (1953).

"D-Xylosamine," M. L. Wolfrom and Kimiko Anno, *J. Amer. Chem. Soc.*, **75**, 1038–1039 (1953).

"Sulfated Nitrogenous Polysaccharides and Their Anticoagulant Activity," M. L. Wolfrom, (Miss) T. M. Shen, and C. G. Summers, *J. Amer. Chem. Soc.*, **75**, 1519 (1953).

"Amino Acids in Cane Juice and Cane Final Molasses," G. N. Kowkabany, W. W. Binkley, and M. L. Wolfrom, *Agr. Food Chem.*, **1**, 84–87 (1953).

"The Uronic Acid Component of Chondroitinsulfuric Acid," M. L. Wolfrom and W. Brock Neely, *J. Amer. Chem. Soc.*, **75**, 2778 (1953).

"Acid Reversion in Relation to Isomaltose as a Starch Hydrolytic Product," A. Thompson, M. L. Wolfrom, and E. J. Quinn, *J. Amer. Chem. Soc.*, **75**, 3003–3004 (1953).

"Chemical Interactions of Amino Compounds and Sugars. VII. pH Dependency," M. L. Wolfrom, Doris K. Kolb, and A. W. Langer, Jr., *J. Amer. Chem. Soc.*, **75**, 3471–3473 (1953).

"Preparation of Crystalline Anhydrous β-Gentiobiose," A. Thompson and M. L. Wolfrom, *J. Amer. Chem. Soc.*, **75**, 3605 (1953).

"Nitrated Aldonic Acids," M. L. Wolfrom and Alex Rosenthal, *J. Amer. Chem. Soc.*, **75**, 3662–3664 (1953).

"Acyclic Acetates of Dialdoses," M. L. Wolfrom and Earl Usdin, *J. Amer. Chem. Soc.*, **75**, 4318–4320 (1953).

Letter to Editor, M. L. Wolfrom, *J. Chem. Educ.*, **30**, 479 (1953).

"Ethyl Trithioorthoglyoxylate," M. L. Wolfrom and Earl Usdin, *J. Amer. Chem. Soc.*, **75**, 4619 (1953).

"Selective Hydroxyl Reactivity in Methyl α-D-Glucopyranoside," M. L. Wolfrom and M. A. El-Taraboulsi, *J. Amer. Chem. Soc.*, **75**, 5350–5352 (1953).

"Chemical Interactions of Amino Compounds and Sugars. VIII. Influence of Water," M. L. Wolfrom and C. S. Rooney, *J. Amer. Chem. Soc.*, **75**, 5435 (1953).

"Polysaccharide Chemistry," by R. L. Whistler and C. L. Smart (book foreword), Academic Press, Inc., New York, p. v (1953).

"Chemical Nature of Heparin," M. L. Wolfrom, *Eng. Expt. Sta. News*, The Ohio State University, **25**, No. 5, 22–25 (1953).

"The Composition of Cane Juice and Cane Final Molasses," W. W. Binkley and M. L. Wolfrom, *Advan. Carbohyd. Chem.*, **8**, 291–314 (1953).

Advances in Carbohydrate Chemistry, Vol. 8, edited by C. S. Hudson, M. L. Wolfrom, S. Peat, M. Stacey, and E. L. Hirst, Academic Press, Inc., New York, 1953.

"Observations on the Crystalline Forms of Galactose," M. L. Wolfrom, Max Schlamowitz, and A. Thompson, *J. Amer. Chem. Soc.*, **76**, 1198 (1954).

"Acid Reversion Products from D-Glucose," A. Thompson, Kimiko Anno, M. L. Wolfrom, and M. Inatome, *J. Amer. Chem. Soc.*, **76**, 1309–1311 (1954).

"Studies on the Odor Fraction of Cane Molasses," M. L. Wolfrom, Wendell W. Binkley, and Florinda Orsatti Bobbio, *El Crisol*, **7**, 35–39 (1954).

"Quantitative Immunochemical Methods as an Aid to Determination of Chemical Structure," Michael Heidelberger, M. L. Wolfrom, and Zacharias Dische, *Fed. Proc.*, **13**, 226–227 (1954).

"Immunological Specificities Involving Galactose," Michael Heidelberger and M. L. Wolfrom, *Fed. Proc.*, **13**, 496–497 (1954).

"Benzylation and Xanthation of Cellulose Monoalkoxide," M. L. Wolfrom and M. A. El-Taraboulsi, *J. Amer. Chem. Soc.*, **76**, 2216–2218 (1954).

"Configuration of the Glycosidic Unions in Streptomycin," M. L. Wolfrom, M. J. Cron, C. W. DeWalt, and R. M. Husband, *J. Amer. Chem. Soc.*, **76**, 3675–3677 (1954).

"Pectic (Poly-D-galacturonic) Hydrazide," M. L. Wolfrom, G. N. Kowkabany, and W. W. Binkley, *J. Amer. Chem. Soc.*, **76**, 4011–4012 (1954).

"Comments on a 'Meso-Carbon Atom'," M. L. Wolfrom, *Proc. Nat. Acad. Sci. U. S.*, **40**, 794–795 (1954).

"Synthesis of Glucothiofuranosides," U. S. Pat. No. 2,691,012 (October 5, 1954), M. L. Wolfrom and S. M. Olin.

"Isomaltose Phenylosazone and Phenylosotriazole," A. Thompson and M. L. Wolfrom, *J. Amer. Chem. Soc.*, **76**, 5173 (1954).

"Observations on the Forms of Allose and Its Phenylosazone," M. L. Wolfrom, J. N. Schumacher, H. S. Isbell, and F. L. Humoller, *J. Amer. Chem. Soc.*, **76**, 5816 (1954).

Advances in Carbohydrate Chemistry, Vol. 9, edited by M. L. Wolfrom, R. S. Tipson, and E. L. Hirst, Academic Press, New York, 1954.

Claude Silbert Hudson [Obituary of], M. L. Wolfrom, *Advan. Carbohyd. Chem.*, **9**, xiii–xviii (1954).

"Synthesis of L-Iduronic Acid and an Improved Production of D-Glucose-6-C^{14}," F. Shafizadeh and M. L. Wolfrom, *J. Amer. Chem. Soc.*, **77**, 2568–2569 (1955).

"Two 3-Epimeric Ketononoses," M. L. Wolfrom and Harry B. Wood, Jr., *J. Amer. Chem. Soc.*, **77**, 3096–3098 (1955).

"Tetraacetates of D-Glucose and D-Galactose," A. Thompson, M. L. Wolfrom, and M. Inatome, *J. Amer. Chem. Soc.*, **77**, 3160 (1955).

"Optical Activity and Configurational Relations in Carbon Compounds," M. L. Wolfrom, *Rec. Chem. Progr.*, **16**, 121–136 (1955).

"The Action of Alkali on D-Fructose," M. L. Wolfrom and J. N. Schumacher, *J. Amer. Chem. Soc.*, **77**, 3318–3323 (1955).

"Color Formation in Sugar Solutions under Simulated Cane Sugar Mill Conditions," M. L. Wolfrom, W. W. Binkley, and J. N. Schumacher, *Ind. Eng. Chem.*, **47**, 1416–1417 (1955).

"Immunochemistry and the Structure of Lung Galactan," Michael Heidelberger, Zacharias Dische, W. Brock Neely, and M. L. Wolfrom, *J. Amer. Chem. Soc.*, **77**, 3511–3514 (1955).

"6-*O*-β-Maltosyl-α-D-glucopyranose Hendecaacetate," A. Thompson and M. L. Wolfrom, *J. Amer. Chem. Soc.*, **77**, 3567–3569 (1955).

William Lloyd Evans [Obituary of], M. L. Wolfrom and Edward Mack, Jr., *J. Amer. Chem. Soc.*, **77**, 4949–4955 (1955).

"Biosynthesis of C^{14}-Labeled Cotton Cellulose from D-Glucose-1-C^{14} and D-Glucose-6-C^{14}," F. Shafizadeh and M. L. Wolfrom, *J. Amer. Chem. Soc.*, **77**, 5182–5183 (1955).

"Starch and Its Derivatives," J. A. Radley, Ed., 3rd Edition (book review), M. L. Wolfrom, *Arch. Biochem. Biophys.*, **58**, 519–520 (1955).

"Cellulose and Cellulose Derivatives," E. Ott, H. M. Spurlin, and Mildred W. Grafflin, Eds., 2nd Edition, Parts I and II (book review), M. L. Wolfrom, *Arch. Biochem. Bioplys.*, **59**, 311–312 (1955).

"Biochemistry of Amino Sugars," by P. W. Kent and M. W. Whitehouse, (book review), *Agr. Food Chem.*, **3**, 920 (1955).

"Polybenzyls from Benzyl Alcohol and Sulfuryl Chloride," R. A. Gibbons, Marian N. Gibbons, and M. L. Wolfrom, *J. Amer. Chem. Soc.*, **77**, 6374 (1955).

"Degradation of Amylopectin to Nigerose," M. L. Wolfrom and A. Thompson, *J. Amer. Chem. Soc.*, **77**, 6403 (1955).

"The Controlled Thermal Decomposition of Cellulose Nitrate. I," M. L. Wolfrom, J. H. Frazer, L. P. Kuhn, E. E. Dickey, S. M. Olin, D. O. Hoffman, R. S. Bower, A. Chaney, Eloise Carpenter, and P. McWain, *J. Amer. Chem. Soc.*, **77**, 6573–6580 (1955).

"Glucose," M. L. Wolfrom, in "The Encyclopedia Americana," New York, 1955, pp. 726–727.

"Glycoside," M. L. Wolfrom, in "The Encyclopedia Americana," New York, 1955, p. 731.

Advances in Carbohydrate Chemistry, Vol. 10, edited by M. L. Wolfrom and R. S. Tipson, Academic Press, Inc., New York, 1955.

"An Evaluation of Hudson's Classical Studies on the Configuration of Sucrose," M. L. Wolfrom and F. Shafizadeh, *J. Org. Chem.*, **21**, 88–89 (1956).

"Dithiocarbonate Esters of Arabinose," M. L. Wolfrom and A. B. Foster, *J. Amer. Chem. Soc.*, **78**, 1399–1403 (1956).

"Ethyl Hydrogen DL-Galactarate and Ethyl DL-Galactarate Lactone, and Their Conversion to a Derivative of α-Pyrone," J. W. W. Morgan and M. L. Wolfrom, *J. Amer. Chem. Soc.*, **78**, 1897–1899 (1956).

"Detection of Carbohydrates on Paper Chromatograms," M. L. Wolfrom and J. B. Miller, *Anal. Chem.*, **28**, 1037 (1956).

"Periodate Oxidation of Cyclic 1,3-Diketones," M. L. Wolfrom and J. M. Bobbitt, *J. Amer. Chem. Soc.*, **78**, 2489–2493 (1956).

"An Anomalous Reaction of Methyl 3,4-*O*-Isopropylidene-β-D-arabinopyranoside 2-*O*-(S-Sodium Dithiocarbonate)," A. B. Foster and M. L. Wolfrom, *J. Amer. Chem. Soc.*, **78**, 2493–2495 (1956).

"Lithium Aluminum Hydride Reduction of 3,4,5,6-Di-*O*-isopropylidene-D-gluconamide and Di-*O*-isopropylidene-galactaramide," J. W. W. Morgan and M. L. Wolfrom, *J. Amer. Chem. Soc.*, **78**, 2496–2497 (1956).

"Biosynthesis of C^{14}-Labeled Cotton Seed Oil from D-Glucose-6-C^{14}," F. Shafizadeh and M. L. Wolfrom, *J. Amer. Chem. Soc.*, **78**, 2498–2499 (1956).

"Problems in Communication: Nomenclature," M. L. Wolfrom, in "Polysaccharides in Biology, Transactions of the First Conference," G. F. Springer, Ed., Josiah Macy, Jr. Foundation, New York, 1956, pp. 9–30.

"Moderne Methoden der Pflanzenanalyse, Vol. II," K. Paesch and M. V. Tracey, Eds. (book review), M. L. Wolfrom, *Arch Biochem. Biophys.*, **62**, 520 (1956).

"Phenylboronates of Pentoses and 6-Deoxyhexoses," M. L. Wolfrom and J. Solms, *J. Org. Chem.*, **21**, 815 (1956).

"Occurrence of the (1 → 3)-Linkages in Starches," M. L. Wolfrom and A. Thompson, *J. Amer. Chem. Soc.*, **78**, 4116–4117 (1956).

"Degradation of Glycogen to Isomaltotriose," M. L. Wolfrom and A. Thompson, *J. Amer. Chem. Soc.*, **78**, 4182 (1956).

"The Controlled Thermal Decomposition of Cellulose Nitrate. II," M. L. Wolfrom, J.

H. Frazer, L. P. Kuhn, E. E. Dickey, S. M. Olin, R. S. Bower, G. G. Maher, J. D. Murdock, A. Chaney, and Eloise Carpenter, *J. Amer. Chem. Soc.*, **78**, 4695–4704 (1956).

"General Determination of Acetyl," Alan Chaney and M. L. Wolfrom, *Anal. Chem.*, **28**, 1614–1615 (1956).

"Cellulose and Cellulose Derivatives," E. Ott, H. M. Spurlin, and Mildred W. Grafflin, Eds., 2nd Edition, Vol. V, part III (book review), M. L. Wolfrom, *Arch. Biochem. Biophys.*, **64**, 520 (1956).

"Carbohydrates: Physical and Chemical Characteristics," M. L. Wolfrom and G. G. Maher, in "Handbook of Biological Data," W. S. Spector, Ed., The National Research Council, W. B. Saunders Co., Philadelphia, Pa., 1956, pp. 6–15.

Advances in Carbohydrate Chemistry, Vol. 11, edited by M. L. Wolfrom and R. S. Tipson, Academic Press, Inc., New York, 1956.

"The Cellodextrins: Preparation and Properties," M. L. Wolfrom, J. C. Dacons, and D. L. Fields, *Tappi*, **39**, 803–806 (1956).

"Synthesis of Inosamine and Inosadiamine Derivatives from Inositol Bromohydrins," M. L. Wolfrom, Jack Radell, R. M. Husband, and G. E. McCasland, *J. Amer. Chem. Soc.*, **79**, 160–164 (1957).

"Reaction of Hydroxylamine with Ethyl α-(2,2,2-Trichloroethylidene)acetoacetate," M. L. Wolfrom, Jack Radell, and R. M. Husband, *J. Org. Chem.*, **22**, 329 (1957).

"A Polymer-Homologous Series of Beta-D-Acetates from Cellulose," M. L. Wolfrom and D. L. Fields, *Tappi*, **40**, 335–337 (1957).

"Acyl Migration in the D-Galactose Structure," M. L. Wolfrom, A. Thompson, and M. Inatome, *J. Amer. Chem. Soc.*, **79**, 3868–3871 (1957).

"Degradation of Glycogen to Isomaltotriose and Nigerose," M. L. Wolfrom and A. Thompson, *J. Amer. Chem. Soc.*, **79**, 4212–4215 (1957).

"Dithioacetals of D-Glucuronic Acid and 2-Amino-2-deoxy-D-galactose," M. L. Wolfrom and K. Onodera, *J. Amer. Chem. Soc.*, **79**, 4737–4740 (1957).

"Derivatives of D-Glucose Containing the Sulfoamino Group," M. L. Wolfrom, R. A. Gibbons, and A. J. Huggard, *J. Amer. Chem. Soc.*, **79**, 5043–5046 (1957).

"Esters," A. Thompson and M. L. Wolfrom, in "The Carbohydrates," W. Pigman, Ed., Academic Press, Inc., New York, 1957, pp. 138–172.

"Glycosides, Simple Acetals, and Thioacetals," M. L. Wolfrom and Alva Thompson, in "The Carbohydrates," W. Pigman, Ed., Academic Press, Inc., New York, 1957, pp. 188–240.

"Chemie der Zucker und Polysaccharide," second edition, by F. Micheel (book review), M. L. Wolfrom, *J. Amer. Chem. Soc.*, **79**, 5330 (1957).

"Glycosidation with Trimethyl Orthoformate and Boron Trifluoride," M. L. Wolfrom, J. W. Spoors, and R. A. Gibbons, *J. Org. Chem.*, **22**, 1513–1514 (1957).

"Reaction of *keto*-Acetates with Diazomethane," M. L. Wolfrom, J. B. Miller, D. I. Weisblat, and A. R. Hanze, *J. Amer. Chem. Soc.*, **79**, 6299–6303 (1957).

"The Action of Diazomethane on the Pentaacetates of *aldehydo*-D-Glucose and *aldehydo*-D-Galactose," M. L. Wolfrom, D. I. Weisblat, Evan F. Evans, and J. B. Miller, *J. Amer. Chem. Soc.*, **79**, 6454–6457 (1957).

Advances in Carbohydrate Chemistry, Vol. 12, edited by M. L. Wolfrom and R. S. Tipson, Academic Press, Inc., New York, 1957.

"Structure, Properties and Occurrence of the Oligosaccharides," F. Shafizadeh and M. L. Wolfrom, in "Encyclopedia of Plant Physiology VI," W. Ruhland, Ed., Springer-Verlag, 1958, pp. 1–46.

"Structure, Properties and Occurrence of the Oligosaccharides," F. Shafizadeh and M. L. Wolfrom, in "Encyclopedia of Plant Physiology VI," W. Ruhland, Ed., Springer-Verlag, 1958, pp. 63–86.

"The Hydrolytic Instability of the Aldonamides," M. L. Wolfrom, R. B. Bennett, and J. D. Crum, *J. Amer. Chem. Soc.*, **80**, 944–946 (1957).

"The Controlled Thermal Decomposition of Cellulose Nitrate. III," M. L. Wolfrom, Alan Chaney, and P. McWain, *J. Amer. Chem. Soc.*, **80**, 946–950 (1958).

"The Controlled Thermal Decomposition of Cellulose Nitrate. IV. C^{14}-Tracer Experiments," F. Shafizadeh and M. L. Wolfrom, *J. Amer. Chem. Soc.*, **80**, 1675–1677 (1958).

"The Reduction of Diazomethyl-*keto* Acetates; A New Route to Osone Derivatives," M. L. Wolfrom and J. B. Miller, *J. Amer. Chem. Soc.*, **80**, 1678–1680 (1958).

"Synthesis of Amino Compounds in the Sugar Series by Penylhydrazone Reduction," M. L. Wolfrom, F. Shafizadeh, J. O. Wehrmüller, and R. K. Armstrong, *J. Org. Chem.*, **23**, 571–575 (1958).

"Structures of Isomaltose and Gentiobiose," M. L. Wolfrom, A. Thompson, and A. M. Brownstein, *J. Amer. Chem. Soc.*, **80**, 2015–2018 (1958).

"Sulfated Aminopolysaccharides," U. S. Pat. 2,832,766 (April 29, 1958), M. L. Wolfrom.

"A Polymer-Homologous Series of Oligosaccharide Alditols from Cellulose," M. L. Wolfrom and D. L. Fields, *Tappi*, **41**, 204–207 (1958).

"Synthesis of Amino Sugars by Reduction of Hydrazine Derivatives," M. L. Wolfrom, F. Shafizadeh, and R. K. Armstrong, *J. Amer. Chem. Soc.*, **80**, 4885–4888 (1958).

"Methyl Glucoside," by G. N. Bollenback (book review), M. L. Wolfrom, *J. Amer. Chem. Soc.*, **80**, 5008 (1958).

"Nitrogen-Containing Polymers Arising from 1,2:5,6-Dianhydro-3,4-O-isopropyl-idene-D-mannitol," M. L. Wolfrom, J. O. Wehrmüller, E. P. Swan, and A. Chaney, *J. Org. Chem.*, **23**, 1556–1557 (1958).

"Paramagentic Resonance Study of Irradiation Damage in Crystalline Carbohydrates," Dudley Williams, J. E. Geusic, M. L. Wolfrom, and Leo J. McCabe, *Proc. Nat. Acad. Sci. U. S.*, **44**, 1128–1136 (1958).

Harold Hibbert (biographical memoir), M. L. Wolfrom, *Nat. Acad. Sci. U. S.*, **32**, 146–180 (1958).

Claude Silbert Hudson (biographical memoir), Lyndon F. Small and M. L. Wolfrom, *Nat. Acad. Sci. U. S.*, **32**, 181–220 (1958).

"Condensation Polymers from Tetra-O-acetylgalactaroyl Dichloride and Diamines," M. L. Wolfrom, Madeline S. Toy, and Alan Chaney, *J. Amer. Chem. Soc.*, **80**, 6328–6330 (1958).

"Glycosides, Natural," M. L. Wolfrom, in "The Encyclopaedia Britannica," Chicago, Ill., 1958, pp. 448–449.

"The Composition of Pyrodextrins," A. Thompson and M. L. Wolfrom, *J. Amer. Chem. Soc.*, **80**, 6618–6620 (1958).

Advances in Carbohydrate Chemistry, Vol. 13, edited by M. L. Wolfrom and R. S. Tipson, Academic Press, Inc., New York, 1958.

"Chitosan Nitrate," M. L. Wolfrom, G. G. Maher, and Alan Chaney, *J. Org. Chem.*, **23**, 1990–1991 (1958).

"The Action of Diazomethane on the Tetraacetates of *aldehydo*-D-(and L)-Arabinose," M. L. Wolfrom, J. D. Crum, J. B. Miller, and D. I. Weisblat, *J. Amer. Chem. Soc.*, **81**, 243–244 (1959).

"The Effect of Ionizing Radiations on Carbohydrates," M. L. Wolfrom, W. W. Binkley, L. J. McCabe, T. M. Shen Han, and A. M. Michelakis, *Radiat. Res.*, **10**, 37–47 (1959).

"Studies on the Biosynthesis and Fragmentation of C^{14}-Labeled Cotton Cellulose and Seed Oil," M. L. Wolfrom, J. M. Webber, and F. Shafizadeh, *J. Amer. Chem. Soc.*, **81**, 1217–1221 (1959).

"The Controlled Thermal Decomposition of Cellulose Nitrate. V. C^{14}-Tracer Experiments," F. Shafizadeh, M. L. Wolfrom, and P. McWain, *J. Amer. Chem. Soc.*, **81**, 1221–1223 (1959).

"The Effect of Ionizing Radiation on Carbohydrates. The Irradiation of Sucrose and Methyl α-D-Glucopyranoside," M. L. Wolfrom, W. W. Binkley, and Leo J. McCabe, *J. Amer. Chem. Soc.*, **81**, 1442–1446 (1959).

"2,4-Dinitrophenyl Ethers of the Alditols," M. L. Wolfrom, B. O. Juliano, Madeline S. Toy, and A. Chaney, *J. Amer. Chem. Soc.*, **81**, 1446–1448 (1959).

"The Sulfation of Chitosan," M. L. Wolfrom and (Mrs.) T. M. Shen Han, *J. Amer. Chem. Soc.*, **81**, 1764–1766 (1959).

Proceedings of the Fourth International Congress of Biochemistry, Vienna, Sept. 1–6, 1958, Vol. I, Symposium I, Carbohydrate Chemistry of Substances of Biological Interest, edited by M. L. Wolfrom, Pergamon Press, London, 1959.

"The Controlled Thermal Decomposition of Cellulose Nitrate. VI. Other Polymeric Nitrates," M. L. Wolfrom, Alan Chaney, and K. S. Ennor, *J. Amer. Chem. Soc.*, **81**, 3469–3473 (1959).

"Ethyl Glycosides of 2-Acetamido-2-deoxy-1-thio-D-galactose," M. L. Wolfrom and Z. Yosizawa, *J. Amer. Chem. Soc.*, **81**, 3474–3476 (1959).

"Synthesis of 2-Amino-2-deoxy-L-arabinose (L-Arabinosamine)," M. L. Wolfrom and Z. Yosizawa, *J. Amer. Chem. Soc.*, **81**, 3477–3478 (1959).

"Synthesis of Amino Sugars by Reduction of Hydrazine Derivatives; D- and L-Ribosamine, D-Lyxosamine," M. L. Wolfrom, F. Shafizadeh, R. K. Armstrong, and T. M. Shen Han, *J. Amer. Chem. Soc.*, **81**, 3716–3719 (1959).

"The Composition of Pyrodextrins. II. Thermal Polymerization of Levoglucosan," M. L. Wolfrom, A. Thompson, and R. B. Ward, *J. Amer. Chem. Soc.*, **81**, 4623–4625 (1959).

"Derivatives of D-Mannitol 1,2,3,5,6-Pentanitrate," M. L. Wolfrom, E. P. Swan, K. S. Ennor, and A. Chaney, *J. Amer. Chem. Soc.*, **81**, 5701–5705 (1959).

"3,4,5-Tri-O-benzoyl-D-xylose Dimethyl Acetal," M. L. Wolfrom and Walter von Bebenburg, *J. Amer. Chem. Soc.*, **81**, 5705–5706 (1959).

"Synthesis of Isomaltose," M. L. Wolfrom and Ian C. Gillam, *Science*, **130**, 1424 (1959).

Advances in Carbohydrate Chemistry, Vol. 14, edited by M. L. Wolfrom and R. S. Tipson, Academic Press, Inc., New York, 1959.

"Ethyl 1-Thio-α-D-galactofuranoside," M. L. Wolfrom, Z. Yosizawa, and B. O. Juliano, *J. Org. Chem.*, **24**, 1529–1530 (1959).

"9-β-Lactosyladenine and 2,6-Diamino-9-β-lactosylpurine," M. L. Wolfrom, P. McWain, F. Shafizadeh, and A. Thompson, *J. Amer. Chem. Soc.*, **81**, 6080–6082 (1959).

"Paramagnetic Resonance Spectra of Free Radicals Trapped on Irradiation of Crystalline Carbohydrates," Dudley Williams, Bernhard Schmidt, M. L. Wolfrom, A. Michelakis, and Leo J. McCabe, *Proc. Nat. Acad. Sci. U. S.*, **45**, 1744–1751 (1959).

"Heparin and Related Substances," M. L. Wolfrom, in "Polysaccharides in Biology," Transactions of the Fourth Conference, May 21–23, 1958, Princeton, New Jersey, G. F. Springer, Ed., Josiah Macy, Jr. Foundation, New York, 1960, pp. 115–158.

"Carbohydrates of the Coffee Bean," M. L. Wolfrom, R. A. Plunkett, and M. L. Laver *J. Agr. Food Chem.*, **8**, 58–65 (1960).

John Ulric Nef (biographical memoir), M. L. Wolfrom, *Nat. Acad. Sci. U. S.*, **34**, 204–227 (1960).

"Preparation of 2,4-Dinitrophenylhydrazine Derivatives of Highly Oxygenated Carbonyl Compounds," M. L. Wolfrom and G. P. Arsenault, *J. Org. Chem.*, **25**, 205–208 (1960).

"Chondroitin Sulfate Modifications. II. Peracetylated Sodium Chondroitin Sulfate A," M. L. Wolfrom and J. W. Spoors, *J. Org. Chem.*, **25**, 308 (1960).

"Recent Advances in the Chemistry of Cellulose and Starch," J. Honeyman, Ed.

(book review), M. L. Wolfrom, *J. Amer. Chem. Soc.*, **82**, 1520 (1960).

"Chondroitin Sulfate Modifications. I. Carboxyl-reduced Chondroitin and Chondrosine," M. L. Wolfrom and Bienvenido O. Juliano, *J. Amer. Chem. Soc.*, **82**, 1673–1677 (1960).

"Chromatographic Separation of 2,4-Dinitrophenylhydrazine Derivatives of Highly Oxygenated Carbonyl Compounds," M. L. Wolfrom and G. P. Arsenault, *Anal. Chem.*, **32**, 693–695 (1960).

"Chondroitin Sulfate Modifications. III. Sulfated and N-Deacetylated Preparations," M. L. Wolfrom and Bienvenido O. Juliano, *J. Amer. Chem. Soc.*, **82**, 2588–2592 (1960).

"Acetals and Dithioacetals of 2-S-Ethyl-2-thio-D-xylose(lyxose)," M. L. Wolfrom and Walter von Bebenburg, *J. Amer. Chem. Soc.*, **82**, 2817–2819 (1960).

"The Controlled Thermal Decomposition of Cellulose Nitrate. VII. Carbonyl Compounds," M. L. Wolfrom and G. P. Arsenault, *J. Amer. Chem. Soc.*, **82**, 2819–2823 (1960).

"Nitration of Unsaturated Alcohols," M. L. Wolfrom, G. H. McFadden, and Alan Chaney, *J. Org. Chem.*, 25, 1079–1082 (1960).

"Nucleosides of Disaccharides; Cellobiose and Maltose," M. L. Wolfrom, P. McWain, and A. Thompson, *J. Amer. Chem. Soc.*, **82**, 4353–4354 (1960).

Advances in Carbohydrate Chemistry, Vol. 15, edited by M. L. Wolfrom and R. S. Tipson, Academic Press, Inc., New York, 1960.

"Carbohydrates," M. L. Wolfrom and F. Shafizadeh, in "Encyclopaedia Britannica," Chicago, Ill., 1960, pp. 828–833.

"Pectin," M. L. Wolfrom, in "Encyclopaedia Britannica," Chicago, Ill., 1960, pp. 429–430.

"Alditol Arsenite Esters," M. L. Wolfrom and M. J. Holm, *J. Org. Chem.*, **26**, 273–274 (1961).

"Composition of Pyrodextrins," M. L. Wolfrom, A. Thompson, and R. B. Ward, *Ind. Eng. Chem.*, **53**, 217–218 (1961).

"Preparation of Polyvinylamine Perchlorate," M. L. Wolfrom and Alan Chaney, *J. Org. Chem.*, **26**, 1319–1320 (1961).

"Reaction of Heparin with H[14]CN," A. B. Foster, M. Stacey, P. J. M. Taylor, J. M. Webber, and M. L. Wolfrom, *Biochem. J.*, **80**, 13–14P (1961).

"A Chemical Synthesis of Isomaltose," M. L. Wolfrom, A. O. Pittet, and I. C. Gillam, *Proc. Nat. Acad. Sci. U. S.*, **47**, 700–705 (1961).

"Methyl 2-Deoxy-2-sulfoamino-β-D-glucopyranoside Trisulfate and the Preparation of Tri-O-acetyl-2-amino-2-deoxy-α-D-glucopyranosyl Bromide," M. L. Wolfrom and T. M. Shen Han, *J. Org. Chem.*, **26**, 2145–2146 (1961).

"Carboxyl-Reduced Heparin," M. L. Wolfrom, J. R. Vercellotti, and G. H. S. Thomas, *J. Org. Chem.*, **26**, 2160 (1961).

"Color Reactions of 2-Amino-2-deoxyalditol Derivatives," D. Horton, J. Vercellotti, and M. L. Wolfrom, *Biochim. Biophys. Acta*, **50**, 358–359 (1961).

"The Chemistry of Natural Products. Vol. V. The Carbohydrates," by S. F. Dyke (book review), M. L. Wolfrom, *J. Amer. Chem. Soc.*, **83**, 2970–2971 (1961).

"Synthesis of Benzyl Vinylcarbamate and 3-O-Vinylcarbamoyl-D-mannitol Pentanitrate," M. L. Wolfrom, G. H. McFadden, and Alan Chaney, *J. Org. Chem.*, **26**, 2597–2599 (1961).

"An Attempted Synthesis of Phenyl Nitrate," Alan Chaney and M. L. Wolfrom, *J. Org. Chem.*, **26**, 2998 (1961).

"Acyclic Sugar Nucleoside Analogs," M. L. Wolfrom, A. B. Foster, P. McWain, W. von Bebenburg, and A. Thompson, *J. Org. Chem.*, **26**, 3095–3097 (1961).

"Methyl Glycoside Formation from β-D-Glucopyranose Pentanitrate," M. L. Wolfrom and Ian C. Gillam, *J. Org. Chem.* **26**, 3564–3565 (1961).

"Ethyl 2-S-Ethyl-1,2-dithio-5-*aldehydo*-α-D-*xylo*(*lyxo*)pentodialdofuranoside," M. L.

Wolfrom, Walter von Bebenburg, and A. Thompson, *J. Org. Chem.*, **26**, 4151–4152 (1961).

"Carbohydrates of the Coffee Bean. II. Isolation and Characterization of a Mannan," M. L. Wolfrom, M. L. Laver, and D. L. Patin, *J. Org. Chem.*, **26**, 4533–4535 (1961).

"The Composition of Pyrodextrins. III. Thermal Polymerization of Levoglucosan," M. L. Wolfrom, A. Thompson, R. B. Ward, D. Horton, and R. H. Moore, *J. Org. Chem.*, **26**, 4617–4620 (1961).

"Synthesis of Amino Sugars by Reduction of Hydrazine Derivatives. 2-Amino-2-deoxy-L-lyxose (L-Lyxosamine) Hydrochloride," D. Horton, M. L. Wolfrom, and A. Thompson, *J. Org. Chem.*, **26**, 5069–5074 (1961).

Advances in Carbohydrate Chemistry, Vol. 16, edited by M. L. Wolfrom and R. S. Tipson, Academic Press, Inc., New York, 1961.

"John Ulric Nef," M. L. Wolfrom, in "Great Chemists," E. Farber, Ed., Interscience Publishers, New York, 1961, pp. 1129–1143.

"Phoebus Aaron Theodor Levene," M. L. Wolfrom, in "Great Chemists," E. Farber, Ed., Interscience Publishers, New York, 1961, pp. 1313–1324.

"Claude Silbert Hudson," M. L. Wolfrom, in "Great Chemists," E. Farber, Ed., Interscience Publishers, New York, 1961, pp. 1535–1550.

"A Disaccharide from Carboxyl-Reduced Heparin," M. L. Wolfrom, J. R. Vercellotti, and D. Horton, *J. Org. Chem.*, **27**, 705 (1962).

"Process for the Preparation of Diglycosylureas," U. S. Pat. 3,023,205 (February 27, 1962), P. R. Steyermark and M. L. Wolfrom.

Methods in Carbohydrate Chemistry, Vol. I, "Analysis and Preparation of Sugars," edited by Roy L. Whistler and M. L. Wolfrom, 13 chapters by M. L. Wolfrom and A. Thompson, 1 chapter by M. L. Wolfrom and Murray Laver, 1 chapter by C. F. Snyder, Harriet L. Frush, H. S. Isbell, A. Thompson, and M. L. Wolfrom, Academic Press, Inc., New York, 1962, pp. 3, 4–8, 8–11, 118–120, 120–122, 171–172, 202–204, 209–211, 316–318, 334–335, 368–369, 448–453, 454–461, 517–519, 524–534.

"Thiosugars. I. Synthesis of Derivatives of 2-Amino-2-deoxy-1-thio-D-glucose," D. Horton and M. L. Wolfrom, *J. Org. Chem.*, **27**, 1794–1800 (1962).

"The Reaction of Free Carbonyl Sugar Derivatives with Organometallic Reagents. I. 6-Deoxy-L-idose and Derivatives," M. L. Wolfrom and S. Hanessian, *J. Org. Chem.*, **27**, 1800–1804 (1962).

"Reaction of Free Carbonyl Sugar Derivatives with Organometallic Reagents. II. 6-Deoxy-L-idose and a Branched-Chain Sugar," M. L. Wolfrom and S. Hanessian, *J. Org. Chem.*, **27**, 2107–2109 (1962).

"5-S-Ethyl-5-thio-L-arabinose Diethyl Dithioacetal," M. L. Wolfrom and T. E. Whiteley, *J. Org. Chem.*, **27**, 2109–2110 (1962).

"Olefinic Structures from Acetylated Phenylhydrazones of Sugars," M. L. Wolfrom, A. Thompson, and D. R. Lineback, *J. Org. Chem.*, **27**, 2563–2567 (1962).

"Simple Pigment Structures Containing Condensed Isoprenoid Units," M. L. Wolfrom and F. Komitsky, Jr., in "Chemistry of Natural and Synthetic Colouring Matters and Related Fields," Academic Press, London, 1962, pp. 287–300.

"Hydrolytic Sensitivity of the Sulfoamino Group in Heparin and in Model Compounds," R. A. Gibbons and M. L. Wolfrom, *Arch. Biochem. Biophys.*, **98**, 374–378 (1962).

"Acyclic Sugar Nucleoside Analogs. II. Sulfur Derivatives," M. L. Wolfrom, P. McWain, and A. Thompson, *J. Org. Chem.*, **27**, 3549–3551 (1962).

"Synthesis of Amino Sugars by Reduction of Hydrazine Derivatives. 5-Amino-3,6-anhydro-5-deoxy-L-idose Derivatives," M. L. Wolfrom, J. Bernsmann, and D. Horton, *J. Org. Chem.*, **27**, 4505–4509 (1962).

Advances in Carbohydrate Chemistry, Vol. 17, edited by M. L. Wolfrom and R. S. Tipson, Academic Press, Inc., New York, 1962.

"Synthesis of 2-Acetamido-2-deoxy-L-xylose," M. L. Wolfrom, D. Horton, and A. Böckmann, *Chem. Ind.* (London), 41 (1963).

"Comparative Hydrolysis Rates of the Reducing Disaccharides of D-Glucopyranose," M. L. Wolfrom, A. Thompson, and C. E. Timberlake, *Cereal Chem.*, **40**, 82–86 (1963).

"The Linkage in a Disaccharide from Carboxyl-reduced Heparin," M. L. Wolfrom, J. R. Vercellotti, and D. Horton, *J. Org. Chem.*, **28**, 278 (1963).

"A Second Disaccharide from Carboxyl-reduced Heparin," M. L. Wolfrom, J. R. Vercellotti, and D. Horton, *J. Org. Chem.*, **28**, 279 (1963).

Methods in Carbohydrate Chemistry, Vol. II, "Reactions of Carbohydrates," edited by Roy L. Whistler and M. L. Wolfrom, 5 chapters by M. L. Wolfrom and A. Thompson, 1 chapter by M. L. Wolfrom, A. Thompson, and E. Pacsu, 1 chapter by M. L. Wolfrom and D. R. Lineback, 1 chapter by M. L. Wolfrom and G. H. S. Thomas, Academic Press, Inc., New York, 1963, pp. 21–23, 24–26, 32–34, 65–68, 211–215, 215–220, 341–345, 427–430.

"Chapter VII. Polysaccharides," D. Horton and M. L. Wolfrom, in "Comprehensive Biochemistry, Vol. 5, Carbohydrates," M. Florkin and E. H. Stotz, Eds., Elsevier Publishing Co., Amsterdam, 1963, pp. 185–232.

Methods in Carbohydrate Chemistry, Vol. III, Cellulose, Roy L. Whistler, Ed., 2 chapters by M. L. Wolfrom and A. Thompson, 2 chapters by M. L. Wolfrom and F. Shafizadeh, Academic Press, Inc., New York, 1963, pp. 143–150, 150–153, 375–376, 377–382.

"Thiosugars. II. Rearrangement of 2-(3,4,6-Tri-O-acetyl-2-amino-2-deoxy-β-D-glucopyranosyl)-2-thiopseudourea," M. L. Wolfrom, D. Horton, and D. H. Hutson, *J. Org. Chem.*, **28**, 845–847 (1963).

"Isopropyl Tetra-O-acetyl-α-D-glucopyranoside; A Synthesis of Kojibiose," M. L. Wolfrom, A. Thompson, and D. R. Lineback, *J. Org. Chem.*, **28**, 860–861 (1963).

"Effect of Maltose on Acid Reversion Mixtures from D-Glucose in Relation to the Fine Structures of Amylopectin and Glycogen," M. L. Wolfrom, A. Thompson, and R. H. Moore, *Cereal Chem.*, **40**, 182–186 (1963).

"Thioglycosides of 3-Amino-3-deoxy-D-mannose," M. L. Wolfrom, D. Horton, and H. G. Garg, *J. Org. Chem.*, **28**, 1569–1572 (1963).

"Macluraxanthone and Two Accompanying Pigments from the Root Bark of the Osage Orange," M. L. Wolfrom, F. Komitsky, Jr., G. Fraenkel, J. H. Looker, E. E. Dickey, P. McWain, A. Thompson, P. M. Mundell, and O. M. Windrath, *Tetrahedron Lett.*, 749–755 (1963).

"The Linkage Sequence in Heparin," M. L. Wolfrom, J. R. Vercellotti, H. Tomomatsu, and D. Horton, *Biochem. Biophys. Res. Commun.*, **12**, 8–13 (1963).

"3,4,6-Tri-O-acetyl-2-O-nitro-α-D-glucopyranosyl Chloride and the Anomeric Tetraacetates of 2-O-Nitro-D-glucopyranose," M. L. Wolfrom, A. Thompson, and D. R. Lineback, *J. Org. Chem.*, **28**, 1930 (1963).

"Chromatography of Sugar Acetates and Methyl Ethers on Magnesol," M. L. Wolfrom, Rosa M. de Lederkremer, and L. E. Anderson, *Anal. Chem.*, **35**, 1357–1359 (1963).

"Tetra-O-acylglycosyl Chlorides from 1-Thioglycosides and Their Conversion to Penta-O-acyl Esters," M. L. Wolfrom and Wolfgang Groebke, *J. Org. Chem.*, **28**, 2986–2988 (1963).

"Glycosyl Halide Derivatives of 3-Amino-3-deoxy-D-mannose," M. L. Wolfrom, H. G. Garg, and D. Horton, *J. Org. Chem.*, **28**, 2989–2992 (1963).

"Conversion of 2-Amino-2-deoxy-1-thio-D-glucose Derivatives into Glycosyl Halide Derivatives. A Tetra-O-acetylglycosylsulfenyl Bromide," D. Horton, M. L. Wolfrom, and H. G. Garg, *J. Org. Chem.*, **28**, 2992–2995 (1963).

"3,4,6-Tri-*O*-acetyl-2-amino-2-deoxy-α-D-galactopyranosyl Bromide Hydrobromide," M. L. Wolfrom, W. A. Cramp, and D. Horton, *J. Org. Chem.*, **28**, 3231–3232 (1963).

"Hydroboration in the Sugar Series," M. L. Wolfrom, K. Matsuda, F. Komitsky, Jr., and T. E. Whiteley, *J. Org. Chem.*, **28**, 3551–3553 (1963).

"Amino Derivatives of Starches. Amination of Amylose," M. L. Wolfrom, Mahmoud I. Taha, and D. Horton, *J. Org. Chem.*, **28**, 3553–3554 (1963).

Advances in Carbohydrate Chemistry, Vol. 18, edited by M. L. Wolfrom and R. S. Tipson, Academic Press, Inc., New York, 1963.

"Hexofuranosyl Nucleosides from Sugar Dithioacetals," M. L. Wolfrom, P. McWain, R. Pagnucco, and A. Thompson, *J. Org. Chem.*, **29**, 454–457 (1964).

"Structural Investigations of Acetylated Sugar Phenylhydrazine Derivatives," M. L. Wolfrom, G. Fraenkel, D. R. Lineback, and F. Komitsky, Jr., *J. Org. Chem.*, **29**, 457–461 (1964).

Methods in Carbohydrate Chemistry, Vol. IV, "Starch," Roy L. Whistler, Ed., 2 chapters by M. L. Wolfrom and N. E. Franks, Academic Press, New York, 1964, pp. 205–251, 269–271.

"Carboxyl-Reduced Heparin. Monosaccharide Components," M. L. Wolfrom, J. R. Vercellotti, and G. H. S. Thomas, *J. Org. Chem.*, **29**, 536–539 (1964).

"Two Disaccharides from Carboxyl-Reduced Heparin. The Linkage Sequence in Heparin," M. L. Wolfrom, J. R. Vercellotti, and D. Horton, *J. Org. Chem.*, **29**, 540–547 (1964).

"Methylation Studies on Carboxyl-Reduced Heparin. 2-Amino-2-deoxy-3,6-di-*O*-methyl-α-D-glucopyranose from the Methylation of Chitosan," M. L. Wolfrom, J. R. Vercellotti, and D. Horton, *J. Org. Chem.*, **29**, 547–550 (1964).

"Osage Orange Pigments. XIII. Isolation of Three New Pigments from the Root Bark," M. L. Wolfrom, E. E. Dickey, P. McWain, A. Thompson, J. H. Looker, O. M. Windrath, and F. Komitsky, Jr., *J. Org. Chem.*, **29**, 689–691 (1964).

"Osage Orange Pigments. XIV. The Structure of Macluraxanthone," M. L. Wolfrom, F. Komitsky, Jr., G. Fraenkel, J. H. Looker, E. E. Dickey, P. McWain, A. Thompson, P. M. Mundell, and O. M. Windrath, *J. Org. Chem.*, **29**, 692–697 (1964).

"A Synthesis of *O*-α-D-Glucopyranosyl-(1→4)-2-amino-2-deoxy-α-D-glucopyranose Hydrochloride," M. L. Wolfrom, H. El Khadem, and J. R. Vercellotti, *Chem. Ind.* (London), 545 (1964).

"A Polymer-Homologous Series of Methyl β-D-Glycosides from Cellulose," M. L. Wolfrom and S. Haq, *Tappi*, **47**, 183–185 (1964).

"Alkaline Hypochlorite Oxidation of Cellulose Analogs," M. L. Wolfrom and W. E. Lucke, *Tappi*, **47**, 189–192 (1964).

"Anomeric Purine Nucleoside Derivatives of 2-Amino-2-deoxy-D-glucose," M. L. Wolfrom, H. G. Garg, and D. Horton, *Chem. Ind.* (London), 930 (1964).

"2-Amino-2-deoxy-L-xylose," M. L. Wolfrom, D. Horton, and A. Böckmann, *J. Org. Chem.*, **29**, 1479–1480 (1964).

"Thin-layer Chromatography on Microcrystalline Cellulose," M. L. Wolfrom, D. L. Patin, and Rosa M. de Lederkremer, *Chem. Ind.* (London), 1065 (1964).

"Isolation and Characterization of Cellulose in the Coffee Bean," M. L. Wolfrom and D. L. Patin, *J. Agr. Food Chem.*, **12**, 376–377 (1964).

"Action of Nitrous Acid on Osazone Acetates. A New Synthesis of Osotriazoles," M. L. Wolfrom, H. El Khadem, and H. Alfes, *J. Org. Chem.*, **29**, 2072 (1964).

"Thio Sugars. III. Synthesis and Rearrangement of 2-(3,4,6-Tri-*O*-acetyl-2-amino-2-deoxy-β-D-galactopyranosyl)-2-thiopseudourea Dihydrobromide and Analogs," M. L. Wolfrom, W. A. Cramp, and D. Horton, *J. Org. Chem.*, **29**, 2302–2305 (1964).

"Carbohydrates: Physical and Chemical Characteristics," M. L. Wolfrom, George G. Maher, and Rinaldo G. Pagnucco, in "Biology Data Book," P. L. Altman and D. S. Dittmer, Eds., Committee on Biological Handbooks, Federation of American Societies for Experimental Biology, Washington, D. C., 1964, pp. 351–367.

"Amidino and Carbamoyl Osazones of Sugars," M. L. Wolfrom, H. El Khadem, and H. Alfes, *J. Org. Chem.*, **29**, 3074–3076 (1964).

"The Polymer-Homologous Series of Acetylated Methyl β-D-Glycosides from Cellulose: Improved Preparation and Extension," M. L. Wolfrom and S. Haq, *Tappi*, **47**, 692–694 (1964).

"Halogen and Nucleoside Derivatives of Acyclic 2-Amino-2-deoxy-D-glucose. I," M. L. Wolfrom, H. G. Garg, and D. Horton, *J. Org. Chem.*, **29**, 3280–3283 (1964).

"Synthetic Confirmation of an Interglycosidic Linkage in Heparin," M. L. Wolfrom, H. El Khadem, and J. R. Vercellotti, *J. Org. Chem.*, **29**, 3284–3286 (1964).

Advances in Carbohydrate Chemistry, Vol. 19, edited by M. L. Wolfrom and R. S. Tipson, Academic Press, Inc., New York, 1964.

"2-Deoxy-D-*arabino*-hexonic Acid 6-Phosphate and Methyl 2-Deoxy-β-D-*arabino*-hexopyranoside 4,6-(Monophenyl phosphate)," M. L. Wolfrom and N. E. Franks, *J. Org. Chem.*, **29**, 3645–3647 (1964).

"The Polymer-Homologous Series of Acetylated Methyl α-D-Glycosides From Cellulose," M. L. Wolfrom and S. Haq, *Tappi*, **47**, 733–734 (1964).

"Osage Orange Pigments. XV. Structure of Osajaxanthone. Synthesis of Dihydroosajaxanthone Monomethyl Ether," M. L. Wolfrom, F. Komitsky, Jr., and J. H. Looker, *J. Org. Chem.*, **30**, 144–149 (1965).

Methods in Carbohydrate Chemistry, Vol. V. General Polysaccharides. Roy L. Whistler, Ed., 1 chapter by M. L. Wolfrom and N. E. Franks, Academic Press, Inc., 1965, pp. 276–280.

"Synthesis of L-*threo*-Pentulose and 3,4,5-Tri-O-benzoyl-1-deoxy-L-*threo*-pentulose," M. L. Wolfrom and R. B. Bennett, *J. Org. Chem.*, **30**, 458–462 (1965).

"The Structure of Osazones," H. El Khadem, M. L. Wolfrom, and D. Horton, *J. Org. Chem.*, **30**, 838–841 (1965).

"Synthesis of Amino Compounds in the Sugar Series by Reduction of Hydrazine Derivatives. Two Epimeric Pairs of 1,2-Diamino-1,2-dideoxyalditols," M. L. Wolfrom and J. L. Minor, *J. Org. Chem.*, **30**, 841–843 (1965).

"Thin-layer Chromatography on Microcrystalline Cellulose," M. L. Wolfrom, D. L. Patin, and Rosa M. de Lederkremer, *J. Chromatogr.*, **17**, 488–494 (1965).

"Osage Orange Pigments. XVI. The Structure of Alvaxanthone," M. L. Wolfrom, F. Komitsky, Jr., and P. M. Mundell, *J. Org. Chem.*, **30**, 1088–1091 (1965).

"Isomerization of Tetra-O-acetyl-1-deoxy-D-*arabino*-hex-1-enopyranose," R. U. Lemieux, D. R. Lineback, M. L. Wolfrom, F. B. Moody, E. G. Wallace, and F. Komitsky, Jr., *J. Org. Chem.*, **30**, 1092–1096 (1965).

"Halogen and Nucleoside Derivatives of Acyclic 2-Amino-2-deoxy-D-glucose. II," M. L. Wolfrom, H. G. Garg, and D. Horton, *J. Org. Chem.*, **30**, 1096–1098 (1965).

"Nucleosides of D-Glucuronic Acid and of D-Glucofuranose and D-Galactofuranose," M. L. Wolfrom and P. McWain, *J. Org. Chem.*, **30**, 1099–1101 (1965).

"Synthesis of D-*lyxo*-Hexulose (D-Tagatose) and 1-Deoxy-D-*lyxo*-hexulose," M. L. Wolfrom and R. B. Bennett, *J. Org. Chem.*, **30**, 1284–1287 (1965).

"2,6-Diamino-2,6-dideoxy-D-mannose Dihydrochloride," M. L. Wolfrom, P. Chakravarty, and D. Horton, *Chem. Commun.*, 143 (1965).

"Extrusion Column Chromatography on Cellulose," M. L. Wolfrom, D. H. Busch, Rosa M. de Lederkremer, Sharon C. Vergez, and J. R. Vercellotti, *J. Chromatogr.*, **18**, 42–46 (1965).

"The Anomeric 9-(2-Amino-2-deoxy-D-glucopyranosyl)adenines," M. L. Wolfrom, H. G. Garg, and D. Horton, *J. Org. Chem.*, **30**, 1556–1560 (1965).

"Products from the Ortho Ester Form of Acetylated Maltose," M. L. Wolfrom and Rosa M. de Lederkremer, *J. Org. Chem.*, **30**, 1560–1563 (1965).

"Two Forms of 2-*O*-(2-Acetamido-3,4,6-tri-*O*-acetyl-2-deoxy-β-D-glucopyranosyl)-1,3-*O*-benzylideneglycerol," W. A. Szarek, M. L. Wolfrom, and H. Tomomatsu, *Chem. Commun.*, 326–327 (1965).

"Amino Derivatives of Starches. 2,6-Diamino-2,6-dideoxy-D-mannose Dihydrochloride," M. L. Wolfrom, P. Chakravarty, and D. Horton, *J. Org. Chem.*, **30**, 2728–2731 (1965).

"Acyclic Sugar Nucleoside Analogs. III," M. L. Wolfrom, W. von Bebenburg, R. Pagnucco, and P. McWain, *J. Org. Chem.*, **30**, 2732–2735 (1965).

"Ethyl 3,4,6-Tri-*O*-acetyl-2-amino-2-deoxy-1-thio-β-D-glucopyranoside," M. L. Wolfrom, W. A. Cramp, and D. Horton, *J. Org. Chem.*, **30**, 3056–3058 (1965).

"Benzylsulfonyl as *N*-Blocking Group in Amino Sugar Nucleoside Synthesis," M. L. Wolfrom and R. Wurmb, *J. Org. Chem.*, **30**, 3058–3061 (1965).

"Osage Orange Pigments. XVII. 1,3,6,7-Tetrahydroxyxanthone from the Heartwood," M. L. Wolfrom and H. B. Bhat, *Phytochemistry*, **4**, 765–768 (1965).

"Synthesis of Diamino Sugars from 1,2-Diamino-1,2-dideoxyalditols. 4,5-Diacetamido-4,5-dideoxy-L-xylose," M. L. Wolfrom, J. L. Minor, and W. A. Szarek, *Carbohyd. Res.*, **1**, 156–163 (1965).

"Amino Derivatives of Starches. Derivatives of 3,6-Diamino-3,6-dideoxy-D-altrose," M. L. Wolfrom, Yen-Lung Hung, and Derek Horton, *J. Org. Chem.*, **30**, 3394–3400 (1965).

"Isomaltose Synthesis Utilizing 2-Sulfonate Derivatives of D-Glucose," M. L. Wolfrom, K. Igarashi, and K. Koizumi, *J. Org. Chem.*, **30**, 3841–3844 (1965).

"Chemical Evidence for the Structure of Starch," M. L. Wolfrom and H. El Khadem, in "Starch: Chemistry and Technology. Vol. I. Fundamental Aspects," R. L. Whistler and E. F. Paschall, Eds., Academic Press, Inc., New York, 1965, pp. 251–278.

"Carbohydrates of the Coffee Bean. IV. An Arabinogalactan," M. L. Wolfrom and D. L. Patin, *J. Org. Chem.*, **30**, 4060–4063 (1965).

Advances in Carbohydrate Chemistry, Vol. 20, edited by M. L. Wolfrom and R. S. Tipson, Academic Press, Inc., New York, 1965.

"Stereoisomere Formen des Äthyl-hemiacetals von *aldehydo*-D-Galactosepentaacetat," M. L. Wolfrom and William H. Decker, *Liebigs Ann. Chem.*, **690**, 163–165 (1965).

Methods in Carbohydrate Chemistry, Vol. V, "General Polysaccharides," Roy L. Whistler, Ed., 1 chapter by M. L. Wolfrom and N. E. Franks, Academic Press, New York, 1965, pp. 276–279.

"A Chemical Synthesis of Panose," M. L. Wolfrom and K. Koizumi, *Chem. Commun.*, 2 (1966).

"Trifluoroacetyl as *N*-Blocking Group in Amino-sugar Nucleoside Synthesis," M. L. Wolfrom and H. B. Bhat, *Chem. Commun.*, 146 (1966).

"2-Amino-2-deoxy-D-xylose and 2-Amino-2-deoxy-D-ribose and Their 1-Thioglycofuranosides," M. L. Wolfrom and M. W. Winkley, *J. Org. Chem.*, **31**, 1169–1173 (1966).

"Configuration of the Glycosidic Linkage of 2-Amino-2-deoxy-D-glucopyranose to D-Glucuronic Acid in Heparin," M. L. Wolfrom, H. Tomomatsu, and W. A. Szarek, *J. Org. Chem.*, **31**, 1173–1178 (1966).

"Quantitative Thin-layer Chromatography of Sugars on Microcrystalline Cellulose," M. L. Wolfrom, Rosa M. de Lederkremer, and Gerhart Schwab, *J. Chromatogr.*, **22**, 474–476 (1966).

"Starch Acetals. *O*-Tetrahydropyran-2-yl and *O*-(1-Alkoxyethyl) Derivatives of Starch," M. L. Wolfrom, S. S. Bhattacharjee, and G. G. Parekh, *Die Stärke*, **18**, 131–135 (1966).

"Amino Derivatives of Starches. Sulfonation Studies on Methyl 3,6-Anhydro-α-D-glucopyranoside and Related Derivatives," M. L. Wolfrom, Yen-Lung Hung, P. Chakravarty, G. U. Yuen, and D. Horton, *J. Org. Chem.*, **31**, 2227–2232 (1966).

"Amino Derivatives of Starches. 2-Amino-3,6-anhydro-2-deoxy-D-mannose," M. L. Wolfrom, P. Chakravarty, and D. Horton, *J. Org. Chem.*, **31**, 2502–2504 (1966).

"Two-Dimensional Thin-layer Chromatography of Amino Acids on Microcrystalline Cellulose," D. Horton, A. Tanimura, and M. L. Wolfrom, *J. Chromatogr.*, **23**, 309–312 (1966).

"Anomeric Nucleosides of the Furanose Forms of 2-Amino-2-deoxy-D-glucose and 2-Amino-2-deoxy-D-ribose," M. L. Wolfrom and M. W. Winkley, *Chem. Commun.*, 533–534 (1966).

"Alkaline Hypochlorite Oxidation of Methyl β-Cellobioside," M. L. Wolfrom and Rosa M. de Lederkremer, *Carbohyd. Res.*, **2**, 426–438 (1966).

"Synthesis of a D-Glucofuranosyl Nucleoside Derivative through an Oxazoline," M. L. Wolfrom and M. W. Winkley, *J. Org. Chem.*, **31**, 3711–3713 (1966).

Advances in Carbohydrate Chemistry, Vol. 21, edited by M. L. Wolfrom and R. S. Tipson, Academic Press, Inc., New York, 1966.

"Bis(phenoxy)phosphinyl as N-Blocking Group in Amino Sugar Nucleoside Synthesis," M. L. Wolfrom, P. J. Conigliaro, and E. J. Soltes, *J. Org. Chem.*, **32**, 653–655 (1967).

"A Chemical Synthesis of Panose and an Isomeric Trisaccharide," M. L. Wolfrom and K. Koizumi, *J. Org. Chem.*, **32**, 656–660 (1967).

"On Sulphate Placement in Heparin," M. L. Wolfrom and P. Y. Wang, *Chem. Commun.*, 241 (1967).

"Osage Orange Pigments. XVIII. Synthesis of Osajaxanthone," M. L. Wolfrom, E. W. Koos, and H. B. Bhat, *J. Org. Chem.*, **32**, 1058–1060 (1967).

"Carbohydrate Nomenclature," M. L. Wolfrom, *J. Chem. Doc.*, **7**, 78–81 (1967).

"Substituted Arylazoethylenes from Aldose Arylhydrazones," H. El Khadem, M. L. Wolfrom, Z. M. El Shafei, and S. H. El Ashry, *Carbohyd. Res.*, **4**, 225–229 (1967).

"Trichloroacetyl and Trifluoroacetyl as *N*-Blocking Groups in Nucleoside Synthesis with 2-Amino Sugars," M. L. Wolfrom and H. B. Bhat, *J. Org. Chem.*, **32**, 1821–1823 (1967).

"Anomeric Purine Nucleosides of the Furanose Form of 2-Amino-2-deoxy-D-ribose," M. L. Wolfrom and M. W. Winkley, *J. Org. Chem.*, **32**, 1823–1825 (1967).

"Polysaccharides from Instant Coffee Powder," M. L. Wolfrom and L. E. Anderson, *J. Agr. Food Chem.*, **15**, 685–687 (1967).

"2,4-Dinitrophenyl as *N*-Blocking Group in Pyrimidine Nucleoside Synthesis with 2-Amino Sugars," M. L. Wolfrom and H. B. Bhat, *J. Org. Chem.*, **32**, 2757–2759 (1967).

"Amino Derivatives of Starches. Amination of 6-*O*-Tritylamylose," M. L. Wolfrom, H. Kato, M. I. Taha, A. Sato, G. U. Yuen, T. Kinoshita, and E. J. Soltes, *J. Org. Chem.*, **32**, 3086–3089 (1967).

"Novel Reaction of a Nitro Sugar with Methanol," M. L. Wolfrom, U. G. Nayak, and T. Radford, *Science*, **157**, 538 (1967).

"A Novel Reaction of a Nitro Sugar with Alcohols," M. L. Wolfrom, U. G. Nayak, and T. Radford, *Proc. Nat. Acad. Sci. U. S.*, **58**, 1848–1851 (1967).

Advances in Carbohydrate Chemistry, Vol. 22, edited by M. L. Wolfrom and R. S. Tipson, Academic Press, New York, 1967.

"Amination of Amylose Oxidized with Dimethyl Sulphoxide–Acetic Anhydride," M. L. Wolfrom and P. Y. Wang, *Chem. Commun.*, 113–114 (1968).

"Reaction of Alkyl Vinyl Ethers with Methyl α-D-Glucopyranoside," M. L. Wolfrom, Anne Beattie, and Shyam S. Bhattacharjee, *J. Org. Chem.*, **33**, 1067–1070 (1968).

"L-Iduronic Acid in Purified Heparin," M. L. Wolfrom, S. Honda, and P. Y. Wang, *Chem. Commun.*, 505–506 (1968).

"Carbohydrates," M. L. Wolfrom, in "The Encyclopaedia Britannica," Chicago, Ill., 1968, pp. 863–868.

"Glycosides, Natural," M. L. Wolfrom, in "The Encyclopaedia Britannica," Chicago, Ill., 1968, pp. 501–502.

"Pectin," M. L. Wolfrom, in "The Encyclopaedia Britannica," Chicago, Ill., 1968, p. 515.

Tables, George G. Maher and M. L. Wolfrom, in "Handbook of Biochemistry," Section D, Carbohydrates, H. A. Sober, Ed., The Chemical Rubber Co., Cleveland, Ohio, 1968, pp. D-1–D-80.

"Reaction of 3,4-Dihydro-2H-pyran with Methyl α-D-Glucopyranoside," M. L. Wolfrom, Anne Beattie, S. S. Bhattacharjee, and G. G. Parekh, *J. Org. Chem.*, **33**, 3990–3991 (1968).

"Anomeric Adenine Nucleosides of 2-Amino-2-deoxy-D-ribofuranose," M. L. Wolfrom, M. W. Winkley, and P. McWain, in "Synthetic Procedures in Nucleic Acid Chemistry," Vol. I, W. W. Zorbach and R. S. Tipson, Eds., John Wiley and Sons, Inc., New York, 1968, pp. 168–171.

"1-(Adenin-9-yl)-1-deoxyl-1-S-ethyl-1-thio-*aldehydo*-D-galactose Aldehydrol," M. L. Wolfrom, P. McWain, and A. Thompson, in "Synthetic Procedures in Nucleic Acid Chemistry," Vol. I, W. W. Zorbach and R. S. Tipson, Eds., John Wiley and Sons, Inc., New York, 1968, pp. 219–223.

"9-(2-Acetamido-3,5,6-tri-O-acetyl-2-deoxy-D-glucosyl)-2,6-dichloro-9H-purine," M. W. Wolfrom, M. W. Winkley, and P. McWain, in "Synthetic Procedures in Nucleic Acid Chemistry," Vol. I, W. W. Zorbach and R. S. Tipson, Eds., John Wiley and Sons, Inc., New York, 1968, pp. 239–241.

"1-(2-Amino-2-deoxy-β-D-glucopyranosyl)thymine," M. L. Wolfrom, H. B. Bhat, and P. McWain, in "Synthetic Procedures in Nucleic Acid Chemistry," Vol. I, W. W. Zorbach and R. S. Tipson, Eds., John Wiley and Sons, Inc., New York, 1968, pp. 323–326.

"Anomeric 2-Amino-2-deoxy-D-glucofuranosyl Nucleosides of Adenine and 2-Amino-2-deoxy-β-D-glucopyranosyl Nucleosides of Thymine and 5-Methylcytosine," *J. Org. Chem.*, **33**, 4227–4231 (1968).

Advances in Carbohydrate Chemistry, Vol. 23, edited by M. L. Wolfrom and R. S. Tipson, Academic Press, Inc., New York, 1968.

"Quantitative Analysis of Gentiobiose and Isomaltose in Admixture, and Its Application to the Characterization of Dextrins," M. L. Wolfrom and G. Schwab, *Carbohyd. Res.*, **9**, 407–413 (1969).

"On the Amination of Amylose," M. L. Wolfrom, K. C. Gupta, K. K. De, A. K. Chatterjee, T. Kinoshita, and P. Y. Wang, *Die Stärke*, **21**, 39–43 (1969).

"Anomeric Forms of 9-(2-Amino-2-deoxy-D-xylofuranosyl)adenine," M. L. Wolfrom, M. W. Winkley, and S. Inouye, *Carbohyd. Res.*, **10**, 97–103 (1969).

"The Isolation of L-Iduronic Acid from the Crystalline Barium Acid Salt of Heparin," M. L. Wolfrom, S. Honda, and P. Y. Wang, *Carbohyd. Res.*, **10**, 259–265 (1969).

"Starch Acetals. Acid Sensitivity and Preferred Site of Reaction," M. L. Wolfrom and S. S. Bhattacharjee, *Die Stärke*, **21**, 116–118 (1969).

"Trifluoroacetyl as an N-Protective Group in the Synthesis of Purine Nucleosides of 2-Amino-2-deoxy Saccharides," M. L. Wolfrom and P. J. Conigliaro, *Carbohyd. Res.*, **11**, 63–76 (1969).

"Reaction of Carbohydrates with Vinyl Ethers; A Differential Hydrolysis," M. L.

Wolfrom, S. S. Bhattacharjee, and Rosa M. de Lederkremer, *Carbohyd. Res.*, **11**, 148–150 (1969).

"Gas–liquid Chromatography in the Study of the Maillard Browning Reaction," M. L. Wolfrom and N. Kashimura, *Carbohyd. Res.*, **11**, 151–152 (1969).

"On the Distribution of Sulfate in Heparin," M. L. Wolfrom, P. Y. Wang, and S. Honda, *Carbohyd. Res.*, **11**, 179–185 (1969).

"Reaction of Alkyl Vinyl Ethers with D-Galactose Diethyl Dithioacetal," M. L. Wolfrom and G. G. Parekh, *Carbohyd. Res.*, **11**, 547–557 (1969).

Advances in Carbohydrate Chemistry and Biochemistry, Vol. 24, edited by M. L. Wolfrom, R. S. Tipson, and D. Horton, Academic Press, Inc., New York, 1969.

"Mono- and Oligo-saccharides," M. L. Wolfrom in "Symposium on Foods: Carbohydrates and Their Roles," H. W. Schultz, R. F. Cain, and R. W. Wrolstad, Eds., Avi Publishing Co., Inc., Westport, Conn., 1969, pp. 12–25.

"Amination of Amylose at the C-2 Position," M. L. Wolfrom and P. Y. Wang, *Carbohyd. Res.*, **12**, 109–114 (1970).

"A Synthetic Heparinoid from Amylose," M. L. Wolfrom and P. Y. Wang, *Carbohyd. Res.*, **18**, 23–37 (1971).

"Halogen and Nucleoside Derivatives of Acyclic 2-Amino-2-deoxy-D-glucose. III," M. L. Wolfrom and P. J. Conigliaro, *Carbohyd. Res.*, **20**, 369–374 (1971).

"Pyrimidine Nucleosides of the Furanose Form of 2-Amino-2-deoxy-D-glucose," M. L. Wolfrom, P. J. Conigliaro, and H. B. Bhat, *Carbohyd. Res.*, **20**, 383–390 (1971).

"Pyrimidine Nucleosides of 2-Amino-2-deoxy-D-galactose," M. L. Wolfrom, H. B. Bhat, and P. J. Conigliaro, *Carbohyd. Res.*, **20**, 375–381 (1971).

"Pyrimidine Nucleosides of the Furanose Form of 2-Amino-2-deoxy-D-xylose," M. L. Wolfrom and P. J. Conigliaro, *Carbohyd. Res.*, **20**, 391–398 (1971).

"Esters," M. L. Wolfrom and Walter A. Szarek, in "The Carbohydrates," Vol. IA, W. Pigman and D. Horton, Eds., Academic Press, Inc., New York, 1971, pp. 217–238.

"Halogen Derivatives," M. L. Wolfrom and Walter A. Szarek, in "The Carbohydrates," Vol. IA, W. Pigman and D. Horton, Eds., Academic Press, Inc., New York, 1971, pp. 239–251.

"Acyclic Derivatives," M. L. Wolfrom, in "The Carbohydrates," Vol. IA, W. Pigman and D. Horton, Eds., Academic Press, Inc., New York, 1971, pp. 355–422.

CONFORMATIONAL ANALYSIS OF SUGARS AND THEIR DERIVATIVES*

By Philippe L. Durette and Derek Horton

Department of Chemistry, The Ohio State University, Columbus, Ohio

I. Introduction

"I held them in every light. I turned them in every attitude. I surveyed their characteristics. I dwelt upon their peculiarities. I pondered upon their conformation."

Edgar Allan Poe   *"Berenice"*

*Preparation of this Chapter was supported by a grant, No. GP-9646, from the National Science Foundation.

The study of the conformations of organic molecules originated in the early eighteen-nineties, when Sachse pointed out[1] that six-membered ring-systems can exist free from bond-angle (Baeyer) strain, in two puckered forms, the flexible or unsymmetric form and the rigid or symmetric (chair) form. Sachse also recognized that the rigid form of cyclohexane can have two monosubstituted

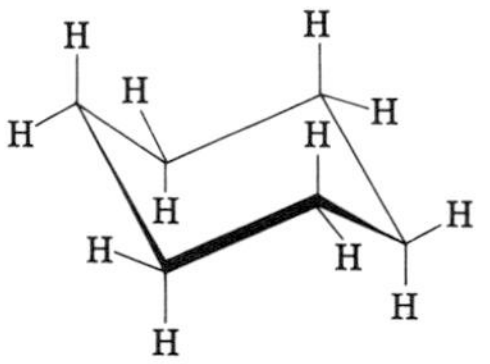

Rigid (chair) form of cyclohexane

Flexible forms of cyclohexane

derivatives existing in dynamic equilibrium with each other and interconvertible by a process of inversion, as illustrated in the following scheme. These fundamental ideas on the exact shapes of mol-

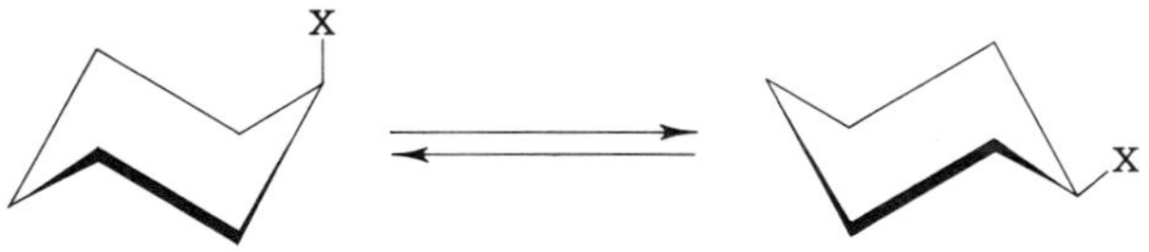

Interconversion of chair forms of a monosubstituted cyclohexane

ecules, however, failed to develop for almost thirty years, until they were reconsidered by Mohr[2] in 1918 and extended by him to ring systems containing more than six atoms.

The first application to pyranoid sugars of the Sachse–Mohr theory of puckered rings was made by Sponsler and Dore[3] in 1926 in their

(1) H. Sachse, Ber., 23, 1363 (1890); Z. Phys. Chem. (Leipzig), 10, 203 (1892).
(2) E. Mohr, J. Prakt. Chem., [2] 98, 315 (1918); [2] 103, 316 (1922); Ber., 55, 230 (1922).
(3) O. L. Sponsler and W. H. Dore, Colloid Symp. Monograph, 4, 174 (1926).

interpretation of the X-ray data for ramie-fiber cellulose. They could explain the observed spacings from the X-ray data if the D-glucopyranose residues were formulated in the rigid or chair form, but not if the residues were in the alternative, flexible form. Haworth[4] further extended these principles in the carbohydrate field by pointing out that each pyranoid sugar can exist in several strainless ring-forms of the Sachse–Mohr type. In order to describe these various shapes, Haworth introduced the term "conformation." "Conformations" were later defined[5] "as the different arrangements in space of the atoms in a single classical organic structure (configuration), the arrangement being produced by the rotation or twisting (but not breaking) of [normal, single] bonds." The significance of conformational analysis in carbohydrate chemistry was thus clearly recognized in 1929 by Haworth, but the lack of experimental data at that time precluded further speculations.

The effect of conformation on the chemical reactivity of organic molecules was first observed by Isbell[6] in the nineteen-thirties. From an investigation of the oxidation of free sugars by bromine, Isbell noted differences in the reactivity of the α and β anomers of the same (D or L) sugar. These differences were considered to arise from an important structural factor, and the observations led to the first attempt to classify sugar derivatives according to the orientation of the α and β (anomeric) hydroxyl groups relative to the strainless pyranoid ring, that is, according to what are now known as the axial and equatorial dispositions. The profound chemical consequences of differences in orientation of substituents on the chair form of cyclohexane were clearly pointed out by Barton[7] in 1950.

The existence of two types of bonds in cyclohexane in its chair conformation was first demonstrated experimentally by Kohlrausch and coworkers[8] from Raman-spectroscopic evidence, and was later corroborated by Hassel[9] from electron-diffraction data and by Pitzer and

(4) W. N. Haworth, "The Constitution of Sugars," Edward Arnold and Co., London, 1929, pp. 90–96.

(5) W. Klyne, *Progr. Stereochem.*, **1**, 36 (1954).

(6) H. S. Isbell, *J. Amer. Chem. Soc.*, **54**, 1692 (1932); *J. Chem. Educ.*, **12**, 96 (1935); H. S. Isbell and W. W. Pigman, *J. Org. Chem.*, **1**, 505 (1937); *J. Res. Nat. Bur. Stand.*, **18**, 141 (1937); H. S. Isbell, *ibid.*, **18**, 505 (1937); **19**, 639 (1937); **20**, 97 (1938).

(7) D. H. R. Barton, *Experientia*, **6**, 316 (1950); *Science*, **169**, 539 (1970).

(8) K. W. F. Kohlrausch, A. W. Reitz, and W. Stockmair, Z. *Phys. Chem.*, **B32**, 229 (1936).

(9) O. Hassel, *Tidsskr. Kjemi, Bergv. Met.*, **3**, 32 (1943).

coworkers[10] from thermodynamic measurements. The six bonds that are parallel to the principal axis were named "axial bonds," and the six bonds that subtend an angle of 109.5° to the axis[11] were named[12] "equatorial bonds."

A basis for predicting the relative stabilities of six-membered ring-systems in different conformations was obtained from studies on cyclohexane derivatives. Hassel[9] recognized that cyclohexane is more stable in the chair form than in the boat form, and that the two chair forms of a monosubstituted cyclohexane[9,13,14] have different energies, the form having the substituent equatorially attached being thermodynamically more stable than the one having the group axially attached. In an extension of these ideas to pyranoid sugars, Hassel and Ottar[15] were able to predict the favored chair form for many sugar derivatives, and thereby firmly established the technique of conformational analysis in the carbohydrate field.

The study of the conformations of cyclic sugars was developed more fully by Reeves in a series of papers, beginning in 1949, dealing with the formation of complexes of sugars and their derivatives in cuprammonia solution.[16-24] Reeves was able to provide experimental evidence to indicate that some pyranoid sugars indeed adopt chair conformations in solution, and that, in such cases, one chair form usually appeared to preponderate over the other. Boat or skew conformations were considered possible if the nonbonded interactions between the substituents in the chair conformation became too large.[17,24] Reeves's investigations have formed the basis of much subsequent work in the conformational analysis of cyclic carbohydrate

(10) C. W. Beckett, K. S. Pitzer, and R. Spitzer, *J. Amer. Chem. Soc.*, **69**, 2488 (1947).
(11) Prior to 1953, the axial bonds had been called[10] "polar." The designations "ϵ" and "κ" had also been used for axial and equatorial bonds, respectively.[9]
(12) D. H. R. Barton, O. Hassel, K. S. Pitzer, and V. Prelog, *Nature*, **172**, 1096 (1953); *Science*, **119**, 49 (1954).
(13) O. Hassel and H. Viervoll, *Tidsskr. Kjemi, Bergv. Met.*, **3**, 35 (1943).
(14) O. Hassel and O. Bastiansen, *Tidsskr. Kjemi, Bergv. Met.*, **6**, 96 (1946).
(15) O. Hassel and B. Ottar, *Acta Chem. Scand.*, **1**, 929 (1947).
(16) R. E. Reeves, *Advan. Carbohyd. Chem.*, **6**, 107 (1951).
(17) R. E. Reeves, *Annu. Rev. Biochem.*, **27**, 15 (1958).
(18) R. E. Reeves, *J. Amer. Chem. Soc.*, **71**, 212 (1949).
(19) R. E. Reeves, *J. Amer. Chem. Soc.*, **71**, 215 (1949).
(20) R. E. Reeves, *J. Amer. Chem. Soc.*, **71**, 1737 (1949).
(21) R. E. Reeves, *J. Amer. Chem. Soc.*, **71**, 2116 (1949).
(22) R. E. Reeves, *J. Amer. Chem. Soc.*, **72**, 1499 (1950).
(23) R. E. Reeves, *J. Amer. Chem. Soc.*, **73**, 957 (1951).
(24) R. E. Reeves and F. A. Blouin, *J. Amer. Chem. Soc.*, **79**, 2261 (1957).

molecules. This field has been the subject of several reviews and monographs.[5,17,25-31a]

The conformational analysis of saturated five- and six-membered heterocycles, presenting data relevant to the carbohydrate field, has also been extensively reviewed.[32-36]

The profound significance of conformational analysis in organic chemistry was especially recognized in 1969, when D. H. R. Barton and O. Hassel shared the Nobel Prize for Chemistry for their pioneering contributions to the basic understanding of the exact shapes of saturated, six-membered ring-systems.

II. METHODS FOR DETERMINATION OF CONFORMATION OF SUGARS

1. X-Ray Crystallography

X-Ray crystal-structure analysis is one of the most precise techniques at present available for determining the conformation of a sugar molecule, as it provides extremely accurate information on bond lengths, conformational (dihedral) and valency angles, and interatomic distances. However, it is to be noted that molecular shapes, as determined by crystallographic methods, refer to molecules held rigidly in the crystal lattice, and do not necessarily correspond to the favored conformation in solution, although such a correspondence is frequently observed.

Structures of considerable complexity can now be solved with increasing ease by X-ray crystallography as a result of improvements in both the direct and vector-analysis methods of phase determin-

(25) E. L. Eliel, N. L. Allinger, S. J. Angyal, and G. A. Morrison, "Conformational Analysis," Interscience Division, John Wiley and Sons, Inc., New York, 1965, Chapter 6.
(26) J. A. Mills, Advan. Carbohyd. Chem., 10, 1 (1955).
(27) R. J. Ferrier and W. G. Overend, Quart. Rev. (London), 13, 265 (1959).
(28) B. Capon and W. G. Overend, Advan. Carbohyd. Chem., 15, 11 (1960).
(29) M. Hanack, "Conformation Theory," Academic Press, New York, 1965, Chapter 7, pp. 324–330.
(30) S. Hirano, Kagaku No Ryoiki, 22, 54 (1968).
(31) S. J. Angyal, Angew. Chem. Int. Ed. Engl., 8, 157 (1969).
(31a) J. F. Stoddart, "Stereochemistry of Carbohydrates," Wiley–Interscience, New York, 1971.
(32) F. G. Riddell, Quart. Rev. (London), 21, 364 (1967).
(33) A. R. Katritzky, in "Topics in Heterocyclic Chemistry," R. N. Castle, Ed., Wiley–Interscience, New York, 1969, pp. 35–55.
(34) C. Romers, C. Altona, H. R. Buys, and E. Havinga, Top. Stereochem., 4, 39 (1969).
(35) E. L. Eliel, Svensk Kem. Tidskr., 81, No. 6/7, 22 (1969).
(36) E. L. Eliel, Accounts Chem. Res., 3, 1 (1970).

ation. The presence of a "heavy atom" is no longer essential. The accuracy and speed of the method has been greatly improved as a result of such technical advances as the advent of automatic diffractometers and computerized data-reduction systems. Therefore, the conformations of a wide range of complex carbohydrate molecules in the crystalline state may now be explored by crystal-structure analysis. The literature up to 1964 has been reviewed[37] and specific applications up to 1970 have been described.[38]

2. Spectroscopic Methods

a. Infrared Spectroscopy.—The application of infrared (i.r.) spectroscopy to structural problems in carbohydrate chemistry has been reviewed.[39-41] I.r. spectra can be recorded for carbohydrates both as solids and dissolved in various solvents, and any conformational differences between the solid state and the liquid or dissolved state are detectable.[42] I.r. spectroscopy can also yield information about intermolecular and intramolecular hydrogen-bonding, and can thus be applied in deducing the conformation of carbohydrates and their derivatives containing free hydroxyl groups. However, the complexity of the i.r. spectra of sugars and their derivatives has rendered the technique of limited applicability in the solution of conformational problems in the carbohydrate field.

A notable exception has been the application of i.r. spectroscopy by Isbell and Tipson to the analysis of the conformations of methyl aldopyranosides,[43] acetylated methyl aldopyranosides,[44] and fully acetylated aldopyranoses[45] in the solid state. Their approach was based on (1) the assumption that the pyranoid sugar derivatives predicted by Reeves to exist in only one of the chair conformations in solution would adopt that same conformation in the crystalline state, and (2) the observation that axially and equatorially attached groups at the anomeric carbon atom give rise to different vibrations.

(37) G. A. Jeffrey and R. D. Rosenstein, *Advan. Carbohyd. Chem.*, **19**, 7 (1964).
(38) G. Strahs, *Advan. Carbohyd. Chem. Biochem.*, **25**, 53 (1970).
(39) W. B. Neely, *Advan. Carbohyd. Chem.*, **12**, 13 (1957).
(40) H. Spedding, *Advan. Carbohyd. Chem.*, **19**, 23 (1964).
(41) R. S. Tipson, "Infrared Spectroscopy of Carbohydrates," National Bureau of Standards Monograph 110, Washington, D. C., 1968.
(42) Compare, H. R. Buys, C. H. Leeuwestein, and E. Havinga, *Tetrahedron*, **26**, 845 (1970) and O. D. Ul'yanova, M. K. Ostrovskii, and Yu. A. Pentin, *Zh. Fiz. Khim.*, **44**, 1013 (1970), for examples in non-carbohydrate systems.
(43) R. S. Tipson and H. S. Isbell, *J. Res. Nat. Bur. Stand.*, **64A**, 239 (1960).
(44) R. S. Tipson and H. S. Isbell, *J. Res. Nat. Bur. Stand.*, **64A**, 405 (1960).
(45) R. S. Tipson and H. S. Isbell, *J. Res. Nat. Bur. Stand.*, **65A**, 249 (1961).

The empirical conformational assignments made from the i.r. evidence were in good agreement with those reported by Reeves.[16,22] For those derivatives expected to exist either in a flexible form or in equilibrium between the two chair forms, the i.r. data did not fit the correlations established for the chair conformers. The failure of the method to differentiate between the two possibilities, or to yield any quantitative information, clearly illustrates the limitations of i.r. spectroscopy in the conformational analysis of sugar molecules.

b. Microwave Spectroscopy.—The microwave spectrum of tetrahydropyran shows that, in the gas phase, the molecule exists in the chair conformation.[46] The potential of this technique for determining the conformations of multisubstituted tetrahydropyran derivatives has not yet been exploited. There would be substantial technical problems involved in attempts to study sugar derivatives in the gas phase.

c. Nuclear Magnetic Resonance Spectroscopy.—Since the pioneering work of Lemieux and coworkers[47] in 1958 on the application of high-resolution, nuclear magnetic resonance (n.m.r.) spectroscopy to the solution of structural problems in the carbohydrate field, this physical method has developed into the most powerful and direct technique for the investigation of the conformational aspects of sugars and their derivatives in solution. The applicability of the n.m.r. technique in the carbohydrate field has been described in several reviews.[48-52]

The value of the n.m.r. method lies in its ability to provide both thermodynamic information regarding the various conformations that sugar molecules may adopt in solution, and kinetic information on molecular rate-processes, such as conformational interconversion and chemical exchange, that take place in solution. As a tool for conformational analysis of mobile systems, it has the particular merit that the perturbations used for measurement of resonances have energies several orders of magnitude lower than those required for inducing significant conformational change. Although the enormous contribution of n.m.r. spectroscopy to the solution of stereochemical problems

(46) V. M. Rao and R. Kewley, *Can. J. Chem.*, **47**, 1289 (1969).
(47) R. U. Lemieux, R. K. Kullnig, H. J. Bernstein, and W. G. Schneider, *J. Amer. Chem. Soc.*, **79**, 1005 (1957); **80**, 6098 (1958).
(48) L. D. Hall, *Advan. Carbohyd. Chem.*, **19**, 51 (1964).
(49) R. W. Lenz and J. P. Heeschen, *J. Polym. Sci.*, **51**, 247 (1961).
(50) L. D. Hall and J. F. Manville, *Advan. Chem. Ser.*, **74**, 228 (1968).
(51) T. D. Inch, *Annu. Rev. NMR Spectrosc.*, **2**, 35 (1969).
(52) J. J. M. Rowe, J. Hinton, and K. L. Rowe, *Chem. Rev.*, **70**, 42 (1970).

has come primarily from the study of proton magnetic resonance, the measurement of resonances of other nuclei, such as that[53–57a] of carbon-13 and[58–66] fluorine-19, has also proved to be well suited for the elucidation of configurational and conformational effects. Such studies have also been enhanced by the use of various specialized instrumental techniques, such as homonuclear, heteronuclear, and internuclear (indor)[67,68] double-resonance, electronic computation, variable temperature applications, nuclear Overhauser effects,[69] and signal enhancement by use of a "computer of average transients" (CAT), or by application of Fourier transformation. Another powerful tool is available in the specific, deshielding effects of certain transition-metal ions, and, in particular, the use of such complexes of the lanthanide elements as tris(dipivalomethanato)europium[III] as an aid in spectral dispersion, although the possible effect of the reagent itself in influencing the conformation needs to be taken into consideration.[70] Useful conformational data can also be obtained[70a] by chlorine-35 quadrupolar resonance spectroscopy.

N.m.r. spectroscopy yields basically two kinds of data: spin–spin coupling-constants and chemical shifts. The relationship between the chemical shift of a proton and its environment in six-membered ring-systems was investigated by Lemieux and coworkers.[47] From a study of the n.m.r. spectra of substituted cyclohexane and tetrahydro-

(53) L. D. Hall and L. F. Johnson, *Chem. Commun.*, 509 (1969).
(54) A. S. Perlin and B. Casu, *Tetrahedron Lett.*, 2921 (1969).
(55) D. E. Dorman and J. D. Roberts, *J. Amer. Chem. Soc.*, **92**, 1355 (1970).
(56) A. S. Perlin, B. Casu, and H. J. Koch, *Can. J. Chem.*, **48**, 2596 (1970).
(57) A. S. Perlin and H. J. Koch, *Can. J. Chem.*, **48**, 2639 (1970).
(57a) W. Voelter, E. Breitmaier, R. Price, and G. Jung, *Chimia*, **25**, 168 (1971).
(58) L. D. Hall and J. F. Manville, *Chem. Ind.* (London), 991 (1965).
(59) L. D. Hall and J. F. Manville, *Carbohyd. Res.*, **4**, 512 (1967).
(60) L. D. Hall and J. F. Manville, *Chem. Ind.* (London), 468 (1967).
(61) L. D. Hall and J. F. Manville, *Can. J. Chem.*, **45**, 1299 (1967).
(62) L. D. Hall and J. F. Manville, *Chem. Commun.*, 37 (1968).
(63) L. D. Hall and L. Evelyn, *Chem. Ind.* (London), 183 (1968).
(64) L. D. Hall and J. F. Manville, *Can. J. Chem.*, **47**, 1 (1969).
(65) L. D. Hall and J. F. Manville, *Can. J. Chem.*, **47**, 19 (1969).
(66) L. D. Hall, P. R. Steiner, and C. Pedersen, *Can. J. Chem.*, **48**, 1155 (1970).
(67) R. Burton, L. D. Hall, and P. R. Steiner, *Can. J. Chem.*, **48**, 2679 (1970).
(68) B. Coxon, *Carbohyd. Res.*, **18**, 427 (1971).
(69) F. A. L. Anet and A. J. R. Bourn, *J. Amer. Chem. Soc.*, **87**, 5250 (1965).
(70) C. C. Hinckley, *J. Amer. Chem. Soc.*, **91**, 5160 (1969); I. Armitage and L. D. Hall, *Chem. Ind.* (London), 1537 (1970); P. Girard, H. Kagan, and S. David, *Bull. Soc. Chim. Fr.*, 4515 (1970); R. F. Butterworth, A. G. Pernet, and S. Hanessian, *Can. J. Chem.*, **49**, 981 (1971); D. Horton and J. K. Thomson, *Chem. Commun.*, in press (1971).
(70a) S. David and L. Guibé, *Carbohyd. Res.*, **20**, 440 (1971).

pyran ring-systems, three conclusions were reached that permitted the assignment of conformation to various aldopyranoses and their derivatives: (1) the anomeric hydrogen atom, being on a carbon atom attached to two oxygen atoms, is strongly deshielded, and usually resonates at lower field than the other ring-protons, (2) axially attached protons usually resonate at higher field than equatorially attached protons in chemically similar environments, and (3) "axial" acetyl methyl protons usually resonate at lower field than "equatorial" acetyl methyl protons. In a subsequent study, empirical rules were derived for estimating long-range, shielding effects that occur with changes in configuration in the aldopyranoses[71] and the corresponding peracetates.[72]

From their n.m.r. investigations on six-membered ring-systems, Lemieux and coworkers[47] also discovered that the vicinal, spin–spin coupling is a function of the dihedral angle. Thus, it was found that the spin–spin coupling-constant between vicinal, antiparallel protons is about two to three times that between gauche-disposed protons. In a related study, it was observed[73] that $J_{antiparallel}$ (9.4 Hz) in 1,4-dioxane was more than three times J_{gauche} (2.7 Hz), again indicating the dependence of vicinal couplings on the geometrical arrangement of the coupled protons. Karplus[74] rationalized the change in coupling constants with dihedral angle by the following relationship, obtained from a valence-bond calculation,

$$J = J_0 \cos^2 \phi + K,$$

where J is the coupling constant between two hydrogen atoms attached to adjacent carbon atoms at a dihedral angle of ϕ. J_0 and K are constants whose values were evaluated as $K = -0.28$ and $J_0 = 8.5$ for $0° \leqslant \phi \leqslant 90°$, and $J_0 = 9.5$ for $90° \leqslant \phi \leqslant 180°$.

The publication of Karplus's equation was immediately followed by a rash of conformational studies on furanoid[75–78] and pyranoid[49] sugar derivatives, based on the relationship between coupling con-

(71) R. U. Lemieux and J. D. Stevens, Can. J. Chem., 44, 249 (1966).

(72) R. U. Lemieux and J. D. Stevens, Can. J. Chem., 43, 2059 (1965).

(73) A. D. Cohen, N. Sheppard, and J. J. Turner, Proc. Chem. Soc., 118 (1958).

(74) M. Karplus, J. Chem. Phys., 30, 11 (1959).

(75) C. D. Jardetzky, J. Amer. Chem. Soc., 82, 229 (1960); 83, 2919 (1961); 84, 62 (1962).

(76) R. U. Lemieux, Can. J. Chem., 39, 116 (1961).

(77) R. J. Abraham, K. A. McLauchlan, L. D. Hall, and L. Hough, Chem. Ind. (London), 213 (1962).

(78) R. J. Abraham, L. D. Hall, L. Hough, K. A. McLauchlan, and H. J. Miller, J. Chem. Soc., 748 (1963).

stant and dihedral angle, and slight modifications thereof.[79-81] These investigations were based on the assumption that exact dihedral angles and precise conformational assignments could be made by direct calculation from the Karplus equation, and thereby they implied that vicinal coupling-constants were a function of the dihedral angle only. Furthermore, the "coupling constants" were frequently the observed splittings, measured under conditions where their accuracy as true couplings is questionable. Karplus[82] was obliged to warn against such indiscriminate use of the equation in the solution of stereochemical problems. He stated that conclusions based on estimations of dihedral angles made to an accuracy of one or two degrees were drawn at great risk, as vicinal coupling-constants depend also on the electronegativity of attached substituents, the carbon–carbon bond length, bond angles, and other molecular properties. A better, semi-theoretical relationship, based on a valence-bond, σ-electron calculation, is given in Karplus's later equation[82]:

$$J_{HH'} = (A + B \cos \phi + C \cos 2\phi) \, (1 - m\Delta X),$$

where A, B, C, and m are constants, ϕ is the dihedral angle, and ΔX is the difference between the electronegativities of the substituent and hydrogen.

By use of the generalized Karplus equation that takes into account the electronegativity of substituents,[82] the variations observed in vicinal coupling-constants for the same nominal dihedral angle of the protons can be rationalized; for example, calculations from the equation indicate that, for antiparallel protons in a pyranoid sugar derivative, the $J_{1,2}$ coupling will be smaller than the $J_{4,5}$ coupling, in agreement with experimental observations.[83] Quantitative differences between the observed couplings and those predicted from the Karplus equation, for systems known to be conformationally homogeneous, were attributed to distortion of the ring by slight flattening from "ideal" geometry.[83]

It is noteworthy that the original Karplus equation was based on coupling constants for acetylated pyranoid sugars[47] that deviate substantially from ideal-chair geometry (that is, $\phi_{aa} \neq 180°$; $\phi_{ee} \neq \phi_{ea} \neq 60°$), as has been demonstrated by X-ray crystal-structure analyses of related systems.[84,84a] Thus, crystalline tri-O-acetyl-β-D-arabinopy-

(79) R. U. Lemieux, J. D. Stevens, and R. R. Fraser, Can. J. Chem., **40**, 1955 (1962).

(80) R. J. Abraham, L. D. Hall, L. Hough, and K. A. McLauchlan, J. Chem. Soc., 3699 (1962).

(81) L. D. Hall, Chem. Ind. (London), 950 (1963).

(82) M. Karplus, J. Amer. Chem. Soc., **85**, 2870 (1963).

(83) P. L. Durette and D. Horton, Org. Magn. Resonance, **3**, 417 (1971).

(84) J. D. Mokren, P. W. R. Corfield, P. L. Durette, and D. Horton, to be published.

(84a) R. U. Lemieux, Abstr. Papers 5th Int. Conf. Carbohydrates (Paris), 1970.

ranosyl bromide[84] has H-2 and H-3 almost antiparallel (176° ± 5°), but H-4 bisects the 5-CH_2 group unequally (H-4–H-5e = 86 ± 5°, H-4–H-5a = 42 ± 5°), and although the H-1–H-2 dihedral angle (63 ± 7°) is close to ideal, the H-3–H-4 angle (40 ± 4°) deviates substantially. The 1-chloride analog gives similar values.

Despite difficulties in estimating exact couplings from theory, the magnitudes of observed, vicinal, spin–spin coupling-constants can often be employed to differentiate between various conformational possibilities and to provide a measure of conformational equilibria in solution.[85–91] For example, if two vicinal hydrogen atoms on a six-membered ring are *trans*, a large value (8–11 Hz) for the spin coupling indicates preponderance of that conformer having the coupled protons diaxial, whereas a small value (1–3 Hz) indicates preponderance of the conformer having the coupled protons diequatorial. By use of "model" couplings determined from compounds known to be conformationally homogeneous, this method has been used[92–99] quantitatively to determine the conformational equilibria of a wide range of aldopentopyranose derivatives in solution (see p. 93).

The magnitude of geminal spin-couplings[100,101] and the observation of long-range coupling[102–104] may also be valuable in the conformational analysis of carbohydrate derivatives. For an extensive series of aldopentopyranose derivatives, it has been shown[83] that the $J_{5e,5a}$ geminal coupling is about 11 Hz when the acyloxy group at C-4 is equatorial (C-4 substituent bisects the H-5e–C-5–H-5a angle),

(85) F. A. L. Anet, *J. Amer. Chem. Soc.*, **84**, 1053 (1962).
(86) H. Booth, *Tetrahedron*, **20**, 2211 (1964).
(87) H. Feltkamp and N. C. Franklin, *J. Amer. Chem. Soc.*, **87**, 1616 (1965); H. Feltkamp, N. C. Franklin, K. D. Thomas, and W. Brügel, *Ann.*, **683**, 64 (1965).
(88) G. E. Booth and R. J. Ouellette, *J. Org. Chem.*, **31**, 544 (1966).
(89) Y. Pan and J. B. Stothers, *Can. J. Chem.*, **45**, 2943 (1967).
(90) G. O. Pierson and O. A. Runquist, *J. Org. Chem.*, **33**, 2572 (1968).
(91) A. J. de Hoog, H. R. Buys, C. Altona, and E. Havinga, *Tetrahedron*, **25**, 3365 (1969).
(92) P. L. Durette and D. Horton, *Chem. Commun.*, 516 (1969).
(93) P. L. Durette and D. Horton, *Chem. Commun.*, 1608 (1970).
(94) P. L. Durette and D. Horton, *Carbohyd. Res.*, **18**, 57 (1971).
(95) P. L. Durette and D. Horton, *J. Org. Chem.*, **36**, 2658 (1971).
(96) P. L. Durette and D. Horton, *Carbohyd. Res.*, **18**, 289 (1971).
(97) P. L. Durette and D. Horton, *Carbohyd. Res.*, **18**, 389 (1971).
(98) P. L. Durette and D. Horton, *Carbohyd. Res.*, **18**, 403 (1971).
(99) P. L. Durette and D. Horton, *Carbohyd. Res.*, **18**, 419 (1971).
(100) J. A. Pople and A. A. Bothner-By, *J. Chem. Phys.*, **42**, 1339 (1965).
(101) A. A. Bothner-By, *Advan. Magn. Resonance*, **1**, 195 (1965).
(102) B. Coxon, *Carbohyd. Res.*, **13**, 321 (1970).
(103) L. D. Hall and L. Hough, *Proc. Chem. Soc.*, 382 (1962).
(104) L. D. Hall, J. F. Manville, and A. Tracey, *Carbohyd. Res.*, **4**, 514 (1967).

whereas its value is over 13 Hz when the C-4 substituent is axial (antiparallel to H-5a); intermediate values observed provide a rough measure of conformational equilibria. Long-range couplings of appreciable (>0.5 Hz) magnitude, between protons separated by four sigma bonds, are observed in certain, geometrically constrained molecules. The optimal stereochemical requirement for this type of coupling in saturated carbohydrate derivatives,[103] and other saturated cyclic systems,[105] is the planar "M" or "W" arrangement of the four sigma bonds.[105] Although small (~0.5 Hz) long-range couplings have been observed between protons having 1,3-axial–equatorial and 1,3-diaxial orientations,[104] the occurrence of long-range couplings for a pyranoid carbohydrate in the range 1.0–2.5 Hz indicates that the protons involved have a 1,3-diequatorial orientation, as in 2,3,4-tri-O-acetyl-1,6-anhydro-β-D-talopyranose[106] (**1**).

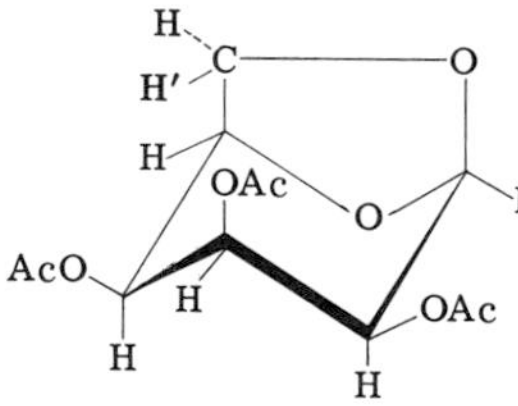

(1)

$$J_{1,3} \approx J_{3,5} \approx J_{4,6} \approx 1.1 \text{ Hz}$$

The n.m.r. spectral method, as a tool for monitoring rate processes, can also, in suitable cases, be used for measurement of the rate of chair–chair conformational inversion of pyranoid sugar derivatives, and thus, of the energy barrier for ring inversion[95,107,108] (see p. 91).

3. Methods Based on Optical Activity

Because optical rotation is sensitive to small changes in molecular structure, measurement of rotatory power possesses great potential as a means of obtaining conformational information for optically active, organic molecules. However, the theoretical relationship between optical rotation and molecular structure is not yet well understood, and it is not possible at present to predict accurately,

(105) S. Sternhell, *Rev. Pure Appl. Chem.*, **14**, 15 (1964).
(106) D. Horton and J. S. Jewell, *Carbohyd. Res.*, **5**, 149 (1967).
(107) N. S. Bhacca and D. Horton, *J. Amer. Chem. Soc.*, **89**, 5993 (1967).
(108) P. L. Durette, D. Horton, and N. S. Bhacca, *Carbohyd. Res.*, **10**, 565 (1969).

from theory alone, the sign and magnitude of the rotation of a particular molecule in a given configuration and conformation. Hence,
attempts at correlating optical rotation and conformation have taken
an empirical or semi-empirical approach.

The first such approach to the interpretation of optical rotation in
terms of the conformational properties of carbohydrates was made by
Whiffen[109] in 1956. He proposed that the observed rotation of an
optically active molecule can be regarded as an algebraic summation
of partial rotatory contributions of various conformational elements
of asymmetry. For these contributions, empirical values were determined that allowed estimation of the net rotations of various
cyclic sugars and cyclitols with fair accuracy. A more extensive
treatment was presented by Brewster,[110] and it was applied to a wide
range of optically active, acyclic and cyclic compounds. The best
correlations between the calculated and the experimental values
were obtained with compounds that do not absorb in the near-ultra-
violet and have predictable, fixed conformations, or for which the
conformational populations can be reliably estimated.[111,112] In the
carbohydrate field, the calculations are quite simple for the poly-
hydroxycyclohexanes,[109,113,114] and differences between the calculated
and observed values for the rotation have been interpreted in terms
of conformational equilibria.[109,114,115] Similar comparisons have been
made for the methyl D-aldopyranosides,[109,114,116] although the lack of
precision in the correlation does not allow a detailed treatment in
terms of conformational populations.

In an attempt to determine an empirical set of values for partial
rotatory contributions of various conformational elements, in systems
devoid of conformational ambiguities, correlations have been made
of the optical rotations of the eight 1,6-anhydro-β-D-hexopyranoses
and their triacetates[117] (see Table I), the 1,6-anhydro-monodeoxy-β-
D-hexopyranoses, and the 2,7-anhydro-β-D-heptulopyranoses and
their acetates.[118] The calculated rotations were found to be in excel-

(109) D. H. Whiffen, *Chem. Ind.* (London), 964 (1956).
(110) J. H. Brewster, *J. Amer. Chem. Soc.*, **81**, 5475, 5483 (1959).
(111) W. D. Celmer, *J. Amer. Chem. Soc.*, **87**, 1797 (1965).
(112) W. D. Celmer, *J. Amer. Chem. Soc.*, **88**, 5028 (1966); *Abstr. Papers Amer. Chem.
Soc. Meeting* (Winter), c13, c14 (1966); *ibid.*, **153**, c19 (1967).
(113) S. Yamana, *Bull. Chem. Soc. Jap.*, **33**, 1741 (1960).
(114) Ref. 25, Chapter 6, pp. 381–394.
(115) G. E. McCasland, S. Furuta, L. F. Johnson, and J. N. Shoolery, *J. Org. Chem.*, **29**,
2354 (1964).
(116) W. Kauzmann, F. B. Clough, and I. Tobias, *Tetrahedron*, **13**, 57 (1961).
(117) D. Horton and J. D. Wander, *J. Org. Chem.*, **32**, 3780 (1967).
(118) D. Horton and J. D. Wander, *Carbohyd. Res.*, **14**, 83 (1970).

TABLE I

Observed and Calculated Rotations of 1,6-Anhydro-β-D-hexopyranoses
and Their Triacetates[a]

Configuration	1,6-Anhydro-β-D-hexopyranoses in water [M]$_D$, degrees		Triacetates in chloroform [M]$_D$, degrees	
	Obsd.	Calcd.	Obsd.	Calcd.
allo	−12,280	−12,400	−20,390	−20,200
altro	−34,506	−35,400	−49,536	−48,700
gluco	−10,724	−10,800	−18,864	−18,500
manno	−20,671	−20,600	−35,597	−36,500
gulo	+8,165	+8,100	+6,365	+6,600
ido	−15,001	−14,900	−21,629	−21,900
galacto	−3,548	−3,500	−1,642	−2,200
talo	−12,960	−13,300	−21,024	−20,200

[a] Data taken from Table II, Ref. 117.

lent agreement with values reported in the literature, and the consistency observed throughout the series indicated that all of these anhydropyranose derivatives exist in chairlike conformations.

In a detailed application of empirical rules for optical rotation to problems of conformational analysis,[119] the optical rotation exhibited by a wide range of pyranoid derivatives, predicted to exist mainly in one conformation, were considered in terms of only "pairwise interactions,"[116] that is, in terms of the dihedral angles defined by carbon and oxygen substituents on adjacent atoms. It was observed that the nature of the bridging atom is at least as significant in determining the rotatory contribution of conformational, asymmetric units as is the nature of the atoms that define the dihedral angle.[119]

O/C = 10°   O/O = 45°   C/C$_O$ > 60°   O/C$_O$ = 115°

Empirically derived rotatory contributions
of conformational asymmetric units

In a further approach to the study of conformational equilibria of carbohydrates, based on the correlation of specific rotation with

(119) R. U. Lemieux and J. C. Martin, *Carbohyd. Res.*, **13**, 139 (1970).

conformation, it was inferred,[120-122] from a comparison of n.m.r.-spectral and optical rotatory data, that the changes in optical rotation occurring upon change of the solvent were the result of a change in the position of the conformational equilibria. The changes in rotation agreed with those anticipated from Brewster's rules.[110] It was thus concluded that simple polarimetry could be effectively employed in the study of the effect of solvent environment on conformation.

The methods so far described involve measurement of the rotation at the D-line of sodium. However, the techniques of optical rotatory dispersion and circular dichroism have shown some utility in providing conformational information about carbohydrates.[123-125] As an example, it was found[125] that molybdate complexes of sugars are suitable for conformational analysis by circular dichroism, and several rules have been postulated relating the number and sign of observed Cotton effects to the orientation of adjacent hydroxyl groups in the chair conformations of pyranoid sugar derivatives. A useful development by Nakanishi and coworkers,[126] termed by them the "aromatic chirality method," permits assignment of relative configurations and favored conformations of benzoates or other aroyl esters of sugars from measurements of circular dichroism. For vicinal bis(p-chlorobenzoates), an antiparallel disposition of the two groups gives rise to a single Cotton effect at the perturbed wavelength, whereas, when they are gauche-disposed, the two groups interact to give a split pattern in the circular-dichroism spectra; there are observed two maxima of opposite sign, situated on each side of the perturbed wavelength. The signs of these maxima are related to the chirality of the screw pattern defined by the two aroyloxy groups along the connecting C-C bond (see Table II). The p-chlorobenzoates absorb at a wavelength more accessible than that for the unsubstituted benzoates. Quantitative variations in the intensities of the split patterns can be correlated with minor deviations in the dihedral angles between the aroyloxy groups.

(120) R. U. Lemieux and A. A. Pavia, *Can. J. Chem.*, **46**, 1453 (1968).

(121) R. U. Lemieux, A. A. Pavia, J. C. Martin, and K. A. Watanabe, *Can. J. Chem.*, **47**, 4427 (1969).

(122) R. U. Lemieux and A. A. Pavia, *Can. J. Chem.*, **47**, 4441 (1969).

(123) N. Pace, C. Tanford, and E. A. Davidson, *J. Amer. Chem. Soc.*, **86**, 3160 (1964).

(124) I. Listowsky, G. Avigad, and S. Englard, *J. Amer. Chem. Soc.*, **87**, 1765 (1965).

(125) W. Voelter, G. Kuhfittig, G. Schneider, and E. Bayer, *Ann.*, **734**, 126 (1970).

(126) N. Harada, H. Sato, and K. Nakanishi, *Chem. Commun.*, 1691 (1970); compare, N. Harada and K. Nakanishi, *J. Amer. Chem. Soc.*, **91**, 3989 (1969).

TABLE II
Correlation of Cotton Effects with Chirality for Vicinal Bis(p-chlorobenzoates)
of Sugar Derivatives in Methanol[126]

Bis(p-chlorobenzoate) of	Chiral relationship of ester groups in $C1(\text{D})^a$ conformation	Cotton effects, $\Delta\epsilon$ (and wavelength, nm)	
Methyl 2,6-dideoxy-α-D-galactopyranoside	(+)-gauche	+17.2 (246)	−7.2 (229)
Methyl 2,6-dideoxy-β-D-galactopyranoside	(+)-gauche	+28.9 (246)	−7.5 (229)
Methyl 2,6-dideoxy-α-D-glucopyranoside	(−)-gauche	−37.1 (246)	+17.1 (229)
Methyl 2,6-dideoxy-β-D-glucopyranoside	(−)-gauche	−29.0 (245)	+14.5 (230)
Methyl 4-O-methyl-6-deoxy-α-D-altropyranoside	$(0)^b$	−10.5 (247)	

^{a} For nomenclature, see p. 76. ^{b} Ester groups antiparallel in the $C1(\text{D})$ conformation.

4. Other Methods

a. Formation of Complexes in Cuprammonia—From studies on the complexing of sugar derivatives in ammoniacal copper(II) solution, Reeves was able to deduce the favored conformation for many methyl D-aldopyranosides and their derivatives.[16,22] The procedure was based on a theory, developed from a study of model compounds,[18] which defined the necessary conditions for formation of a complex between cuprammonia and two hydroxyl groups on a pyranoid ring. The actual structures of the complexes are not known, but the steric requirements for their formation are quite well established. Complexes can form only when the hydroxyl groups are *syn*-diaxial or are on adjacent carbon atoms oriented at a projected dihedral angle of approximately 0° or 60°. At dihedral angles of approximately 120° or 180°, no such complex can be formed.

Formation of a complex was first detected from optical rotatory measurements.[127] However, since complexes can form without appreciable change in optical rotation (as with vicinal-eclipsed or *syn*-diaxial hydroxyl groups), the question of possible sites for complexing was re-investigated by observing changes in the electrical conductivity of the cuprammonia solution. By consideration of the electrical conductivity and optical rotatory properties of the methyl D-aldopyranosides and their derivatives in cuprammonia solution, Reeves assigned conformations to the glycopyranosides studied.

(127) R. E. Reeves, *J. Biol. Chem.*, **154**, 49 (1944).

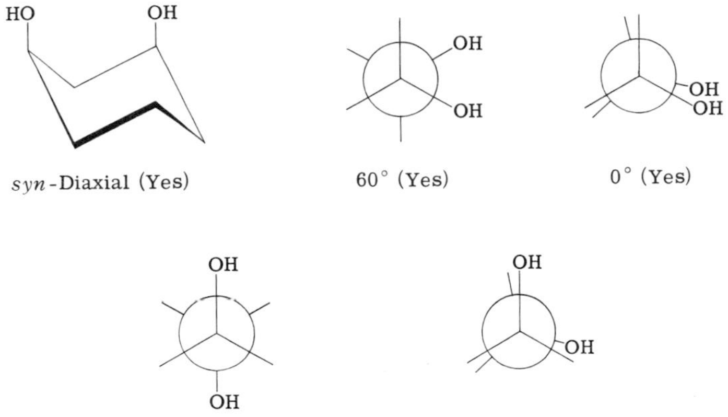

syn-Diaxial (Yes)   60° (Yes)   0° (Yes)

180° (No)   120° (No)

Stereochemical requirements for formation
of cuprammonia complexes

A criticism of Reeves's method is that formation of a complex might strongly affect the position of any equilibrium existing between conformers. Indeed, a complex having a conformation intrinsically less favored might form, even though complexing might not be possible with the preponderant form at equilibrium. Hence, an indication of complexing in such a situation could lead to an erroneous conclusion as to the favored conformation of the non-complexed derivative. In practice, however, this consideration applies to only a few of the examples examined by Reeves, with the result that most of his conformational assignments have been found correct when established later by other physical methods. The criticism does not apply when negative evidence, from lack of formation of a complex, is used for excluding a particular, possible conformation where a complex could be formed.

b. Dipole Moments.—The complexity of the total electronic situation in polysubstituted sugar molecules has greatly limited the potential of the comparison of calculated with measured values of dipole moments as a tool for conformational analysis in the carbohydrate field. Nevertheless, the method has proved useful in several instances, as in the measurement[128] of the dipole moment of 1,3:2,4-di-O-methylene-L-threitol to show that the diacetal exists in the so-called "O-inside" conformation and not in the alternative, "H-inside" conformation. The finding was corroborated by n.m.r.-spectral evidence.

(128) R. U. Lemieux and J. Howard, *Can. J. Chem.*, **41**, 393 (1963).

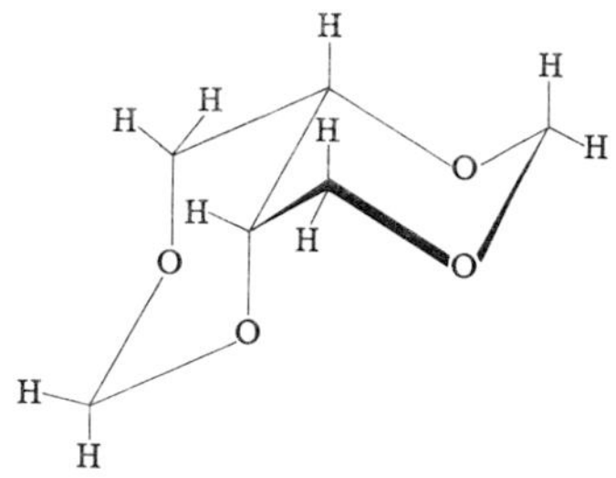

"O-Inside" 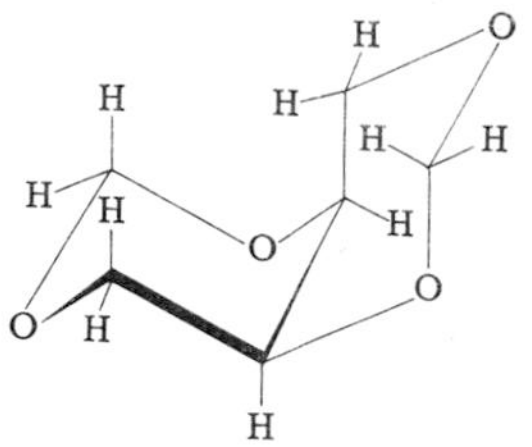   "H-Inside"

1, 3 : 2, 4-Di-O-methylidene-L-threitol

Dipole moment (debyes)		
Calculated	4. 2	1. 9
Observed	4. 0	

The correlation between vicinal coupling-constants and dipole moments has been applied to conformational problems concerning derivatives of 2-alkoxytetrahydropyran,[91,129] the fundamental skeleton of all pyranoid sugar derivatives. The observation of a linear relationship between the squares of the electric-dipole moments and the vicinal coupling-constants, as the conformational equilibrium constant was varied by changing the alkoxyl substituent, indicated[91,129] that the equilibrium mixture is composed of only two conformers in appreciable proportion.

In another example, as a model study for an investigation of the conformations of certain D-glucosides and polysaccharide derivatives,[130] the comparison of observed dipole moments with values calculated from bond-moment data was used to demonstrate that, in carbon tetrachloride solution, tetrahydropyran exists in a chair conformation, and that 2-chlorotetrahydropyran exists in that chair conformation having the C–Cl bond axially disposed.

c. Theoretical Approaches.—Several theoretical studies on the conformations of the aldopyranoses have been reported. By calculating the extent of atomic (Van der Waals) overlap of non-bonded atoms, the relative stabilities of the chair conformations of the aldopyranoses were estimated.[131] Although the approach considered the exact shape of the pyranoid ring, it did not take into account the effect

(129) A. J. de Hoog and H. R. Buys, *Tetrahedron Lett.*, 4175 (1969).
(130) J. M. Eckert and R. J. W. LeFèvre, *J. Chem. Soc.* (B), 855 (1969).
(131) G. R. Barker and D. F. Shaw, *J. Chem. Soc.*, 584 (1959).

of hydrogen bonding on the effective size of the hydroxyl groups, nor the effect of polar interactions on the favored conformation. The conformational assignments are in general agreement with those of Reeves for the methyl D-aldopyranosides,[16] but some of the conclusions, such as the indication that α-D-ribopyranose favors the $1C$(D) conformation, appear irregular.

Sundararajan and Rao[132] made potential-energy calculations for the aldohexopyranoses and aldopentopyranoses. Polar interactions, intermolecular forces, hydrogen-bonding interactions, and solvent effects were ignored completely, and the non-bonded (Van der Waals) interaction energies were determined from the potential-energy functions of Kitaigorodsky; the calculations indicated that, for the sixteen α- and β-D-aldohexopyranoses and most of the eight α- and β-D-aldopentopyranoses, the chair conformer having the lower energy is the $C1$(D) form. However, as polar factors are significant in influencing the conformation adopted, a valid, theoretical model for predicting favored conformation would, at least, have to include terms for polar interactions. Their deriving of conclusions that are at variance with experimental data is most probably attributable to failure to consider such polar interactions. Better agreement with experiment was later obtained from a consideration both of non-bonded and electrostatic interactions,[133] and the approach was successfully extended to acetylated derivatives (V. S. R. Rao, personal communication).

A very similar approach, considering only Van der Waals interactions, was used by Szafranek[134] for calculating the relative energies of the two chair conformers of several D-aldohexopyranoses. The results obtained were consistent with those of Reeves.[16] However, the agreement may be fortuitous, as this treatment, too, neglected polar interactions.

A mathematical method has been described[135,136] for determining, from independent variables that include the contribution of valence-angle deformation, the potential barrier for conformational inversion in β-D-glucopyranose. The energy required for the conversion of a chair form to a boat form was calculated to be 13.5 kcal.mole^{-1}, and the energy required for the reverse process, 6.5 kcal.mole^{-1}. Ex-

(132) P. R. Sundararajan and V. S. R. Rao, *Tetrahedron,* **24,** 289 (1968).

(133) V. S. R. Rao, K. S. Vijayalakshmi, and P. R. Sundararajan, *Carbohyd. Res.,* **17,** 341 (1971).

(134) J. Szafranek, *Zesz. Nauk. Wyzsz. Szk. Pedagog. Gdansku: Mat., Fiz., Chem.,* **7,** 165 (1967).

(135) K. M. Grushetskiĭ and A. F. Bestsennyi, *Zh. Strukt. Khim.,* **8,** 332 (1967).

(136) K. M. Grushetskiĭ and A. F. Bestsennyi, *Zh. Strukt. Khim.,* **9,** 870 (1968).

perimental values[95,107,108] obtained by n.m.r. spectroscopy for conformational inversion in certain aldopentopyranose derivatives are
in the range of 10–11 kcal.mole⁻¹ (see p. 91).

(see p. 91)

III. CONFORMATIONS OF ACYCLIC SUGARS AND THEIR DERIVATIVES

1. General Considerations

By rotation about single bonds, it is possible for an acyclic sugar
derivative to adopt numerous conformational states. In the liquid
state or in solution, at room temperature, the energy barriers separating these various conformations for an acyclic sugar are generally
small enough for rapid conformational interconversion to occur,
whereas, in the crystalline state, lattice forces usually hold the
molecule in a single, fixed conformation.

The most useful way to depict the various conformational states of
an acyclic sugar chain is by means of a formalized drawing that shows
the relative dihedral angles of the substituents at each end of a carbon–carbon bond. Two such types of representation are the sawhorse
and Newman (Böeseken) formulas,[137] as illustrated for erythritol.
The zigzag type of representation shown has the advantage that it
can be employed to display the relative orientations of groups along
a chain of several atoms that lie approximately in a plane (for example,
the plane of the paper).

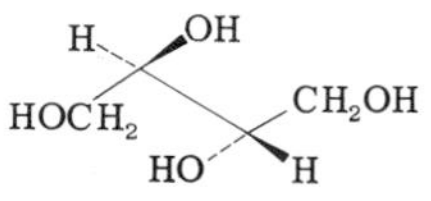

| Zigzag | Sawhorse | Newman |

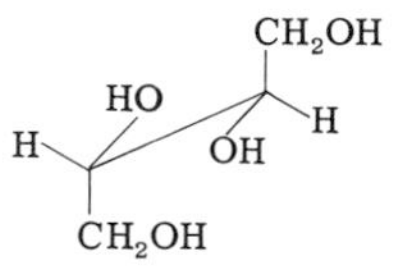

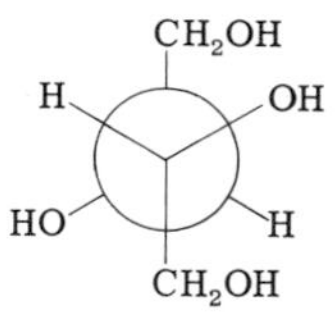

Low-energy (staggered) conformations of erythritol

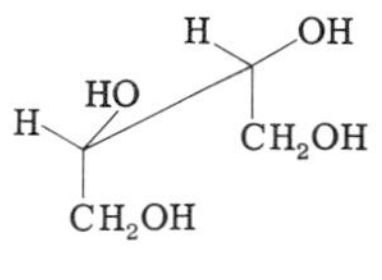

Sawhorse 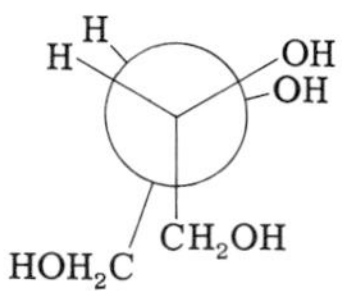   ≡   Newman

High-energy (eclipsed) conformations of erythritol

(137) M. S. Newman, *J. Chem. Educ.*, **32**, 344 (1955); similar formulas were used earlier
by J. Böeseken and R. Cohen, *Rec. Trav. Chim. Pays-Bas*, **47**, 839 (1928).

2. Acyclic Derivatives in the Crystalline State

From crystal-structure analyses of unsubstituted alditols, it has been demonstrated[138] that, in the crystalline state, such acyclic compounds tend to adopt a conformation having an extended, planar, zigzag carbon-chain, unless this arrangement leads to a parallel, 1,3-steric interaction between oxygen atoms, in which case a more favorable conformation is obtained by rotation through 120° about a single carbon–carbon bond, to give a nonlinear (bent) chain. Thus, the planar, zigzag conformation is adopted in the solid state by DL-arabinitol,[139] galactitol[140] (2), D-mannitol[141] (3), and the D-ara-

(2)   (3)

binonate ion,[142] none of which have parallel 1,3-interactions between oxygen atoms in this conformation. In contrast, the ribitol chain in crystalline riboflavine hydrobromide monohydrate[143] exists in the bent-chain conformation, presumably because the planar, zigzag conformation has an unfavorable 1,3-interaction between two oxygen atoms. Similar bent conformations have been observed in the solid state for ribitol,[144] xylitol[145] (4), D-glucitol[146] (5), and D-iditol,[147] all of which would have a 1,3-interaction in the extended-chain arrangement. For the D-gluconate ion,[148] however, the steric strain of such a parallel interaction is apparently relieved, in the solid state, by bond-angle deformation, because the X-ray data indicate an essentially planar, zigzag conformation for the ion.

(138) G. A. Jeffrey and H. S. Kim, *Carbohyd. Res.*, **14**, 207 (1970).
(139) F. D. Hunter and R. D. Rosenstein, *Acta Crystallogr.*, **B24**, 1652 (1968).
(140) H. M. Berman and R. D. Rosenstein, *Acta Crystallogr.*, **B24**, 435 (1968).
(141) H. M. Berman, G. A. Jeffrey, and R. D. Rosenstein, *Acta Crystallogr.*, **B24**, 442 (1968); H. S. Kim, G. A. Jeffrey, and R. D. Rosenstein, *ibid.*, **B24**, 1449 (1968).
(142) S. Furberg and S. Helland, *Acta Chem. Scand.*, **16**, 2373 (1962).
(143) N. Tanaka, T. Ashida, Y. Sasada, and M. Kakudo, *Bull. Chem. Soc. Jap.*, **40**, 1739 (1967); .**42**, 1546 (1969).
(144) H. S. Kim, G. A. Jeffrey, and R. D. Rosenstein, *Acta Crystallogr.*, **B25**, 2223 (1969).
(145) H. S. Kim and G. A. Jeffrey, *Acta Crystallogr.*, **B25**, 2607 (1969).
(146) Y. J. Park, G. A. Jeffrey, and W. C. Hamilton, *Acta Crystallogr.*, **B**, in press.
(147) N. Azarnia, M. S. Shen, and G. A. Jeffrey, *Acta Crystallogr.*, **B**, in press.
(148) C. D. Littleton, *Acta Crystallogr.*, **6**, 775 (1953).

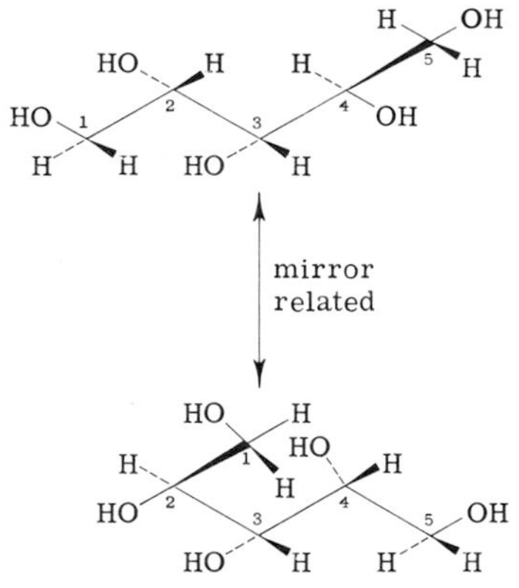

mirror related

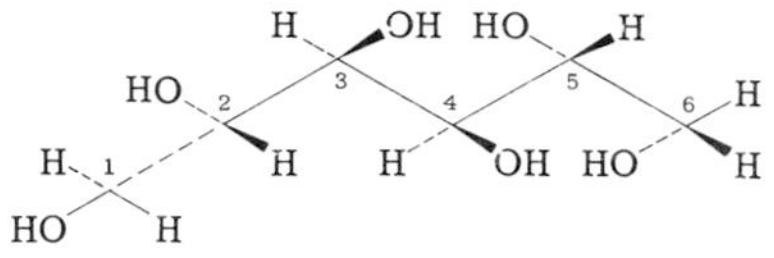

(5)

(4)

The relationship between configuration and conformation observed for the unsubstituted alditols in the crystalline state can be used to predict, with significant confidence, the favored conformations of other acyclic sugar derivatives whose structures have not yet been determined experimentally.

3. Acyclic Derivatives in Solution

Before the advent of n.m.r. spectroscopy as a tool for solving stereochemical problems in carbohydrate chemistry, it was generally believed that acyclic sugar derivatives in solution exist in the extended, planar, zigzag conformation. Such a planar, zigzag structure was invoked for the alditols in order to explain the favored formation of certain rings in cyclic acetals of the alditols[149] and to rationalize the relative rates of glycol fission by periodate at the various carbon–carbon bonds in the alditols.[150] The conclusions reached in both instances as to the favored conformation of the alditols can scarcely be considered valid, because, in the first example, the equilibrium between the various possible acetals is a function only of the free-energy content of each acetal[26] and not necessarily of the conformation of the alditol precursor, and, in the second, the rate of oxidation by periodate depends on the free energy of the cyclic intermediates and not on that of the initial conformation.

A conformational assignment from n.m.r. spectroscopy for an acyclic sugar chain in solution was reported[151] in 1965. Analysis of the proton

(149) S. A. Barker, E. J. Bourne, and D. H. Whiffen, *J. Chem. Soc.*, 3865 (1952).
(150) J. C. P. Schwarz, *J. Chem. Soc.*, 276 (1957); D. H. Hutson and H. Weigel, *ibid.*, 1546 (1961).
(151) D. Horton and M. J. Miller, *J. Org. Chem.*, **30**, 2457 (1965).

spin-couplings in the n.m.r. spectrum of 2-(D-*arabino*-tetraacetoxy-butyl)quinoxaline (**6**) in chloroform-*d* showed that, for the poly-

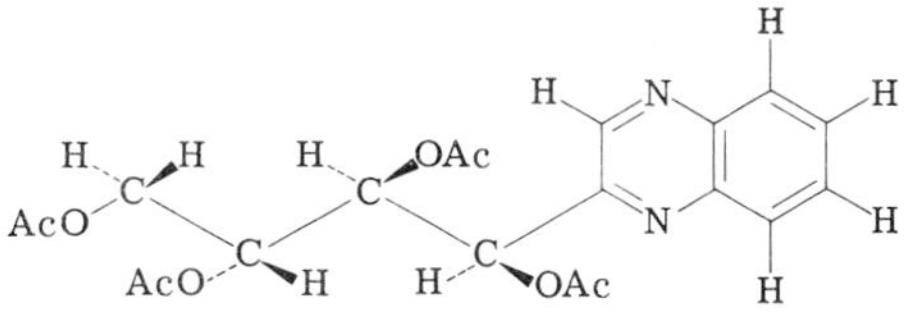

(6)

acetoxyalkyl chain of this acyclic sugar derivative, the rotameric state most highly populated is the planar, zigzag arrangement, corresponding to attainment of minimum non-bonded interactions between the small–medium–large sets of groups at the ends of each carbon–carbon bond. A similar model was later used[152] in interpreting the n.m.r. spectra of certain sugar phenylosotriazoles and their acetates.

In a more extensive study of the effect of configuration on this conformational model for acyclic sugar derivatives in solution, a detailed analysis by n.m.r. spectroscopy was made[153] of seven, unsubstituted, sugar osotriazoles in methyl sulfoxide. It was found that the planar, zigzag arrangement of carbon atoms in the polyhydroxyalkyl chain is the favored conformation, except where such an arrangement would lead to an eclipsed 1,3-interaction of hydroxyl groups, as in the L-*xylo* configuration (this interaction was pointed out by Schwarz[154] in 1964 and is analogous[155] to an axial–axial interaction in a 1,3-disubstituted cyclohexane). The spin-coupling data for this derivative were interpreted in terms of a favored conformation (**7**) generated from the

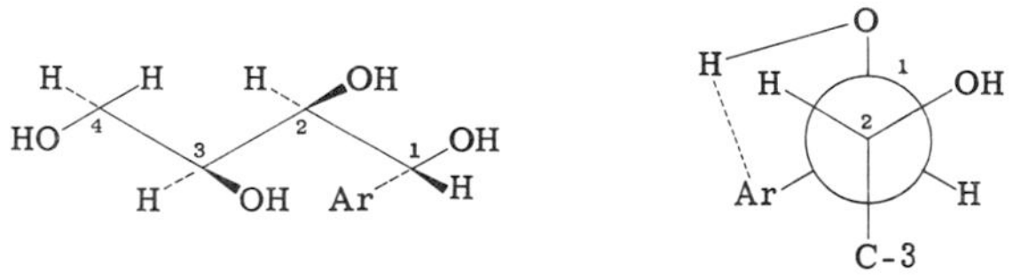

(7)

planar, zigzag form by rotation through one-third of a revolution about the C-1–C-2 bond in order to remove the parallel, 1,3-interaction between the hydroxyl groups on C-1 and C-3.

(152) G. G. Lyle and M. J. Piazza, *J. Org. Chem.*, **33**, 2478 (1968).
(153) H. S. El Khadem, D. Horton, and T. F. Page, Jr., *J. Org. Chem.*, **33**, 734 (1968).
(154) J. C. P. Schwarz, cited in footnote 8 of Ref. 153.
(155) A. B. Dempster, K. Price, and N. Sheppard, *Chem. Commun.*, 1457 (1968).

The importance of parallel 1,3-interactions in determining the favored conformational state of acyclic sugar derivatives in solution was also apparent from n.m.r.-spectral studies[156] of polyhydroxyalkyl-quinoxalines in methyl sulfoxide, and of their acetates in carbon tetrachloride.

The conformations of the peracetylated aldose diethyl dithioacetals in chloroform solution have been examined by n.m.r. spectroscopy,[157] and the term "sickle" was introduced to designate the conformation generated from the extended, planar, zigzag form by rotation through 120° about an internal carbon–carbon bond. Such a sickle arrangement is adopted in solution by the D-xylose (8) and D-ribose (9) diethyl

(8)

(9)

dithioacetals, whose extended conformation would have a pair of parallel, eclipsed acetoxyl groups on C-2 and C-4.

Similar observations have been made with the fully acetylated diphenyl dithioacetals of D- and L-arabinose, D-xylose, and D-ribose,[158] the *aldehydo*-aldose peracetates having the D-*ribo*, D-*arabino*, D-*xylo*, D-*lyxo*, and L-*galacto* configurations[159] and analogous dimethyl acetals,[159a] the tetraacetoxy-*trans*-1-nitro-1-hexenes having the D-*arabino*, D-*xylo*, the D-*ribo* configurations,[160] the pentononitrile tetraacetates,[161] a series of acetylated arylosotriazoles,[162] and various aldose dithioacetals,[163] and acetylated alditols.[164]

Related investigations[165] have further suggested that, for those derivatives adopting the sickle arrangement, the particular carbon–

(156) W. S. Chilton and R. C. Krahn, *J. Amer. Chem. Soc.*, **90**, 1318 (1968).
(157) D. Horton and J. D. Wander, *Carbohyd. Res.*, **10**, 279 (1969).
(158) D. Horton and J. D. Wander, *Carbohyd. Res.*, **13**, 33 (1970).
(159) D. Horton and J. D. Wander, *Carbohyd. Res.*, **15**, 271 (1971).
(159a) J. Defaye, D. Gagnaire, D. Horton, and M. Muesser, *Carbohyd. Res.*, in press (1972).
(160) J. M. Williams, *Carbohyd. Res.*, **11**, 437 (1969).
(161) W. W. Binkley, D. R. Diehl, R. W. Binkley, *Carbohyd. Res.*, **18**, 459 (1971).
(162) H. El Khadem, D. Horton, and J. D. Wander, to be published.
(163) D. Horton and J. D. Wander, to be published.
(164) S. J. Angyal, D. Gagnaire, and R. Le Fur, *Carbohyd. Res.*, in press (1972).
(165) J. B. Lee and B. F. Scanlon, *Tetrahedron*, **25**, 3413 (1969).

carbon bond that undergoes rotation depends on the relative size of the groups attached to the carbon atoms bearing the two substituents in 1,3-interaction. That the location of the *gauche* arrangement in the sickle conformation depends on the size of the attached substituents was also deduced[166] from an analysis of the n.m.r. spectra of some methyl tetra-*O*-acetyl-5-hexulosonates and peracetylated *keto*-hexuloses. By correlation of data from different reports, various substituent groups can be arranged in increasing order of their tendency to occupy a *gauche* position in the *xylo* and *ribo* configurations[166]; the sequence for the *ribo* configuration differs from that for the *xylo* configuration (as shown in the following lists), as would be anticipated from the different steric arrangements in the two series.

ribo Configuration: $CN > CO_2Me > COCH_2OAc$
$$> CH{=}CHNO_2 >> CH_2OAc > CH(SEt)_2$$

xylo Configuration: $CH(SEt)_2 > CH{=}CHNO_2 > CH_2OAc >$
$$CHOAcCH_2OAc \approx COCH_2OAc$$

Decreasing order for tendency of groups to occupy a gauche disposition[166]

For molecules in the solid state, it is possible that lattice forces may profoundly affect the process of adoption of a particular conformation, and thus the favored conformation of a molecule in the crystalline state is not *necessarily* the same as that adopted when the molecule is surrounded by a sheath of solvent molecules. Nevertheless, there is a high degree of correlation between the results of X-ray crystallography of alditols and their derivatives (see p. 69) and the conformations adopted by various types of acyclic sugar derivatives in solution.

Considerations of conformational stability for acyclic sugar derivatives may be of use in predicting the outcome of kinetically controlled, ring-forming reactions of the sugars. Thus, the tendency of the four pentose dialkyl dithioacetals to form 2,5-anhydrides when they are treated with one molecular proportion of *p*-toluenesulfonyl chloride in pyridine can be correlated with their conformational stability and their ability to attain the transition state for cyclization.[167]

(166) S. J. Angyal and K. James, *Aust. J. Chem.*, **23**, 1223 (1970).
(167) J. Defaye and D. Horton, *Carbohyd. Res.*, **14**, 128 (1970); compare, J. Defaye, D. Horton, and M. Muesser, *ibid.*, **20**, 305 (1971).

IV. Conformations of Pyranoid Sugars and Their Derivatives

1. General Considerations

The conformational analysis of pyranoid sugars and their derivatives may be expected to follow the broad principles established with cyclohexane derivatives,[168] but also to exhibit several major points of difference.[25,32,169] The symmetry of the ring is decreased by replacement of a methylene group by an oxygen atom,[16] and the latter exerts a stereo-electronic influence on the conformation that differs from that of a methylene group. Pyranoid sugars and their derivatives may be formulated in two, energetically nonequivalent, chairlike conformations, interconversion between which involves an energy barrier corresponding to that for a conformer having five ring-atoms in a plane, and also in a "flexible cycle" of skew forms[17,170–172] interconvertible through energy barriers corresponding to the boat forms. At conformational equilibrium, the various conformers will be populated to extents depending on their relative free-energy content, according to the classical thermodynamic distribution; the concentrations of those conformers corresponding to the energy barriers will be infinitesimally small.

The relative, free-energy contents of the various conformers possible are determined by the net effect of various stereo-electronic factors, including (1) steric (Van der Waals and London) interactions, (2) bond-torsional (Pitzer) strain, (3) bond-angle (Baeyer) strain, (4) electronic factors involving interaction of dipoles, (5) the effects of solvation and hydrogen bonding, and (6) crystal-lattice forces. Experimental evidence, mainly from n.m.r. spectroscopy,[48,71,72,173–176] indicates that many pyranoid sugars, especially aldohexopyranoses

(168) Ref. 25, Chapter 2.
(169) D. Horton, in "Handbook of Biochemistry and Biophysics," H. C. Damm (Ed.), World Publishing Co., Cleveland, Ohio, 1966, pp. 128–131.
(170) H. S. Isbell and R. S. Tipson, *Science*, **130**, 793 (1959); the skew ring-shape was apparently first discussed by P. C. Henriquez, *Koninkl. Ned. Akad. Wetenschap. Proc., Ser. B*, **37**, 532 (1934); compare R. D. Stolow, *J. Amer. Chem. Soc.*, **81**, 5806 (1959).
(171) H. S. Isbell and R. S. Tipson, *J. Res. Nat. Bur. Stand.*, **64A**, 171 (1960).
(172) R. Bentley, *J. Amer. Chem. Soc.*, **82**, 2811 (1960).
(173) D. Horton and W. N. Turner, *J. Org. Chem.*, **30**, 3387 (1965).
(174) C. V. Holland, D. Horton, M. J. Miller, and N. S. Bhacca, *J. Org. Chem.*, **32**, 3077 (1967).
(175) C. V. Holland, D. Horton, and J. S. Jewell, *J. Org. Chem.*, **32**, 1818 (1967).
(176) N. S. Bhacca, D. Horton, and H. Paulsen, *J. Org. Chem.*, **33**, 2484 (1968).

and their monocyclic derivatives, exhibit a high degree of conformational homogeneity in one of the two chairlike conformations; convincing evidence for nonchair conformers as favored forms has been presented only for systems having another ring or rings fused to the pyranoid ring.[102,177] For monocyclic, aldopentopyranoid derivatives in solution, conformational homogeneity in a single chair form is, however, the exception rather than the rule, and the chair–chair conformational equilibria have been measured for several series of derivatives as a function of their stereochemistry.[92–99] The energy barrier to chair–chair interconversion in certain pyranoid sugar derivatives has been measured, and found to differ little from that observed for unsubstituted tetrahydropyran.[95,107,108]

2. Conformational Representation and Terminology

The Haworth perspective formulas[178] express more closely than the Fischer projection formulas the actual bond-lengths in the cyclic sugars and their derivatives. The convention of establishing one plane for the atoms in the ring, and planes above and below the ring for the substituents, facilitates the recognition of configurational relationships. However, the Haworth formulas were not intended to suggest that the rings are planar in the actual molecule. By simple rotation about single bonds, molecules of a cyclic sugar, having a given structure and configuration, may adopt an infinite number of relative orientations (conformations[4]) of the component atoms in three-dimensional space.[179] For molecules in a crystalline solid, a single conformation is generally adopted, and its geometry can be determined accurately by crystallographic techniques. For freely mobile molecules (as in solution or in the molten state), a conformational equilibrium is established. The thermal energy in sugar molecules at room temperature is normally sufficient to establish a freely mobile system in conformational equilibrium, unless the molecule is restricted from assuming more than one conformation.

The two chairlike conformations of pyranoid sugars and their derivatives represent potential-energy minima as compared with other theoretically possible conformations. The idealized depictions given for the two possible chairlike forms of β-D-glucopyranose show the axial and equatorial dispositions of the bonds to the various substituents.

(177) C. Cone and L. Hough, Carbohyd. Res., 1, 1 (1965).
(178) H. D. K. Drew and W. N. Haworth, J. Chem. Soc., 2303 (1926).
(179) Compare Ref. 25, Chapters 2, 4, and 6.

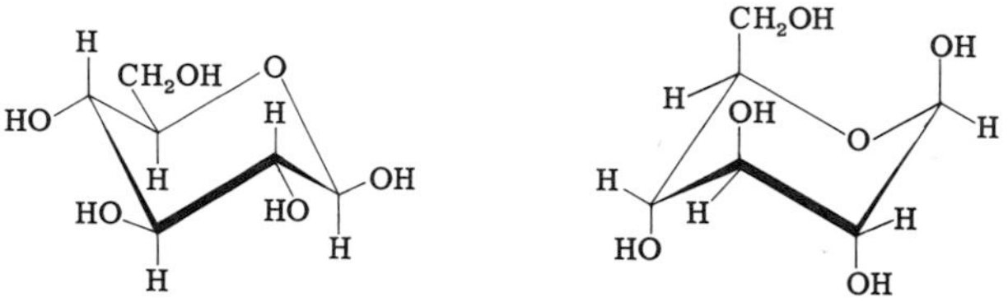

All-equatorial conformation   All-axial conformation

β-D-Glucopyranose

There exists as yet no officially approved system of nomenclature for naming the conformers of sugars. A suitable structural formula can be used to express conformation without ambiguity, but it is frequently desirable to refer to a particular conformation by use of an appropriate symbol as part of a name, in comparisons in a table, or in a textual description. In practical terms, a major issue centers around the naming of the two chairlike forms of pyranoid sugars and their derivatives. A question of logic arises as to whether an assigned symbol should be independent of chirality (in the spirit of the α,β nomenclature for anomers), or whether it should refer to a particular ring-shape and require the independent specification of chirality before the conformational symbol becomes meaningful. By the first system, a specific symbol is used to represent a given axial–equatorial disposition of substituent groups, and this symbol applies to *both* enantiomers; the chirality must then be specified before a unique conformational structure for a single enantiomer can be drawn. The second system allows the structure to be drawn directly from the symbol, but leads to *different* symbols for mirror-image conformations (that is, enantiomorphs in chair conformations having the *same* axial–equatorial dispositions of substituents).

The system of conformational terminology that has thus far been utilized most widely was introduced by Reeves,[16,19] and is based on the second system mentioned, that is, it assigns symbols for specified ring-shapes, as shown. The symbols are defined as *C1* and *1C* for

C1   *1C*

the ring-shapes shown, with the understanding that the ring numbering runs clockwise (because the unsubstituted rings are superposable).
For these symbols to be meaningful, the chirality of the molecule
must be specified, because, as shown for the two chairlike conformers of α-D-glucopyranose and α-L-glucopyranose, the symbol
is changed when mirror-image conformers are named.

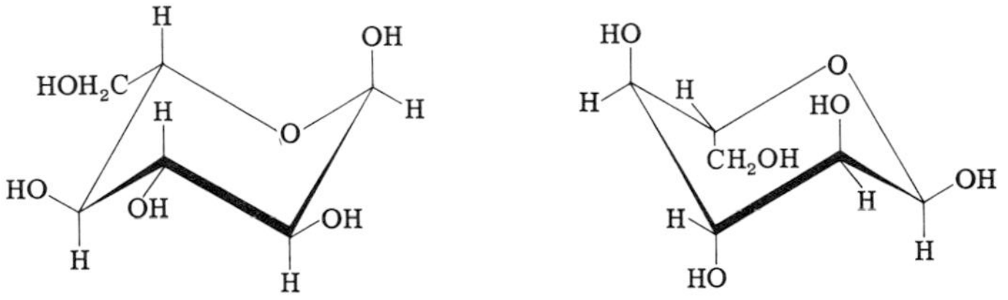

1C   C1

α-L-Glucopyranose

------------------------MIRROR PLANE----------------------

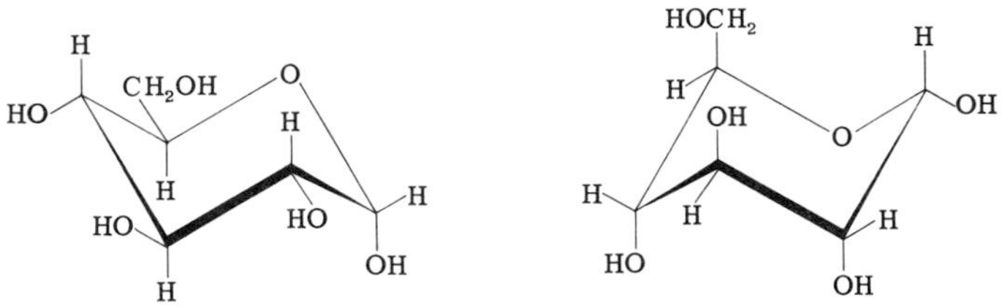

C1   1C

α-D-Glucopyranose

By this system, the conformational symbol has no meaning unless
the chirality of the molecule to which it refers is specified. If the
chiral symbol is not given in the description of the compound, the
appropriate chiral designation must be specified, together with the
conformational symbol. Thus, in order to indicate a particular axial–
equatorial relationship of the substituent groups in a sugar without
specifying its chirality, two symbols must be employed; for example,
α-glucopyranose in the *C1*(D) or *1C*(L) conformation. This terminology, although quite explicit, is rather inconvenient to use in conjunction with such widely utilized physical methods as n.m.r. spectroscopy and X-ray crystallography, which give information on the
relative dispositions of the groups in the molecule, but seldom give
chiral information directly.

Various alternatives to the Reeves system for the nomenclature of conformers of pyranoid sugars have been proposed.[170,171,180,181] The system of Isbell and Tipson[170,171] embodies the principle of achiral conformational symbols. The symbols CE and CA are used to designate the two pyranoid chair conformations. For the purpose of assigning the conformational symbol, the particular anomeric configuration of the molecule under consideration is disregarded. That chair conformation which, in the α anomeric form, would have the bond to the anomeric group axial is assigned the symbol CA, and that chair conformation which, in its α anomeric form, would have the bond to the anomeric group equatorial is given the symbol CE. The use of these

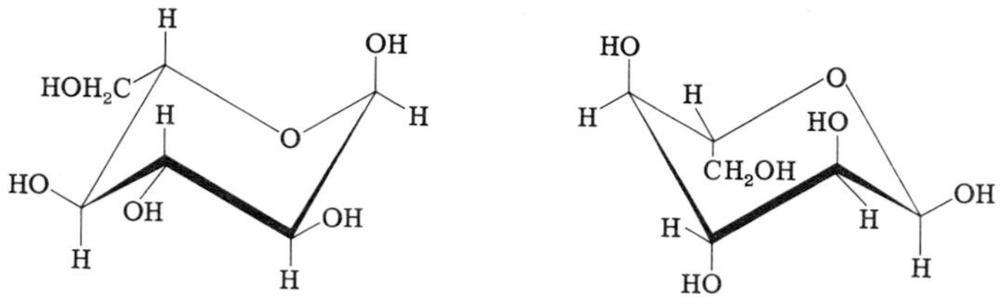

CA   CE

α-L-Glucopyranose

------------------MIRROR PLANE--------------

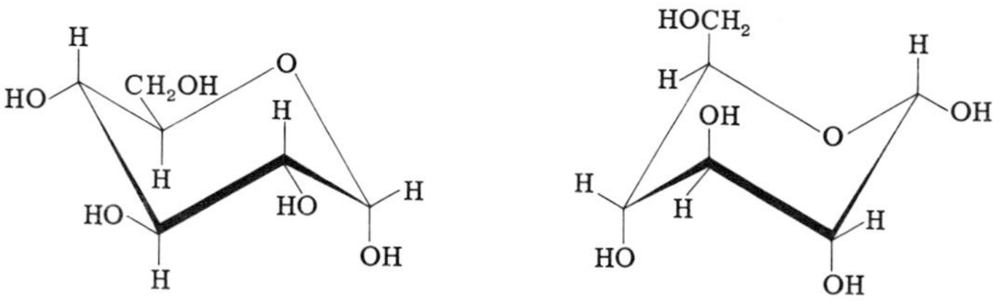

CA   CE

α-D-Glucopyranose

symbols is illustrated with reference to the conformational designation of both chairlike conformations of α-D-glucopyranose and α-L-glucopyranose. It is seen that mirror-image conformations receive the *same* symbol, thus following the system used for the anomeric designators α and β (which also remain the same regardless of chirality).

The Isbell–Tipson system of conformational nomenclature has been much less widely used in the literature than the Reeves system, al-

(180) R. D. Guthrie, *Chem. Ind.* (London), 1593 (1958); D. F. Shaw, *Tetrahedron Lett.*, 1 (1965).
(181) J. Szejtli, *Acta Chim. Acad. Sci. Hung.*, **61**, 57 (1969).

though there are many examples in which the Reeves system has been used ambiguously because the chiral series was not specified. The system adopted in this Chapter is that of Reeves, with the chiral designators added whenever the chirality is not evident from the context; the four symbols used are $C1(\text{D})$, $1C(\text{D})$, $C1(\text{L})$, and $1C(\text{L})$; the use of such symbols was first introduced in this Series.

In addition to the chair conformations, the pyranoid ring may also be formulated in a flexible cycle,[170-172] where it is also free from bond-angle strain. The cycle includes six boat and six skew forms, each boat form being a maximum-energy transition-state between two skew forms (see Fig. 1). The high-energy transition state in passing

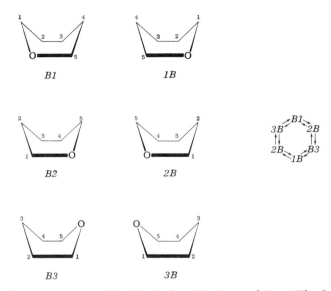

FIG. 1.—Boat Forms in the Flexible Cycle of the Pyranoid Ring. (The forms are interrelated by a process of pseudorotation as shown, and between each pair of boat forms in the cycle there is a skew form of lower potential energy. The symbols are those of Reeves.[16])

from one of the chair conformers into the flexible cycle or into the other chair conformer may be regarded as a structure having five of the six ring-atoms approximately coplanar (see Fig. 2).

For monocyclic sugars, non-bonded interactions render the skew, and, to a greater extent, the boat, forms less stable than the chair forms. The boat forms make an insignificant contribution to the conformational equilibrium of monocyclic pyranoid sugars at room

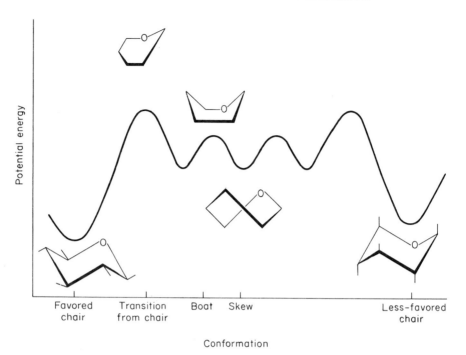

Fig. 2.—Potential-energy Diagram (Somewhat Idealized) for a Monocyclic Pyranoid Sugar as a Function of the Conformation.

temperature.[25] The energy required for conversion from a chair form into the flexible cycle is small compared with that needed for most chemical reactions, and thus the skew and boat forms may be intermediates in reactions of the sugars. In addition, the skew or boat form may be the favored conformation for certain pyranoid sugar derivatives that are conformationally "locked" by bridging substituents. For example, a 1,4-anhydride bridge locks a pyranoid ring-system in a boat-like conformation.

An example of a sugar derivative held rigidly in a skew conformation is 3-O-benzoyl-1,2,4-O-benzylidene-α-D-ribopyranose[102] (**10**).

(10)

A generalized system of conformational nomenclature, based on the Isbell–Tipson symbols,[170,171] that has been suggested[171,182] gives a letter symbol (C, chair; S, skew; and so on) to describe the shape of the ring, and denotes the molecular shape by defining a plane that contains as many ring atoms as possible (four for a chair, boat, or skew form), and identifies the orientation of the remaining ring atoms as above or below this plane by superscript or subscript numbers. This system could be adapted, according to the way in which the terms are defined, to conform to either the achiral (Isbell–Tipson type) or chiral (Reeves type) concept of conformational symbols, and furthermore, it can be used with furanoid and other ring systems, as well as with pyranoid rings.

3. Pyranoid Sugar Derivatives in the Crystalline State

The structures of a number of pyranoid sugar derivatives in the crystalline state have been determined by X-ray crystallography. In all examples reported, the pyranoid ring has been found to be in a chair shape, but the true shape deviates somewhat from ideal geometry. This deviation is evident from the dihedral angles observed around the ring (51–62°, as compared with 56–62° for an idealized, pyranoid ring),[183] and the distortion is usually the result of flattening in that portion of the ring involving C-2 and C-3. From a comparison of the X-ray data, it may also be seen that the actual dihedral angles in the various pyranoid sugars investigated differ from one sugar to another. This variation may be ascribed to the changes in the type and relative orientation of substituents on the ring and to differences in the inter- and intra-molecular forces of attraction existing in the crystalline lattice.

The monocyclic pyranoid sugars whose structures have been determined by X-ray or neutron diffraction are listed in Table III.

A comparison of the geometries of *epi*- and *myo*-inositol in the crystalline state with that of an ideal, strain-free model revealed that

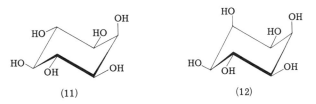

(11)   (12)

(182) J. C. P. Schwarz, cited in L. Hough and A. C. Richardson, "Rodd's Chemistry of Carbon Compounds," S. Coffey (Ed.), Elsevier, Amsterdam, Vol. I, part F, 1967, p. 90.
(183) H. M. Berman and S. H. Kim, *Acta Crystallogr.*, **B24**, 897 (1968).

TABLE III: Conformations of Monocyclic, Pyranoid Sugar Derivatives in the Crystalline State as Determined by X-Ray Crystallography

Pyranoid sugar or sugar residue	Derivatives studied	Conformation	References
α-D-Glucopyranose	sodium bromide complex, dihydrate	$C1$(D)	184
	sucrose[a]	$C1$(D)	185
	free sugar	$C1$(D)	186
	free sugar[a]	$C1$(D)	187
α-L-Rhamnopyranose	free sugar, monohydrate	$C1$(D)	188
2-Deoxy-β-D-erythro-hexopyranose	free sugar, monohydrate	$1C$(L)	189
β-D-Glucopyranose	free sugar	$1C$(D)	190
β-D-Lyxopyranose	free sugar	$C1$(D)	191,192
β-L-Arabinopyranose	free sugar	$C1$(D)	193,194
β-D-Arabinopyranosyl bromide (and chloride)	2,3,4-tri-O-acetyl-	$C1$(L)	195,196
β-Cellobiose	free sugar	$1C$(D)	84
β-Maltose	monohydrate	$C1$(D)	192,197
α-Lactose	monohydrate	$C1$(D)	197a
β-D-Glucopyranuronic acid	dihydrates of K and Rb salts	$C1$(D)	197b
Methyl α-D-altropyranoside	free glycoside	$C1$(D)	198
Methyl α-D-galactopyranoside	6-bromo-6-deoxy-	$C1$(D) (distorted)	198a
Methyl α-D-galactopyranoside	monohydrate	$C1$(D)	199a
α-D-Glucopyranosyl phosphate	dihydrate of dipotassium salt	$C1$(D)	199b
Methyl β-D-xylopyranoside	free glycoside	$C1$(D)	200
α-D-Glucopyranose	2-acetamido-2-deoxy-	$C1$(D)	201
Methyl β-cellobioside	methanolate	$C1$(D)	202
Methyl β-D-maltopyranoside	free glycoside	$C1$(D)	202a
α-L-Sorbopyranose	free sugar	$1C$(L)	203
β-D-Fructopyranose	free sugar	$1C$(D)	204
α-D-Tagatopyranose	free sugar	$C1$(D)	205
Methyl α-D-glucopyranoside	free glycoside	$C1$(D)	206
	4,6-dichloro-4,6-dideoxy-	$C1$(D)	183
1-O-(α-D-Glucopyranosyl)glycerol		$C1$(D)	207
	6-deoxy-6-sulfo-	$C1$(D)	208

[a] Determined by neutron diffraction.

the presence of axially attached hydroxyl groups on the ring has only a small effect on the conformation.[209] In *myo*-inositol (11), which has one hydroxyl group axially attached, the cyclohexane ring is somewhat flattened, as indicated by the mean dihedral angle between the C–C bonds of 56.8° ("ideal" value, 60°). The ring flattening is increased only very slightly in *epi*-inositol (12), which has two *syn*-diaxial hydroxyl groups (mean dihedral angle of 56.3°). The effect of the 1,3-interaction is to increase the non-bonded distance between the diaxial O-2 and O-6 from the "ideal" value of 2.50 Å to 2.96 Å, mainly by distortion of valency angles. These results support[209] the assumption that intermolecular hydrogen-bonding plays only a minor

(184) C. A. Beevers and W. Cochran, *Proc. Roy. Soc., Ser. A*, **190**, 257 (1947); C. A. Beevers, T. R. R. McDonald, J. H. Robertson, and F. Stern, *Acta Crystallogr.*, **5**, 689 (1952).
(185) G. M. Brown and H. A. Levy, *Science*, **141**, 921 (1963).
(186) T. R. R. McDonald and C. A. Beevers, *Acta Crystallogr.*, **5**, 654 (1952).
(187) G. M. Brown and H. A. Levy, *Science*, **147**, 1038 (1965).
(188) R. C. G. Killean, W. G. Ferrier, and D. W. Young, *Acta Crystallogr.*, **15**, 911 (1962).
(189) H. M. McGeachin and C. A. Beevers, *Acta Crystallogr.*, **10**, 227 (1957).
(190) S. Furberg, *Acta Chem. Scand.*, **14**, 1357 (1960).
(191) W. G. Ferrier, *Acta Crystallogr.*, **13**, 678 (1960); **16**, 1023 (1963).
(192) S. S. C. Chu and G. A. Jeffrey, *Acta Crystallogr.*, **B24**, 830 (1968).
(193) A. Hordvik, *Acta Chem. Scand.*, **15**, 1781 (1961).
(194) A. Hordvik, *Acta Chem. Scand.*, **20**, 1943 (1966).
(195) A. Hordvik, *Acta Chem. Scand.*, **15**, 16 (1961).
(196) S. H. Kim and G. A. Jeffrey, *Acta Crystallogr.*, **22**, 537 (1967).
(197) R. A. Jacobson, J. A. Wunderlich, and W. N. Lipscomb, *Acta Crystallogr.*, **14**, 598 (1961); C. J. Brown, *J. Chem. Soc.* (A). 927 (1966).
(197) (a) G. J. Quigley, A. Sarko, and R. H. Marchessault, *J. Amer. Chem. Soc.*, **92**, 5834 (1970); (b) C. A. Beevers and H. N. Hansen, *Acta Crystallogr.*, **B27**, 1323 (1971).
(198) G. E. Gurr, *Acta Crystallogr.*, **16**, 690 (1963); S. Furberg, H. Hammer, and A. Mostad, *Acta Chem. Scand.*, **17**, 2444 (1963).
(198a) B. M. Gatehouse and B. J. Poppleton, *Acta Crystallogr.*, **B27**, 871, (1971).
(199) (a) J. H. Robertson and B. Sheldrick, *Acta Crystallogr.*, **19**, 820 (1965); (b) B. M. Gatehouse and B. J. Poppleton, *Acta Crystallogr.*, **B27**, 654 (1971).
(200) C. A. Beevers and G. H. Maconochie, *Acta Crystallogr.*, **18**, 232 (1965).
(201) C. J. Brown, G. Cox, and F. J. Llewellyn, *J. Chem. Soc.* (A), 922 (1966).
(202) L. N. Johnson, *Acta Crystallogr.*, **21**, 885 (1966).
(202a) J. T. Ham and D. G. Williams, *Acta Crystallogr.*, **B26**, 1373 (1970).
(203) S. S. C. Chu and G. A. Jeffrey, *Acta Crystallogr.*, **23**, 1038 (1967).
(204) S. H. Kim and R. D. Rosenstein, *Acta Crystallogr.*, **22**, 648 (1967).
(205) R. D. Rosenstein, *Am. Crystallographic Assoc., Abstr.*, 1968, Summer Meeting.
(206) S. Takagi and R. D. Rosenstein, *Carbohyd. Res.*, **11**, 156 (1969).
(207) R. Hoge and J. Trotter, *J. Chem. Soc.* (A), 267 (1968).
(208) Y. Okaya, *Acta Crystallogr.*, **17**, 1276 (1964).
(209) G. A. Jeffrey and H. S. Kim, *Carbohyd. Res.*, **15**, 310 (1970).

role in affecting the conformation of the cyclitols, the primary factors being the intramolecular, non-bonded interactions between hydroxyl groups.

4. Pyranoid Sugar Derivatives in Solution

a. Determination of Favored Conformation.—In 1947, Hassel and Ottar[15] concluded that the chair shape is thermodynamically the most stable conformation that the pyranoid ring can assume. Experimental justification was subsequently provided by Reeves in a series of extensive investigations on the formation of complexes from pyranoid sugars in cuprammonia solution,[16,17,19–23] and the results indicated that the pyranoid ring does indeed adopt a chair shape when this is structurally possible.

Neglecting contributions from boat or skew forms to the conformational population, Reeves[22] predicted the favored chair conformation for a wide range of methyl D-aldopyranosides in aqueous solution. Only methyl α-D-idopyranoside, methyl 4,6-O-benzylidene-α- and -β-D-idopyranoside, and methyl α- and β-D-arabinopyranoside appeared to exist in the $1C(D)$ conformation. The methyl D-altro- and D-lyxo-pyranosides exhibited behavior intermediate between that expected for the two chair forms, and it was, therefore, concluded that these glycosides exist in solution in an equilibrium composed of appreciable proportions of each chair conformer. However, it was later acknowledged[17,24] that boat or skew conformations might be possible were the nonbonded interactions between the axial groups in the chair conformation to become too large.

The conformations of the aldopyranoses in aqueous solution have been investigated by n.m.r. spectroscopy[71,210,211] (see Table IV). The chair form favored by the various aldohexopyranoses appears to be controlled by the tendency of the 5-(hydroxymethyl) group to assume the equatorial orientation. Hence, all the β-D anomers exist preponderantly in the $C1(D)$ conformation, as the alternative, $1C(D)$ conformation appears to be highly destabilized by the syn-diaxial interaction between the 1-hydroxyl and 5-(hydroxymethyl) groups. Although this interaction is not found in the $1C(D)$ conformation of the α-D anomers, most of the α-D anomers nevertheless adopt the $C1(D)$ conformation, presumably because of a favorable anomeric effect (see p. 103). Solutions of α-D-idopyranose and α-D-altropyranose appear to contain appreciable proportions of each chair conformer at equilibrium.

(210) S. J. Angyal and V. A. Pickles, unpublished data cited in Ref. 229.
(211) M. Rudrum and D. F. Shaw, *J. Chem. Soc.*, 52 (1965).

TABLE IV

Favored Conformations of D-Aldopyranoses in Aqueous Solution[a]

| Aldopyranose | Conformation | | Estimated free-energies, kcal. mole^{-1} | |
	By summation of interaction energies[229]	By n.m.r. spectroscopy[71,210,211]	$C1$	$1C$
α-D-Allose	$C1$	$C1$	3.9	5.35
β-D-Allose	$C1$	$C1$	2.95	6.05
α-D-Altrose	$1C \rightleftarrows C1$	$1C \rightleftarrows C1$	3.65	3.85
β-D-Altrose	$C1$	$C1$	3.35	5.35
α-D-Galactose	$C1$	$C1$	2.85	6.3
β-D-Galactose	$C1$	$C1$	2.5	7.75
α-D-Glucose	$C1$	$C1$	2.4	6.55
β-D-Glucose	$C1$	$C1$	2.05	8.0
α-D-Gulose	—	$C1$	4.0	4.75
β-D-Gulose	$C1$	$C1$	3.05	5.45
α-D-Idose	$1C \rightleftarrows C1$	$1C \rightleftarrows C1$	4.35	3.85
β-D-Idose	—	$C1$	4.05	5.35
α-D-Mannose	$C1$	$C1$	2.5	5.55
β-D-Mannose	$C1$	$C1$	2.95	7.65
α-D-Talose	$C1$	$C1$	3.55	5.9
β-D-Talose	—	$C1$	4.0	8.0
α-D-Arabinose	$1C$	$1C$	3.2	2.05
β-D-Arabinose	—	$1C \rightleftarrows C1$	2.9	2.4
α-D-Lyxose	$1C \rightleftarrows C1$	$1C \rightleftarrows C1$	2.05	2.6
β-D-Lyxose	$C1$	$C1$	2.5	3.55
α-D-Ribose	$1C \rightleftarrows C1$	$1C \rightleftarrows C1$	3.45	3.55
β-D-Ribose	$1C \rightleftarrows C1$	$1C \rightleftarrows C1$	2.5	3.1
α-D-Xylose	$C1$	$C1$	1.95	3.6
β-D-Xylose	$C1$	$C1$	1.6	3.9

[a] Data taken from Table I, Ref. 31.

The conformations adopted in solution by the eight aldopentopyranoses is determined by the relative configurations of the hydroxyl groups. Thus, α- and β-D-arabinopyranose favor the $1C$ conformation, α-D-lyxopyranose and α-D-ribopyranose appear to contain substantial proportions of each chair form, and β-D-lyxopyranose, β-D-ribopyranose, α-D-xylopyranose, and β-D-xylopranose exist preponderantly in the $C1$ conformation. In methyl sulfoxide, however, α-D-lyxopyranose exists almost entirely in the $C1$ conformation, and in this solvent, β-D-ribopyranose appears to contain an appreciable proportion of the alternative, $1C$ conformer.[212]

The favored conformations of several ketohexopyranoses have been determined from a study of long-range couplings of hydroxyl

(212) W. Mackie and A. S. Perlin, *Can. J. Chem.*, **44**, 2039 (1966).

protons.[213] α-D-Sorbopyranose and α-D-tagatopyranose were observed to exist in the $C1$(D) conformation in methyl sulfoxide solution, and β-D-tagatopyranose in the $1C$(D) conformation. In the crystalline state, α-L-sorbopyranose[204] (**13**) was found to be in the $1C$(L) form, and α-D-tagatopyranose[206] (**14**) in the $C1$(D) form.

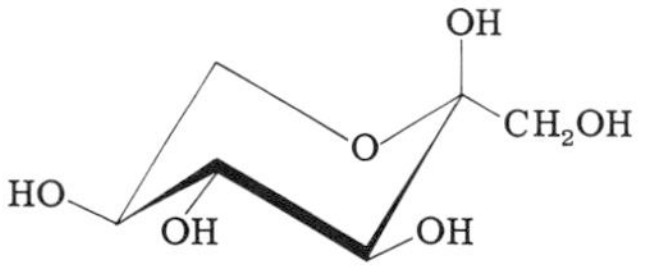

α-L-Sorbopyranose-*1C*   α-D-Tagatopyranose-*C1*
(13)   (14)

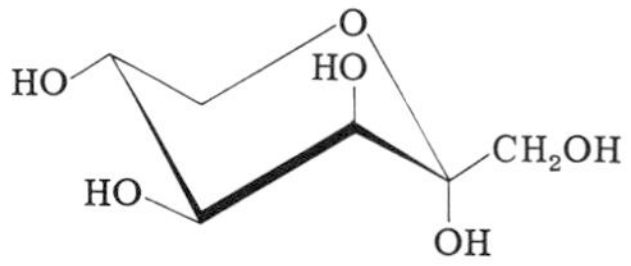

Favored conformations of ketohexopyranoses in the solid state[204,206] and in methyl sulfoxide solution[213]

The introduction of substituents and the effects of solvation profoundly influence the conformational equilibria of monocyclic pyranoid sugars and their derivatives, because of the changes in steric and electronic interactions that result. The conformations of numerous polysubstituted pyranoid sugars in various solvents have been investigated, mostly by application of the n.m.r. spectroscopic method. Six aldopentopyranose tetraacetates were examined[72] in chloroform solution near room temperature; they were considered to exist almost entirely in the $C1$(D) conformation, except for the β-D-*ribo* derivative, which appeared to contain substantial proportions of each chair form, and the β-L-*arabino* derivative[214,215] which was estimated to contain mostly the $C1$(L) conformation. 2-Deoxy-β-D-*erythro*-pentopyranose triacetate was considered to exist mainly in the $1C$ conformation.

β-D-Ribopyranose tetrabenzoate in chloroform solution near room temperature was reported[216] to contain the $1C$ and $C1$ forms in a ratio of ~2:1, whereas the α-D anomer appeared to exist mainly in the $C1$ conformation. The anomeric effect (see p. 103) was invoked as the principal factor determining these differences in conformational distribution. 2-Deoxy-β-D-*erythro*-pentopyranose tribenzoate, whose

(213) J. C. Jochims, G. Taigel, A. Seeliger, P. Lutz, and H. E. Driesen, *Tetrahedron Lett.*, 4363 (1967).
(214) R. U. Lemieux and N.-J. Chü, *Abstr. Papers Amer. Chem. Soc. Meeting*, **133**, 31N (1958).
(215) R. U. Lemieux, in "Molecular Rearrangements," Part 2, P. de Mayo (Ed.), Interscience Division, John Wiley and Sons, New York, 1964, pp. 735–743.
(216) B. Coxon, *Tetrahedron*, **22**, 2281 (1966).

1C conformation does not have a *syn*-diaxial interaction between benzoyloxy groups at C-2 and C-4, favors the *1C* conformation strongly.[217]

N.m.r.-spectral data for α-D-lyxopyranose tetraacetate in acetone-d_6 indicated that the *C1* form is the major conformer present at equilibrium near room temperature.[107]

The favored conformations of a number of peracetylated 1-thioaldopyranose derivatives have been examined by use of 220-MHz n.m.r. spectroscopy.[174] For the D-aldohexose derivatives investigated, the *C1*(D) conformation was strongly favored, most probably because of the large steric effect of the acetoxymethyl group; with the D-aldopentoses, the possibility that the alternative chair conformation was present in appreciable proportion at equilibrium with the (apparently favored) forms having the 1-SAc group equatorially attached could not be discounted; the conformational equilibria were measured in a subsequent study (see p. 91).[99]

In an extension to aldohexose peracetates, the favored form of α-D-idopyranose pentaacetate in solution[176] in acetone or chloroform was found to be the *C1*(D) conformer, having four substituents axially and one equatorially attached. α-D-Altropyranose pentaacetate had previously been reported[218] also to adopt the *C1*(D) conformation, having three substituents axial and two equatorial. Consideration of the observed proton spin-couplings allowed the exclusion of the boat or skew forms as major contributors to the conformational populations for both compounds near room temperature.

The conformations of 18, substituted D-galactopyranoid derivatives have been determined from their n.m.r. spectra.[219] As expected, all of the compounds were found to adopt the *C1*(D) conformation.

The effect of an anomeric halogen atom on the conformations of monocyclic pyranoid compounds has been examined in detail.[94] For a range of poly-*O*-acylated aldopyranosyl halides, each in the thermodynamically more stable anomeric form, it was shown[173] that the halogen atom is axially attached in the favored conformation, even when this means that three of the four substituents are axially attached and a *syn*-diaxial interaction exists, as observed with tri-*O*-acetyl-β-D-ribopyranosyl bromide. The data were interpreted as indicating that the large anomeric effect (see p. 103) of a halogen atom at C-1 plays an overriding role in determining the conformation favored.

(217) R. J. Cushley, J. F. Codington, and J. J. Fox, *Carbohyd. Res.*, **5**, 31 (1967).
(218) B. Coxon, *Carbohyd. Res.*, **1**, 357 (1966).
(219) H. Libert, I. Schuster, and L. Schmid, *Chem. Ber.*, **101**, 1902 (1968); W. Sibral, H. Libert, and L. Schmid, *Monatsh. Chem.*, **99**, 884 (1968).

Similar behavior was observed[216] for the perbenzoylated α- and β-D-ribopyranosyl halides, for which the operation of the anomeric effect was again considered to be the principal factor in determining the chair conformation favored. This phenomenon has also been observed[220] with some chlorosulfated α- and β-D-aldopentopyranosyl chlorides. Crystallographic analysis of tri-O-acetyl-β-D-arabinopyranosyl bromide (and chloride)[84] (see also p. 59) showed the halogen atom in axial orientation. The molecule deviates from ideal-chair geometry by slight flattening at C-4 and slight additional puckering at C-1. The acetate-methyl groups lie almost antiparallel to the respective carbon atoms of the ring.

When the conformation of tri-O-acetyl-β-D-xylopyranosyl chloride, the thermodynamically less stable anomeric form, was investigated by n.m.r. spectroscopy,[175] it was found that, in chloroform or acetone solution, the chair conformer having all four substituents axially attached [the $1C(D)$ conformer] is favored. This observation would be totally unexpected from a consideration of steric factors alone

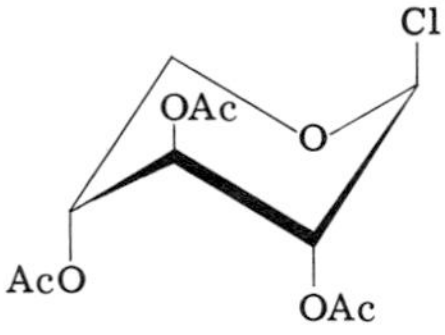

$1C(D)$

All-axial conformation favored for tri-O-acetyl-β-D-xylopyranosyl chloride[175]

(see p. 99). In a related study,[59] tri-O-acetyl- and tri-O-benzoyl-β-D-xylopyranosyl fluoride were also found to adopt the $1C$ form as the favored conformation. Examination of various fully esterified pentopyranosyl fluorides by proton and fluorine-19 n.m.r. spectroscopy showed that all such derivatives favor that chair conformation having the fluorine substituent attached axially.[65]

The conformations of pyranoid sugars containing an amide nitrogen atom in place of the ring-oxygen atom have been investigated. For example, 1,2,3,4-tetra-O-acetyl-5-[(benzyloxycarbonyl)amino]-5-deoxy-α-D-xylopyranose (15) and methyl 5-acetamido-5-deoxy-2,3,4-tri-O-methyl-α-D-xylopyranoside (16) assume the $C1$ conformation.[221] Other 5-acetamido-5-deoxypentopyranose derivatives were also found to favor that chair conformation having the C-1 substituent attached axially. Thus, 1,2,3,4-tetra-O-acetyl-5-[(benzyloxycarbonyl)-

(220) H. J. Jennings, Can. J. Chem., 47, 1157 (1969).
(221) H. Paulsen and F. Leupold, Carbohyd. Res., 3, 47 (1966).

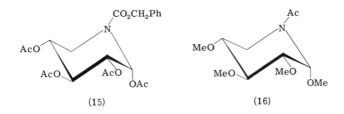

(15)   (16)

amino]-5-deoxy-β-L-arabinopyranose exists in the $C1$(L) conformation
(**17**) and the α-D-*ribo* and α-D-*lyxo* analogs adopt the $C1$(D) conformation, whereas the β-D-*ribo* derivative (**18**) favors the $1C$(D) con-

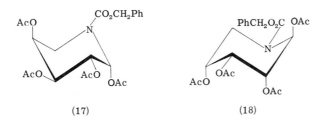

(17)   (18)

formation.[222] This tendency of the anomeric substituent to occupy
the axial position was ascribed to destabilization of the chair conformer having the aglycon attached equatorially, as a result of strong,
steric interaction with the N-acyl group of the ring-nitrogen atom.[222]

The favored conformations in solution of a wide range of pentopy-
ranoid derivatives, including several complete configurational series,
are now well established[92-99] as a result of quantitative measurements
of conformational equilibria between the two chair forms (see the
following Section).

b. Conformational Equilibria.—In the absence of direct evidence,
the apparent conformational homogeneity of many pyranoid sugar
derivatives, as observed by n.m.r. spectroscopy, may be ascribed to
rapid conformational inversion, so that a time-averaged spectrum is
observed in which only the favored conformer makes a substantial
contribution; alternatively, if interconversion is slow on the n.m.r.
time-scale, superimposed spectra of the two chair conformers would
be obtained, but the proportion of the minor conformer might be so
small that its signals are lost in the "background noise" of the spectrum. This question was resolved by n.m.r.-spectral studies[107,108] on
β-D-ribopyranose tetraacetate in solution at low temperature, which

(222) H. Paulsen and F. Leupold, *Chem. Ber.*, **102**, 2804 (1969).

provided direct confirmation that this aldopentopyranose tetraacetate, and, presumably, other monocyclic pyranoid sugars and their derivatives, indeed exist in rapid conformational equilibrium and that the spectrum observed at room temperature is a time-average of the spectra of the separate conformers.

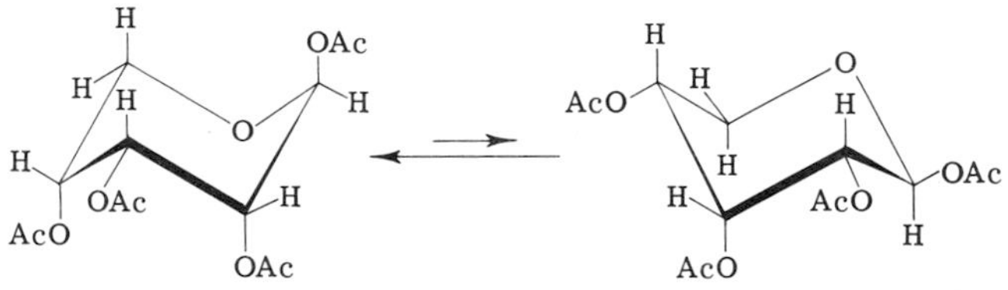

Conformational equilibrium of β-D-ribopyranose tetraacetate at −84°

At room temperature, the spin-coupling data observed for β-D-ribopyranose tetraacetate in various solvents suggest that a conformational equilibrium may be involved; in particular the $J_{1,2}$ spin-coupling (4.8 Hz in acetone-d_6) appears intermediate between the values expected for the diequatorial and the diaxial dispositions of H-1 and H-2, as would be found in the $1C(D)$ and $C1(D)$ conformations, respectively. When the temperature of the solution in acetone-d_6 was progressively lowered, the pattern observed at 220 MHz changed little until near −60°, when a broadening of certain signals commenced rather abruptly. At lower temperatures, the spectrum became resolved into a new pattern that could be interpreted as resulting from the superposition of the separate spectra of the individual conformers. The original, time-averaged, doublet signal for H-1 became separated into a narrow, low-field signal for an equatorial proton at C-1 [H-1e in the $1C(D)$ conformation] and a wide doublet at higher field for an axial proton at C-1 [H-1a in the $C1(D)$ conformation]. At low temperatures, the equilibrium lies approximately 2:1 in favor of the triaxial $1C(D)$ form, indicating that the steric destabilization of three axially attached acetoxyl groups and a syn-diaxial interaction are counteracted by other factors (presumably electronic factors).

From the temperature ("coalescence temperature") at which the transition between the low-temperature pattern and the time-averaged pattern takes place, in a system of two equilibrating species, it is possible[223,224] to determine the frequency of interconversion at that

(223) H. S. Gutowsky and C. H. Holm, J. Chem. Phys., **25**, 1228 (1956); J. A. Pople, W. G. Schneider, and H. J. Bernstein, "High-Resolution Nuclear Magnetic Resonance," McGraw-Hill Book Co., Inc., New York, 1959, p. 223.

(224) H. Shanan-Atidi and K. H. Bar-Eli, J. Phys. Chem., **74**, 961 (1970).

temperature, and thus to estimate the energy of activation for the interconversion. It was found that, at $-60°$, the rate of inversion of the $C1(D)$ to the $1C(D)$ conformer of β-D-ribopyranose tetraacetate is about 117 times per second, corresponding to a free energy of activation of $\Delta G^+ = 10.3 \pm 0.3$ kcal.mole^{-1}, and the rate of inversion of the $1C(D)$ to the $C1(D)$ conformer is about 57 times per second corresponding a ΔG^+ value of 10.6 ± 0.3 kcal.mole^{-1}. These values are very similar to those determined for the ring-inversion of cyclohexane,[225] and of unsubstituted tetrahydropyran.[226] This observation excludes the possibility that "passing interactions" during "ring inversion" of a polysubstituted tetrahydropyran ring-system might greatly raise the barrier to inversion relative to that of unsubstituted tetrahydropyran; similarly, extrapolation to substituted cyclohexane systems would appear reasonable.

Examples of pyranoid derivatives in which a "conformational freeze-out"[227] of the type observed with β-D-ribopyranose tetraacetate can be observed directly are rare. Apart from the obvious problems of (a) freezing of the solvent (precluding studies in water solution), (b) limited solubility of derivatives in suitable solvents, and (c) insufficient spectral resolution, dispersion, or definition at low temperatures, there is observed for most derivatives a progressive trend toward greater conformational homogeneity at low temperatures, with the result that signals of only one conformer are observed, even below the temperature of conformational freeze-out. From a study of fifty aldopentopyranose derivatives, it was found[92-99] that, in their n.m.r. spectra, only three showed a detectable proportion of the minor chair conformer below the temperature of conformational freeze-out. In general, the aldohexopyranoses and their derivatives appear to exhibit even greater conformational homogeneity at low temperatures. However, it is possible to determine conformational populations at temperatures at which inversion is rapid, provided that suitable n.m.r. parameters for the individual conformers are available for use in conjunction with time-averaged values.

For β-D-ribopyranose tetraacetate, the $J_{1.2}$ spin-couplings, J_e and J_a, for the individual $1C(D)$ and $C1(D)$ conformers are available directly from the n.m.r. spectrum measured at low temperature, and thus

(225) F. A. L. Anet, M. Ahmad, and L. D. Hall, *Proc. Chem. Soc.*, 145 (1964); F. A. Bovey, F. P. Hood, III, E. W. Anderson, and R. L. Kornegay, *ibid.*, 146 (1964).

(226) G. Gatti, A. L. Segre, and C. Morandi, *J. Chem. Soc. (B)*, 1203 (1967).

(227) L. W. Reeves and K. O. Strømme, *Can. J. Chem.*, 38, 1241 (1960); F. R. Jensen and C. H. Bushweller, *J. Amer. Chem. Soc.*, 88, 4279 (1966); C. H. Bushweller, J. Golini, G. U. Rao, and J. W. O'Neil, *ibid.*, 92, 3055 (1970).

the $J_{1,2}$ spin-coupling observed (J_{obs}) in the time-averaged spectrum can be related[108] to the mole fraction (N_e) of the $1C(D)$ form (having H-1 equatorially attached) and the mole fraction (N_a) of the $C1(D)$ form (having H-1 axially attached) by the equation

$$J_{obs} = N_e J_e + N_a J_a.$$

From the values of N_e and N_a thus determined, it is possible to calculate the standard, free-energy difference ($\Delta G°$), and the equilibrium constant (K), for the interconversion H-eq ⇌ H-ax by the standard thermodynamic relationship:

$$\Delta G° = -RT\ln K = -RT\ln\frac{N_a}{N_e}.$$

By this method, it was found[108] that a solution of β-D-ribopyranose tetraacetate in acetone-d_6 at room temperature contains ~45% of the $1C(D)$ conformer and ~55% of the $C1(D)$ conformer, thus indicating that the equilibrium constant for the $1C(D)$ ⇌ $C1(D)$ process increases with rising temperature [negative enthalpy change for $1C(D) \rightarrow C1(D)$].

In principle, for determining conformational populations, a method based on chemical shifts instead of coupling constants might be useful, but, in practice, such an approach has been found[108] to give erratic and inconsistent results. It cannot be assumed that, even for pure conformers, chemical shifts are necessarily independent of temperature.

The method of averaging of coupling constants, as applied for determining the conformational equilibrium of β-D-ribopyranose tetraacetate, utilizes for the pure conformers spin-coupling values obtained directly from the spectrum measured at low temperature. In an extension to the full series of D-aldopentopyranose tetraacetates,[92,95] it was not found possible to observe, in six of the other seven examples, the spectrum of the minor conformer; the spectrum of a single conformer, only, was observed at low temperature. It was found that α-D-xylopyranose tetraacetate is essentially all in the $C1(D)$ conformation over a wide range of temperatures, and that the spin couplings observed in its spectrum did not change with temperature. β-D-Arabinopyranose tetraacetate was found to exist almost exclusively in the $1C(D)$ conformation at room temperature, and a slight change in the spin couplings observed for this molecule as the temperature was lowered indicated that the population becomes exclusively $1C(D)$ at low temperatures. These two compounds were used[95] to provide, for the pure, individual conformers, "model" coupling-constants for protons disposed similarly through the series of the

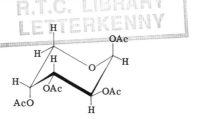

α-D-Xylopyranose tetraacetate

[Exclusively *C1* (D) over a wide range of temperature]

β-D-Arabinopyranose tetraacetate

[Almost entirely *1C* (D) at room temperature, becoming exclusively so at low temperature]

eight stereoisomers. Because analysis of the $J_{1,2}$ couplings is only useful for the four examples having the *trans*-arrangement of H-1 and H-2, the calculations were based, instead, on the coupling of H-4 with the *trans*-disposed proton at C-5; these protons are antiparallel in the $C1$(D) conformation and diequatorial in the $1C$(D) conformation. The results for the eight derivatives in acetone-d_6 at room temperature are summarized in Fig. 3.

The only aldopentopyranose tetraacetates for which conformational freeze-outs were observed at temperatures down to −100° were the β-D-*ribo* and the β-D-*lyxo* derivatives.[95]

The general approach employing the method of averaging of spin couplings has also been applied with the eight D-aldopentopyranose tetrabenzoates,[95] the eight methyl tri-*O*-acetyl-D-aldopentopyranosides and several of their *O*-benzoylated analogs,[98] the four 1,2-*trans*-tri-*O*-acetyl-D-aldopyranosyl benzoates[97] and thiolacetates,[99] the four 1,2-*trans*-tri-*O*-benzoyl-D-aldopentopyranosyl acetates,[97] several acetylated and benzoylated D-aldopentopyranosyl halides,[94] and the methyl, ethyl, isopropyl, and *tert*-butyl β-D-ribopyranoside triacetates and tribenzoates.[96] The "model" couplings used in each series were taken from those members of the series that were found to be essentially homogeneous in a single conformation. The conformational populations determined, and the free-energy values for the $1C$(D) $\rightleftarrows$ $C1$(D) equilibrium, are listed in Table V (see p. 96).

By inspection of the data in Table V, certain significant generalizations emerge. For the aldopentopyranoid derivatives listed, a conformational equilibrium in solution between the two chair forms, with an appreciable proportion (10% or more) of the less-favored chair form, is the rule rather than the exception. Only in the α-*xylo* series is the $C1$(D) conformer favored overwhelmingly for a range of substituents at C-1, and only in the β-*arabino* series is the $1C$(D) conformer favored very strongly for a range of substituents at C-1. It is

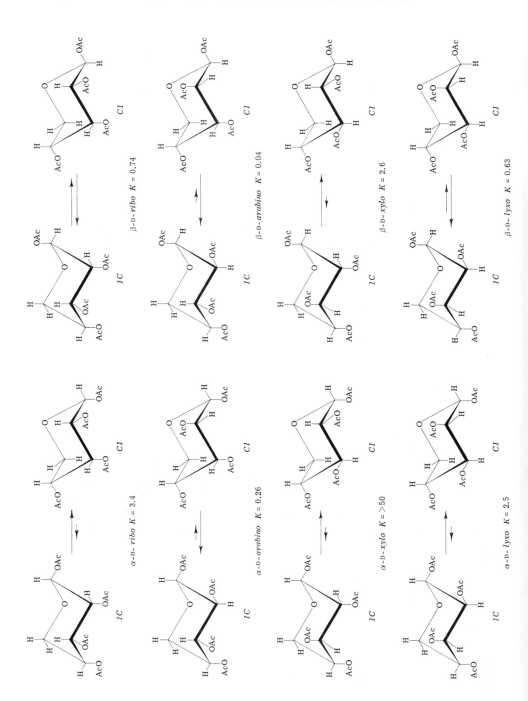

therefore clear that the classical concept that steric factors control the conformation assumed is *untenable*; for example, tri-*O*-acetyl-β-D-xylopyranosyl chloride exists to ~80% in the all-axial form, and the corresponding benzoylated analog favors the tetra-axial arrangement almost exclusively.[94] Although this behavior may be partially explicable by invoking the polar effect that favors the axial disposition of an electronegative substituent on C-1 (anomeric effect, see p. 103), a single polar contribution of this type would need to have a magnitude unreasonably large in order to outweigh the combined steric destabilizations of the four axial substituents, on the assumption that these can be treated additively (see p. 99).

The nature of the solvent and its polarity do not appear to affect in any regular way the position of the conformational equilibria; for example,[93,95] β-D-xylopyranose tetrabenzoate, which exists in acetone-d_6 (dielectric constant, ϵ, = 20.7) as a 1:1 mixture of the $C1$(D) and $1C$(D) conformations, shows practically the same conformational population in a range of solvents, including benzene-d_6 (ϵ = 2.3), toluene-d_8 (ϵ = 2.4), chloroform-d (ϵ = 4.8), pyridine-d_5 (ϵ = 12.3), hexachloroacetone, and methyl sulfoxide-d_6 (ϵ = 48.9). Similar results, showing negligible dependence of conformational populations on the polarity of the solvent, were obtained with β-D-ribopyranose tetraacetate,[92,95] tri-*O*-acetyl-β-D-xylopyranosyl chloride,[94] and some simpler derivatives of tetrahydropyran.[121,122] In contrast, as the polarity of the solvent was increased, there was observed[98] with methyl 2,3,4-tri-*O*-benzoyl-β-D-xylopyranoside a broad trend in favor of that conformation having the C-1 substituent equatorially attached.

The conformational effect of the aglycon in alkyl glycopyranosides has been examined with the methyl, ethyl, isopropyl, and *tert*-butyl tri-*O*-acetyl-β-D-ribopyranosides, and the corresponding benzoates.[96] In each series, there was a negligible change in conformational population as between the methyl, ethyl, and isopropyl derivatives, but the *tert*-butyl derivatives contained somewhat less of that conformer [$1C$(D)] having the aglycon axially attached, presumably because there is loss of some rotational freedom about the C-1–O-1 bond in the *tert*-butyl derivatives as a result of steric hindrance.

The conformational population of a given aldopentopyranoid derivative is not independent of the substituents at O-2, O-3, and O-4. For example, for the foregoing methyl tri-*O*-acetyl-β-D-ribopyranosides, the acetylated derivatives exist to ~60% in the $1C$(D) form

---

Fig. 3.—Conformational Equilibria[95] for the Aldopentopyranose Tetraacetates in Acetone-d_6 at 31°.

TABLE V

Conformational Equilibria of Aldopentopyranose
Derivatives in Acetone-d_6 Solution at 31°

D-Aldopentopyranose derivative	Equilibrium data[b]			$\Delta G^\circ_{31°}$, kcal.mole^{-1} for $1C$(D) $\rightleftarrows$ $C1$(D)	References
	% $C1$	% $1C$	$K = C1/1C$		
Acetylated pyranosyl halides					
α-xylo, bromide[a]	>98	<2	>50	<−2.4	94
chloride[a]	>98	<2	>50	<−2.4	94
α-lyxo, bromide[a]	96	4	24	−1.9 ±1.0	94
chloride[a]	91	9	9.6	−1.4 ±0.5	94
β-ribo, bromide[a]	5	95	0.05	+1.8 ±0.9	94
chloride[a]	6	94	0.08	+1.5 ±0.6	94
β-arabino, bromide[a]	3	97	0.03	+2.1 ±1.1	94
chloride[a]	2	98	0.02	+2.4 ±1.2	94
β-xylo, chloride[a]	21	79	0.26	+0.81 ±0.32	94
Benzoylated pyranosyl halides					
α-xylo, chloride[a]	>98	<2	>50	<−2.4	94
β-xylo, chloride[a]	2	98	0.02	+2.4 ±1.2	94
β-xylo, chloride	16	84	0.19	+1.0 ±0.39	94
β-ribo, bromide[a]	2	98	0.02	+2.4 ±1.2	94
Tetraacetates					
α-ribo	77	23	3.4	−0.74 ±0.33	95
β-ribo	43	57	0.74	+0.18 ±0.26	95
α-arabino[a]	21	79	0.26	+0.81 ±0.34	95
β-arabino	4	96	0.04	+1.9 ±1.0	95
α-xylo	>98	<2	>50	<−2.4	95
β-xylo	72	28	2.6	−0.58 ±0.30	95
α-lyxo	71	29	2.5	−0.55 ±0.30	95
β-lyxo	39	61	0.63	+0.28 ±0.27	95
Tetrabenzoates					
α-ribo	73	27	2.7	−0.60 ±0.29	95
β-ribo	23	77	0.30	+0.72 ±0.31	95
α-arabino[a]	30	70	0.43	+0.51 ±0.28	95
β-arabino	5	95	0.05	+1.8 ±0.9	95
α-xylo	>98	<2	>50	<−2.4	95
β-xylo	50	50	0.98	+0.01 ±0.21	95
α-lyxo	74	26	2.8	−0.63 ±0.30	95
β-lyxo	22	78	0.29	+0.76 ±0.32	95
β-ribo (tetra-p-toluate)	28	72	0.39	+0.57 ±0.29	95

(continued)

Table V (*continued*)

D-Aldopentopyranose derivative	Equilibrium data[b]			$\Delta G^{\circ}_{31^{\circ}}$, kcal.mole^{-1} for $1C(\text{D}) \rightleftarrows C1(\text{D})$	References
	% C1	% 1C	K = C1/1C		
Tri-O-acetylpentopyranosyl benzoates					
β-*ribo*	44	56	0.77	+0.16 ±0.28	97
α-*arabino*[a]	27	73	0.36	+0.62 ±0.31	97
β-*xylo*	61	39	1.6	−0.28 ±0.27	97
α-*lyxo*	72	28	2.6	−0.58 ±0.30	97
Tri-O-benzoylpentopyranosyl acetates					
β-*ribo*	22	78	0.29	+0.76 ±0.32	97
α-*arabino*[a]	28	72	0.39	+0.57 ±0.29	97
β-*xylo*	53	47	1.1	−0.08 ±0.26	97
α-*lyxo*	73	27	2.7	−0.60 ±0.29	97
Methyl peracetylated pyranosides					
α-*ribo*	65	35	1.8	−0.36 ±0.29	98
β-*ribo*	39	61	0.63	+0.28 ±0.28	98
α-*arabino*[a]	17	83	0.20	+0.98 ±0.40	98
β-*arabino*	3	97	0.03	+2.1 ±1.1	98
α-*xylo*	>98	<2	>50	<−2.4	98
β-*xylo*	81	19	4.3	−0.89 ±0.37	98
α-*lyxo*	83	17	5.0	−0.95 ±0.39	98
β-*lyxo*	58	42	1.4	−0.20 ±0.28	98
Methyl perbenzoylated pyranosides					
β-*ribo*	20	80	0.25	+0.85 ±0.36	98
β-*arabino*	5	95	0.05	+1.8 ±0.9	98
α-*xylo*	>98	<2	>50	<−2.4	98
β-*xylo*	74	26	2.8	−0.63 ±0.32	98
α-*lyxo*	86	14	6.4	−1.10 ±0.44	98
Alkyl tri-O-acetyl-β-D-ribopyranosides					
methyl	39	61	0.63	+0.29 ±0.29	96
ethyl	39	61	0.63	+0.29 ±0.29	96
isopropyl	38	62	0.60	+0.31 ±0.29	96
tert-butyl	46	54	0.85	+0.10 ±0.28	96

(*continued*)

TABLE V (continued)

D-Aldopentopyranose derivative	Equilibrium data[b]			$\Delta G^{\circ}_{31^{\circ}}$, kcal.mole^{-1} for $1C$(D) $\rightleftarrows$ $C1$(D)	References
	% C1	% 1C	K = C1/1C		
Alkyl tri-O-benzoyl-β-D-ribopyranosides					
methyl	20	80	0.25	+0.86 ±0.37	96
ethyl	19	81	0.23	+0.90 ±0.38	96
isopropyl	22	78	0.28	+0.78 ±0.35	96
tert-butyl	26	74	0.35	+0.64 ±0.33	96
1-Thioaldopentopyranose tetraacetates					
β-ribo	66	34	2.0	−0.41 ±0.28	99
α-arabino[a]	32	68	0.46	+0.47 ±0.29	99
β-xylo	72	28	2.6	−0.58 ±0.30	99
α-lyxo	64	36	1.8	−0.35 ±0.24	99

[a] In chloroform-d. [b] The limits of accuracy for these values may be determined from the uncertainty limits given by the ΔG° values.

(axial bond to the glycosidic group), whereas the corresponding benzoates exist to ~80% in the form having the substituent on C-1 axially attached.[93,95] Similar behavior, in which the apparent axial-directing effect of the substituent on C-1 is enhanced by changing the substituents at C-2, -3, and -4 from acetates to benzoates, is observed for all of the other configurations except the α-ribo series, although the quantitative magnitude of the enhancement varies; it is large for the β-ribo, β-xylo, β-lyxo, and α-arabino series, but less for the α-lyxo derivatives. Attractive interactions between syn-diaxial acyloxy groups, of higher magnitude in the benzoates than in the acetates, were invoked to rationalize these observed results.[93,95]

5. Empirical Approaches to Prediction of Favored Conformation

In the first attempt to devise a rationalization of the observed conformational behavior of aldopyranose derivatives, Reeves assigned arbitrary "instability" values to certain structural features of pyranoid sugars in their chair conformations.[6,17,22] According to this model, the following factors introduced instability into a chair conformer: (1) an axially attached substituent (other than hydrogen), to which was assigned one unit of instability; (2) an axial hydroxyl group at C-2 when the C–O bond at C-2 bisects the two C–O bonds at C-1 (Δ2 condition), 0.5 unit; and (3) an axial hydroxymethyl group at C-5 syn to a hydroxyl group at C-1 or C-3 (the Hassel–Ottar effect[15]), 2.5 units. From a summation of the "instability factors" for each chair form, Reeves predicted the chair conformation that would be the more stable. When the difference between the two chair forms was less than one unit, it was postulated that both forms would be present in appreciable proportions. With the exception of methyl α-D-gulopyranoside and methyl β-D-mannopyranoside, an excellent correlation was obtained between the predicted and experimental results, although this is not surprising, because the values for the "instability factors" were deliberately chosen to give good agreement.

In an attempt to correlate conformational stability and the physical properties of some methyl O-methylaldopyranosides, Kelly[228] modified Reeves's arbitrary "instability factors" as follows: (1) axial hydroxyl group, 1 unit; (2) Δ2 factor, 2.5 units; (3) the Hassel–Ottar effect, 2.5 units; and (4) an axial hydroxymethyl group, 2.0 units. With these values, improved agreement between the predicted and the experimentally determined conformations was obtained.

Because the Reeves–Kelly scheme neglects several important non-bonded interactions, such as those between syn-diaxial hydroxyl groups and vicinal, gauche substituents, and because polar interactions are neglected altogether, the scheme may, at best, be considered to represent a qualitative attempt to predict the conformational behavior of pyranoid sugars in solution.

In a more quantitative effort, Angyal[31,229] estimated the approximate free-energy content of each chair form for the D-aldopentopyranoses and D-aldohexopyranoses by summation of quantitative free-energy terms for conformational interaction-energies of various groups, also taking into account the value of the anomeric effect (see next Section) for the hydroxyl group. Inherent in these calculations were the assumptions that the pyranoid ring has the same geometry as cyclohexane and that the various non-bonded interactions are independent of one another. The approximate nature of the results was, however, clearly recognized by the author. Not only were interactions between axial groups considered, but also those between vicinal, gauche substituents. The values used for the non-bonded interactions were determined, in aqueous solution at room temperature, from the equilibria of cyclitols with their borate complexes[230] and from the anomeric equilibria of pyranoid sugars.[229–231] The following values

(228) R. B. Kelly, Can. J. Chem., 35, 149 (1957).
(229) S. J. Angyal, Aust. J. Chem., 21, 2737 (1968).
(230) S. J. Angyal and D. J. McHugh, Chem. Ind. (London), 1147 (1956).
(231) S. J. Angyal, V. A. Pickles, and R. Ahluwahlia, Carbohyd. Res., 1, 365 (1966).

for the interactions, as illustrated, were used in the calculations: (1) an axial hydrogen *syn* to an axial hydroxyl group, 0.45 kcal.mole^{-1}; (2) an axial hydrogen *syn* to an axial hydroxymethyl (or methyl) group, 0.9 kcal.mole^{-1}; (3) two *syn*-diaxial hydroxyl groups, 1.5 kcal.mole^{-1}; (4) an axial hydroxyl group *syn* to an axial hydroxymethyl (or methyl) group, 2.5 kcal.mole^{-1}; (5) two vicinal, gauche, hydroxyl groups, 0.35 kcal.mole^{-1}; and (6) a hydroxyl group gauche to a hydroxymethyl (or methyl) group, 0.45 kcal.mole^{-1}. All other interactions were considered to be less than 0.1 kcal.mole^{-1} and were, therefore, neglected. The value for the anomeric effect of the hydroxyl group was taken as being 0.85 kcal.mole^{-1} when no hydroxyl group was present on C-2, as 1.0 kcal.mole^{-1} when an axial hydroxyl group was present on C-2, and as 0.55 kcal.mole^{-1} when the hydroxyl group on C-2 was equatorially attached. The Reeves Δ2 effect was regarded as constituting part of the anomeric effect.

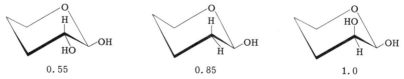

Estimated values for non-bonded interaction energies (kcal. mole^{-1}) in aqueous solution at room temperature, according to Angyal (see text)

Estimated values for the anomeric effect (kcal. mole^{-1}) of the hydroxyl group in aqueous solution, according to Angyal (see text)

The estimated free-energies for both chair conformations of the free aldopyranoses were then calculated by summation of the various steric interactions and the anomeric effect of the anomeric hydroxyl group. The energies calculated are given in Table IV (see p. 85), together with the conformations predicted and those indicated by experiment. When the free-energy difference between the two chair forms was less than 0.7 kcal.mole^{-1}, both conformers were considered to be present in comparable amounts at equilibrium.

The estimated conformational tendencies were found to be in excellent agreement with those observed by n.m.r. spectroscopic studies of the aldopyranoses in aqueous solution.[71,210,211] The n.m.r. data indicated that, for several of the free sugars, the *1C* and *C1* conformers are probably in equilibrium, because the spin-coupling data gave values intermediate between those anticipated for each single conformer. Taking into consideration the difference in the anomeric effect of the methoxyl and the hydroxyl group and the limitations of the cuprammonia method for determining conformations, the conformational assignments were found to be compatible with those found by Reeves for the methyl aldopyranosides.[16,22]

In a further quantitative check of the validity of the calculated free-energy values, the predicted α,β anomeric equilibria of the aldopyranoses in aqueous solution[31,229] were found to be in reasonable agreement with those determined experimentally by studies of their mutarotation,[229] oxidation by bromine,[232] and n.m.r. spectra.[71,210,211,233] In addition, the position of the equilibria between aldohexopyranoses (and 3-deoxy-aldohexopyranoses) and their 1,6-anhydrides, and between heptuloses and their 2,7-anhydrides, as determined by gas–liquid chromatography,[234] were in good agreement with data estimated from conformational interaction-energies.

From data accumulated on the anomeric equilibria of certain peracetylated aldopyranoses in 1:1 acetic acid–acetic anhydride, with perchloric acid as the catalyst, and an estimated value for the conformational equilibrium of β-L-arabinopyranose tetraacetate in chloroform solution, Lemieux and Chü estimated values for the various non-bonded interactions present in a pyranoid ring.[214,215] Although the experimental details of this work have not yet been published, the following interaction-energies were proposed in a review article[215]: an axial hydrogen atom opposing an axial acetoxyl group, 0.18 kcal.mole^{-1}; two *syn*-diaxial acetoxyl groups, 2.08 kcal.

(232) H. S. Isbell, *J. Res. Nat. Bur. Stand.*, **66A**, 233 (1962).
(233) S. J. Angyal and V. A. Pickles, *Carbohyd. Res.*, **4**, 269 (1967).
(234) S. J. Angyal and K. Dawes, *Aust. J. Chem.*, **21**, 2747 (1968).

mole^{-1}; and two vicinal, gauche acetoxyl groups, 0.55 kcal.mole^{-1}. The first of these values appears small compared with the value of 0.36 kcal.mole^{-1} determined for acetoxycyclohexane in carbon disulfide.[235] The conformational properties of the peracetylated aldopentopyranoses[72] were found to be in qualitative agreement with those expected from these non-bonded interaction-energies, taken in conjunction with the magnitude proposed for the anomeric effect (see next Section) of the acetoxyl group.[214,215] However, predictions from these estimates agree poorly with the anomeric equilibrium actually observed for D-altropyranose pentaacetate,[218] and these estimates would predict for α-D-idopyranose pentaacetate (19) a conformation that is the opposite of that found[176] experimentally.

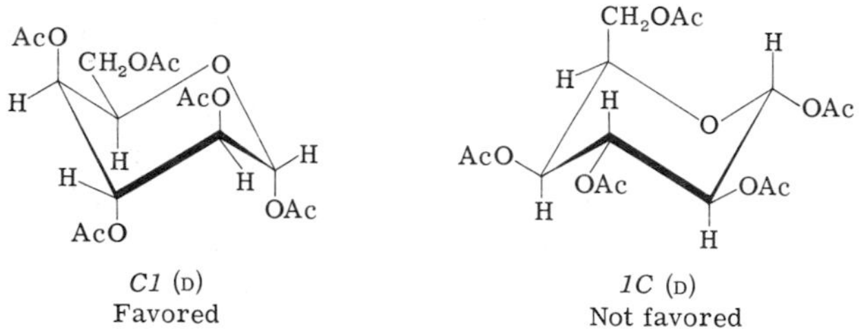

C1 (D)
Favored

1C (D)
Not favored

(19)

The large preponderance of the C1(D) chair form at equilibrium could not have been predicted from the concept of additive conformational free-energies,[214,215,236] even were the estimated anomeric effect of the acetoxyl group taken into account.[214,215] On the basis of published estimates[214,215,235,236] for these energies, both chair conformers would be expected to be present in substantial proportion, the 1C(D) conformation being somewhat favored. Attractive interactions between axial acetoxyl groups and other groups in the molecule, or significant repulsive interactions between vicinal equatorial groups, were postulated as being possible factors involved in controlling the position of the conformational equilibrium for this aldohexose peracetate,[176] and the validity of the model[214,215,236] for additive free-energies in these systems was questioned.[176] Likewise, the all-axial, 1C(D) conformation observed[175] for tri-O-acetyl-β-D-xylopyranosyl chloride could not be rationalized on the basis of

(235) F. R. Jensen, C. H. Bushweller, and B. H. Beck, J. Amer. Chem. Soc., 91, 344 (1969).
(236) Ref. 25, Chapter 2, pp. 44 and 52; Chapter 6, pp. 356 and 371–378.

additive interaction-energies, without postulation of an unreasonably large value for the anomeric effect of the chlorine atom.

From a consideration of detailed results on the conformational equilibria of aldopentopyranose derivatives, it has been pointed out[92–99] that a more sophisticated model is required before conformational populations can be reliably predicted, at least with acylated derivatives. Even with adjustment of the original parameters in order to take revised values for the anomeric equilibrium of D-lyxopyranose tetraacetate and the conformational equilibrium of β-D-arabinopyranose tetraacetate[95] into account, the observed data cannot be accommodated within the framework of this model, except on a very broad, qualitative basis. Other possible factors that should be considered[93–95] include polar contributions from substituents other than that on C-1, attractive interactions between *syn*-diaxial acyloxy groups, non-bonded interactions between atoms that have unshared pairs of electrons,[122] repulsive interactions between gauche-vicinal groups, the effect of solvent pressure, and differences between the molar volume of conformers.

6. The Anomeric Effect

It has been found that the α-D is more stable than the β-D anomer in anomerically equilibrated mixtures of methyl D-glucopyranosides,[237] penta-O-acetyl-D-glucopyranoses,[238] and peracetylated D-glucopyranosyl halides,[173,239] even though the substituent on C-1 is undoubtedly axially attached in the α-D anomer.

Results for the anomeric equilibria of the pentopyranose tetra-acetates (see Table VI), taken in conjunction with their conformational equilibria (see Fig. 3, p. 94), also indicate that that anomer having the substituent on C-1 axially attached is the more stable. This predisposition of a polar substituent at C-1 of a pyranoid ring to assume the axial orientation, contrary to expectations based merely on steric considerations, has been termed[214,215] the "anomeric effect." The phenomenon has been attributed by Edward[240] to an unfavorable dipole–dipole interaction between the carbon–oxygen bonds on the

(237) C. L. Jungius, Z. *Phys. Chem.*, **52**, 97 (1905).
(238) W. A. Bonner, *J. Amer. Chem. Soc.*, **73**, 2659 (1951); **81**, 1448 (1959).
(239) R. U. Lemieux, *Advan. Carbohyd. Chem.*, **9**, 1 (1954); L. J. Haynes and F. H. Newth, *ibid.*, **10**, 207 (1955).
(240) J. T. Edward, *Chem. Ind.* (London), 1102 (1955).

TABLE VI

Anomeric Equilibria of D-Aldopentopyranose Tetraacetates at 27° in 1:1
Acetic Anhydride–Acetic Acid, 0.1M in Perchloric Acid[95]

Anomeric pairs[a]	Equilibrium constant, $K = \beta/\alpha$	$\Delta G°$, kcal.mole^{-1}, for $\alpha \rightleftarrows \beta$ at 27°
Tetra-O-acetyl-α,β-D-ribopyranose	3.4	-0.73 ± 0.03
Tetra-O-acetyl-α,β-D-arabinopyranose	5.4	-1.01 ± 0.03
Tetra-O-acetyl-α,β-D-xylopyranose	0.23	$+0.89 \pm 0.03$
Tetra-O-acetyl-α,β-D-lyxopyranose	0.20	$+0.98 \pm 0.05$

[a] Similar values for the first three anomeric pairs listed were given earlier in ref. 215.

ring and the bond from the anomeric carbon atom to the equatorial, polar substituent.

Lemieux and Chü[214,215] have interpreted the effect in terms of an electrostatic interaction between the C-1 to substituent and C-5–O-5 bonds.[241] The interaction was later alluded to[242] as the "rabbit-ear effect," and was considered to arise from a repulsion of the electric dipoles engendered by the parallel disposition of electron pairs occupying non-bonding orbitals of the ring hetero-atom and the electronegative atom bonded directly to the anomeric carbon atom.[35,242,243] By this rationale, the anomeric effect would be expected to be in-

versely proportional to the dielectric constant of the solvent, and directly proportional to the magnitude of the dipole moment of the C-1–substituent bond.

(241) J. T. Edward, P. F. Morand, and I. Puskas, Can. J. Chem., 39, 2069 (1961).
(242) R. O. Hutchins, L. D. Kopp, and E. L. Eliel, J. Amer. Chem. Soc., 90, 7174 (1968).
(243) (a) E. L. Eliel and C. A. Giza, J. Org. Chem., 33, 3754 (1968); (b) M. A. Kabayama and D. Patterson, Can. J. Chem., 36, 536 (1958).

The validity of the Lemieux–Chü interpretation was entirely supported, and the picturesque "rabbit-ear" idea shown untenable, by a theoretical study that used an *ab initio* (Hartree–Fock) calculation, with fluoromethanol as the model compound.[244] The calculations showed that the stable conformation has the C–F bond *trans* to one "electron pair" and *gauche* to another, whereas the conformation in which the C–F bond bisects the "electron pairs" is the energy maximum. Similar interpretations were established for hydrazine, hydroxylamine, and hydrogen peroxide. The concept can be stated in general terms as a destabilization of a conformation that places a polar (C–X) bond eclipsed between two electron pairs.

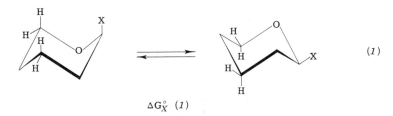

Destabilized   Stabilized

Quantitatively, the anomeric effect has been expressed[245] as the sum of the free-energy difference for the process shown in equation

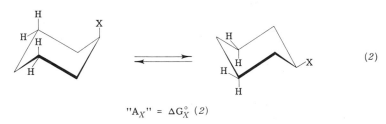

$$\Delta G_X^\circ \ (1)$$

$$(1)$$

(1) and the conformational free-energy or[246] "A-value" for an axial substituent X in cyclohexane, as shown in equation (2); that is, the anomeric effect $= \Delta G_X^\circ \ (1) + \text{"}A_X\text{"}$.

$$"A_X" = \Delta G_X^\circ \ (2)$$

$$(2)$$

(244) S. Wolfe, A. Rauk, L. M. Tel, and I. G. Csizmadia, *J. Chem. Soc. (B)*, 136 (1971).
(245) C. B. Anderson and D. T. Sepp, *Tetrahedron*, **24**, 1707 (1968).
(246) S. Winstein and N. J. Holness, *J. Amer. Chem. Soc.*, **77**, 5562 (1955).

This expression of the anomeric effect is based on the assumption that the steric influence of the substituent on the tetrahydropyran ring is the same as that on cyclohexane. However, the assumption may not be generally true, because the C–O bond is shorter than the C–C bond, and this leads to a larger *syn*-diaxial interaction in the heterocycle, as has been demonstrated experimentally in 1,3-dioxane systems.[243a,247]

The anomeric effect has been found to depend not only on the nature of the solvent but also on the presence and the configuration of substituents at the other positions of the pyranoid ring. Thus, the anomeric effect of the hydroxyl group in the aldopyranoses in aqueous solution was estimated[229] to be 0.55 kcal.mole^{-1} when the hydroxyl group on C-2 is equatorial, 1.0 kcal.mole^{-1} when it is axial, and 0.85 kcal.mole^{-1} when there is no hydroxyl group on C-2. It should be mentioned that the "A-value" of 0.9 kcal.mole^{-1} for the hydroxyl group in aqueous solution (estimated from the results of borate complexing[230]) employed in these calculations appears small, in view of its value in carbon disulfide[248] (0.97 kcal.mole^{-1}) as determined by an accurate, direct method (low-temperature, n.m.r.-spectral integration). Its magnitude should be even larger in water, because the steric influence of the group should be increased by the effects of solvation. Therefore, 1.25 kcal.mole^{-1} is probably[85] closer to the true value, and thus the estimates previously given for the anomeric effect of the hydroxyl group should be revised to 0.9, 1.35, and 1.20 kcal.mole^{-1}, respectively.

Estimates of the magnitude of the anomeric effect of other polar groups have been reported: (*a*) OMe: 1.2 kcal.mole^{-1} in the methyl aldopentopyranosides[249] and 1.4 kcal.mole^{-1} in the methyl aldohexopyranosides[250] in 1% methanolic hydrogen chloride, 1.3 kcal.mole^{-1} in 2-methoxy-4-methyltetrahydropyran in *p*-dioxane[245] and 0.9 kcal. mole^{-1} in aqueous methanol[245]; (*b*) OAc: 1.3 kcal.mole^{-1} in peracetylated aldopentopyranoses and 1.5 kcal.mole^{-1} in the peracetylated aldohexopyranoses in 1:1 acetic acid–acetic anhydride,[214,215] 1.35 kcal.mole^{-1} in 2-acetoxy-4-methyltetrahydropyran in acetic acid[245]; (*c*) Cl: ~2 kcal.mole^{-1} in tetra-*O*-acetyl-D-glucopyranosyl chloride in acetonitrile[251] and 2.7 kcal.mole^{-1} in neat 2-chlorotetrahydropyran[252];

(247) E. L. Eliel and M. C. Knoeber, *J. Amer. Chem. Soc.*, **90**, 3444 (1968); E. L. Eliel and F. W. Nader, *ibid.*, **92**, 3050 (1970).
(248) C. H. Bushweller, J. A. Beach, J. W. O'Neil, and G. U. Rao, *J. Org. Chem.*, **35**, 2086 (1970).
(249) C. T. Bishop and F. P. Cooper, *Can. J. Chem.*, **41**, 2743 (1963).
(250) V. Smirnyagin and C. T. Bishop, *Can. J. Chem.*, **46**, 3085 (1968).
(251) R. U. Lemieux and J. Hayami, *Can. J. Chem.*, **43**, 2162 (1965).
(252) C. B. Anderson and D. T. Sepp, *J. Org. Chem.*, **32**, 607 (1967).

and (d) Br: 3.2 kcal.mole^{-1}, or greater, in neat 2-bromotetrahydro-pyran.[252]

The magnitude of the anomeric effect also appears to increase with the extent of methylation of the hydroxyl groups.[212] Thus, at anomeric equilibrium in aqueous solution, D-mannose contains 64% of the α-D-pyranose form; 2,3-di-O-methyl-D-mannose, 80%; and 2,3,4,6-tetra-O-methyl-D-mannose, 86%.

In acylated pyranoid sugar derivatives, the net axial-directing influence of a substituent at C-1 (presumably the resultant of a positive, polar contribution and a negative, steric contribution) has been shown[93-99] not to be independent of the total stereochemistry. Thus, by keeping the substituents at C-2, 3, and 4 constant in an aldopento-pyranose system, and varying the nature of the substituent on C-1, it was found that the axial-directing effect falls in the order Br ≈ Cl >> OMe > OBz ≈ OAc > SAc in both the β-ribo and the α-lyxo series. On the other hand, for the β-xylo series, the order is Cl > OBz > SAc ≈ OAc > OMe, and a similar order is observed in the α-arabino series. The axial-directing influence of the halogeno groups is consistently strong,[94] presumably because the C–halogen bond-moments are larger than the C–O bond-moment, but the influence of other groups appears to depend on whether or not the axially attached substituent on C-1 has a syn-axial group at C-3. If such a syn-diaxial arrangement is present, the directing influence of the SAc group is relatively strong (possibly as a result of London attraction), and that of the OMe group is weak; if it is not, the directing influence of the SAc group is relatively weak, and that of the OMe group, relatively strong.[98,99] These observations appear to be best explained[92-99,176] by a concept of attractive interactions between syn-diaxial acyloxy groups and a polar stabilization of acyloxy groups axially attached to carbon atoms other than C-1 (especially C-3).

The enhancement of the axial-directing effect of the C-1 substituent when the ring substituents are changed from acetates to benzoates has already been discussed (see p. 98).

From an n.m.r.-spectral study of some pyridinium α-glycopyrano-sides, it was discovered that a quaternized nitrogen atom at the anomeric center favors the equatorial orientation to an extent greater than that predicted from steric considerations alone.[253] Thus, N-(tetra-O-acetyl-α-D-glucopyranosyl)-4-methylpyridinium bromide (20) in deuterium oxide was considered to exist in a conformation, close to the 1C(D) conformation, having the anomeric substituent equatorial.

(253) R. U. Lemieux and A. R. Morgan, Can. J. Chem., 43, 2205 (1965).

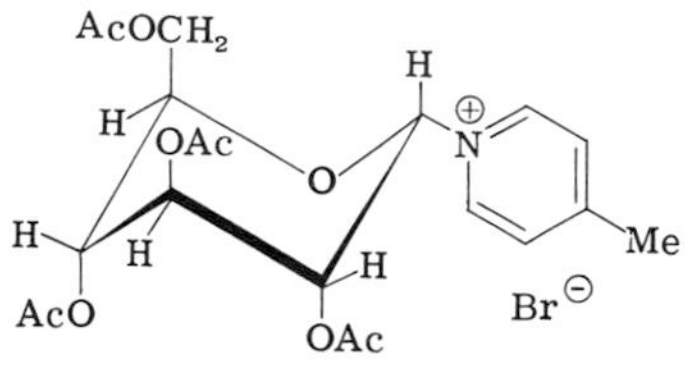

(20)

This unexpected shift in conformation was attributed to a strongly destabilizing electrostatic interaction, resulting from the establishment of a positively charged atom in axial orientation at the anomeric center. This phenomenon, termed[253] the "reverse anomeric effect," was further substantiated from a study of the effects of protonation and N-methylation on the conformational populations of N-glycosyl-imidazoles.[254] N.m.r. and X-ray data indicated, however, a shift from the $C1(\text{D})$ to a boat-like conformation, and *not* toward the alternative $1C(\text{D})$ conformation as had previously been reported.[253]

The reverse anomeric effect has been invoked to explain the observed conformational behavior of other α-D-glycosylamine derivatives in the form of their salts.[255] For example, 9-α-D-mannopyrano-syladenine hydrochloride (**21**) has been assigned[255] the $1C(\text{D})$

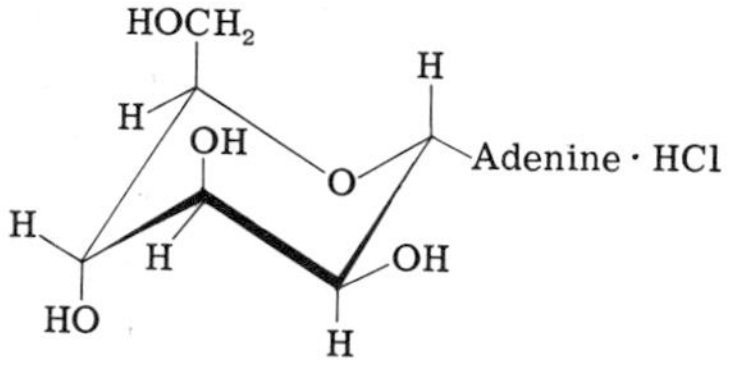

(21)

conformation in methyl sulfoxide near room temperature from the observation that $J_{1,2}$ is 8.3 Hz. In addition, it was observed that other α-glycosylamines having a tertiary nitrogen atom also favor that chair conformation having the aglycon in equatorial orientation, indicating that the magnitude of the anomeric effect of an amino group is relatively small. The bulkiness of the anomeric substituent was also considered to be a factor in the destabilization of the conformer having the aglycon attached axially.

(254) R. U. Lemieux and S. S. Saluja, *Abstr. Papers ACS—CIC Meeting (Toronto)*, c33 (1970).
(255) K. Onodera, S. Hirano, and F. Masuda, *Carbohyd. Res.*, **7**, 27 (1968).

This conformational change has also been detected in a pento-pyranoid system.[256] Thus, although 1-*O*-acetyl-3-benzamido-2,4-di-*O*-benzoyl-3-deoxy-3-*C*-(ethoxycarbonyl)-β-D-ribopyranose (**22**) adopts the *1C*(D) conformation in chloroform solution, the corresponding nucleoside derivative **23** exists preponderantly in the *C1*(D) con-formation.

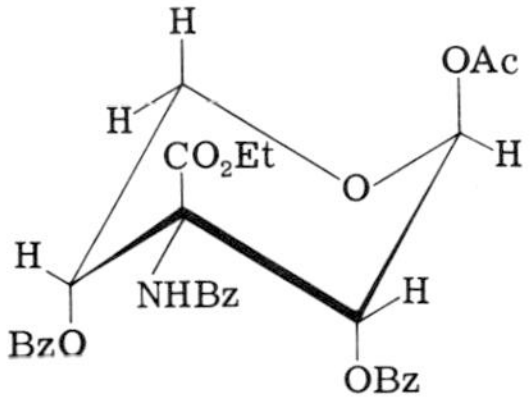

(22) 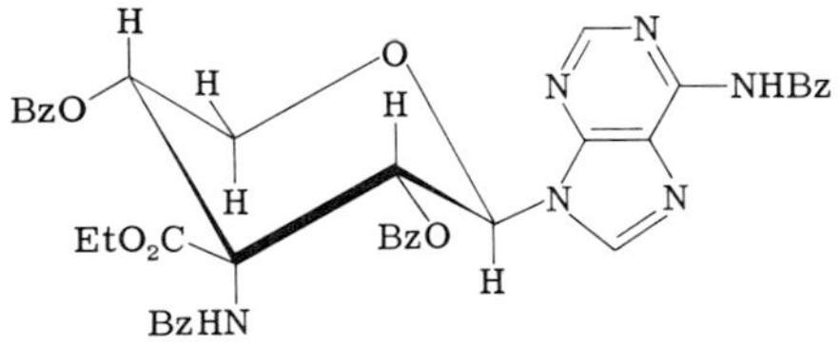   (23)

V. Conformations of Furanoid Sugars and Their Derivatives

1. General Considerations

Furanoid ring-systems do not adopt a planar conformation, because of the excessive torsional (Pitzer) and Van der Waals strain that would exist in a planar ring—strain caused by eclipsing of orbitals and substituent groups on adjacent ring-atoms. The destabilization generated by such eclipsing outweighs the stabilization gained by relief of bond-angle (Baeyer) strain that would occur were the ring to be planar. Furanoid sugar derivatives usually adopt either an "envelope" conformation (which has been designated[81] by the sym-bol "V") having four atoms, including the ring-oxygen atom, almost coplanar, and the remaining ring-atom exoplanar, or a "twist" form (which has been designated[81] by the symbol "*T*") having three ad-jacent atoms coplanar, and one of the remaining atoms above and the

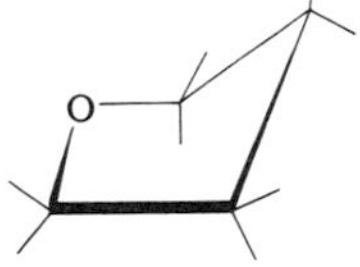

Envelope    Twist

(256) H. Yanagisawa, M. Kinoshita, S. Nakada, and S. Umezawa, *Bull. Chem. Soc. Jap.*, **43**, 246 (1970).

other below the plane. (For cyclopentane, these two conformational states have been termed[257] C_s and C_2, respectively, from the terminology of point-group symmetry.) In principle, for any furanoid ring, there exist ten possible envelope-forms and ten possible twist-forms. The energy barriers separating these various conformations are presumed to be small enough to allow rapid interconversion at room temperature by the process of "pseudorotation."

2. Furanoid Sugar Derivatives in the Crystalline State

That furanoid rings do not adopt planar shapes was first demonstrated for a crystalline sugar derivative by X-ray and neutron diffraction with the observations that (a) the β-D-fructofuranose residue of sucrose, in crystals of sucrose sodium bromide dihydrate[184] and of the free sugar,[185] and (b) the D-ribofuranosyl residue of cytidine,[258] exist in an envelope conformation. These two crystal-structure analyses did not, however, give very accurate detail. The non-planarity of the furanoid ring in crystalline sugar derivatives has been repeatedly confirmed by more precise analyses of the structures of the D-ribose and 2-deoxy-D-*erythro*-pentose moieties of such biologically important derivatives as the β-D-nucleosides.[259–263] Crystal-structure determinations of certain α-D-nucleosides,[264] and various other furanoid sugar derivatives,[185,265,266] have also been reported. In the examples studied, the conformation has been found to be of the envelope type, with four ring atoms approximately in one plane, and the fifth atom, either C-2 or C-3 (the atoms *meta* to the ring-oxygen atom[267]), out of the plane by 0.5–0.6 Å. Certain of the structures approached, to a greater or lesser extent, the character of a twist conformation. The furanoid compound methyl 1,2,3,5-tetra-O-acetyl-β-D-galactofuranuronate has been shown to exist in an envelope conformation, but in this instance it is C-1 that is the out-of-plane atom.[267a]

(257) J. E. Kilpatrick, K. S. Pitzer, and R. Spitzer, *J. Amer. Chem. Soc.*, **69**, 2483 (1947).
(258) S. Furberg, *Acta Crystallogr.*, **3**, 325 (1950).
(259) M. Spencer, *Acta Crystallogr.*, **12**, 59 (1959).
(260) M. Sundaralingam, *J. Amer. Chem. Soc.*, **87**, 599 (1965); M. Sundaralingam and L. H. Jensen, *J. Mol. Biol.*, **13**, 930 (1965).
(261) D. W. Young, P. Tollin, and H. R. Wilson, *Acta Crystallogr.*, **B25**, 1423 (1969).
(262) M. Sundaralingam, *Biopolymers*, **7**, 821 (1969).
(263) M. Sundaralingam, *Acta Crystallogr.*, **21**, 495 (1966); S. T. Rao and M. Sundaralingam, *J. Amer. Chem. Soc.*, **91**, 1210 (1969); **92**, 4963 (1970).
(264) D. C. Rohrer and M. Sundaralingam, *J. Amer. Chem. Soc.*, **92**, 4950, 4956 (1970).
(265) R. Parthasarathy and R. E. Davis, *Acta Crystallogr.*, **23**, 1049 (1967).
(266) P. Groth and H. Hammer, *Acta Chem. Scand.*, **22**, 2059 (1968).
(267) R. U. Lemieux and R. Nagarajan, *Can. J. Chem.*, **42**, 1270 (1964).
(267a) J. P. Beale, N. C. Stephenson, and J. D. Stevens, *Chem. Commun.*, 25 (1971).

The conformation of the furanoid moiety in nucleic acids and other carbohydrate derivatives, as determined by X-ray or neutron diffraction, has been discussed elsewhere.[37,260,262,268]

3. Furanoid Sugar Derivatives in Solution

The conformations of furanoid sugar derivatives in solution have been investigated by n.m.r. spectroscopy. The establishment of precise conformations for five-membered ring-systems by the n.m.r. method is, however, limited by the fact that the energy barriers between the numerous conformational states possible are quite small (3–4 kcal.mole^{-1}).[257,269,270] Conformational interconversion is sufficiently rapid that a time-averaged spectrum of the conformers can be expected over the accessible temperature-range of the instruments at present available. The barrier to conformational inversion in *pyranoid* sugar derivatives is of a magnitude (~10.5 kcal.mole^{-1}) very similar to that for cyclohexane, and the barrier in *furanoid* sugar derivatives can thus be expected to differ little from that in cyclopentane. Nevertheless, from application of the Karplus equation relating vicinal spin-couplings to dihedral angle,[74] the n.m.r. spectra of a wide range of furanoid sugar derivatives have been interpreted in terms of either an envelope or a twist conformation.[75,76,81,249,271-274] However, in view of the ease of conformational interconversion, and the limitations to application of the Karplus equation for the determination of favored conformation of furanoid ring-systems,[66,84,275] it seems more reasonable, until a more accurate method for relating vicinal coupling-constants to dihedral angles becomes available, to treat the study of the conformations of furanoid sugar derivatives in solution in terms of conformational equilibria between the various, rapidly interconverting, puckered forms, and *not* in terms of a unique shape and precise dihedral angles for each particular furanoid ring. Some of the later investigations on the conformations of nucleosides[276] and pentofuranosyl fluoride derivatives[66] have taken this approach.

As observed for corresponding molecules in the solid state, furanoid

(268) H. R. Wilson, A. Rahman, and P. Tollin, *J. Mol. Biol.*, **46**, 585 (1969).
(269) K. S. Pitzer and W. E. Donath, *J. Amer. Chem. Soc.*, **81**, 3213 (1959).
(270) J. B. Hendrickson, *J. Amer. Chem. Soc.*, **83**, 4537 (1961).
(271) D. Gagnaire and P. Vottero, *Bull. Soc. Chim. Fr.*, 2779 (1963).
(272) G. Casini and L. Goodman, *J. Amer. Chem. Soc.*, **86**, 1427 (1964).
(273) R. J. Cushley, J. F. Codington, and J. J. Fox, *Can. J. Chem.*, **46**, 1131 (1968).
(274) J. D. Stevens and H. G. Fletcher, Jr., *J. Org. Chem.*, **33**, 1799 (1968).
(275) T. D. Inch and P. Rich, *J. Chem. Soc. (C)*, 1784 (1968).
(276) F. E. Hruska, A. A. Grey, and I. C. P. Smith, *J. Amer. Chem. Soc.*, **92**, 214, 4088 (1970); **93**, 1765 (1971).

sugar derivatives adopt non-planar ring-shapes in solution. Because the interaction energy between the ring-oxygen atom and an eclipsed carbon atom is smaller than that between two carbon atoms, the oxygen atom usually occupies the least-puckered part of the furanoid ring, with C-2 or C-3, or both, exoplanar. At conformational equilibrium, the various conformers will populate to extents depending on their relative free-energies, as determined by a combination of steric and electronic interactions. It appears that the preponderant conformer is usually the one having maximal staggering of bulky substituents and the C-1 substituent *quasi*-axial[66,274] (favorable anomeric effect). For example, the furanoid moiety of nucleocidin (**24**)

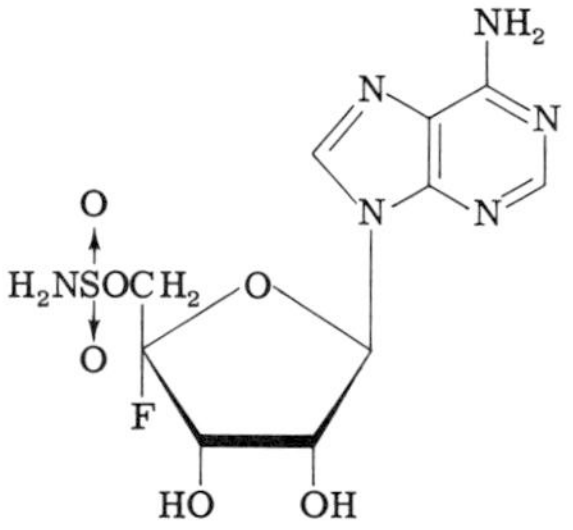

(24)

favors an envelope conformation having C-3 above the plane of the other ring-atoms.[66] However, because of uncertainties in the conformational assignments, the relative importance of steric and polar factors in determining favored conformations for furanoid sugar derivatives has not yet been fully ascertained, and thus, a quantitative treatment is not yet warranted.

The conformational behavior of furanoid sugar derivatives has been discussed[267,277] on the basis that, in general, "furanoside rings will tend to be puckered in such a manner as to have a carbon *meta* to the ring oxygen (either the 2- or the 3-carbon for aldofuranosides; either the 3- or 4-carbon for ketofuranosides) farthest out of the mean plane of the ring."[267] Thus, the envelope conformation for the furanoid moieties in 2,1'-anhydro-(1-*O*-α-D-fructofuranosyl-β-D-fructofuranose) (**25**) can account for the behavior observed on oxidation with periodate, because the hydroxyl groups of the two α-glycol groups define dihedral angles of ~ 150° and 75° in the α- and β-D-furanoid rings, respectively.

(277) Ref. 215, p. 709.

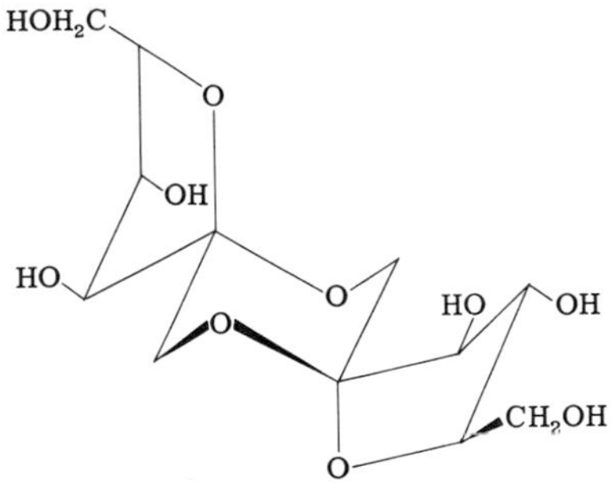

(25)

VI. Conformations of Septanoid Sugar Derivatives

The crystal-structure analysis of sugar derivatives having a seven-membered ring have been reported.[278] The X-ray data for 5-*O*-(chloroacetyl)-1,2:3,4-di-*O*-isopropylidene-α-D-glucoseptanose (**26**) showed

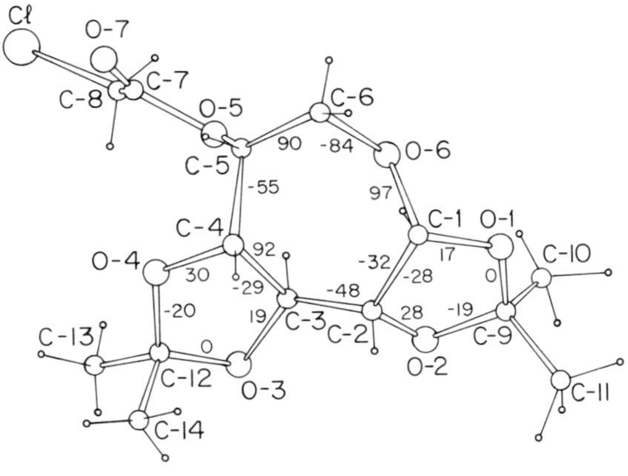

(26)

Crystal-structure diagram for 5-*O*-(chloroacetyl)-1,2:3,4-di-*O*-isopropylidene-α-D-glucoseptanose (**26**). (The numbers given along the bonds are the projected dihedral angles between the atoms attached to the bonded atoms.)

that the seven-membered (septanoid) ring adopts a shape that lies between the chair and twist-chair forms. However, this intermediate conformation may not be typical of septanoid compounds, as it is

(278) (a) J. Jackobs and M. Sundaralingam, *Chem. Commun.*, 157 (1970); (b) J. P. Beale, N. C. Stephenson, and J. D. Stevens, *Chem. Commun.*, 484 (1971).

influenced by the rigidity imposed by the two dioxolane rings. These two rings adopt an almost "ideal" envelope conformation, C-2 being out of the plane of the 1,2-ring and C-4 exoplanar to the 3,4-ring.[278a] In another crystallographic study on septanoid derivatives,[278b] it was shown that, in ethyl 2,3:4,5-di-O-isopropylidene-1-thio-β-D-glucoseptanoside, a twist-chair shape is adopted by the seven-membered ring, whereas the dioxolane rings adopt twist shapes. A monocyclic septanoid derivative, namely, methyl 2,3,4,5-tetra-O-acetyl-β-D-glucoseptanoside, was found[278b] to have a conformation somewhat twisted from an ideal chair (C^1_5 having C-5 below, and C-1 and C-2 above, a plane defined by C-3, C-4, C-6, and O-6.

The conformations of septanoid sugar derivatives in solution have been investigated by n.m.r. spectroscopy. Preliminary evidence points to a chair-like conformation for the septanoid molecule in solution.[279] Such a chair conformation has been observed for D-*xylo*-4,5,6-trihydroxy-1-methyl-4,5,6,7-tetrahydro-1H-1,2-diazepin (**27**) in

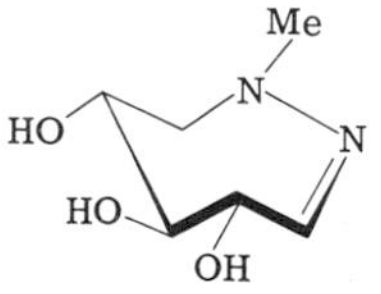

(27)

deuterium oxide.[280] In a related study, the seven-membered, 1,3-dioxepan ring in 1,3:2,5:4,6-tri-O-ethylidene-D-mannitol in chloroform solution was found[281] to exist mainly in a twist-chair shape.

VII. CONFORMATIONS OF SUGAR DERIVATIVES HAVING FUSED-RING SYSTEMS

1. Pyranoid Sugar Derivatives Having Fused Rings

Fusion of a second ring to a pyranoid ring often limits the number of possible conformations that the pyranoid part may adopt. For example, *trans*-fusion of a benzylidene acetal ring at C-4 and C-6 of a hexopyranose system effectively prevents attainment of the 1C(D) conformation, although skew or boat conformations may be possible. Thus, a comparative study[282] of the n.m.r. spectra of 36 methyl 4,6-O-

(279) J. D. Stevens, *Chem. Commun.*, 1140 (1969), and unpublished results cited in Ref. 278a.
(280) H. Paulsen and G. Steinert, *Chem. Ber.*, **103**, 475 (1970).
(281) T. B. Grindley, J. F. Stoddart, and W. A. Szarek, *J. Chem. Soc. (B)*, 623 (1969).
(282) B. Coxon, *Tetrahedron*, **21**, 3481 (1965).

benzylidene-α-D-aldohexopyranosides showed that, in all examples, the pyranoid part exists exclusively in the $C1(D)$ conformation. Even the presence of a 1,2,3-triaxial system of substituents, as in the α-D-*altro* configuration, does not result in any conformational modifica-

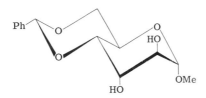

Favored conformation of methyl
4,6-*O*-benzylidene-α-D-altropyranoside

tion toward the skew or boat forms. The tendency of the pyranoid part to adopt the $C1(D)$ conformation in these systems was attributed to the anomeric effect of the methoxyl group and the greater thermo-dynamic stability of the chair form as compared to that of the skew or boat forms. That the $C1(D)$ conformation is the favored form of the pyranoid part of methyl 4,6-*O*-benzylidene-α-D-glycopyranosides and of other similar derivatives having the same configuration at C-4 is supported by the results of i.r.-spectral,[40] chemical,[283] and other n.m.r.-spectral studies.[284–287] The crystal-structure analysis[288] of methyl 4,6-*O*-benzylidene-2-*O*-(*p*-bromophenylsulfonyl)-3-cyano-3-deoxy-α-D-altropyranoside (**28**) showed that the pyranoid ring as-

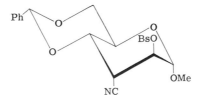

Bs = *p*-Bromophenylsulfonyl

(28)

(283) C. B. Barlow and R. D. Guthrie, *J. Chem. Soc.* (C), 1194 (1967).
(284) R. D. Guthrie and L. F. Johnson, *J. Chem. Soc.*, 4166 (1961).
(285) Y. Ali, A. C. Richardson, C. F. Gibbs, and L. Hough, *Carbohyd. Res.*, 7, 255 (1968).
(286) C. B. Barlow, E. O. Bishop, P. R. Carey, R. D. Guthrie, M. A. Jensen, and J. E. Lewis, *Tetrahedron*, 24, 4517 (1968); C. B. Barlow, E. O. Bishop, P. R. Carey, and R. D. Guthrie, *Carbohyd. Res.*, 9, 99 (1969).
(287) E. L. Albano, D. Horton, and J. H. Lauterbach, *Carbohyd. Res.*, 9, 149 (1969).
(288) B. E. Davison and A. T. McPhail, *J. Chem. Soc.* (B), 660 (1970).

sumes a slightly flattened chair-shape. The flattening is caused by a strong, 1,3-diaxial interaction, between the 1-methoxyl group and the 3-cyano group, that displaces them to a separation of 3.03 Å.

The formation of a 1,6-anhydro bridge in the aldohexopyranoses provides another example of conformational restriction. The results of n.m.r.-spectral[103,104,106,289] and optical-rotatory[117] studies strongly support the conclusion that all of the 1,6-anhydro-β-D-hexopyranoses and their triacetates adopt the $1C(D)$ conformation in solution, even for the D-*gluco* structure, which has all substituents axially attached. Conformational uniformity in the $1C(D)$ form for the 1,6-anhydro-β-D-aldohexopyranose structure was also indicated by the close agreement between empirically calculated and experimentally determined optical rotations for the 1,6-anhydro-deoxy-β-D-hexopyranoses and their diacetates and for the 2,7-anhydro-β-D-heptulopyranoses and their tetraacetates[118] (see p. 61). An X-ray crystal-structure analysis of 1,6-anhydro-β-D-glucopyranose[290] showed that the pyranoid part also exists in the all-axial, $1C(D)$ conformation in the crystalline state, but is somewhat distorted by the anhydro ring, which causes compression of C-1 toward C-5 and extension of C-2 away from C-4 (see Fig. 4). The two *syn*-axial oxygen atoms, on C-2 and C-4, are considerably farther apart (3.30 Å) than in an unstrained, pyranoid ring-system, for which the separation calculated is ~2.5 Å.

The crystal structure of a pyranoid ring-system containing a three-atom bridge, namely, N-(p-bromophenyl)-α-D-ribopyranosylamine 2,4-benzeneboronate (**29**), has been reported.[291] The presence of the

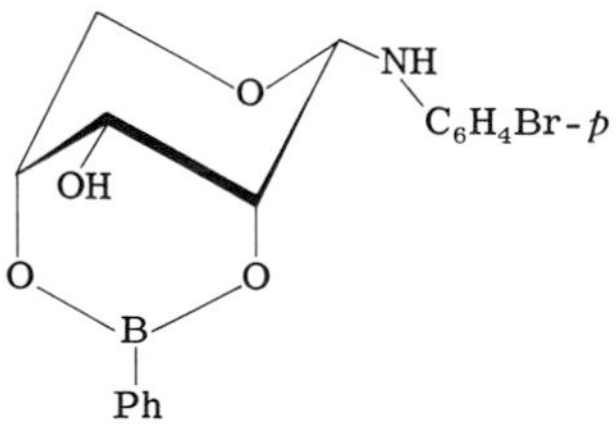

(29)

third atom in the bridge leaves the pyranoid ring undistorted, except for a slight compression of the ring angle at C-3.

The presence of a 3,6-anhydro ring in aldohexopyranose deriv-

(289) K. Heyns and J. Weyer, *Ann.*, **718**, 224 (1968).
(290) Y. J. Park, H. S. Kim, and G. A. Jeffrey, *Acta Crystallogr.*, **B27**, 220 (1971).
(291) H. Shimanouchi, N. Saito, and Y. Sasada, *Bull. Chem. Soc. Jap.*, **42**, 1239 (1969).

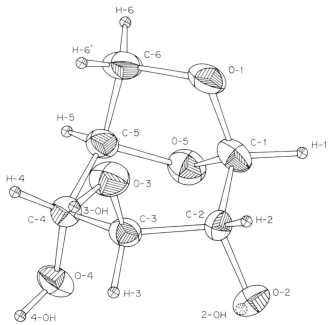

FIG. 4.—Conformation of 1,6-Anhydro-β-D-glucopyranose as Determined by X-Ray Crystallography.

atives also restricts the conformational behavior of the pyranoid part. Thus, the bicyclic part in 3,6-anhydro-α,α-trehalose hexaacetate appears, from the evidence of chemical shifts of the acetate groups, to adopt the $1C(\text{D})$ conformation.[292] Analysis of spin couplings indicated that both of the pyranoid moieties in 3,6:3',6'-dianhydro-α,α-trehalose in solution adopt a slightly distorted $1C(\text{D})$ conformation.[293]

When a five-membered ring is *cis*-fused to a pyranoid ring, the chair form of the latter ring cannot remain completely undistorted. It was at one time assumed that the 1,3-dioxolane ring of a vicinal, cyclic acetal was planar, and would cause a pyranoid ring fused to it to adopt a boat conformation. However, n.m.r. studies have convincingly established that such acetal rings, *cis*-fused to furanoid rings[77,80] and pyranoid rings[294,295] are nonplanar, and that the angle

(292) E. R. Guilloux, F. Percheron, and J. Defaye, *Carbohyd. Res.*, **10**, 267 (1969).
(293) G. Birch, C. K. Lee, and A. C. Richardson, *Carbohyd. Res.*, **16**, 235 (1971).
(294) L. D. Hall, L. Hough, K. A. McLauchlan, and K. G. R. Pachler, *Chem. Ind.* (London), 1465 (1962).
(295) A. S. Perlin, *Can. J. Chem.*, **41**, 399 (1963).

between the bridgehead hydrogen atoms is about 40°. Simple diox-olane derivatives are likewise nonplanar.[79,296] The conformations that are possible for cyclic acetals of sugars have been discussed.[297]

There has been considerable interest in, and controversy over, the conformational influence that a *cis*-fused, vicinal, 5-membered ring may have on the chair form of cyclohexane derivatives[298,299] and py-ranoid sugar derivatives.[177,294,300-305] In the first such study,[294] it was concluded from n.m.r. studies that the pyranoid ring of (4-O-acetyl-2,3-O-isopropylidene-α-D-lyxopyranosyl)bis(ethylsulfonyl)methane (30) is distorted only slightly by the *cis*-fused acetal ring. Studies on

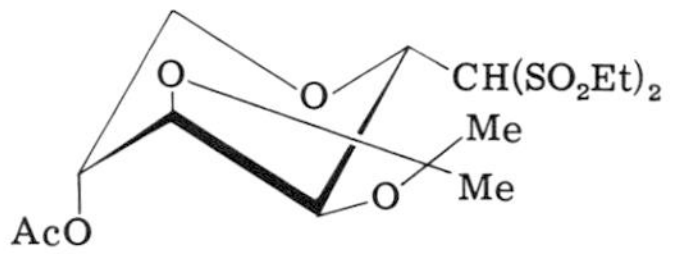

(30)

hydrogen bonding indicated[298] that *cis*-fusion of one isopropylidene acetal ring also causes only slight distortion of the chair conformation of an inositol. In contrast to these observations, the spin-coupling data measured for various 1,2-O-alkylidene aldopyranoid derivatives were interpreted[300] in terms of a skew conformation, and it was sug-gested[299] that a cyclohexane ring is considerably distorted by a *cis*-fused 2,2-dimethyldioxolane ring. On the other hand, a crystal-structure analysis[302] of 1,2-O-(1-aminoethylidene)-α-D-glucopyranose (31), previously assigned a skew conformation from n.m.r. data,[300]

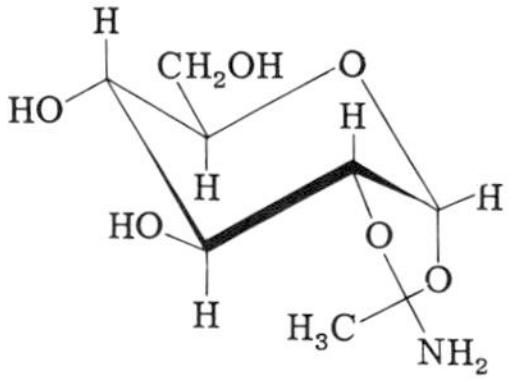

(31)

(296) F. A. L. Anet, *J. Amer. Chem. Soc.*, **84**, 747 (1962).
(297) R. S. Tipson, H. S. Isbell, and J. E. Stewart, *J. Res. Nat. Bur. Stand.*, **62**, 257 (1959).
(298) S. J. Angyal and R. M. Hoskinson, *J. Chem. Soc.*, 2991 (1962).
(299) R. U. Lemieux and J. W. Lown, *Can. J. Chem.*, **42**, 893 (1964).
(300) B. Coxon and L. D. Hall, *Tetrahedron*, **20**, 1685 (1964).
(301) R. U. Lemieux and A. R. Morgan, *Can. J. Chem.*, **43**, 2199 (1965).
(302) J. Trotter and J. K. Fawcett, *Acta Crystallogr.*, **21**, 366 (1966).
(303) R. G. Rees, A. R. Tatchell, and R. D. Wells, *J. Chem. Soc.* (C), 1768 (1967).
(304) R. U. Lemieux and D. H. Detert, *Can. J. Chem.*, **46**, 1039 (1968).
(305) T. Maeda, K. Tori, S. Satoh, and K. Tokuyama, *Bull. Chem. Soc. Jap.*, **42**, 2635 (1969).

indicated that it actually exists in a distorted, flattened $C1(D)$ conformation, and it was pointed out[302] that the n.m.r. data observed[300] are compatible with the conformation determined for the solid state. 3,4,6-Tri-O-acetyl-1,2-O-isopropylidene-α-D-glucopyranose (32) has been shown to have a distorted, flattened $C1(D)$ conformation, and 1,2:4,5-di-O-isopropylidene-β-D-fructopyranose has been demonstrated[306] to exist, in the crystalline state, in a conformation intermediate between $1C(D)$ (33) and $H_0^2(D)$.

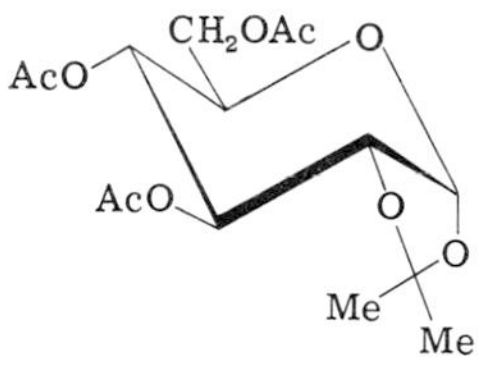

(32) 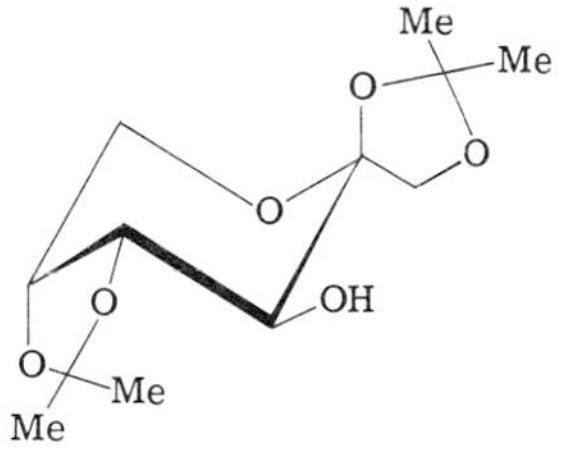   (33)

The precise conformational tendency of pyranoid sugar derivatives having a single, *cis*-fused acetal ring remains unresolved. Various assignments of skew conformations,[300,307] made on the basis of spin couplings, may need further interpretation that takes into account, for the determination of skew conformations of carbohydrates, the model parameters obtained[102] from an analysis, by computer methods, of the n.m.r. spectrum of a carbohydrate molecule (10) (see p. 80) that is rigidly locked in a skew conformation.

Although there is some uncertainty as to the conformational effect of one *cis*-fused, 5-membered acetal ring, evidence from both i.r.- and n.m.r.-spectral studies clearly indicates that the presence of two such *cis*-fused rings forces the cyclohexane or pyranoid ring to adopt a skew conformation. Thus, the pyranoid ring in derivatives of 1,2:3,4-di-O-isopropylidene-β-L-arabinose and -α-D-galactopyranose[177] (34), and of 2,3:4,5-di-O-isopropylidene-β-D-fructopyranose[305] (35), and the

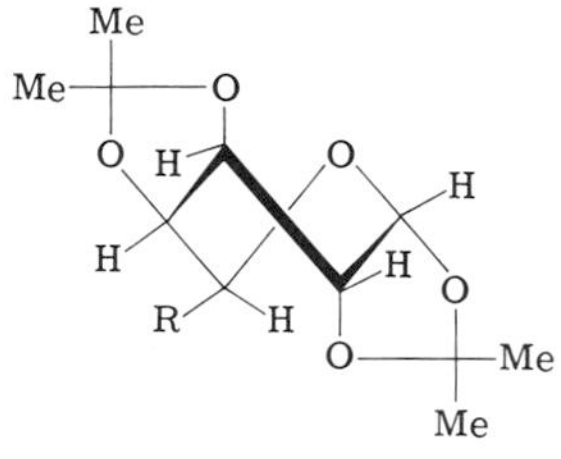

(34) 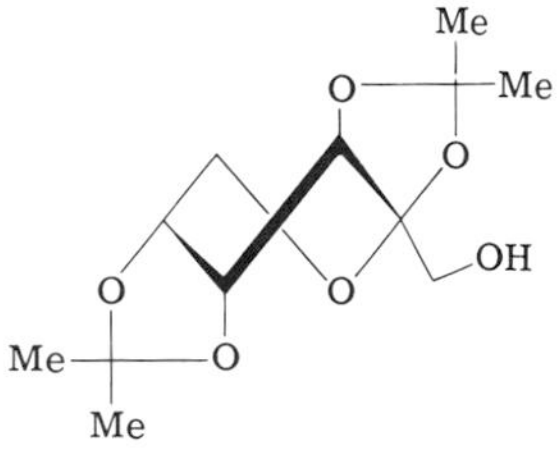   (35)

(306) S. Takagi and R. D. Rosenstein, *Inter. Union Crystallogr., Abstr.*, XV–61 (1969).
(307) A. E. El-Ashmawy and D. Horton, *Carbohyd. Res.*, 3, 191 (1966).

cyclohexane ring in 1,2:3,4-di-O-isopropylidene-5,6-dithio-*neo*-inositol,[308] and di-O-isopropylidene-*muco*-, -*epi*-, and -*cis*-inositols,[298] adopt non-chair shapes close to the skew form.

In a later study, the conformations of aldohexopyranoid derivatives containing both 1,2- and 4,6-cyclic acetal substituents were investigated.[309] From a computer analysis of the coupling constants, measured for a series of 1,2:4,6-di-O-benzylidene-α-D-glucopyranoses (**36**), it

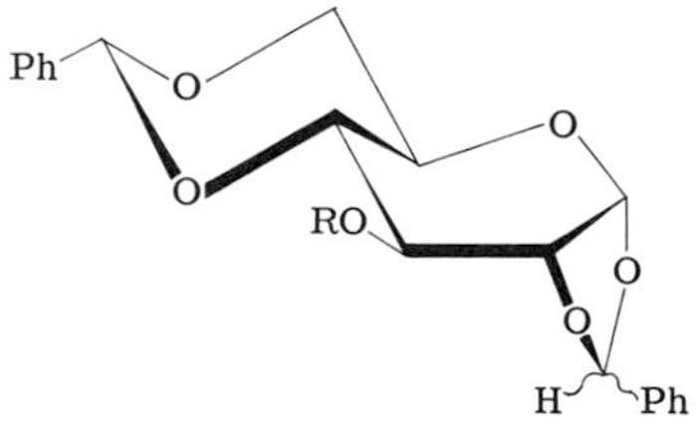

(36)

was concluded that these compounds adopt conformations in which the *m*-dioxane and pyranoid rings have chair and flattened-chair forms, respectively. The smaller conformational distortion of the pyranoid ring in the di- than in the mono-benzylidene acetals[300,303,304] was attributed to the tendency of the 4,6-O-benzylidene substituent to increase the rigidity of the pyranoid ring.

2. Furanoid Sugar Derivatives Having Fused Rings

A vicinal acetal ring *cis*-fused to a furanoid ring forces the furanoid part to adopt a twist conformation. This conformational influence was first demonstrated from a study of the conformations of various derivatives of 1,2-O-isopropylidene-α-D-gluco- and -α-L-idofuranose[77] and 1,2-O-isopropylidene-α-D-xylofuranose.[80] It was considered that the furanoid part adopts a twist conformation in which C-2 is below and C-3 above the plane defined by the other atoms, and that the 3,6-anhydro ring in 3,6-anhydro-1,2-O-isopropylidene-α-D-glucofuranose[80] exists in an envelope form having C-4 above the plane defined by the other four ring-atoms. Similar results were obtained from a study of the conformations of 3,6-anhydro-β-L-idofuranose derivatives.[310] Introduction of a 4,6- or 2,3-O-isopropylidene group in a series of L-sorbofuranose derivatives appears to lock the furanoid

(308) G. E. McCasland, S. Furuta, A. Furst, L. F. Johnson, and J. N. Shoolery, *J. Org. Chem.*, **28**, 456 (1963).
(309) B. Coxon, *Carbohyd. Res.*, **14**, 9 (1970).
(310) L. D. Hall and P. R. Steiner, *Can. J. Chem.*, **48**, 2439 (1970).

part in a symmetric, twist conformation, as, for example, in 1,2:4,6-di-*O*-isopropylidene-α-L-sorbofuranose.[311]

In a study related to the original work on 1,2-*O*-isopropylidenealdo-furanoses,[77,80] the conformations of a number of 1,2:3,5-di-*O*-benzylidene-α-D-glucofuranose derivatives (**37**) in solution were investi-

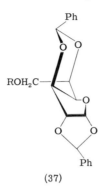

(37)

gated[312] by n.m.r. spectroscopy. It was found[312] that the furanoid part of the 1,2-monobenzylidene acetals has a symmetrical conformation approximately the same as that in the 1,2-*O*-isopropylidenealdofuranose derivatives. However, the spin-coupling data for the dibenzylidene acetals indicated that the conformation of the furanoid part is markedly different from that in the monobenzylidene acetals. Thus, fusion of the 3,5-*O*-benzylidene ring to the 1,2-*O*-benzylidene-α-D-glucofuranose system results in a distortion of the symmetric twist conformation toward that envelope conformation (V^3) in which C-3 is above the plane of the other ring atoms. The *m*-dioxane part of the dibenzylidene acetal adopts that chair conformation in which C-6 of the sugar is attached axially. In a subsequent n.m.r.-spectral study of the conformations of the diastereoisomeric 1,2-*O*-isopropylidene-3,5-*O*-(methoxymethylidene)-6-*O*-*p*-tolylsulfonyl-α-D-glucofuranoses, by using computer analysis,[313] it was observed that the furanoid ring of these diacetals is less distorted toward the envelope form than is the case for the dibenzylidene acetals.[312]

The preparation of several furanoid sugar derivatives having a fused oxetane ring has been reported,[314] as, for example, 3,5-anhydro-

(311) T. Maeda, K. Tori, S. Satoh, and K. Tokuyama, *Bull. Chem. Soc. Jap.*, **41**, 2495 (1968).
(312) B. Coxon, *Carbohyd. Res.*, **8**, 125 (1968).
(313) B. Coxon, *Carbohyd. Res.*, **12**, 313 (1970).
(314) J. G. Buchanan and E. M. Oakes, *Tetrahedron Lett.*, 2013 (1964); *Carbohyd. Res.*, **1**, 242 (1965); R. L. Whistler, T. J. Luttenegger, and R. M. Rowell, *J. Org. Chem.*, **33**, 396 (1968).

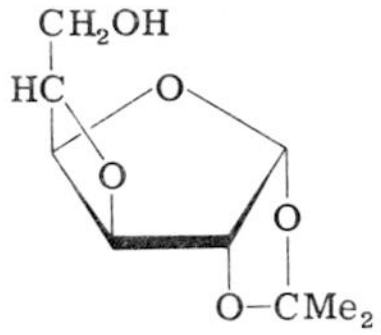

CH$_2$OH
HC
O
O
O
O—CMe$_2$

(38)

1,2-O-isopropylidene-β-L-idofuranose (**38**). Although the shape of the four-membered ring was not determined, it most likely has a non-planar shape allowing relief of torsional strain.

VIII. Conformations of Unsaturated, Cyclic Sugar Derivatives

When four of the atoms of the pyranoid ring are constrained to coplanarity, as by a double bond or an epoxide ring, two half-chair forms are possible, one of which will usually be favored. When five

$H\,^5_4$ $H\,^4_5$

Half-chair conformations

of the six atoms are held coplanar, as by an unsaturated lactone ring, the single exoplanar atom leads to a shape that has been termed a "sofa" conformation.[315]

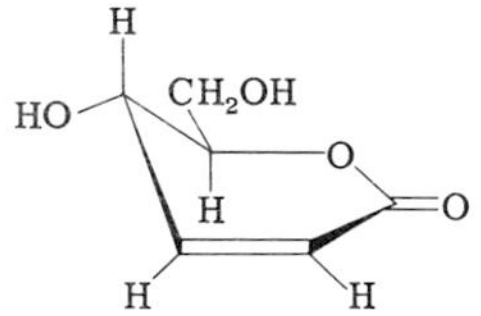

H
HO   CH$_2$OH
O
H   O
H   H

Sofa conformation
(for an unsaturated, pyranoid lactone)

The first conformational assignment for an unsaturated carbohydrate derivative was made[316] from an analysis of the n.m.r. spectrum of 3,4,6-tri-O-acetyl-1,5-anhydro-D-hex-1-enitol (tri-O-acetyl-D-glucal) in chloroform solution. It was shown that this 1,2-unsaturated, pyranoid

(315) E. M. Philbin and T. S. Wheeler, *Proc. Chem. Soc.*, 167 (1958).
(316) L. D. Hall and L. F. Johnson, *Tetrahedron*, **20**, 883 (1964).

sugar adopts that half-chair $[H^4_5(D)]$ conformation having C-4 above and C-5 below the plane defined by C-1, C-2, C-3, and O-5. The 2-acetoxy derivative of tri-O-acetyl-D-glucal was also shown to exist in the $H^4_5(D)$ conformation.[317]

From a more extensive n.m.r.-spectral study of various acylated, 2,3-unsaturated aldose derivatives, it was further observed that the anomeric effect is a dominant factor in determining the conformations of these derivatives.[318] Thus, tetra-O-acetyl-3-deoxy-α-D-*erythro*-hex-2-enopyranose (**39**) adopts the H^0_4 conformation, whereas the corresponding β-D anomer (**40**) exists in the alternative H^4_0 conformation

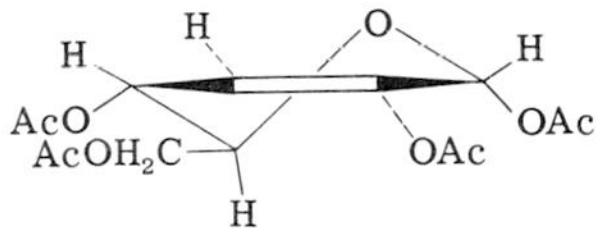

(39) 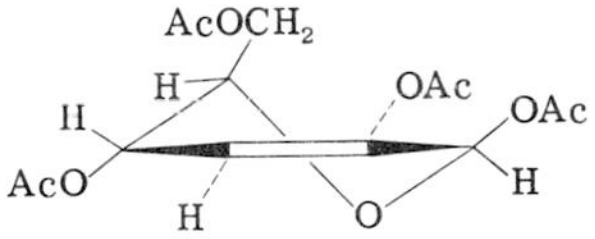   (40)

having the bonds to all substituents *quasi*-axial. This observation was confirmed[319] by n.m.r.-spectral measurements at 100 MHz.

Consideration of the anomeric equilibria of various acetylated, 2,3-unsaturated, pyranoid sugars[318] led to the conclusion that polar, allylic substituents favor the *quasi*-axial orientation (0.8 kcal.mole^{-1} for the acetoxyl group). Such an allylic effect was found to exert significant control over the conformational tendency and configurational equilibria of unsaturated, pyranoid sugar derivatives. Thus, tri-O-acetyl-2-hydroxy-D-xylal (**41**) adopts the $H^5_4(D)$ conformation. In

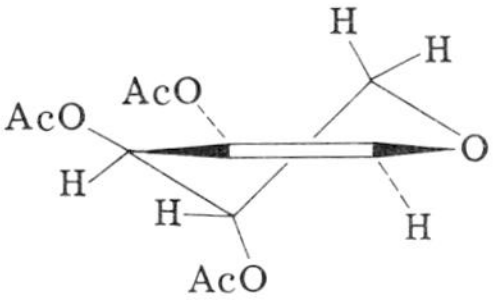

(41)

contrast to these observations, it was considered[320] that both anomers of methyl 3-deoxy-2,4,6-tri-O-methyl-D-*erythro*-hex-2-enopyranoside

(317) R. J. Ferrier, W. G. Overend, and G. H. Sankey, *J. Chem. Soc.*, 2830 (1965).
(318) R. J. Ferrier and G. H. Sankey, *J. Chem. Soc.* (C), 2345 (1966).
(319) R. U. Lemieux and R. J. Bose, *Can. J. Chem.*, 44, 1855 (1966).
(320) E. F. L. J. Anet, *Carbohyd. Res.*, 1, 348 (1966).

(**42**) exist in the H_4^0 conformation, indicating that the steric influence of the 5-(methoxymethyl) substituent, and not the operation of the

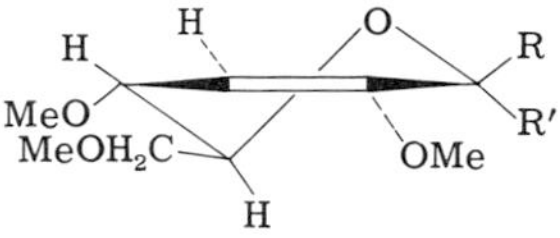

α: R = H, R′ = OMe
β: R′ = H, R′ = OMe

(42)

allylic and the anomeric effects, determines the conformations of these 2,3-unsaturated derivatives. However, in the absence of such a group at C-5, the operation of the anomeric effect appears to be the controlling factor in determining the conformation, as observed[321] for the methyl 3,4-dichloro-4-deoxy-D-*glycero*-pent-2-enopyranosides (**43**).

α

β

(43)

The n.m.r. spectral parameters of several 6-chloro-9-(2,3-dideoxy-glyc-2-enopyranosyl)purines in chloroform-*d* or benzene d_6 have been interpreted in terms of an equilibrium between the $H_5^0(D)$ and $H_0^5(D)$ conformations.[321a]

N.m.r. studies on pyranoid sugar derivatives containing an epoxide ring indicate[322,323] that the compounds adopt a half-chair conformation similar to that observed for unsaturated, pyranoid sugars. From an analysis of the spin-coupling data for some methyl 2,3- and 3,4-anhydroaldopyranosides,[323] it was found that the particular half-chair form adopted is determined by the anomeric effect and the tendency of the 5-(hydroxymethyl) or 5-methyl group to assume the equatorial

(321) B. Coxon, H. J. Jennings, and K. A. McLauchlan, *Tetrahedron*, **23**, 2395 (1967).
(321a) M. Fuertes, G. Garcia-Muñoz, R. Madroñero, M. Stud, and M. Rico, *Tetrahedron*, **26**, 4823 (1970).
(322) D. H. Buss, L. Hough, L. D. Hall, and J. F. Manville, *Tetrahedron*, **21**, 69 (1965).
(323) J. G. Buchanan, R. Fletcher, K. Parry, and W. A. Thomas, *J. Chem. Soc. (B)*, 377 (1969).

orientation. The only exception noted was methyl 3,4-anhydro-α-D-arabinopyranoside (**44**), which, in deuterium oxide solution, exists

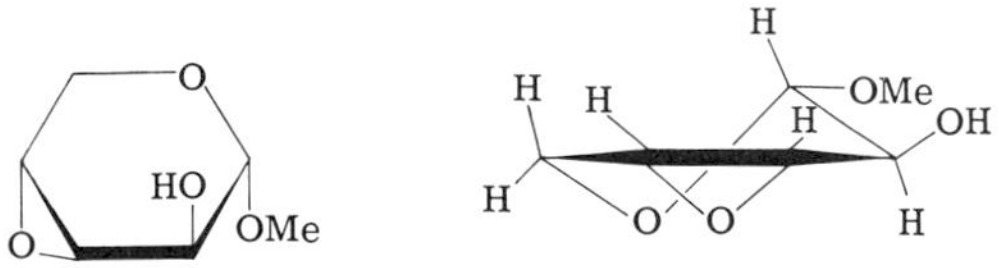

(44)

for the most part in that half-chair conformation having the anomeric substituent equatorially attached. This exception was attributed to a large electrostatic repulsion (estimated to be 1.7 kcal.mole^{-1}) between the axially attached methoxyl group and the epoxide-ring oxygen atom in the alternative half-chair conformation. The allylic effect (axial predisposition for the allylic group) noted for the unsaturated sugar analogs[318] did not appear to be important for these derivatives of anhydro sugars.

The anomeric effect was also found to play a significant role in determining the conformation of methyl 2,3-anhydro-α- and -β-L-ribopyranosides and the corresponding 2,3-unsaturated derivatives.[324] Both anomers adopt that half-chair conformation having the bond to the 1-methoxyl group *quasi*-axial.

(324) R. U. Lemieux, K. A. Watanabe, and A. A. Pavia, *Can. J. Chem.*, **47**, 4413 (1969).

CYCLIC ACYLOXONIUM IONS IN CARBOHYDRATE CHEMISTRY*

By Hans Paulsen

Institut für Organische Chemie und Biochemie, Universität Hamburg,
Bundesrepublik Deutschland

I. Introduction

1. Scope

Intramolecular substitution-reactions that take place from secondary sulfonates on sugar rings by so-called "neighboring-group

---

*Translated from the German by D. Horton.

reactions" constitute a valuable and important method for selectively introducing new substituents onto sugar rings. These reactions are frequently successful even when a direct, non-assisted SN2 reaction cannot be made to proceed. The driving force of the neighboring-group reaction is considered to result from the intermediate formation of a cyclic intermediate of largely cationic character. A detailed treatment of this type of reaction has been provided by Goodman[1] in an earlier Volume in this Series.

The neighboring-group reactions studied most intensively have been those of acyloxy groups, which, by the definition of Winstein,[2] constitute complex neighboring-groups. Here, the intermediate formation of a dioxolanylium ion is invoked. The formation of such a ring between C-1 and C-2 of aldoses determines, as shown by Tipson,[3] the stereochemical outcome of glycosylation reactions from glycosyl halides. This type of reaction, involving acyloxy neighboring-groups, has already been discussed by Lemieux.[4]

Neighboring-group reactions of acyloxy groups involve a synchronous process whereby the nucleophilic oxygen atom of the carbonyl group becomes attached at the same time as the nucleophilic leaving-group is split off. The intermediate dioxolanylium ion is exceedingly reactive. It is always a debatable point whether full charge-separation takes place before the attack of a nucleophile on the dioxolanylium system. It is probable that, at least in the majority of neighboring-group reactions involving acyloxy groups, the attack on the dioxolanylium ion by a nucleophile from the solvent medium takes place while the ion is still in the form[1,4] of an "intimate ion-pair."

Reactions are described in the article presented here whereby, in the absence of a nucleophile and in the presence of a difficultly polarizable, complex anion, the dioxolanylium and dioxanylium ring-systems can be generated selectively in carbohydrate derivatives. As intermolecular, nucleophilic attack is not possible in such cases, there is a resultant, intramolecular attack by substituents, within the sugar derivative, that react nucleophilically, at least so far as is permitted by the stereochemistry of the molecule. Many rearrangement reactions in a carbohydrate matrix occur in this way. Discussion of these rearrangement reactions constitutes the essential content of the present Chapter.

(1) L. Goodman, *Advan. Carbohyd. Chem.*, **22**, 109 (1967).
(2) S. Winstein, and R. Boschan, *J. Amer. Chem. Soc.*, **72**, 4669 (1950).
(3) R. S. Tipson, *J. Biol. Chem.*, **130**, 55 (1939).
(4) R. U. Lemieux, *Advan. Carbohyd. Chem.*, **9**, 1 (1954).

2. Preparation of Acyloxonium Salts

Meerwein[5-7] was the first to succeed in obtaining dioxolanylium ions of type **2**, sufficiently stabilized as salts with non-polarizable anions that they could be isolated crystalline. The compounds can be prepared by splitting out of an anion from cyclic ortho esters or acetals wherein the required ring-system is already present. The ortho ester **1** reacts[6] with antimony pentachloride or boron trifluoride, with splitting out of $\ominus OR$, to give **2**. Acetals (**3**) from aldehydes can be converted, by hydride abstraction with triphenylmethyl or triethyloxonium fluoroborate,[5,7] into salts (**2**); this reaction proceeds well only with acetals of the 1,3-dioxolane type (**3**) that have little steric hindrance. With acetals of the 1,3-dioxane type, formed from aldehydes, the reaction of hydride abstraction is not, as a rule, possible.[8] In all such reactions, the anion involved is either $SbCl_6^\ominus$ or $BF_4^\ominus$.

In the following discussion, the dioxolanylium (**2**) and dioxanylium ring-systems will be referred to simply as "acyloxonium derivatives," as the carboxonium ion of the ring is formally derived from a carboxylic acid. Thus,[8] system **2** is an acetoxonium salt when $R = Me$, and a benzoxonium salt when $R = Ph$.

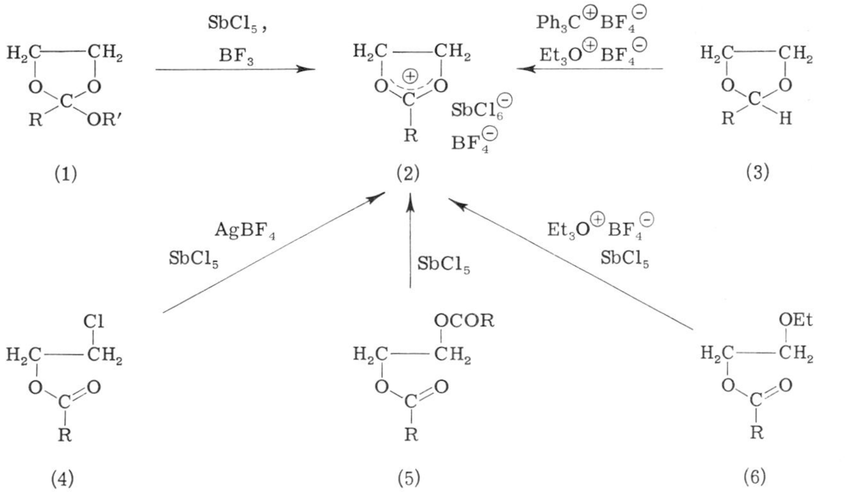

(5) H. Meerwein, V. Hederich, and K. Wunderlich, *Arch. Pharm.* (Weinheim), **291**, 541 (1958).

(6) H. Meerwein, K. Bodenbenner, P. Borner, F. Kunert, and K. Wunderlich, *Ann. Chem.*, **632**, 38 (1960).

(7) H. Meerwein, V. Hederich, H. Morschel, and K. Wunderlich, *Ann. Chem.*, **635**, 1 (1960).

(8) H. Paulsen, H. Behre, and C.-P. Herold, *Fortschr. Chem. Forsch.*, **14**, 472 (1970).

Acetals formed from ketones and 1,2-diols are of major significance in carbohydrate chemistry, and can be converted into acyloxonium salts only in special cases. Acetals (7) from acetophenone
react[9] with the highly reactive ketonium salts (8) to give benzoxonium

(7)   (8)   (9)   (10)

salts (9), with transfer of the methyl group to give 10. Acetals from
acetone and 1,2-diols react with antimony pentachloride and boron
trifluoride to give acetoxonium salts, but only in poor yield; at least
50% of the compound is converted into undesired, alkylated product,
which reacts further in as yet unpredictable ways.[9]

The method used most frequently for preparation of acyloxonium
salts is the neighboring-group reaction. The esterified chlorohydrin
(4) reacts[5,6] with antimony pentachloride (or, better, with silver
fluoroborate) to give the salt 2. The fluoro or bromo analogs react
similarly with boron trifluoride or silver fluoroborate. Even the
ether 6 can be transformed into 2 by use of antimony pentachloride
or triethyloxonium fluoroborate.[6]

The simplest approach to 2 is by reaction of the readily accessible
diol ester (5) with antimony pentachloride.[10,11] Ester groups often
give adducts with antimony pentachloride. The acyloxy group is
strongly complexed in these adducts, and it can thus be readily detached as its anion by a neighboring-group process. In consequence,
in the reaction with antimony pentachloride, the acetate anion is,
for example, a better leaving-group than the chloride anion. Thus,
1,2-diacetoxy-3-chloropropane (11a) reacts to give exclusively the
chlorine-containing, acetoxonium salt (12a), and not the desired
acetoxonium salt (13a) of glycerol.[11] On treatment with antimony
pentachloride, the corresponding benzoyl derivative (11b) gives[11]
a 3:1 mixture of 12b and 13b; this result indicates that, under the
conditions used, the benzoate ion is a poorer leaving-group than the
acetate ion, although it still is a better one than the chloride ion.

(9) S. Kabusz, Angew. Chem., 80, 81 (1968); Angew. Chem. Intern. Ed. Engl., 7, 64
    (1968).
(10) H. Paulsen and H. Behre, Angew. Chem., 81, 905 (1969); Angew. Chem. Intern.
    Ed. Engl., 8, 886 (1969).
(11) H. Paulsen, and H. Behre, Chem. Ber., 104, 1264 (1971).

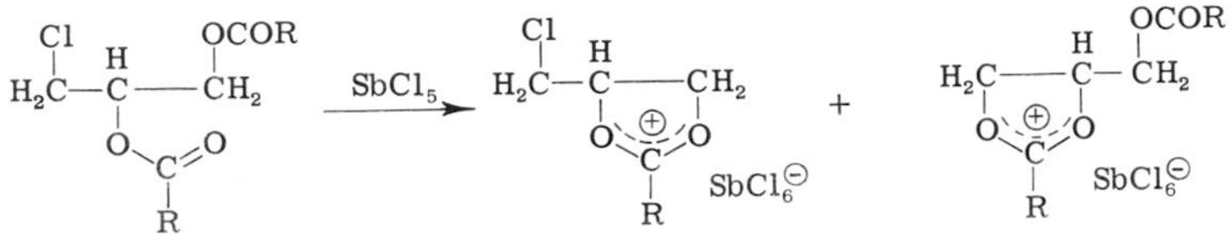

(11)

(12)

(13)

(a) R = Me
(b) R = Ph

By use of the reaction 5 → 2, the reactivities of different acyloxy groups were compared.[8,11] The esters (5) having acetyl (R = Me), propionyl (R = Et), and pivaloyl (R = CMe$_3$) groups react the most rapidly (5 min at 20° in dichloromethane) to give the salts 2. Aroyl analogs (5) having R = Ph, p-McC$_6$H$_4$, or p-MeOC$_6$H$_4$ react much more slowly. All methods described for preparing the 5-membered, cyclic acyloxonium salts (2) lend themselves likewise to preparation of six-membered-ring, acyloxonium salts, if the corresponding 1,3-propanediol derivatives are employed as the starting materials. However, six-membered-ring, acyloxonium salts, being generally formed less readily, require the use of longer reaction-times or higher temperatures for their formation.[8,11] Attempts to prepare seven-membered-ring, acyloxonium derivatives have so far not been successful.

Evidence that the reaction of the diol ester 5 with antimony pentachloride genuinely involves a neighboring-group reaction is provided by the stereospecificity of this reaction with esters of cyclic diols. trans-1,2-Cyclopentanediol diacetate (14) and trans-1,2-cyclohexanediol diacetate (17) react[8,11] with antimony pentachloride to give the cis-acetoxonium salts 15 and 18, whereas the cis-diol esters (16 and 19) merely give difficultly soluble adducts from which unchanged starting material can be recovered after hydrolytic treatment.

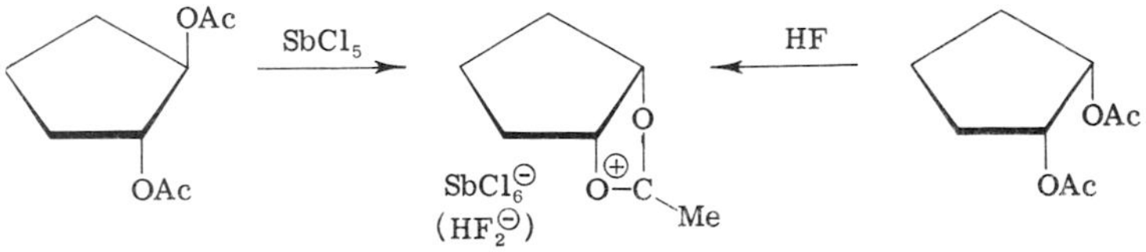

(14)

(15)

(16)

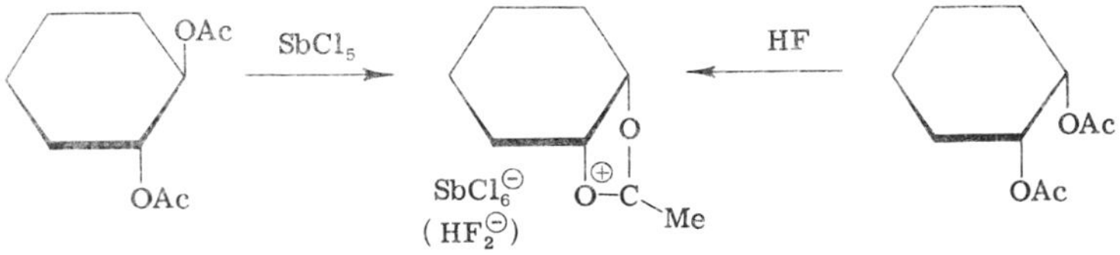

(17)

(18)

(19)

The cyclopentane derivative **14** clearly reacts more rapidly than the cyclohexane derivative **17**. In **14**, the acetoxyl groups can readily approach the orientation favored for rearside, nucleophilic attack, and the neighboring-group reaction is thus facilitated, whereas, in **17**, both acetoxyl groups are initially in a diequatorial orientation and must be brought into diaxial orientation to permit the neighboring-group reaction. *trans*-1,3-Diacetoxycyclopentane and *trans*-1,3-diacetoxycyclohexane both react with antimony pentachloride to give six-membered-ring, acetoxonium salts. Understandably, in this situation, the cyclohexane derivative reacts the faster, because of stereochemical considerations.[8,11]

A further method permits stereoselective conversion of *cis*-1,2-diol diacetates into acetoxonium derivatives; it involves reaction with liquid hydrogen fluoride.[12] Thus, *cis*-1,2-diacetoxycyclopentane (**16**) reacts rapidly in liquid hydrogen fluoride to give the ion **15**. *trans*-1,2-Diacetoxycyclopentane (**14**), on the other hand, is converted[11] only very slowly (72 hours) into **15**. Similarly, *cis*-1,2-diacetoxycyclohexane (**19**) is rapidly converted into the ion **18**, whereas the corresponding *trans* derivative does not react.[12] Such acyclic derivatives as **5** react relatively slowly, to give the corresponding ion (**2**). Pinacol, in acetic anhydride and in the presence of 70% perchloric acid, can be converted into the acetoxonium perchlorate.[13] This reaction is, presumably, comparable to the reaction with hydrogen fluoride.

In the reaction with hydrogen fluoride, one molecule of acetic acid is split out from one molecule of the diacetyl derivative; the exact mechanism and the nature of the anion involved have not yet been clarified. Possibly, the HF_2^- ion functions as the anion. The reaction can readily be followed by nuclear magnetic resonance (n.m.r.) spectroscopy, and, if the reaction can be performed in the same Teflon tube used for the n.m.r. measurements,[12] hydrogen fluoride can serve as the solvent medium. Excellent n.m.r. spectra of the acyloxonium derivatives are thereby obtained, although isolation of these derivatives as salts from the hydrogen fluoride involves difficulties. In the simplest examples, such as the ions **15** and **18**, the corresponding fluoroborate salts can be obtained by introduction of boron trifluoride into the hydrogen fluoride.[11,12]

These acyloxonium salts show interesting n.m.r. spectra. Because of the positive charge, the signals of protons on the dioxolanylium ring are shifted to low field with respect to signals of protons on the side

(12) C. Pedersen, *Tetrahedron Lett.*, 511 (1967).
(13) G. N. Dorofeenko and L. V. Mesheritskaya, *Zh. Obshch. Khim.*, **38**, 1192 (1968).

chains. The magnitude of this downfield shift is, therefore, a measure of how this positive charge is distributed on the ring. Several studies of the extent of this shift, in variously substituted acyloxonium salts, have been reported.[14-17]

3. Reactions of Acyloxonium Ions

Acyloxonium ions are ambident cations; that is to say, they can react with anions in two ways.[18] The *cis* pathway (**21** → **22**) leads to the product of kinetic control, and is the pathway generally adopted if there is a large gain in energy by bond formation to give **22**. If, however, an equilibrium between **22** and **21** is achieved, the thermodynamically favored, *trans* product (**20**) is formed.[18]

Nucleophilic agents that favor reaction by the *cis* pathway are NaOR (giving orthoesters),[6,17,19] sodium cyanide (giving orthocyanides),[6] and sodium borohydride (giving acetals).[20,21] The *cis* pathway is also adopted on hydrolysis with water. The partially esterified ortho acid thereby formed is unstable, and splits spontaneously to give[15] the acyclic, *cis*-diol derivative **23**. In the reaction by the *cis*

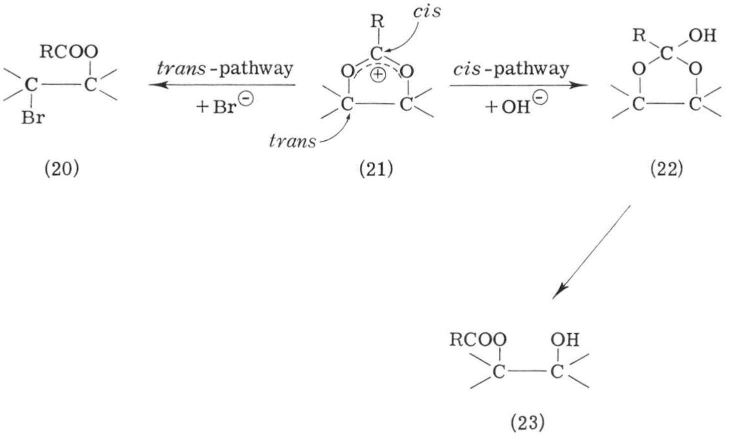

(14) H. Hart and D. A. Tomalia, *Tetrahedron Lett.*, 3383, 3389 (1966).
(15) C. B. Anderson, E. C. Friedrich, and S. Winstein, *Tetrahedron Lett.*, 2037 (1963).
(16) H. Hart and D. A. Tomalia, *Tetrahedron Lett.*, 1347 (1967).
(17) M. Beringer and S. A. Galton, *J. Org. Chem.*, **32**, 2630 (1967).
(18) S. Hünig, *Angew. Chem.*, **76**, 400 (1964); *Angew. Chem. Intern. Ed. Engl.*, **3**, 548 (1964).
(19) S. Kabusz, *Angew. Chem.*, **78**, 940 (1966); *Angew. Chem. Intern. Ed. Engl.*, **5**, 896 (1966).
(20) J. G. Buchanan and A. R. Edgar, *Chem. Commun.*, 29 (1967).
(21) H. Paulsen and C.-P. Herold, *Chem. Ber.*, **103**, 2450 (1970).

pathway, the dioxolane structure is retained, and therefore the
cis-1,2-diol arrangement is obtained.

Nucleophilic agents that favor reaction by the *trans* pathway are
potassium acetate in acetic acid, which leads to formation of *trans*-
1,2-diol diacetates,[15] and lithium chloride and tetrabutylammonium
bromide, which afford[6,15] the corresponding *trans*-1,2-chlorohydrin
or bromohydrin acetates (**20**). In reactions by the *trans* pathway,
the dioxolanylium ring is opened by rearside attack by the nucleophile, so that inversion of configuration at the reaction center takes
place; consequently, a *trans* product is obtained from a *cis*-diol
precursor.

The direction of opening of the dioxolanylium ring in unsym-
metrical molecules is determined by the nature of the other groups
present. With acyloxonium salts (**25**) of diols constrained by a conformationally fixed decalin system, both reaction-pathways proceed
stereoselectively.[22–24] By the *cis* pathway (**25** → **26**) that is involved

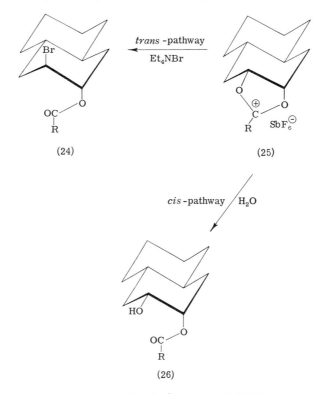

(22) J. F. King and A. D. Allbutt, *Tetrahedron Lett.*, 49 (1967).
(23) J. F. King and A. D. Allbutt, *Can. J. Chem.*, **47**, 1445 (1969).
(24) J. F. King and A. D. Allbutt, *Can. J. Chem.*, **48**, 1754 (1970).

on hydrolysis with water, the preponderant product obtained is that having the axial hydroxyl group acetylated and the equatorial one un-substituted.[24] With reactions by the *trans* pathway (**25** → **24**), the nucleophile shows the expected, favored attack on the equatorially substituted carbon atom of **25**, and this leads[23] to the *trans*-diaxially substituted product **24**.

Under solvolytic conditions, intramolecular, SN2 reactions of secondary sulfonates on sugar rings normally proceed with neigh-boring-group assistance. The acyloxonium intermediates react at once by the *cis* pathway with a nucleophile (generally water) from the solvent medium, to give *cis*-diol derivatives.[1] A neighboring-group reaction *via* a 1,2-acyloxonium ion of a cyclic aldose derivative may, however, proceed by either of the reaction pathways according to the conditions of the reaction. Thus, tetra-O-acetyl-β-D-glucopy-ranosyl bromide reacts, *via* an acetoxonium intermediate, with (*a*) alcohols in pyridine by the *cis* pathway, giving a 1,2-ortho ester, and (*b*) alcohols and silver carbonate by the *trans* pathway, giving β-D-glucosides.[4,25] The attacking nucleophile in the latter process enters at C-1. In the absence of a reactive, external nucleophile, the *trans* pathway can also occur intramolecularly if a sufficiently nu-cleophilic, vicinal substituent is present in favorable stereochemical disposition. These intramolecular reactions, likewise, take place with configurational inversion.

II. REARRANGEMENT OF ACYLOXONIUM IONS IN POLYOL SYSTEMS

1. Rearrangement of Acyloxonium Ions in 1,2,3-Triols

Esters of acyclic 1,2,3-triols, and esters of cyclic (1,3/2)-triols react, *via* a neighboring-group reaction, with antimony pentachloride to give acyloxonium salts in good yield. Thus, glycerol triesters (**27**)

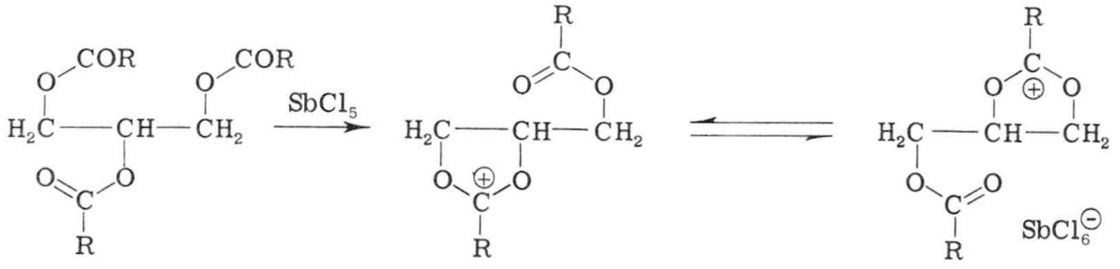

(27)   (28)   (29)

(a) R = Me
(b) R = CMe$_3$

(25) E. Pacsu, *Advan. Carbohyd. Chem.*, **1**, 77 (1945).

give the salts **28,** the rate of the reaction being dependent on the nature of the acyl group; for example, acetates and pivalates react more rapidly than benzoates.[8,11]

Acyloxonium ions from glycerol, of type **28,** are of interest as they can be isomerized into the corresponding ions (**29**) that result from a neighboring-group reaction (intramolecular, nucleophilic ring-opening by the *trans* pathway, see Section I,3) through attack of the carbonyl oxygen-atom of the vicinal acyl group. Because these two ions are structurally identical, this process may be considered to be a valence isomerism of the free acyloxonium cation.

It is possible to detect such rearrangements by n.m.r. spectroscopy. Thus, the n.m.r. spectrum of the acetoxonium salt **28a** at room temperature shows an acetoxonium-methyl signal (CH_3 group on the dioxolanylium ring) that resonates at low field because of the effect of the positive charge, together with a normal, acetyl-methyl group signal. The reaction **28a** $\rightleftarrows$ **29a** leads to interconversion of the acetoxonium methyl group and the normal methyl group. When this process occurs rapidly (on the n.m.r. time-scale), both methyl-group signals merge into a single, time-averaged signal. A solution of **28a** in nitromethane-d_3 shows separate methyl-group signals at room temperature, but a single, coalesced signal at 105°, indicating that the interconversion **28a** $\rightleftarrows$ **29a** takes place rapidly at the higher temperature, and proving that valence isomerism occurs in the acyloxonium cation of glycerol.[10,26]

As acyloxonium derivatives of polyols, for example, **28,** are relatively sensitive, decomposition reactions tend to occur during high-temperature n.m.r. studies, and it is desirable to verify that the coalescence of peaks observed is reversed when the temperature is lowered. For the coalescence temperature to be low, the free energy of activation of the rearrangement should be low, and the frequency separation of the signals that coalesce should be small, because, by the Eyring equation, a relationship exists between the coalescence temperature and this frequency separation.[27]

These conditions are singularly well fulfilled for the pivaloxonium derivative **28b.** The pivaloyl group is, as already indicated, at least as effective a neighboring group as the acetyl group. The singlet signal for the *tert*-butyl group on the dioxolanylium ring of **28b** is not shifted as far to low field as are those of the methyl-group protons in the methyl analog, because the methyl groups of the *tert*-butyl group

(26) H. Paulsen and H. Behre, *Chem. Ber.,* **104,** 1281 (1971).
(27) M. T. Rogers and J. C. Woodbrey, *J. Phys. Chem.,* **66,** 540 (1962).

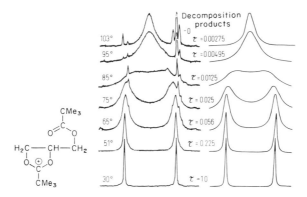

FIG. 1.—Temperature Dependence of the N.M.R. Spectrum of **28b** ⇄ **29b** (*tert*-Butyl Group Signals Only). [The experimental spectra are given on the left and the calculated ones on the right. Line-shape calculations compared with the observed spectra give the following energy parameters for the rearrangement process in nitromethane: E_a 20.1 ±0.3 kcal.mol^{-1}, ΔG^+ 18.55 ±0.06 kcal.mol^{-1}, and ΔS^+ 3.2 ±1.2 cal.deg.$^{-1}$]

are one carbon atom farther away from the positive charge, and thus, the frequency separation of the coalescing signals is considerably smaller; the signal-coalescence for the rearrangement **28b** ⇄ **29b** is observed[26] at a temperature as low as 87°. On the left-hand side of Fig. 1, a part of the n.m.r. spectrum of **28b** is given; it shows the appearance of the *tert*-butyl signals at various temperatures. On the right is given the theoretically computed spectrum.[26] As the observed coalescence indicates, the interconversion **28b** ⇄ **29b** proceeds rapidly in relation to the n.m.r. time-scale.

Acyloxonium derivatives of (1,2/3)-cyclopentanetriol can be expected to undergo an analogous equilibration reaction. The (1,3/2)-triol ester (**30**) can thus, according to which acyloxy group is split off with antimony pentachloride, react by two pathways, to give the acyloxonium ion **31** or **33**. The acetyl derivative **30a** reacts exclusively to give **31a**, isolated in 70% yield.[10,26] In contrast, the pivaloyl derivative **30b** gives[10,26] a 43:57 mixture of **31b** and **33b**; the latter can be separated by fractional recrystallization, and can be used for preparation of (1,2,3/0)-cyclopentanetriol.[26] That rapid equilibrium exists between the ions **31a** and **31b** and the respective isomers **32a** and **32b** is indicated (as for the equilibrium **28** ⇄ **29**) by a coalescence of the appropriate signals in the n.m.r. spectra measured at elevated temperatures. This rearrangement is not possible with the all-*cis* ion **33b**, and the n.m.r. spectrum remains unaltered at elevated temperatures.[26]

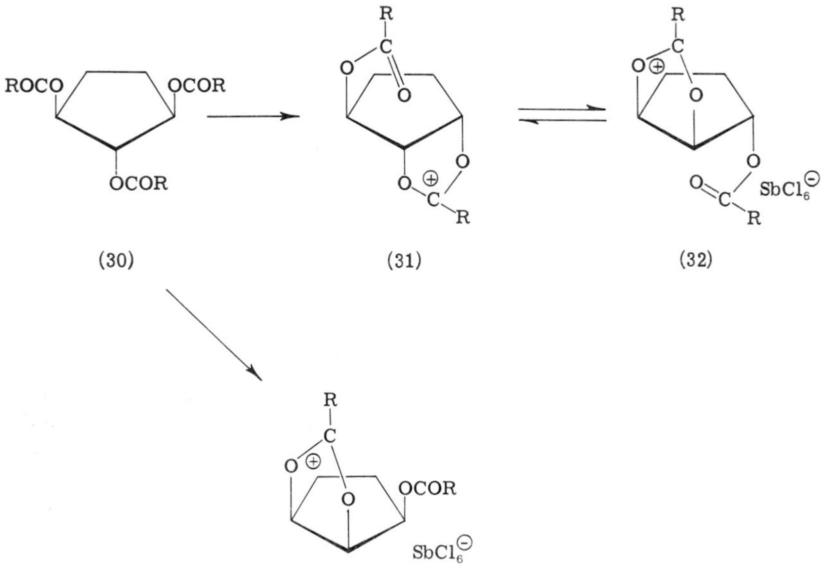

(30)   (31)   (32)

(33)

(a) R = Me
(b) R = CMe$_3$

The formation of acyloxonium derivatives (**35**) of (1,2/3)-cyclo-hexanetriol from tri-O-acetyl-(1,3/2)-cyclohexanetriol and antimony pentachloride does not proceed[26] in a manner analogous to the pathway **30** → **31**. In this triol system, in its favored conformation, all three substituents are equatorially disposed. The neighboring-group reaction leading to splitting out of an acetate anion requires, however, a diaxial orientation of the substituents reacting. It is understandable that this reaction should proceed considerably more slowly, because of the expenditure of energy required for conversion of the triol into its all-axial conformation.

From the acetal **34**, the fluoroborate salt of the ion **35** can be obtained by use of triphenylmethyl fluoroborate to split out hydride

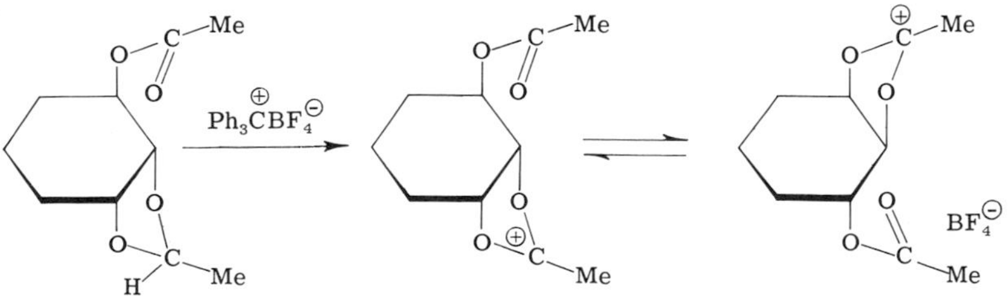

(34)   (35)   (36)

TABLE I

Free Energies of Activation ($\Delta G^{\ddagger}$) of the Acyloxonium Rearrangement in 1,2,3-Triols[26]

Compound (equilibrium)	Signal observed	T_c (K)	$\Delta \nu$ (Hz)	$\Delta G^{\ddagger}$ (kcal. mol^{-1})	Solvent medium
28a ⇄ 29a	Me	~378	53.5	18.7	CD$_3$NO$_2$
28b ⇄ 29b	CMe$_3$	360	21.8	18.4	CD$_3$NO$_2$
	CH$_2$	370	54.5	18.3	
31a ⇄ 32a	Me	365	54.0	18.0	CD$_3$NO$_2$
31b ⇄ 32b	CMe$_3$	348	22.3	17.8	CD$_3$NO$_2$
35 ⇄ 36	Me	~368	43.0	18.4	CD$_3$CN

ion.[26] The acetoxonium ion **35** shows, likewise, a rapid, reversible rearrangement **35** ⇄ **36**, but the thermal stability of this compound is so low that coalescence of the methyl-group signals cannot be observed at elevated temperature.

The free energy of activation ($\Delta G^{\ddagger}$) for the rearrangement of acyloxonium cations of different triols can be calculated, by use of the Eyring relationship, from the coalescence temperature (T_c) and frequency separation ($\Delta \nu$) measured by n.m.r. spectroscopy.[27] The values determined are given in Table I.

From Table I, it may be seen that the open-chain compounds (**28**) show the highest $\Delta G^{\ddagger}$ values. The values are 0.6–0.7 kcal.mol^{-1} lower for the cyclopentanetriol derivatives (**31**), presumably because of steric facilitation of the neighboring-group reaction. The facilitation is no longer so significant in the cyclohexanetriol system (**35**), and, consequently, the $\Delta G^{\ddagger}$ value is distinctly higher. The $\Delta G^{\ddagger}$ values for the pivaloxonium derivatives lie approximately 0.2–0.3 kcal.mol^{-1} lower than for the corresponding acetoxonium derivatives. Probably, the pivaloyl group is slightly the more effective in neighboring-group participation.[26]

2. Cyclic Acyloxonium Rearrangements in Cyclopentanepentols

From the results given in Table I, it may be seen that the acyloxonium ion rearrangement proceeds the most readily in the (1,2/3)-cyclopentanetriol system. This type of study has been extended to polyhydroxycyclopentane derivatives that are capable of such rearrangements. Both tetra-O-acetyl- and tetra-O-pivaloyl-(1,3/2,4)-cyclopentanetetrol[28,29] (**38a** or **38b**) react with antimony pentachloride

(28) A. Hasegawa and H. Z. Sable, *J. Org. Chem.*, **31**, 4161 (1966).
(29) H. Z. Sable, T. Anderson, B. Tolbert, and T. Posternak, *Helv. Chim. Acta*, **46**, 1157 (1963).

to give non-homogeneous acyloxonium salts.[30] The major product in the salts from **38a** is the hexachloroantimonate of the ion **39**. Similarly, the major product from **38b** is the analog of **39** having CMe_3 groups in place of the methyl groups.

The acetal **37** of (1,2,4/3)-cyclopentanetetrol[29] predictably reacts, by hydride abstraction, with triphenylmethyl fluoroborate to give[30] the fluoroborate salt of the ion **39**. The ion **39** can rearrange reversibly to give the ion **40**, and, subsequently, the ion **41**; this rearrangement,

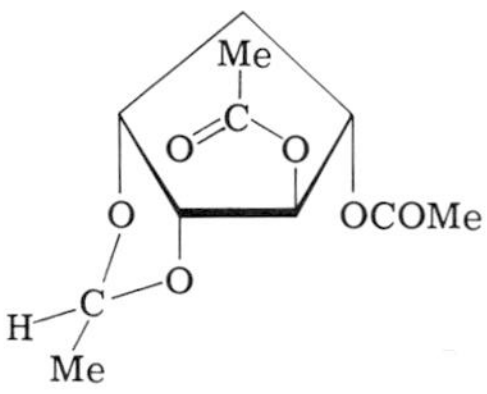

(37)

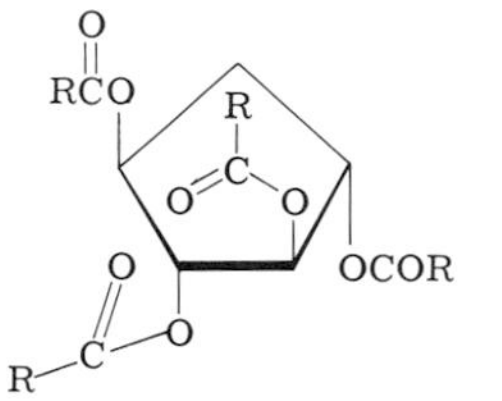

(a) R = Me
(b) R = CMe₃

(38)

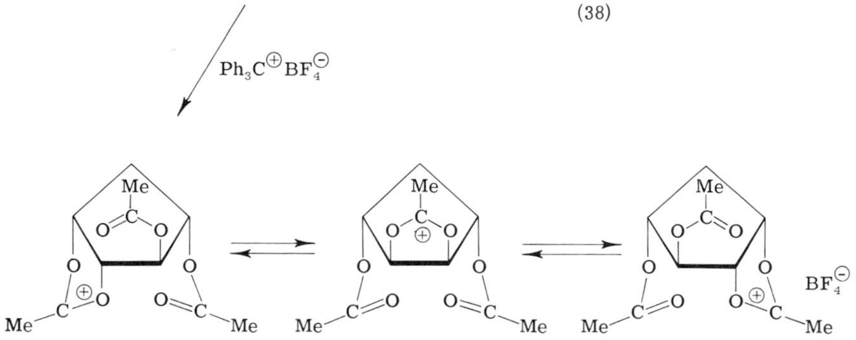

$Ph_3C^{\oplus}BF_4^{\ominus}$

(39)   (40)   (41)

also, can be followed by n.m.r. spectroscopy. At room temperature, separate signals can be observed for acetoxonium-methyl and acetyl-methyl protons, in the intensity ratio of 1:2. On heating, a line-broadening of these methyl-group signals is observed, although the temperature for coalescence of these signals cannot be determined, because decomposition reactions interfere.[30]

The (1,2,4/3,5)-cyclopentanepentol system is especially interesting. In liquid hydrogen fluoride, its pentaacetate readily gives rise to an acetoxonium derivative (**43**) having a well-resolved n.m.r. spectrum

(30) H. Paulsen and H. Behre, *Chem. Ber.*, **104**, 1299 (1941).

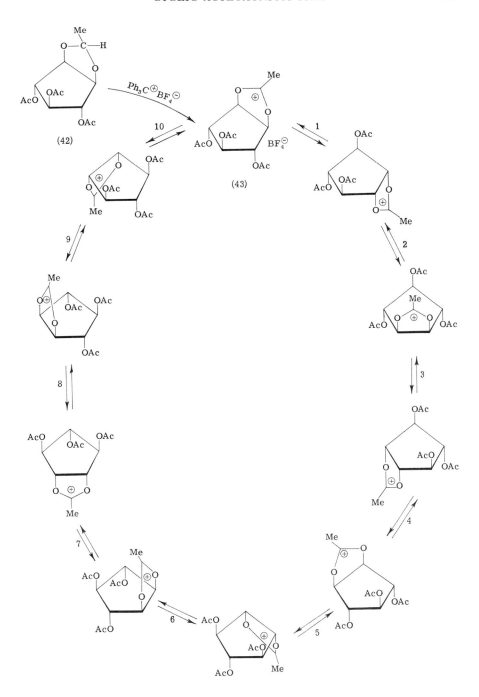

that is in good agreement with the structure assigned.[30] Obviously, only the cis-1,2-diacetoxy system of the pentaacetate reacts; however, attempts to isolate 43 as a salt from the solution were unsuccessful, and studies on the solution at elevated temperatures were unrewarding.[30]

A fluoroborate salt of the ion 43 can be obtained from the acetal 42 by treatment with triphenylmethyl fluoroborate.[30,31] As indicated in the cycle of reactions, the cation 43 can rearrange successively, by neighboring-group reactions, into subsequent ions; all of the ions are structurally identical with 43. The formula scheme indicates that, after each rearrangement step, the stereochemistry is so disposed that a new neighboring-group reaction can follow. The cycle is completed after ten steps, and the starting point is reached.

The actual occurrence of this cycle of rearrangements is supported by n.m.r. spectral evidence. The n.m.r. spectrum of 43 at room temperature shows the acetoxonium-methyl and acetyl-methyl signals in the ratio of 1:3. On heating, all of the methyl-group signals are greatly broadened, indicating that, as required by the cycle, all of the methyl groups are involved in the rearrangement reactions.[30,31] Because of limitations on the thermal stability, the temperature for full coalescence was not reached, but, by extrapolation, coalescence may be estimated as taking place at 85–90°; this indicates a ΔG^+ value of about 18 kcal.mol^{-1} for the cyclic rearrangement. This value is of the same order of magnitude as that observed for the rearrangement 31a $\rightleftarrows$ 32a of the acetoxonium cation from (1,2/3)-cyclopentanetriol.

It was further proved that, together with the rearrangement cycle, side reactions occur that involve 1,3-neighboring-group reactions and configurational inversion.[30] Such processes lead to other isomers of the cyclopentanepentol system. Analysis of an isomerized and hydrolyzed reaction-mixture showed that only 3% of (1,2,3/4,5)-cyclopentanepentol[32] and 0.5% of (1,2,3,4/5)-cyclopentanepentol[32] were formed; the first of these isomers can arise by a 1,3-neighboring-group reaction.

3. Cyclic Acyloxonium Rearrangements in Pentaerythritol

Various investigations on the preparation, from esters of 1,2- and 1,3-diols, of five- and six-membered acyloxonium derivatives have

(31) H. Paulsen and H. Behre, Angew. Chem., 81, 906 (1969); Angew. Chem. Intern. Ed. Engl., 8, 887 (1969).
(32) T. Posternak and G. Wolczunowicz, Naturwissenschaften, 55, 82 (1968).

shown that the six-membered are formed with considerably more difficulty than the five-membered derivatives (see Section I,2). A direct comparison of the stability of the 1,3-dioxolanylium and the 1,3-dioxanylium ring can be demonstrated by the reaction of 1,2,4-butanetriol tripivalate with antimony pentachloride, whereby either the five-membered (**44**) or the six-membered (**45**) ring might be

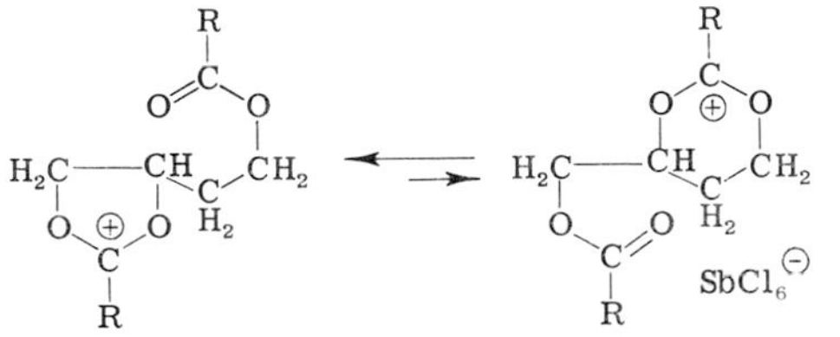

(44) ⇌ (45)

R = CMe$_3$

formed. A salt whose n.m.r. spectrum indicates the structure **44** is obtained, and no component corresponding to **45** can be detected.[33,34] The five-membered ring is thus considerably more stable and, although an equilibrium **44** ⇌ **45** probably exists, the proportion of **45** is so small as to be undetectable by current n.m.r. spectroscopy.

1,3,5-Pentanetriol tripivalate reacts[33,34] with antimony pentachloride to give the dioxanylium salt **46**. The cation of **46** can rearrange into the ion **47**, which remains structurally and energetically

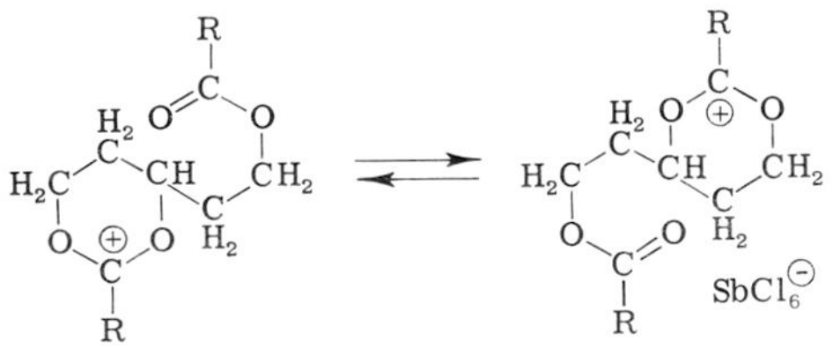

(46) ⇌ (47)

R = CMe$_3$

identical with **46**. The rapid equilibration **46** ⇌ **47** in the salt can be demonstrated by n.m.r. spectroscopy. At room temperature in nitromethane-d_3, there are observed two, separate *tert*-butyl signals ($\Delta\nu$ 20.2 Hz) which merge at 110°, indicating[33,34] a $\Delta G^{\pm}$ value of 19.3

(33) H. Paulsen, H. Meyborg, and H. Behre, *Angew. Chem.*, **81**, 907 (1969); *Angew. Chem. Intern. Ed. Engl.*, **8**, 888 (1969).
(34) H. Paulsen, H. Meyborg, and H. Behre, *Chem. Ber.*, in press.

kcal.mol⁻¹; this value is 0.9–1.0 kcal.mol⁻¹ higher than that for the acyloxonium ion rearrangement **28b** ⇌ **29b** for the pivaloxonium cation of glycerol.

(1,3/5)-Cyclohexanetriol tripivalate (**48**) reacts with antimony pentachloride to give, in 71% yield, the pivaloxonium salt **49** as the sole reaction-product.[35] The all-*cis* analog is not formed. The cation **49** has a structure for which is possible a cycle of rearrangements analogous to those occurring in the cyclopentanepentol series (see Section II, 2). In six successive steps, starting with **49**, the 1,3-dioxolanylium ring alternates around the cyclohexane ring. The results of n.m.r. spectral studies at elevated temperature indicate that this rearrangement does not proceed sufficiently rapidly to permit direct demonstration by this method; the n.m.r. spectrum of **49** appears to be independent of temperature.[35] Alternative proof for this rearrangement is, however, provided by isotope-labeling studies and by the Forsén-Hoffman double-resonance technique.[36]

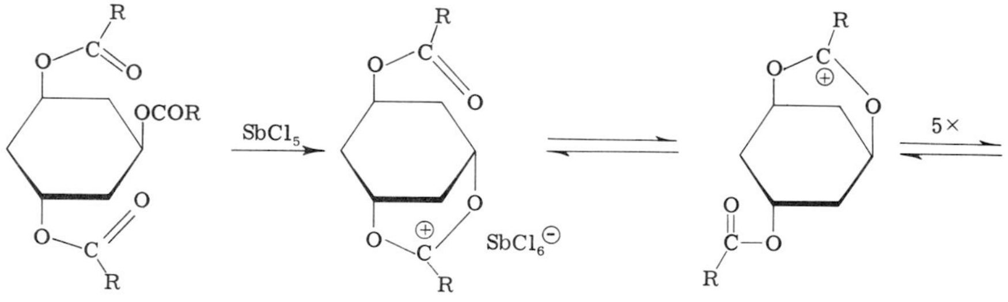

(48)   (49)   (50)

R = CMe₃

The greater difficulty of the rearrangement **49** ⇌ **50** is understandable on conformational grounds. N.m.r. spectroscopy shows that **49** exists in that chair conformation having the dioxanylium ring 1,3-diaxial and the 5-pivaloyloxy group equatorial on the cyclohexane ring.[35] For a neighboring-group reaction to occur to give **50**, the molecule must first assume a boat conformation (energetically less-favored).

A further example of a cyclic rearrangement involving a steady-state interconversion of dioxanylium rings has been satisfactorily established. Pentaerythritol tetrapivalate (**51**) is converted[33,34] by antimony pentachloride in dichloromethane at 40° into the mono-

(35) H. Paulsen, and H. Behre, unpublished results.
(36) S. Forsén and R. A. Hoffman, *J. Chem. Phys.*, **39**, 2892 (1963).

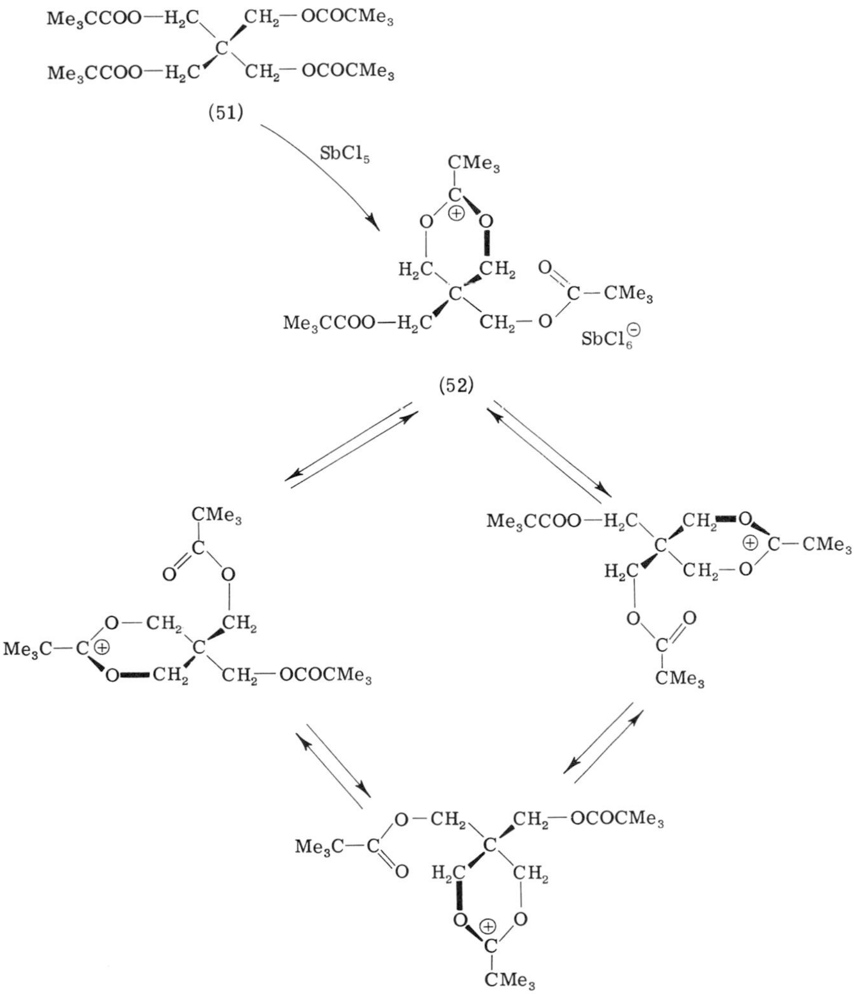

Me₃CCOO—H₂C, CH₂—OCOCMe₃

Me₃CCOO—H₂C, CH₂—OCOCMe₃

(51)

SbCl₅

(52)

pivaloxonium salt **52**. The dioxanylium ring in **52** can be opened by a neighboring acyl group to give a new ion structurally identical with **52**. This rearrangement can proceed further, as indicated in the cyclic scheme, and, after four successive steps, the starting point is reached.

The salt **52** is relatively stable thermally, and can be studied by variable-temperature, n.m.r. spectroscopy. At room temperature in nitromethane-d_3, the two expected *tert*-butyl signals ($\Delta\nu$ 20.2 Hz), in the intensity ratio of 1:2, are observed, and these signals coalesce at 110°. The methylene groups of the dioxanylium ring, and the normal methylene groups, give separate signals ($\Delta\nu$ 54.5 Hz) in the intensity ratio of 1:1 at room temperature, and these signals coalesce at 122°. These data establish the occurrence of the cyclic rearrangement of **52**, and a $\Delta G^{\neq}$ value of 19.3 kcal.mol^{-1} was calculated.[33,34] This value is comparable with that for the pivaloxonium rearrangement in the 1,3,5-pentanetriol system.

III. REARRANGEMENT OF ACYLOXONIUM IONS IN MONOSACCHARIDES

1. Rearrangement of D-Glucose into D-Idose

Esters of monosaccharides may be expected to undergo conversion into acyloxonium salts according to the general principles set forth in Section I,2. However, as they are polyfunctional compounds, monosaccharides can exhibit numerous side-reactions that lead to mixtures of products, together with significant decomposition-reactions manifested by strong darkening of the reaction mixtures. In consequence, it is not possible to obtain acetoxonium derivatives by treating methyl aldohexopyranoside tetraacetates or 1,6-anhydroaldohexose triacetates with antimony pentachloride in the manner already described, even when *trans*-diacetoxyl groups are present.[35]

Fortunately, the C-1 substituent in numerous aldose derivatives is an exceptionally good leaving-group, and it is thus possible to perform reactions selectively at C-1. Consideration of this factor clarifies understanding of the reaction with antimony pentachloride; it permits use of the reaction for the preparation of acyloxonium derivatives of aldoses, and study of their rearrangements.

Tetra-O-acetyl-β-D-glucopyranosyl chloride (**56**) reacts with antimony pentachloride by a 1,2-neighboring-group reaction, with splitting out of chloride ion, and initial generation of an acetoxonium derivative (**59**) having the D-*gluco* configuration.[37] Similarly, penta-O-acetyl-β-D-glucopyranose (**57**) gives **59** by the action of antimony pentachloride, through nucleophilic attack on C-1 by the 2-acetoxyl group in **57**, with loss of an acetate anion from C-1. Penta-O-acetyl-α-D-glucopyranose does not react with antimony pentachloride to give **59**, because neighboring-group assistance in splitting out of the 1-acetoxyl group is lacking; the reaction leads merely to a difficultly soluble adduct.[21] However, tetra-O-acetyl-α-D-glucopyranosyl

(37) H. Paulsen, W.-P. Trautwein, F. Garrido Espinosa, and K. Heyns, *Chem. Ber.*, **100**, 2822 (1967).

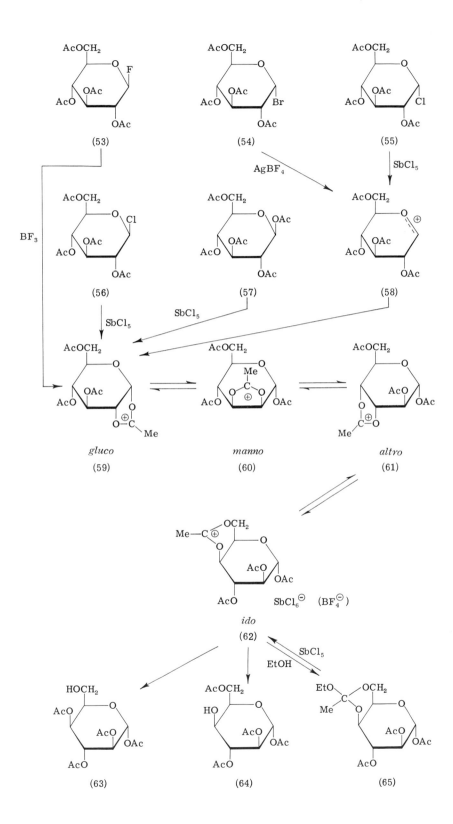

chloride (**55**) reacts with antimony pentachloride, to give **59**, presumably because the chlorine atom can so readily be removed as chloride ion that neighboring-group assistance is not essential for its abstraction.[37] The carboxonium ion (**58**), thereby formed initially, at once undergoes closure of the acetoxonium ring to give the (more stable) ion **59**.

A hexachloroantimonate salt of the D-*gluco* acetoxonium ion is isolable only if the reaction is performed at low temperature ($-10°$) in a nonpolar solvent (such as carbon tetrachloride) from which the salt is precipitated, by virtue of its insolubility, as fast as it is formed. Under these conditions, the ion **59** has no time to undergo successive acyloxonium rearrangements to **60, 61,** and **62,** because the rate of precipitation is greater than the rate of rearrangement. The salt thereby obtained contains over 90% of the D-*gluco* derivative.[21]

In contrast, if **56, 57,** or **55** in dichloromethane is treated at 20° with antimony pentachloride, a rapid acyloxonium rearrangement of the D-*gluco* ion **59** ensues, first by neighboring-group participation to give the D-*manno* ion (**60**), and, subsequently, to the D-*altro* ion (**61**), and then to the D-*ido* ion (**62**). In the solution, an equilibrium **59** ⇄ **60** ⇄ **61** ⇄ **62** exists between the ions. The equilibrium composition contains[21] approximately 54–55% of the D-*gluco* ion (**59**), 13–14% of the D-*manno* ion (**60**), 6–9% of the D-*altro* ion (**61**), and 21–22% of the D-*ido* ion (**62**). The hexachloroantimonate of the D-*ido* acetoxonium ion (**62**) is soluble with difficulty in dichloromethane or 1,2-dichloroethane, and this salt crystallizes more readily than the corresponding salts of the other three ions (**59, 60,** and **61.**) In consequence, only the D-idose salt (**62**) crystallizes out from this reaction mixture.[21,37] Because, as indicated in Section II, the acyloxonium rearrangements generally take place rapidly under these conditions, the *ido* derivative (**62**) is re-formed in the solution by equilibration (**59** ⇄ **60** ⇄ **61** ⇄ **62**) as it crystallizes from the solution (principle of Dimroth), and the yield (73%) of isolated product is considerably higher than the equilibrium concentration of **62** in the solution. The *ido* configuration of the product was established unambiguously from its n.m.r. spectrum;[37] the content of the *ido* derivative in the salt, as crystallized, is about 94%.

This reaction provides an excellent method for preparing D-idose from D-glucose. Hydrolysis of the salt **62** proceeds with *cis* opening of the acetoxonium ring, to give a mixture of the 1,2,3,4-tetraacetate (**63**) and 1,2,3,6-tetraacetate (**64**) of α-D-idopyranose, from which **64** can be isolated crystalline. The yield of **64** can be further increased,[37] because **63** readily undergoes acyl migration to give **64**. Peracetylation of the mixed tetraacetates gives homogeneous penta-O-acetyl-α-D-idopyranose. This compound strongly favors the $C1$(D) conformation, in which all four of the acetoxyl groups on the ring are axially attached and only the acetoxymethyl group is oriented equatorially.[38] This observed conformation may be attributed to (a) the unfavorable anomeric effect[39] in the alternative conformation, and (b) a very small interaction[39] between syn-diaxial acetoxyl groups in the favored conformation.

The reverse rearrangement (from the D-ido to the D-$gluco$ system) is likewise effected readily. The pure D-ido salt (**62**) gives α-D-glucopyranose pentaacetate on treatment with acetic anhydride or acetic acid.[37] In these solvents, a rapid equilibration **62 $\rightleftarrows$ 61 $\rightleftarrows$ 60 $\rightleftarrows$ 59** can be established before the acyloxonium ion captures an anion. The ion **59** evidently reacts the fastest with acetate, to give an ortho acid derivative which, by cis opening, leads to penta-O-acetyl-α-D-glucopyranose. In pyridine–ethanol, the D-ido salt **62** reacts at once to give the non-rearranged product, namely, the ortho ester **65** resulting from the cis pathway.[37] With antimony pentachloride, compound **65** is transformed back into the D-ido salt **62,** which is nevertheless converted into α-D-glucose pentaacetate by the action of acetic anhydride or acetic acid.

The equilibrium **59 $\rightleftarrows$ 60 $\rightleftarrows$ 61 $\rightleftarrows$ 62** may conveniently be observed if the pure D-$gluco$ salt (**59**) on the one hand, and the pure D-ido salt (**62**) on the other, are brought into equilibration in nitromethane at $-20°$ and the composition of the mixture is monitored as a function of time. The first step of the analysis is effected, as in analysis of all mixtures of salts of sugar acyloxonium ions, by hydrolysis of the mixtures with water; by this procedure, all acyloxonium ions undergo cis opening. The resultant mixture of partially acetylated sugars, after peracetylation, is determined qualitatively and quantitatively by gas–liquid chromatography.[21] The results of the equilibration are shown in Fig. 2. The curves on the left-hand side indicate the progress of the reaction starting from the D-$gluco$ acetoxonium salt (**59**), and those on the right indicate the results obtained when the D-ido acetoxonium salt (**62**) is the starting material. It is noteworthy that, starting from either side, the concentration of the original salt declines rapidly, and leads eventually to the equilibrium mixture of all four sugar derivatives.[21]

The final equilibrium-compositions have also been measured[21] in various solvents and at different temperatures (see Table II). The

(38) N. S. Bhacca, D. Horton, and H. Paulsen, *J. Org. Chem.*, **33**, 2484 (1968).
(39) R. U. Lemieux and A. A. Pavia, *Can. J. Chem.*, **47**, 4441 (1969).

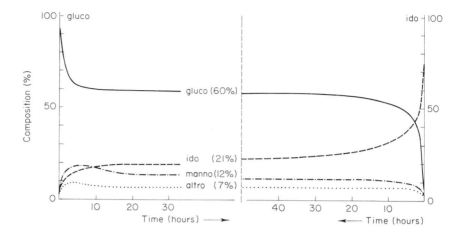

Fig. 2.—Equilibration of the D-*gluco* Acetoxonium Salt (**59**) (left) and the D-*ido* Acetoxonium Salt (**62**) (right) in Nitromethane at −20°. [Content of D-*gluco* ion (**59**) (——), D-*ido* ion (**62**) (– – –), D-*manno* ion (**60**) (·–·–·–·), and D-*altro* ion (**61**) (········).]

equilibrium distribution is, to a limited extent, solvent-dependent. Thus, in acetonitrile (as compared with nitromethane), the concentration of the D-*gluco* derivative is higher, and that of the D-*ido* derivative is correspondingly lower. At higher temperatures (above −20°), side reactions leading to unidentified decomposition-products become significant. The proportion of the D-*manno* derivative (**60**) seems to be somewhat greater at 40°.

TABLE II

Equilibrium Compositions of the Acetoxonium Ions 59 ⇄ 60 ⇄ 61 ⇄ 62 and of the Corresponding Benzoxonium Ions

Solvent medium	Temperature (°C)	Percent composition				Other compounds	Anion
		gluco	*manno*	*altro*	*ido*		
Acetoxonium ions							
CH_3NO_2	−20	60	12	7	21	—	$SbCl_6^{\ominus}$
CH_3CN	−20	72	9	6	12	—	$SbCl_6^{\ominus}$
CH_3NO_2	−20	61	14	6	19	—	$BF_4^{\ominus}$
CH_3CN	−20	75	8	6	10	—	$BF_4^{\ominus}$
CH_3NO_2	+40	54	17	9	15	5	$SbCl_6^{\ominus}$
CH_3CN	+40	65	14	7	9	5	$SbCl_6^{\ominus}$
$C_2H_4Cl_2$	+40	55	14	9	21	1	$SbCl_6^{\ominus}$
Benzoxonium ions							
CH_3NO_2	−20	63	16	5	13	3	$SbCl_6^{\ominus}$
CH_3CN	−20	77	14	3	5	1	$SbCl_6^{\ominus}$
$C_2H_4Cl_2$	+20	33	29	13	20	5	$SbCl_6^{\ominus}$

To explore the dependence of the equilibrium on the nature of the anion, the fluoroborate salt of **59** was prepared. It was obtained at low temperature (0°) either by the reaction of tetra-*O*-acetyl-β-D-glucopyranosyl fluoride (**53**) with boron trifluoride in carbon tetrachloride, or by treatment of tetra-*O*-acetyl-α-D-glucopyranosyl bromide (**54**) with silver fluoroborate in nitromethane.[21] In ether, compound **54** and silver fluoroborate give a mixture of tetra-*O*-acetyl-α- and β-D-glucopyranosyl fluoride, because the ether complexes with the boron trifluoride, and so a fluoroborate anion is no longer available.[40] On dissolution of the fluoroborate salt of the D-*gluco* acetoxonium ion (**59**) in nitromethane or acetonitrile, an acyloxonium-ion rearrangement, is observed with consequent equilibration, between **59** ⇄ **60** ⇄ **61** ⇄ **62**, that is not significantly different from that observed with the hexachloroantimonate salt (see Table II); the rearrangement and equilibration thus appear to be essentially independent of the nature of the anion.[21]

The results of the equilibration studies indicated that the D-*gluco* ion (**59**) is by far the most stable of the four ions. The stability of the acetoxonium ions appears to be determined by the relative influence of the following factors[21]: (*1*) the stereochemistry of the bicyclic ring-systems; (*2*) the effect of the ring-oxygen atom; and (*3*) the differences in stability between five- and six-membered, cyclic acetoxonium ions. The most significant influence, the stereochemistry of the system, is also the most difficult to estimate, because the pyranoid ion very probably assumes a twisted conformation, and a precise conformational analysis of bicyclic acetoxonium ions has not thus far been developed. Because of the effect of the neighboring, ring-oxygen atom, the D-*gluco* 1,2-acetoxonium ion (**59**) appears to be the favored form, but, for other configurations, this tendency can readily be outweighed by steric compensation, so that, for the acetoxonium ions derived from D-xylose and D-galactose (see Sections III,2 and 3), the 1,2-acetoxonium ion is no longer the most stable form. A five-membered, acetoxonium form is regarded as being more stable than a six-membered analog (see Section III,3). This effect seems to be a determining factor in the proportionately low stability of the D-idose 4,6-acetoxonium ion (**62**).

Penta-*O*-benzoyl-β-D-glucopyranose or, better, tetra-*O*-benzoyl-α-D-glucopyranosyl chloride in carbon tetrachloride can be converted at 75° with antimony pentachloride into a mixture of benzoxonium salts that contains 38% of the D-*gluco*, 25% of the D-*manno*, 10% of the D-*altro*, and 20% of the D-*ido* derivatives; these results indicate

(40) K. Igarashi, T. Honma, and J. Iriswa, *Carbohyd. Res.*, **13**, 49 (1970).

a corresponding acyloxonium rearrangement in the benzoyl derivatives. The benzoxonium salt is freely soluble in dichloromethane and
dichloroethane, and a favored crystallization of the D-*ido* salt does
not occur. By hydrolysis of the benzoxonium salt, and perbenzoyla-
tion of the resultant mixture of tetrabenzoates, it is possible to isolate
penta-*O*-benzoyl-α-D-idopyranose in 15% yield from the mixture by
virtue of its extremely low solubility in the system.[21]

The mixture of benzoxonium salts is subject,[21] in various solvent
media, to an equilibration reaction similar to that involving the
acetoxonium salts **59** and **62**. The equilibrium proportions of the
four benzoxonium ions (see Table II) correspond approximately to
those found for the four acetoxonium ions **59, 60, 61,** and **62**. The
content of the D-*gluco* ion is somewhat greater, and that of the D-*ido*
ion is proportionately lower, but, evidently, a change in the nature
of the acyl group does not lead to substantial alteration in the equilibrium composition.

The D-*ido* acetoxonium salt (**62**) exhibits both of the reactions
characteristic of an ambident cation (see Section I,3). With ethanol–
pyridine, **62** reacts by the *cis* pathway to give the orthoester **65**. It is
noteworthy that, in pyridine solution, no equilibration takes place
between the D-*ido* salt (**62**) and the D-*gluco* salt (**59**). After hydrolysis
of solutions of these salts and subsequent further acetylation, the pure
pentaacetates of D-idose or D-glucose, respectively, are recovered.[21]
It may be supposed that, in pyridine solution, a salt such as **62** is
converted at once into the non-isolable pyridinium orthoester (**67**),
whereby each further acetoxonium rearrangement is prevented. In
pyridine–ethanol, the salt **62** evidently[21] reacts initially to give the
intermediate **67**, which reacts further with ethanol to give the orthoester **65**.

At low temperatures, and by a rapid reaction that prevents a reverse rearrangement, the D-*ido* salt **62** reacts in acetonitrile with sodium borohydride by the *cis* pathway, to give the ethylidene derivative **66**, and with lithium bromide by the *trans* pathway, to give[21] the 6-bromo-6-deoxy-D-idose derivative **68**. Under similar conditions, treatment with lithium azide did not provide a 6-azido-6-deoxy-D-idose derivative.

2. Rearrangement of Xylose and Arabinose

On treatment in carbon tetrachloride with antimony pentachloride, tri-*O*-acetyl-β-D-xylopyranosyl chloride (**69**), tetra-*O*-acetyl-β-D-xylopyranose (**70**), and tri-*O*-acetyl-α-D-xylopyranosyl chloride (**71**) give an acetoxonium salt that precipitates directly from the solution. This salt consists of a mixture of the D-*xylo*, D-*lyxo*, and D-*arabino* derivatives (**72**, **73**, and **74**, respectively). None of these three compounds crystallize out preferentially.[41] Evidently, the D-*xylo* derivative **72** is formed initially, and is then transformed by reversible acyloxonium rearrangements into **73** and **74**. Salts are not isolated when the reaction is conducted in dichloromethane, because of their high solubility in this solvent.

As the salt is precipitated at once from carbon tetrachloride, its composition is dependent on the reaction temperature and the nature of the starting material,[41,42] because precipitation of the salt and interconversion leading to equilibration (**72** ⇄ **73** ⇄ **74**) occur concurrently. At low temperatures (0°), the salt precipitates before complete equilibration is achieved, and this salt consequently contains little of the D-*arabino* derivative **74**. At a reaction temperature of 25°, with rapidly rearranging derivatives, equilibrium is generally reached, so that the composition of the salt corresponds to that of the equilibrium mixture. At higher temperatures, side reactions cannot be excluded.

The salt mixtures, obtained in comparable yields from **69, 70,** and **71** (see Table III), have different compositions. The acyloxonium rearrangement proceeds at clearly different rates for the three compounds, as was recognized from the progressively diminished content of D-*arabino* derivative in the product. The β-D chloride **69** reacts the fastest, and the α-D chloride **71**, the slowest. The former has

(41) H. Paulsen, F. Garrido Espinosa, W.-P. Trautwein, and K. Heyns, *Chem. Ber.*, **101**, 179 (1968).

(42) H. Paulsen, C.-P. Herold, and F. Garrido Espinosa, *Chem. Ber.*, **103**, 2463 (1970).

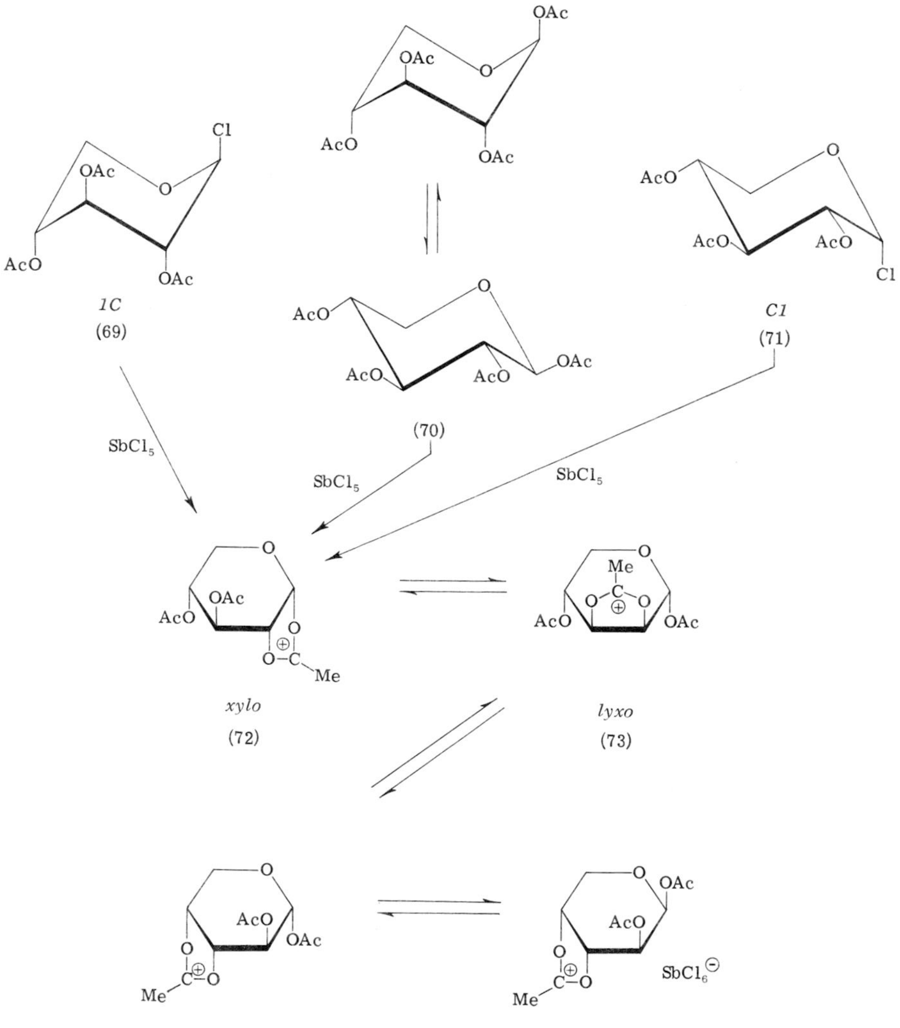

xylo
(72)

lyxo
(73)

α-arabino
(74)

β-arabino
(75)

been shown by n.m.r.-spectral studies to adopt the $1C$(D) conformation, which has four axial substituents[41,43] (see This Volume, Chapter 2, p. 88), a conformation favorable for neighboring-group and rearrangement reactions. The α-D chloride (71) undoubtedly adopts the $C1$(D) conformation,[43] which is not favorable for rearrangement

(43) D. Horton, C. V. Holland, and J. S. Jewell, *J. Org. Chem.*, **32**, 1818 (1967); P. L. Durette and D. Horton, *Carbohyd. Res.*, **18**, 57 (1971).

TABLE III

Composition of Acetoxonium Salts Prepared from 69, 70, and 71, and Proportions of the Acetoxonium Ions $72 \rightleftarrows 73 \rightleftarrows 74 \rightleftarrows 75$ at Equilibrium

Starting material	Solvent medium	Temperature (°C)	xylo (%)	lyxo (%)	α- + β-arabino (%)	Other products (%)	Remarks
69	CCl$_4$	+25	35	7	58	—	isolated salt
70	CCl$_4$	+25	55	7	36	—	isolated salt
71	CCl$_4$	+25	66	7	27	—	isolated salt
Salt from 69	CH$_3$NO$_2$	+25	28	10	57	5	equilibration
Salt from 69	CH$_3$CN	+25	55	12	31	2	equilibration

reactions; furthermore, the initial splitting out of chloride ion is not assisted by a neighboring-group reaction. The β-D tetraacetate **70** exists to about 80% in the $C1$(D) and 20% in the $1C$(D) conformation,[44] and its reactivity[42] lies, predictably and demonstrably (see Table III), between that of the two chlorides **69** and **71**. Hydrolysis of the acetoxonium-salt mixture obtained from **69** at 25° allows the isolation of crystalline 1,2,4-tri-O-acetyl-α-D-arabinopyranose.[41]

Equilibration studies of systems in nitromethane and in acetonitrile have been conducted[42] with the acetoxonium-salt mixture obtained from **69**. As indicated in Table III, in nitromethane, the α- and β-D-*arabino* ions (**74** and **75**) are favored to the extent of 57%. The equilibrium composition corresponds approximately to the composition of the salt isolated from **69** at 25°. In acetonitrile, in contrast, the D-*xylo* ion (**72**) is favored to the extent of 55%. The content (10–12%) of the D-*lyxo* ion (**73**) in the solution thus resembles that of the D-*manno* ion (**60**) in the corresponding rearrangement in the D-glucose series.

Surprisingly, the equilibration studies consistently show a substantial proportion of the β-D-*arabino* ion (**75**), as indicated by a corresponding proportion of tetra-O-acetyl-β-D-arabinopyranose obtained after hydrolysis and subsequent analysis of the product mixture. The β-D ion (**75**) cannot be formed through acetoxonium rearrangement of **72**, but must arise through subsequent anomerization of the α-D ion (**74**) formed initially, because the acetoxonium salt functions as an effective catalyst for anomerization.[42] If solutions of anomerically pure aldopentopyranose tetraacetates in nitromethane are treated with 2-methyl-1,3-dioxolan-2-ylium hexachloroantimonate (**2**, R=Me), a rapid onset of anomeric equilibration is observed. Under these conditions, tetra-O-acetyl-D-arabinopyranose gives[42] an α:β ratio of 9:91, and tetra-O-acetyl-D-xylopyranose, an α:β ratio of 96:4.

A furanoid acyloxonium salt can be prepared from 2,3,5-tri-O-benzoyl-β-D-arabinofuranosyl chloride (**76**). In carbon tetrachloride, compound **76** reacts with antimony pentachloride to give a crystalline benzoxonium salt that contains 95–99% of the D-*arabino* derivative **77**. Rearrangement to **78** and **79** is negligible; the D-*arabino* ion **77** is by far the most stable ion in the equilibrating system **77** $\rightleftarrows$ **78** $\rightleftarrows$ **79**, as indicated by equilibration studies on the salt in nitromethane and in acetonitrile. Together with 87–89% of **77**, there is present only 4–6% of the D-*xylo* ion (**79**) having a six-membered benzoxonium ring, and the content of D-*lyxo* ion (**78**) is negligible. In this example

(44) P. L. Durette and D. Horton, *Chem. Commun.*, 516 (1969); *J. Org. Chem.*, **36**, 2658 (1971).

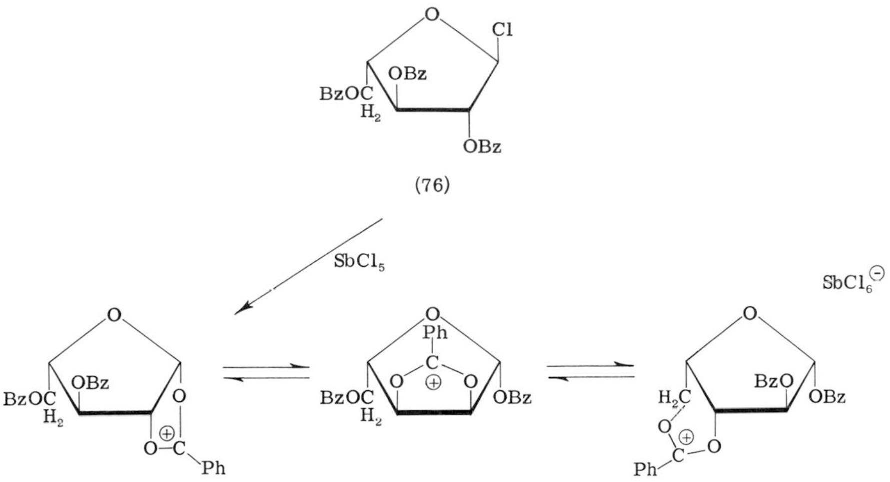

(76)

SbCl$_5$

SbCl$_6^{\ominus}$

arabino

(77)

lyxo

(78)

xylo

(79)

Bz = COPh

from the furanose series, the 1,2-acyloxonium ion (which has the ring-oxygen atom adjacent) is stabilized even more strongly than a corresponding D-*gluco* pyranoid ion.

3. Rearrangement of Derivatives of D-Glucose Substituted at C-6

A range of D-glucose derivatives substituted at C-6 can likewise be transformed into acetoxonium salts.[42] These compounds resemble the D-xylose analogs, inasmuch as here, also, only two rearrangement steps are possible, from the D-*gluco* ion (83) to the D-*manno* ion (84) and the D-*altro* ion (85). When such compounds in carbon tetrachloride at 50° are treated with antimony pentachloride, the salts are immediately precipitated out.[42] Under these conditions, it is evident that the equilibria 83 ⇄ 84 ⇄ 85 are established, and, in all cases, mixtures of the D-*gluco*, D-*manno*, and D-*altro* salts are obtained (see Table IV). The reaction has been investigated[42] starting with the 6-deoxy-D-glucose derivatives 80a and 82a, with the methyl D-glucuronate derivatives 81b and 82b, with the 6-O-methyl-D-glucose derivative 81c, with the 6-deoxy-6-iodo-D-glucose derivatives 80d and 81d, and with the 6-O-p-tolylsulfonyl-D-glucose derivatives 80e, 81e, and 82e.

From Table IV, it may be seen that the D-*gluco* ion is by far the most stable ion (53–74% of the mixture) for all of the C-6 substituted

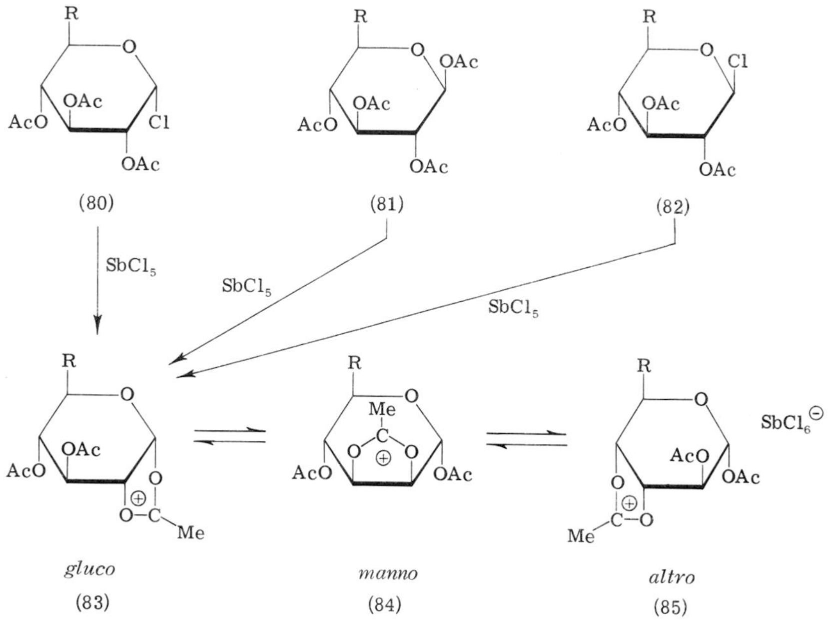

(a) R = Me
(b) R = CO₂Me
(c) R = CH₂OMe

(d) R = CH₂I
(e) R = CH₂OTs

derivatives studied. A comparable ratio between the D-*gluco*, D-*manno*, and D-*altro* ions is observed in the D-glucose–D-idose rearrangement if, in this rearrangement, the content of the D-*ido* ion is not taken into account (see Table IV, last entry). With regard to the different equilibrium composition in the D-xylose–D-arabinose

TABLE IV

Composition of Acetoxonium Salts 83 ⇄ 84 ⇄ 85 Obtained, in Carbon Tetrachloride, from D-Glucose Derivatives Modified at C-6

C-5 Substituent R in 83, 84, and 85	Temperature (°C)	*gluco* (%)	*manno* (%)	*altro* (%)	Other products (%)
—CH₃	+40	68	10	22	—
—CO₂Me	+50	53	20	20	7
—CH₂OMe	+50	74	12	9	5
—CH₂I	+50	61	9	20	10[a]
—CH₂OTs	+50	74	12	9	5
—CH₂OAc[b]	+50	56	28	16	—

[a] Penta-*O*-acetyl-β-D-glucopyranose produced in the reaction.
[b] These values were calculated from the D-glucose–D-idose rearrangement (see Section III, 1, p. 150) by neglecting the content of the D-*ido* ion and taking as 100% the sum for the D-*gluco*, D-*manno*, and D-*altro* ions.

rearrangement (see Section III, 2, p. 156), it may be inferred that the especial stability of the D-*gluco* 1,2-acetoxonium ion is the consequence of the steric influence of the C-5 substituent.

4. Rearrangement of Idose, Galactose, and Altrose

Tetra-O-acetyl-α-D-idopyranosyl chloride (**86**) in carbon tetrachloride at 50° reacts[45] with antimony pentachloride to give an acetoxonium salt that is a mixture of the hexachloroantimonates of the ions **87, 88,** and **89**. The composition of the salt, which, from its mode of

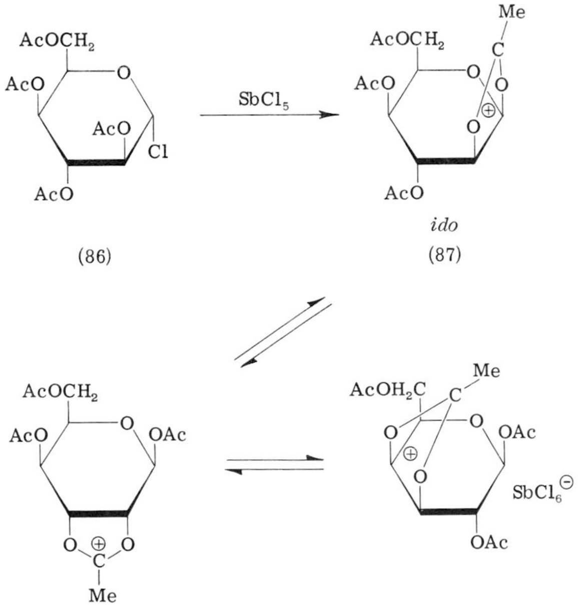

preparation, would appear to contain an equilibrium mixture of the three ions, was found to be 26% of the D-*ido* ion (**87**), 3% of the D-*gulo* ion (**88**), and 71% of the D-*galacto* ion (**89**). The proportions of products in this D-idose system are in a sense comparable to those obtaining in the D-xylose system. In the starting material **86**, as in **69**, all of the substituents on C-1, C-2, C-3, and C-4 are axially attached. The composition of an acetoxonium salt produced from **69**

(45) F. Garrido Espinosa, W.-P. Trautwein, and H. Paulsen, *Chem. Ber.*, **101**, 191 (1968).

under analogous conditions is[42,45] 35% of D-*xylo* ion (**72**), 7% of D-*lyxo* ion (**73**), and 58% of D-*arabino* ion (**74** + **75**); these proportions are entirely analogous to those in the homomorphous D-idose series.

From stereochemical considerations, 1,2-acetoxonium ions of D-galactose and D-altrose can undergo only a single rearrangement step, leading from the D-*galacto* ion (**91**) to the D-*talo* ion (**92**), and from the D-*altro* ion (**94**) to the D-*allo* ion (**95**). Tetra-*O*-acetyl-β-D-

galacto *talo*

(90) (91) (92)

altro *allo*

(93) (94) (95)

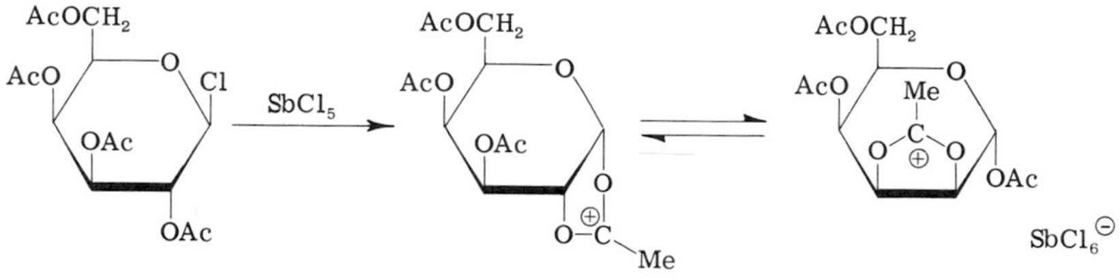

galactopyranosyl chloride (**90**), penta-*O*-acetyl-β-D-galactopyranose, or tetra-*O*-acetyl-α-D-galactopyranosyl chloride react with antimony pentachloride to give a mixture of acetoxonium salts that contains 54–61% of the D-*talo* and 39–46% of the D-*galacto* derivative.[46] Equilibration studies at −20° in nitromethane indicate existence of an equilibrium between 35% of the D-*galacto* ion (**91**) and 65% of the D-*talo* ion (**92**). In acetonitrile at −20°, the proportions are reversed,[42] with 72% of the D-*galacto* ion (**91**) and 28% of the D-*talo* ion (**92**); the solvent-dependence of the equilibrium position is especially significant in this example.

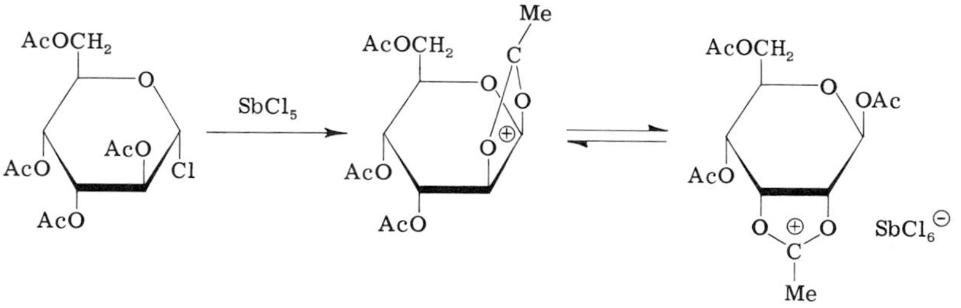

(46) H. Paulsen, F. Garrido Espinosa, and W.-P. Trautwein, *Chem. Ber.*, **101**, 186 (1968).

The acyloxonium rearrangement provides a good preparative access to D-talose from D-galactose.[46] Thus, tetra-O-acetyl-β-D-galactopyranosyl chloride (90) is treated with antimony pentachloride to give the mixture of salts 91 and 92, which is hydrolyzed and the mixture of products acetylated. Penta-O-acetyl-α-D-talopyranose, which is present as the pure α-D anomer, crystallizes out from the mixture, whereas the D-galactose pentaacetate does not. The latter exists as a difficultly crystallizable, α,β mixture, because hydrolysis of 91 and acetylation of the product leads to an anomeric mixture.

Tetra-O-acetyl-α-D-altropyranosyl chloride (93) reacts[42] with antimony pentachloride to give a mixture of salts containing 54% of the D-*altro* derivative (94) and 39% of the D-*allo* derivative (95). This ratio contrasts with that observed for the equilibrium 91 ⇄ 92, and shows that the 1,2-O-acetoxonium ion is favored.

5. Rearrangement of Amino Sugars

Studies on monosaccharides have indicated that, in general, acyloxonium rearrangements lead to a rapidly equilibrated mixture of the participating ions. The stabilities of the various cyclic ions differ somewhat, but, generally, not to such an extent that the equilibrium lies fully on the side of a single ion. In most instances, all ions participating are present in appreciable proportion, so that precipitation of a salt gives a mixture of configurational isomers. This factor leads, in preparative applications, to a difficult separative problem. Isolation of the pure, rearranged monosaccharide is possible if the acyloxonium salt of one pure isomer in the equilibrium mixture crystallizes out (as in the preparation of D-idose from D-glucose), or if the mixture of isomers formed by hydrolysis is readily separated (as in the preparation of D-talose from D-galactose).

Studies have been made to determine whether, by introduction of other substituents into the molecule, it would be possible so to influence the equilibrium between the various acyloxonium ions that one of these would be favored to the greatest extent. The acetamido group has been found to be such a substituent.

3-Acetamido-1,2,4,6-tetra-O-acetyl-3-deoxy-β-D-glucopyranose (96) or 3-acetamido-2,4,6-tri-O-acetyl-3-deoxy-α-D-glucopyranosyl chloride reacts with antimony pentachloride to give the D-*gluco* ion (97); this exists in equilibrium with the D-*manno* ion (98) only, as the nitrogenous substituent is an extremely poor leaving-group that cannot be lost by further neighboring-group reactions. However, the D-*manno* ion 98 contains an oxazolinium ring-system that is more stable than the acetoxonium ring in **97**, and thus, the equilibrium lies fully on the side of the D-*manno* ion (**98**). After hydrolysis of the reaction mixture, only the D-*manno* oxazoline derivative (**101**) resulting from deprotonation of **98** is obtained. Acid hydrolysis of **101** gives 3-amino-3-deoxy-D-mannose, isolable as the hydrochloride.[47]

Treatment of 2,3,4-tri-*O*-acetyl-6-benzamido-6-deoxy-α-D-glucopyranosyl chloride (**99**) with antimony pentachloride in 1,2-dichloroethane leads to an acyloxonium derivative. Between the ions **102** ⇌ **103** ⇌ **104** ⇌ **105** is established an equilibrium in which the

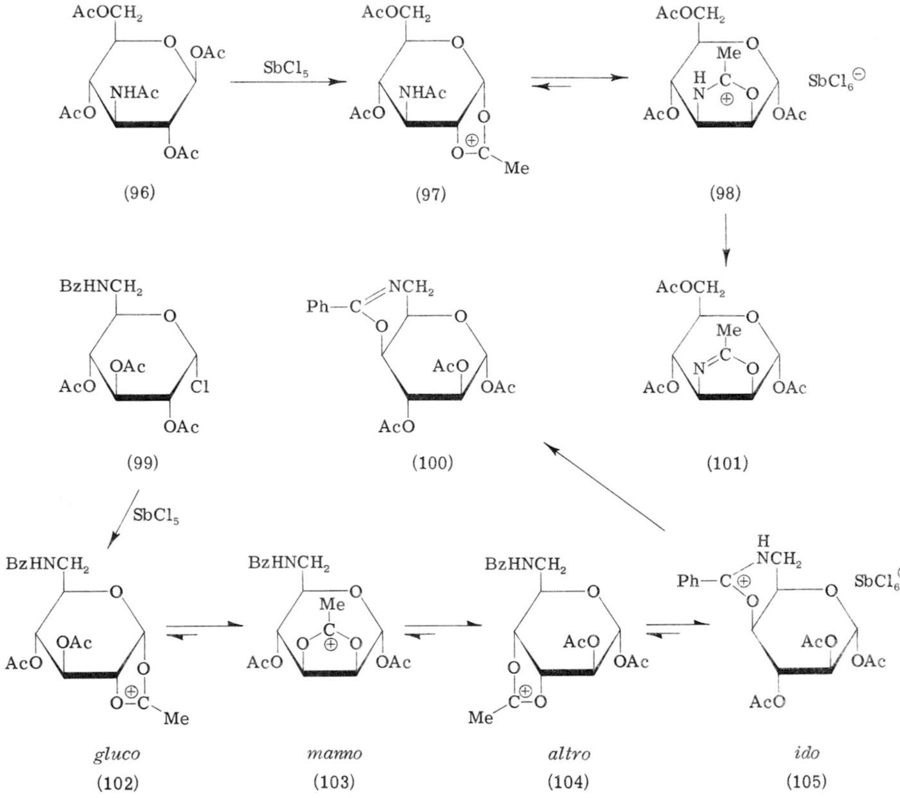

(96)   (97)   (98)

(99)   (100)   (101)

gluco   manno   altro   ido
(102)   (103)   (104)   (105)

Bz = COPh

dihydro-oxazinium ion (**105**) greatly preponderates. The reaction solution contains 30% of unreacted **99**, and 70% of **105** which crystallizes out as its hexachloroantimonate.[47] By treatment of the salt with

(47) H. Paulsen and C.-P. Herold, *Chem. Ber.*, **104**, 1311 (1971).

aqueous sodium hydrogen carbonate, the free dihydro-oxazine (**100**) can be obtained; it is convertible into other derivatives of 6-amino-6-deoxy-D-idose. Treatment of 6-acetamido-2,3,4-tri-*O*-acetyl-6-deoxy-α-D-glucopyranosyl chloride with antimony pentachloride affords a reaction mixture that, likewise, contains 30% of unreacted halide and 70% of the corresponding D-*ido* ion analogous to **105,** but the salt does not crystallize out and is consequently difficult to isolate.[47]

These examples in the 3-amino- and 6-amino-D-glucose systems demonstrate that the equilibrium proportions of the ions can be so influenced by appropriate substituents that only one form is strongly favored, thus facilitating isolation of this form.

6. Reaction of N-Bromosuccinimide with Benzylidene Acetals

As Hanessian[48] and Hullar and coworkers[49] have demonstrated, 4,6-benzylidene acetals of the type **106** in carbon tetrachloride react under anhydrous conditions with *N*-bromosuccinimide to give 4-*O*-benzoyl-6-bromo-6-deoxy derivatives (**108**) in good yield. As

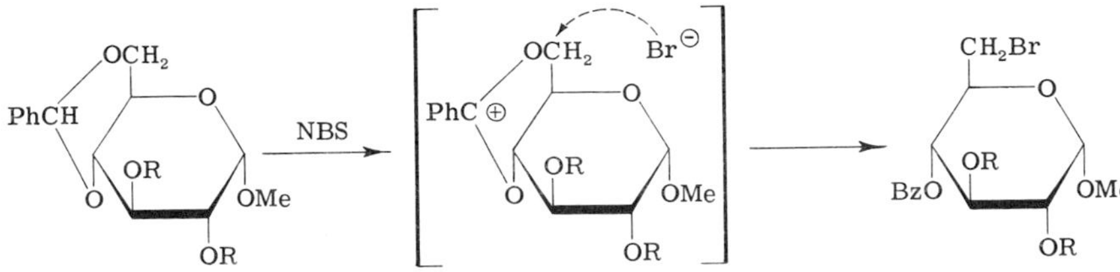

(106)   (107)   (108)

(a) R = H
(b) R = Bz = COPh
NBS = *N*-Bromosuccinimide

the most probable mechanism of this reaction, it has been proposed[48-54] that a benzoxonium intermediate (**107**) is formed and then opened by attack by bromide ion at C-6 by the *trans* pathway, to give **108**. On steric grounds, the attack by bromide must be at C-6 (not at C-4). By such a mechanism, the benzylidene acetal group in **106** must first be brominated by the reagent to give a (non-isolable)

(48) S. Hanessian, *Carbohyd. Res.*, **2**, 86 (1966).
(49) D. L. Failla, T. L. Hullar, and S. B. Siskin, *Chem. Commun.*, 716 (1966).
(50) S. Hanessian and N. R. Plessas, *J. Org. Chem.*, **34**, 1035 (1969).
(51) S. Hanessian and N. R. Plessas, *J. Org. Chem.*, **34**, 1045 (1969).
(52) S. Hanessian and N. R. Plessas, *J. Org. Chem.*, **34**, 1053 (1969).
(53) S. Hanessian and N. R. Plessas, *J. Org. Chem.*, **34**, 2163 (1969).
(54) T. L. Hullar and S. B. Siskin, *J. Org. Chem.*, **35**, 225 (1970).

bromo ortho ester,[55] which gives the benzoxonium ion **107** by splitting out of bromide ion. It is debatable whether this step involves total charge-separation to give the free benzoxonium ion before attack by bromide ion at C-6 takes place. Both steps could occur either synchronously or by way of an ion-pair.[50] It is noteworthy that the 2,3-dibenzoate (**106b**) gives[54] the product (**108b**) in higher yield than in the reaction in which the 2,3-dihydroxy derivative **106a** gives **108a**. Were the reaction to proceed by way of a free benzoxonium ion (**107b**), the 3-benzoyloxy group would be suitably situated for an intramolecular, *trans* opening-reaction in competition with attack at C-6 by bromide ion; rearrangement products formed by such a competing reaction were not, however, detected. Similar observations were made[50,56] with a 2,3-di-*O*-acetyl D-*manno* analog of **106**. A mechanism exclusively involving radicals in the whole reaction cannot be excluded. The observed dependence of the reaction on light[50] and the presence of peroxides[50,54] suggests that the first step of the bromination proceeds by a free-radical process.

The *N*-bromosuccinimide reaction has rapidly achieved adoption as a valuable method for transforming 4,6-benzylidene acetals of hexoses into 6-bromo-6-deoxy derivatives;[48–54,56,57] the latter are readily convertible into 6-deoxy and 6-amino-6-deoxy derivatives.[51] The reaction proceeds satisfactorily with a range of other substituents present at C-3 in the starting acetal (**106**); for example, −OH,[50,51,54] −OMs,[50,51] −OTs,[54] −OAc,[50,56] −OBz,[50,54] −OCH$_3$,[50] −Cl,[51] and −N$_3$,[51] and NHAc[51] or NHTs[54] at C-2. Likewise, 2,3-epoxides in similar systems react without difficulty.[50,51] The reaction has been applied to 4,6-*O*-benzylidene derivatives to give 6-bromo-6-deoxy-hexose derivatives in the D-*gluco*, D-*galacto*, D-*altro*, and D-*manno* configurational series,[50,51,54,56] as well as to corresponding derivatives of the disaccharides α,α-trehalose[50] and sophorose.[50] One limitation is found with *O*-benzyl derivatives, which can undergo bromination of the benzyl methylene group.[50]

On conversion into the supposed benzoxonium intermediate, methyl 2,3-*O*-benzylidene-5-*O*-(methylsulfonyl)-β-D-ribofuranoside (**109**) would offer two sites on the resultant dioxolanylium ring for attack by bromide. These two sites appear equally favored, because, on treatment of **109** with *N*-bromosuccinimide, a 1:1 mixture of the bromides **111** and **112** is obtained.[52]

(55) K. Heyns, W.-P. Trautwein, F. Garrido Espinosa, and H. Paulsen, *Chem. Ber.*, **99**, 1183 (1966).
(56) D. Horton and A. E. Luetzow, *Carbohyd. Res.*, **7**, 101 (1968).
(57) G. B. Howarth, W. A. Szarek, and J. K. N. Jones, *Chem. Commun.*, 62 (1968).

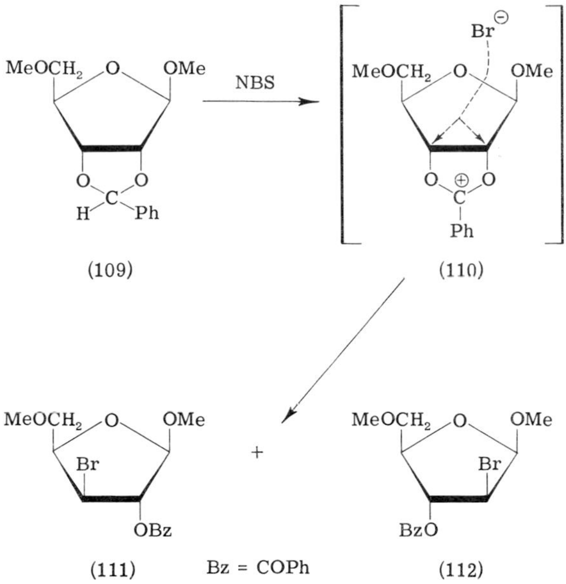

(109) (110)

(111) Bz = COPh (112)

Methyl 2-O-benzoyl-4,6-O-benzylidene-β-D-arabinopyranoside (113) presumably reacts with N-bromosuccinimide to give the benzoxonium ion 114 that, by a neighboring-group reaction, could rearrange to the ion 117. The observed products from 113 were the 4-bromo derivative 115 and a 3-bromo derivative, in 1:2 ratio. Unfortunately, it was not established whether the 3-bromo derivative has the structure 116 or 118, or whether it is a mixture of the two.[52] A 2-bromo derivative 119, whose presence would have been indicative

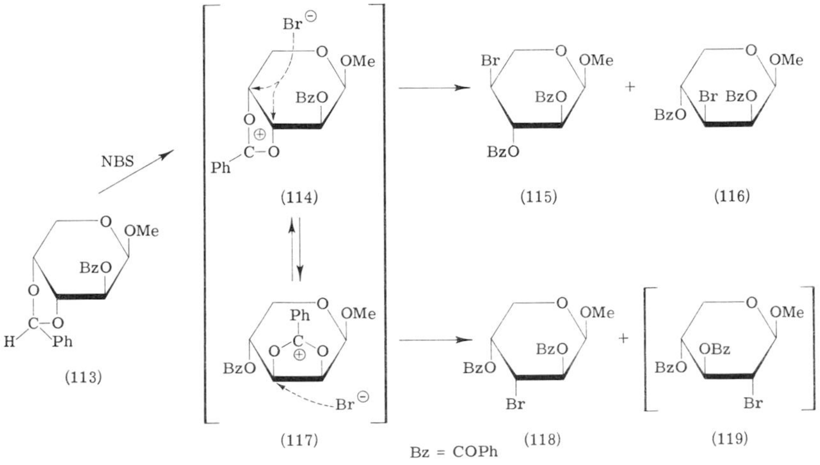

(114) (115) (116)

(113)

(117) (118) (119)

Bz = COPh

of the rearrangement **114** ⇄ **117**, was not detected in the reaction mixture. The observed higher yield of a 3-bromo derivative is scarcely an adequate proof of an equilibration **114** ⇄ **117**, as the reaction of N-bromosuccinimide with methyl 3,4-O-benzylidene-2,6-dichloro-2,6-dideoxy-α-D-altropyranoside and with methyl 2-azido-3,4-O-ben-zylidene-6-chloro-2,6-dideoxy-α-D-altropyranoside gives preponder-antly a 4-O-benzoyl-3-bromo derivative.[53]

The rearrangement of a benzoxonium intermediate can be convincingly demonstrated in the reaction of 3,5-O-benzylidene-1,2-O-isopropylidene-α-D-glucofuranose (**120**) with N-bromosuccini-mide, whereby the 6-bromo derivative **124** and the 3,6-anhydride **123** are obtained.[52] It may be presumed[52] that the benzoxonium ion **121** rearranges into the ion **122** through intramolecular attack by

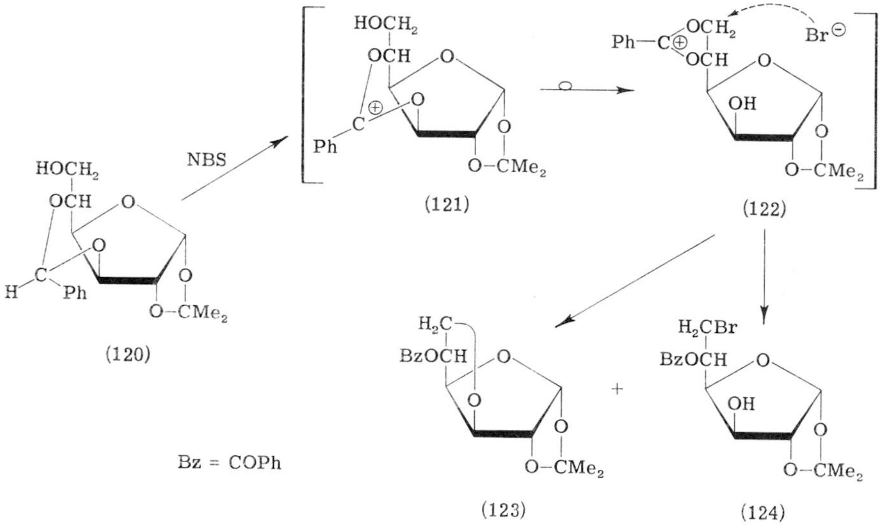

Bz = COPh

(120)   (121)   (122)   (123)   (124)

the 6-hydroxyl group. The ion **122** can either suffer attack by bromide ion at C-6 to give **124**, or the 3-hydroxyl group can attack at C-6 by the *trans* pathway to give the anhydride **123**.

With the dibenzylidene acetal **124a**, a twofold reaction with N-bromosuccinimide is possible.[57a] Thus, compound **124a** reacts with two molecular proportions of N-bromosuccinimide to form the di-bromide **124b**, which, on hydrogenation, yields **124c**, a derivative of tyvelose.

An interesting reaction pattern was found for 2′,3′-O-benzylidene-uridine.[57b] With 2.2 molecular proportions of N-bromosuccinimide,

(57a) M. Haga, M. Chonan, and S. Tejima, *Carbohyd. Res.*, **16**, 486 (1971).
(57b) M. M. Ponpipom and S. Hanessian, *Carbohyd. Res.*, **17**, 248 (1971).

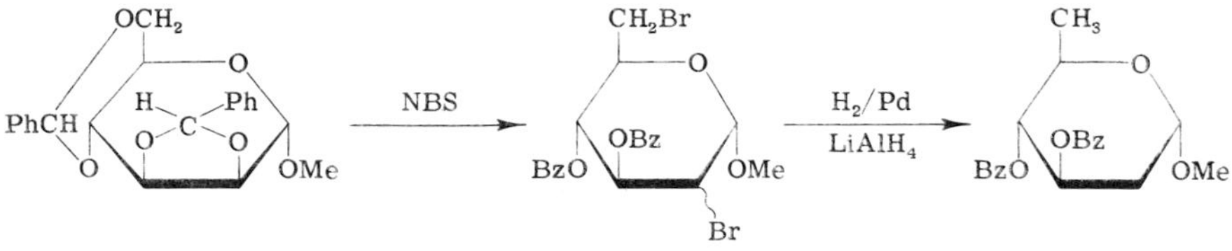

(124a)   (124b)   (124c)

compound **124d** yields the dibromide **124i** which, on reduction, gives the deoxy nucleoside **124j**. On the other hand, reaction of the 5-acetate **124e** leads to the anhydronucleoside **124h**.

It may be supposed that, initially, the acetals **124d** and **124e** are converted into benzoxonium intermediates which, by intramolecular, nucleophilic attack of the nucleoside base at C-2, rearrange to form the protonated intermediates **124f** and **124g**, respectively. Deprotonation of **124g** then leads to **124h**, whereas nucleophilic attack by a bromide ion at C-2 of **124f** yields the product **124i** *via* the *trans* pathway.

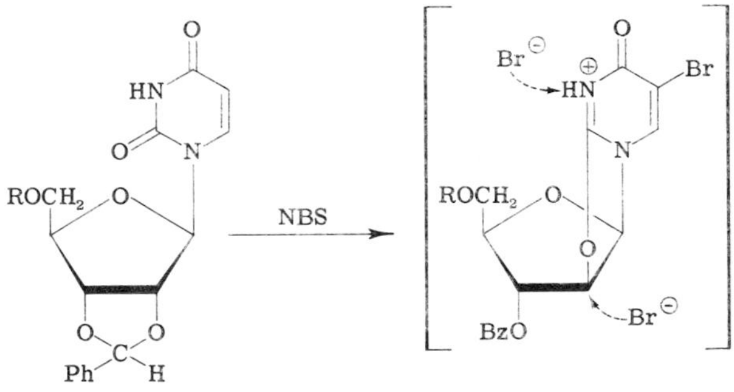

(124d) R = H
(124e) R = Ac

(124f) R = H
(124g) R = Ac

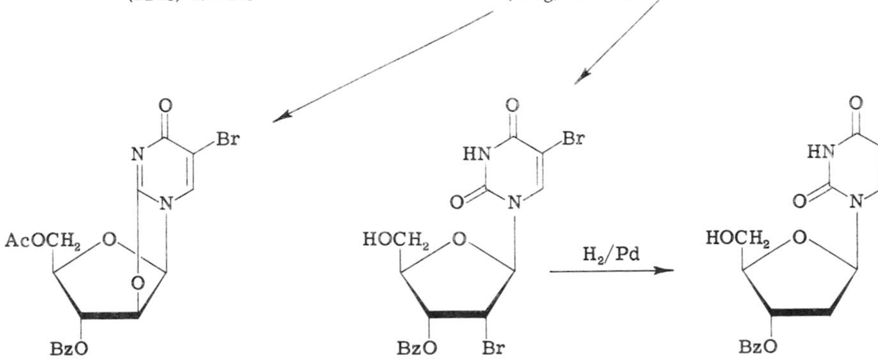

(124h)   (124i)   (124j)

Bz = COPh

An acetoxonium fluoroborate (126) has been prepared by Buchanan[20] by treating methyl 3,5-O-(ethyl orthoacetyl)-β-L-arabinopyranoside (125) with boron trifluoride in benzene. An excess of lithium borohydride rapidly converts 126 into the acetal 127 containing 90% of the *endo* form. Evidently, acid-catalyzed ring-opening of epoxides having neighboring, *trans*-disposed, acetoxyl groups generally proceeds *via* an acetoxonium intermediate, as exemplified by the opening of the epoxides 128 and 129 by way of the corresponding acetoxonium ions 126 and 130, respectively. In both examples,

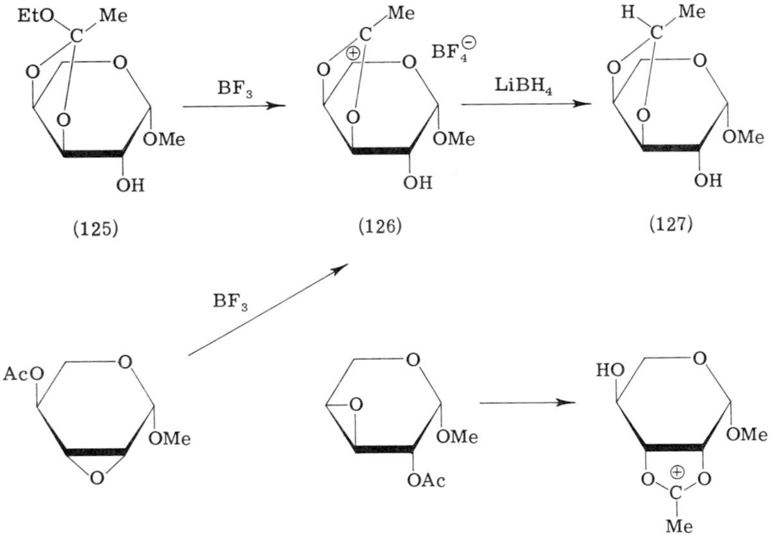

net inversion at C-3 is observed, as would be expected from hydrolytic *cis*-opening of the acetoxonium ring.[58,59] Under strictly anhydrous conditions in acetic acid–acetic anhydride, the products of *trans* opening[60,60a] are also found, in low yield. The intermediate ion 126 formed from 128 can be trapped, and isolated, by treating 128 in benzene with boron trifluoride[20]; with lithium borohydride, it is converted into the acetal 127.

(58) J. G. Buchanan and R. Fletcher, *J. Chem. Soc.* (C), 1926 (1966).
(59) J. G. Buchanan and R. Fletcher, *J. Chem. Soc.*, 6316 (1965).
(60) J. G. Buchanan, A. R. Edgar, and R. Fletcher, *Abstr. Papers Amer. Chem. Soc. Meeting*, 153, C 51 (1967).
(60a) J. G. Buchanan, J. Conn, A. R. Edgar, and R. Fletcher, *J. Chem. Soc.* (C), 1515 (1971).

IV. REARRANGEMENT OF ESTERS OF POLYOLS IN LIQUID HYDROGEN FLUORIDE

1. Rearrangement of Cyclitols

Hedgley and Fletcher[61,62] found that isomerizations take place when solutions of cyclitol acetates or 1,5-anhydroalditol acetates in liquid hydrogen fluoride are left to stand. These isomerizations evidently occur *via* acetoxonium ions, because, as indicated in Section I,2 (p. 132), such ions are quite readily formed from *cis*-diol diacetates in liquid hydrogen fluoride.

When a suitable stereochemical arrangement is present, the acyloxonium ions react further, through intramolecular, neighboring-group reactions, to give the corresponding isomerization products. As a rule, acyloxonium salts are not isolable from the liquid hydrogen fluoride (see Section I,2, p. 132), and yet, in the monosaccharide series (see Section V, p. 176), the acyloxonium derivatives formed as intermediates in hydrogen fluoride can be detected by n.m.r. spectroscopy. The isomerizations in hydrogen fluoride frequently give complex reaction-mixtures, as, in this medium, partial or total deacetylation takes place simultaneously. When the equilibrium composition is favorable, the reaction can be used for preparation of a single isomer.

Isomerizations in liquid hydrogen fluoride exhibit a noteworthy stereoselectivity. Generally, isomerization occurs with a *cis-trans*-1,2,3-triol triester of the type **131**, whereby the configuration of the acyloxy group at the central carbon atom is inverted,[61,62] so that **131** is isomerized reversibly into compound **132**. From studies on model compounds (see Section I,2, p. 132), it may be presumed that, initially, the *cis*-1,2-diol diester system in **131** reacts in hydrogen fluoride by loss of one molecule of acid per molecule, to give the acetoxonium ion **133**. This ion rearranges to **134**, to an extent dependent on the stabilities of the ions in the equilibrium **133** ⇌ **134**, so that subsequent hydrolysis yields products from the unrearranged ion (**133**) and the rearranged one (**134**). Starting from **132**, it is also possible to effect the reverse isomerization by way of the ion **134**. These steps throw light on the observed stereospecificity of the rearrangement. A mechanism, suggested at the outset,[61,62] that invokes an initial, seven-membered, carboxonium ion that is opened by *trans* attack,

(61) E. J. Hedgley and H. G. Fletcher, Jr., *J. Amer. Chem. Soc.*, **84**, 3726 (1962).
(62) E. J. Hedgley and H. G. Fletcher, Jr., *J. Amer. Chem. Soc.*, **85**, 1615 (1963).

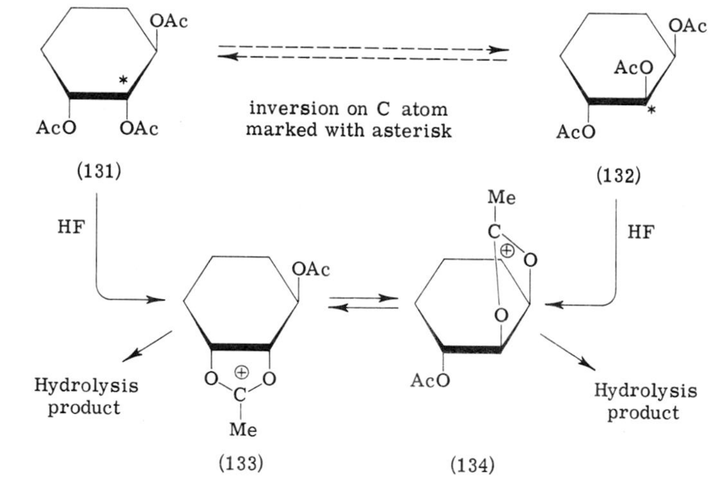

(131)

inversion on C atom
marked with asterisk

(132)

HF

HF

Hydrolysis
product

Hydrolysis
product

(133)

(134)

stands in contradiction to observations[8,12] described in Section I,2 (see p. 132).

Isomerization of numerous inositol derivatives can be effected by this type of reaction. After the action of hydrogen fluoride, and subsequent hydrolysis, *myo*-inositol hexaacetate (**135b**) gives a mixture of *myo*-inositol (**135a**), DL-*chiro*-inositol (**136a**), and *muco*-inositol (**137a**), with **137a** preponderating (over 70%) and *myo*-inositol (**135a**)

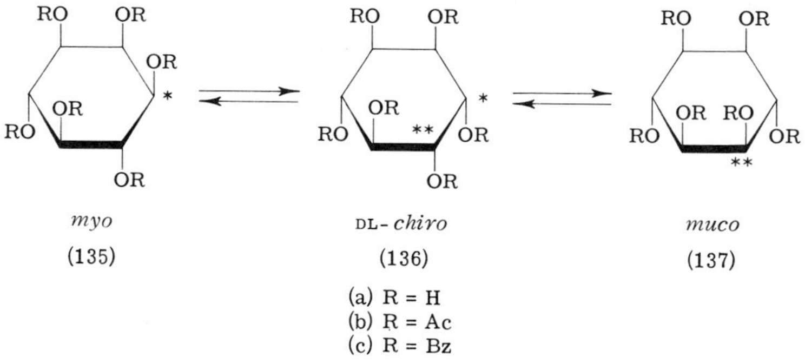

myo

(135)

DL-*chiro*

(136)

muco

(137)

(a) R = H
(b) R = Ac
(c) R = Bz

present in low proportion.[61] The configurational inversion, following the general process **131** → **132**, proceeds first to DL-*chiro*-inositol, which subsequently reacts to give *muco*-inositol. The reversibility of the reaction is demonstrated by the fact that the action of hydrogen fluoride on either hexa-*O*-benzoyl-DL-*chiro*-inositol (**136c**) or hexa-*O*-acetyl-*muco*-inositol (**137b**), with subsequent hydrolysis, leads[61] in each case to a mixture of **135a**, **136a**, and **137a**.

The high proportion of *muco*-inositol formed in the isomerization is surprising. The isomerization of *myo*-inositol with acetic acid–sulfuric acid (see Section VI, p. 190) gives[63,64] an equilibrium mixture of *myo*-, DL-*chiro*-, and *muco*-inositol in the ratios of 54:41:5. These ratios correspond to the relative thermodynamic stabilities of the three inositols as determined on the basis of intramolecular, steric interactions of the substituents,[64,65] whereas the isomerization in hydrogen fluoride gives the thermodynamically least-stable isomer (*muco*) in the highest yield. This discrepancy can be rationalized if it is postulated that the acetoxonium ion in the *muco*-inositol system is further stabilized as a di-cation of structure **138**, formed because

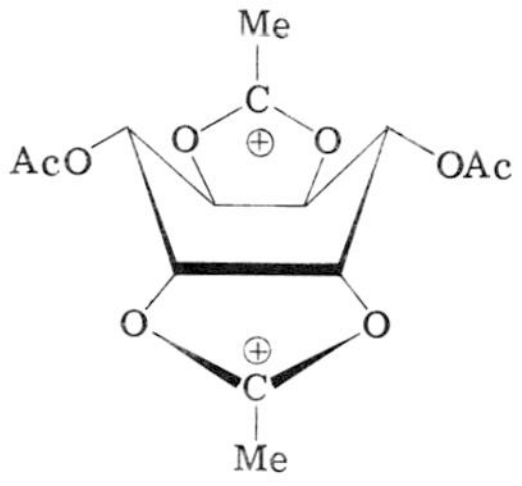

(138)

of the presence of a second, *cis*-diol diester group.[66] Such a di-cation (**138**) could exist in an extremely favorable boat form having all substituents quasi-equatorial; it would certainly be more stable than a similarly possible di-cation from DL-*chiro*-inositol. This reasoning is strongly supported by n.m.r.-spectral studies on the isomerization of hexa-*O*-acetyl-*myo*-inositol (**135b**) in hydrogen fluoride solution.[66]

The behavior of (1,2/3,4)-cyclopentanetetrol derivatives in hydrogen fluoride shows that di-cations can be formed. Thus, dissolution of tetra-*O*-acetyl-(1,2/3,4)-cyclopentanetetrol[28,29] in hydrogen fluoride gives only the di-cation **139**, as indicated by n.m.r.-spectral studies on the solution.[11] However, pentaerythritol tetraacetate reacts in hydrogen fluoride to give the mono-cation **140** only, and a *spiro*-linked di-cation formed from **140** is not obtained.[11]

The reaction of hexa-*O*-acetyl-*epi*-inositol (**141b**) with hydrogen fluoride, and subsequent hydrolysis, gives[61] a mixture of *epi*-inositol (**141a**) and *allo*-inositol (**142**). From penta-*O*-acetyl-5-*O*-methyl-

(63) S. J. Angyal, P. A. J. Gorin, and M. E. Pitman, *Proc. Chem. Soc.*, 337 (1962).
(64) S. J. Angyal, P. A. J. Gorin, and M. E. Pitman, *J. Chem. Soc.*, 1807 (1965).
(65) S. J. Angyal and D. J. McHugh, *Chem. Ind.* (London), 1147 (1956).
(66) H. Paulsen, and H. Höhne, unpublished results.

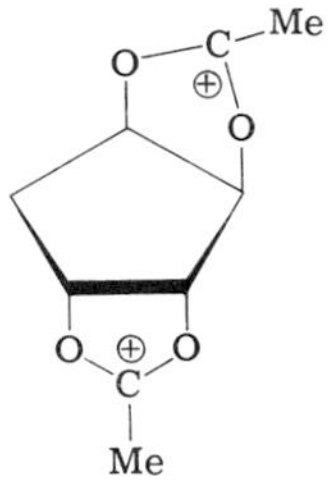

O—C—Me
O
O—C—O
Me

(139)

AcO   OAc
H$_2$C   CH$_2$
H$_2$C   CH$_2$
O   O
C
Me

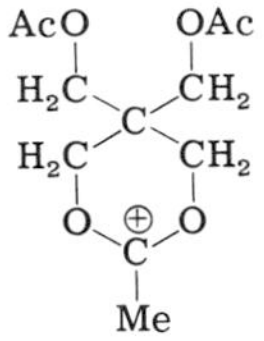

(140)

D-*chiro*-inositol (**145b**), under similar conditions, 3-O-methyl-*muco*-inositol (**146**) and 5-O-methyl-DL-*chiro*-inositol (**145a** + **147**) are obtained.[61] The D-*chiro*-inositol derivative (**145b**) evidently isomerizes to the *muco* derivative (**146**), and this, in turn, isomerizes to the derivative **147**, so that, besides **146**, the racemate (**145a** + **147**) results. By isomerization in hydrogen fluoride, penta-O-acetyl-1-O-methyl-L-*chiro*-inositol (**143b**) gives L-1-O-methyl-*myo*-inositol (**144**) in 30% yield.[61]

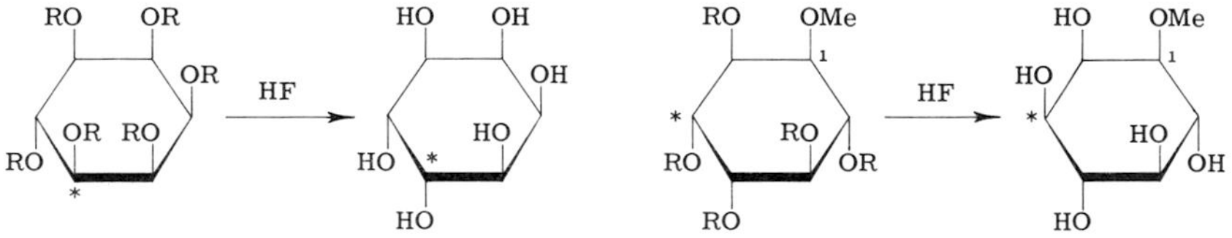

RO   OR
OR
OR RO
RO
*

$\xrightarrow{\text{HF}}$

epi
(141)

HO   OH
OH
HO
HO   *
HO

allo
(142)

RO   OMe
1
RO
RO   OR
RO
*

$\xrightarrow{\text{HF}}$

L-*chiro*
(143)

HO   OMe
1
HO
*
HO
HO
HO

myo
(144)

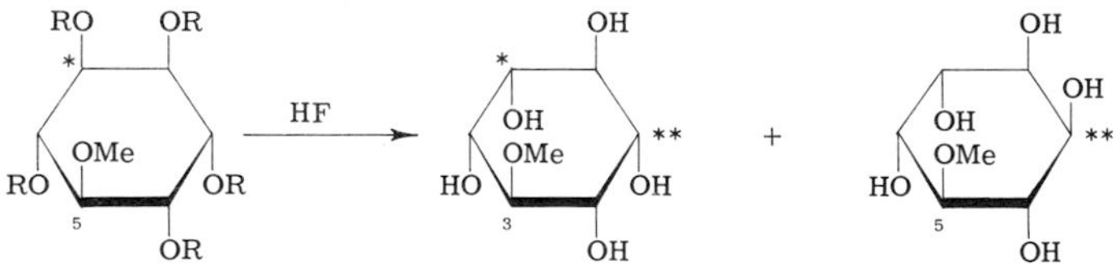

RO   OR
*
OMe
RO   OR
OR
5

$\xrightarrow{\text{HF}}$

D-*chiro*
(145)

OH
*
OH
OMe
HO   OH
OH
3

**

muco
(146)

+

OH
OH
OH
OMe
HO   OH
5

**

L-*chiro*
(147)

(a) R = H
(b) R = Ac

Isomerization in hydrogen fluoride is effective for rearrangement of the readily obtainable triacetate of methyl *rac*-4-epishikimate (**148b**). The methyl ester (**149**) of *rac*-shikimic acid can be isolated[67]

(67) R. Grewe and S. Kersten, *Chem. Ber.*, **100**, 2546 (1967).

crystalline in 80% yield, together with 16% of **148a**. By isomerization, in hydrogen fluoride, of the reduction product (**150b**) of quinic acid, the rearranged product **151** can be obtained.[68]

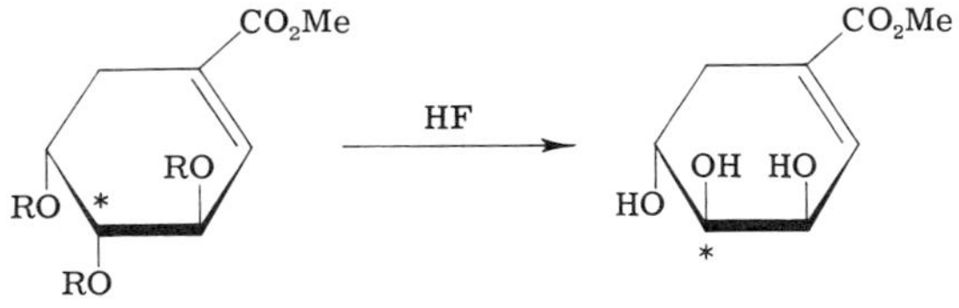

(148)   HF   (149)

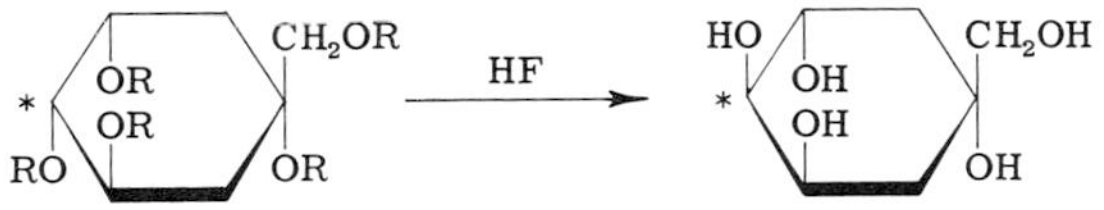

(150)   HF   (151)

(a) R = H
(b) R = Ac

2. Rearrangement of Anhydroalditols and Alditols

Isomerization of 1,5-anhydroalditols in hydrogen fluoride follows the same principle as that obtaining with the cyclitols, in that a *cis-trans* grouping (**131**) is rearranged to give the system **132**. Tetra-*O*-acetyl-1,5-anhydro-D-glucitol (**152**) contains no *cis*-disposed acetoxyl groups, and thus does not undergo isomerization in hydrogen fluoride; indeed, it is stable in this solvent.[62] Tetra-*O*-acetyl-1,5-anhydro-D-galactitol (**153b**) in hydrogen fluoride undergoes the anticipated isomerization at C-3, and, after hydrolysis, equal amounts of 1,5-anhydro-D-galactitol (**153a**) and 1,5-anhydro-D-gulitol (**154**)

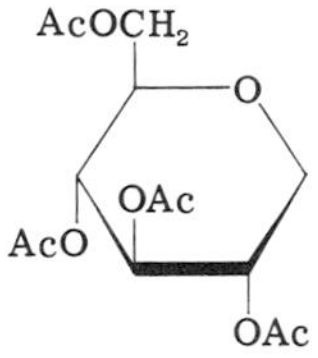

gluco
(152)

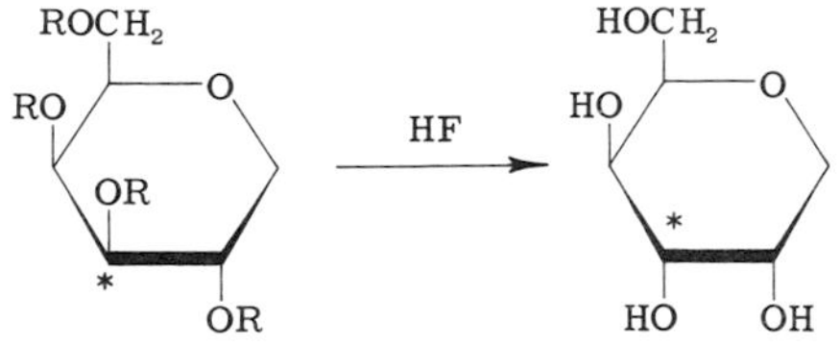

galacto
(153)   HF

gulo
(154)

(a) R = H
(b) R = Ac

(68) P. A. J. Gorin, *Can. J. Chem.*, **41**, 2417 (1963).

are obtained.[62] Likewise, tetra-O-acetyl-1,5-anhydro-D-mannitol
(**156b**) isomerizes at C-3 by the action of hydrogen fluoride. The
yield of D-*manno* derivative (**156a**) obtained after hydrolysis is considerably higher (84%) than that of the D-*altro* product (**157**), which

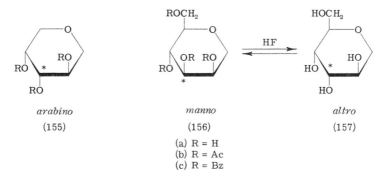

	arabino	*manno*	*altro*
	(155)	(156)	(157)

(a) R = H
(b) R = Ac
(c) R = Bz

is obtained in only 16% yield.[62] In the case of 1,5-anhydro-tri-O-ben-
zoyl-D-arabinitol (**155c**), the anticipated rearrangement at C-3 leads,
by virtue of the symmetry of this compound, to an identical structure.
Treatment of **155c** with hydrogen fluoride, followed by deacetylation,
gives[62] only the product **155a**.

The reaction of peracetylated 1,4-anhydroalditols with hydrogen
fluoride is complex, and leads not only to deacetylated products
but, frequently, to isomerized, open-chain derivatives. Thus, on
treatment with hydrogen fluoride, and subsequent hydrolysis,
di-O-acetyl-1,4-anhydroerythritol gives a mixture of 1,4-anhydroery-
thritol, erythritol, and a threitol. 1,4-Anhydrothreitol is not formed.[69]

1,4-Anhydro-tri-O-benzoyl-D-ribitol (**158b**) reacts with hydrogen
fluoride to give, after hydrolysis, a mixture of 1,4-anhydro-D-ribitol
(**158a**), 1,4-anhydro-L-lyxitol (**159**), DL-arabinitol, xylitol, and ribitol.[69]

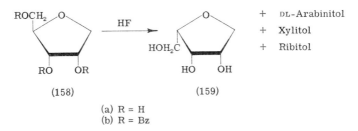

+   DL-Arabinitol
+   Xylitol
+   Ribitol

(158)   (159)

(a) R = H
(b) R = Bz

Even more complex is the corresponding reaction of tri-O-acetyl-
1,4-anhydro-D-arabinitol (**160b**), which gives a mixture of 1,4-anhydro-

(69) E. J. Hedgley and H. G. Fletcher, Jr., *J. Amer. Chem. Soc.*, **86**, 1576 (1964).

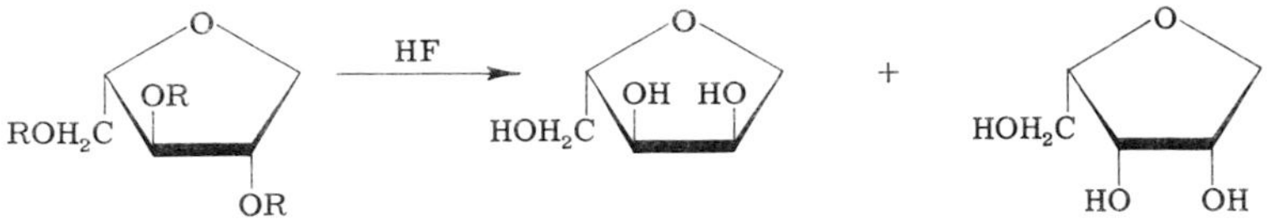

(160)   (161)   (162)

(a) R = H
(b) R = Ac

D-arabinitol (**160a**), 1,4-anhydro-L-ribitol (**161**), 1,4-anhydro-L-lyxitol (**162**), an arabinitol, and ribitol. Tetra-O-acetyl-1,4-anhydro-D-glucitol likewise reacts unpredictably in hydrogen fluoride; the products found after hydrolysis were 1,4-anhydro-D-glucitol, 1,4-anhydro-D-mannitol, galactitol, D-glucitol, an iditol, and a mannitol.[69]

The sole compound found to react with hydrogen fluoride to give a single product is tri-O-acetyl-1,4-anhydro-D-xylitol (**163b**), which leads, after hydrolysis, to 1,4-anhydro-D-ribitol (**164**), isolated crys-

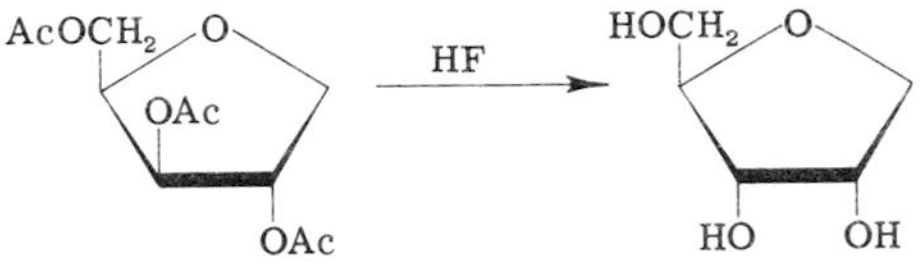

(163)   (164)

(a) R = H
(b) R = Ac

talline.[69] In this reaction, it is conceivable that a 3,5-acetoxonium ion (dioxanylium ring), is formed, and that this is opened at C-3 through a neighboring-group reaction of the 2-acetoxyl group, thereby giving rise to the product having the *ribo* configuration. It is however, more probable, that it is the *trans*-2,3-diol diacetate system that reacts, because, in the cyclopentane system (and only in this system), a slow reaction of a *trans*-1,2-diacetoxyl grouping has been observed in hydrogen fluoride. Di-O-acetyl-*trans*-1,2-cyclopentanediol (**14**) (see p. 132) reacts with hydrogen fluoride in 72 hours to give the acetoxonium ion (**15**), with conversion of the *trans*-diol into a *cis*-diol.[11] By such a pathway, compound **164** could arise from **163b**, and, similarly, **161** and **162** could arise from **160b**. In compound **158b**, the formation of a benzoxonium salt from the *cis*-disposed 2,3-substituents is straightforward.

The action of hydrogen fluoride on acyclic alditol acetates gives deacetylation products and complex mixtures of isomers. This is understandable, because, in such flexible systems, it is possible for

intermediate 1,3-dioxolanylium and 1,3-dioxanylium cations to rearrange in numerous ways. The action of hydrogen fluoride on penta-
O-acetyl-DL-arabinitol gives a mixture of DL-arabinitol, xylitol, and
ribitol.[70] Similarly, hexitol hexaacetates give, after the hydrolytic
step, various isomerization products; hexa-O-acetylgalactitol and
hexa-O-acetyl-D-mannitol give rise to galactitol, a glucitol, an iditol,
and a mannitol; and hexa-O-acetyl-D-glucitol gives a glucitol and an
iditol, but galactitol and a mannitol were not detectable.[70]

V. Rearrangement of Esters of Monosaccharides in Liquid Hydrogen Fluoride

1. Rearrangement of Pentoses and Hexoses

The basic principles observed with simple model compounds
(see Section I,2, p. 132) and cyclitols (see Section IV,1, p. 169),
according to which, esters of cis-1,2-diols react with hydrogen fluoride
to give acyloxonium ions, and esters of cis-trans-1,2,3-triols give a
cis-acyloxonium ion that can undergo acyloxonium rearrangement
leading to configurational inversion at the middle carbon atom, are
also valid for esters of monosaccharides. However, esters of aldoses
have a reactive leaving-group at C-1 that can, on the one hand, react
with hydrogen fluoride by substitution, to give aldosyl fluoride
derivatives, or, on the other hand, be split off directly, to give 1,2-
acyloxonium derivatives, the starting point for further acyloxonium
rearrangements. It is understandable that a complex sequence of
possible reaction-pathways may result.

The reactions of esters of monosaccharides in hydrogen fluoride
have been studied intensively by Pedersen. Besides making analyses
of the rearranged products, it has been found of great value to monitor
the progress of the reactions by n.m.r. spectroscopy. Measurements
can be made directly on the reaction mixtures in hydrogen fluoride
by use of a Teflon tube.[12] The acyloxonium ions formed as intermediates can be detected, and their n.m.r. spectra analyzed.

Tetra-O-benzoyl-β-L-arabinopyranose (enantiomorph of **165b**) is
isomerized by hydrogen fluoride after 64 hours, to give 36% of
2,4-di-O-benzoyl-β-L-ribopyranosyl   fluoride   (enantiomorph   of
167).[71] This reaction with D-arabinose derivatives has been examined
by n.m.r. spectroscopy.[72] Tetra-O-acetyl-α-D-arabinopyranose (**165a**)

(70) E. J. Hedgley and H. G. Fletcher, Jr., *J. Amer. Chem. Soc.*, **86**, 1583 (1964).
(71) C. Pedersen and H. G. Fletcher, Jr., *J. Amer. Chem. Soc.*, **82**, 945 (1960).
(72) C. Pedersen, *Acta Chem. Scand.*, **22**, 1888 (1968).

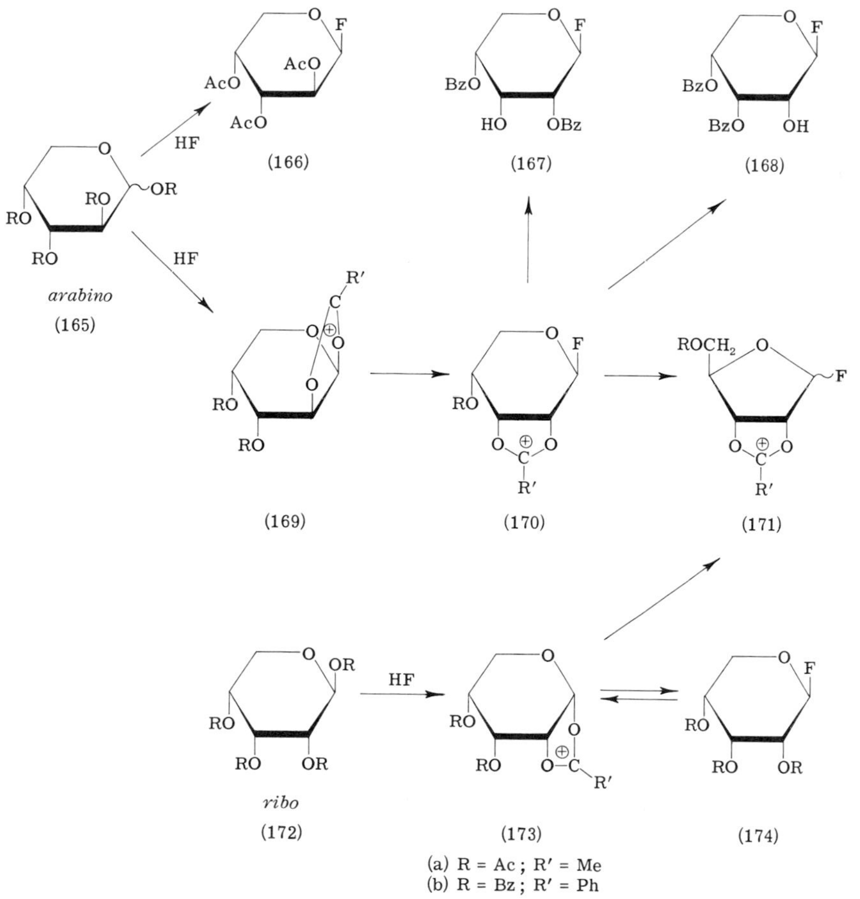

(a) R = Ac ; R′ = Me
(b) R = Bz ; R′ = Ph

reacts rapidly in hydrogen fluoride to give[72] the isolable β-D fluoride **166**. When the solution is kept for 80 hours at 20°, there follows a series of rearrangements, presumably by way of **169a** and **170a**, leading finally to the furanoid ion **171a**, the formation of which can by recognized from its known n.m.r. spectrum.[73]

Dissolution of tetra-*O*-benzoyl-β-D-arabinopyranose (**165b**) in hydrogen fluoride immediately gives the benzoxonium ion derivative **169b**, and not a fluoro sugar,[72] but, if the solution is processed directly, tri-*O*-benzoyl-α-D-arabinopyranosyl fluoride is obtained.[71,72] If the solution of **165b** in hydrogen fluoride is kept for 24 hours at 0°, the ion **169b** undergoes an acyloxonium-ion rearrangement, to give the

(73) N. Gregersen and C. Pedersen, *Acta Chem. Scand.*, **22**, 1307 (1968).

D-*ribo* derivative **170b**. Hydrolysis of this solution gives 2,4-di-O-ben-zoyl-β-D-ribopyranosyl fluoride (**167**) and the 3,4-di-O-benzoyl analog (**168**) in 35 and 22% yields, respectively.[71,72] If the solution of **170b** is kept still longer, namely, for 6 days at 20°, a ring contraction occurs,[72] by a process not yet established, to give the ion **171b**.

Tetra-O-acetyl-β-D-ribopyranose (**172a**) reacts at once in hydrogen fluoride to give[72] tri-O-acetyl-β-D-ribopyranosyl fluoride (**174a**). Upon longer reaction (5 days at 20°), compound **172a** is transformed by way of the ion **173a** into the furanoid ion **171a** as the final product.[72] In contrast, dissolution of tetra-O-benzoyl-β-D-ribopyranose (**172b**) in hydrogen fluoride gives the benzoxonium ion **173b** directly,[72] demonstrating that acyloxonium-ion formation also occurs readily from vicinal, *trans*-diacyloxy groups when C-1 is involved.[72] Processing of the solution of **173b** leads[74] to tri-O-benzoyl-β-D-ribopyranosyl fluoride (**174b**), but dissolution of **174b** in hydrogen fluoride at once gives back[72] the benzoxonium ion **173b**; evidently, the benzoylated fluoride **174b** is not stable in hydrogen fluoride, and dissociates spontaneously to the ion **173b**. When the solution of **173b** is kept for a longer period (6 days at 20°), the principal end-product observed is the ring-contraction product **171b,** evidently a product having high stability. Following hydrolysis of this solution, 52% of D-ribo-furanose derivatives could be isolated, comprised of 12% of 3,5-di-O-benzoyl-α-D-ribofuranosyl fluoride, 32% of a mixture of 2,5- and 3,5-di-O-benzoyl-β-D-ribofuranosyl fluoride, and 8% of tri-O-benzoyl-β-D-ribofuranosyl fluoride.[72]

Tetra-O-acetyl-α-D-xylopyranose in hydrogen fluoride is converted rapidly into tri-O-acetyl-α,β-D-xylopyranosyl fluoride, and this product subsequently undergoes further transformations that have not yet been clarified.[75] In contrast, tetra-O-benzoyl-α- or β-D-xylopy-ranose (**175**), when dissolved in hydrogen fluoride, at once shows[75] the n.m.r. spectrum of the benzoxonium ion **176**. If this solution is rapidly processed, tri-O-benzoyl-α,β-D-xylopyranosyl fluoride (**177**) is obtained.[76] An ion-pair of the cation **176** and HF_2^- probably exists in the solution; these unite, to form **177**, as the solution is concentrated.

When the solution of the benzoxonium ion **176** in hydrogen fluoride is kept for 20 hours, the compound is completely transformed, and the n.m.r. spectrum indicates[75,76] that the principal product is the di-cation **183**. The ion is quite stable, because no significant further

(74) C. Pedersen and H. G. Fletcher, Jr., *J. Amer. Chem. Soc.*, **82**, 941 (1960).
(75) C. Pedersen, unpublished results.
(76) C. Pedersen, *Acta Chem. Scand.*, **17**, 1269 (1963).

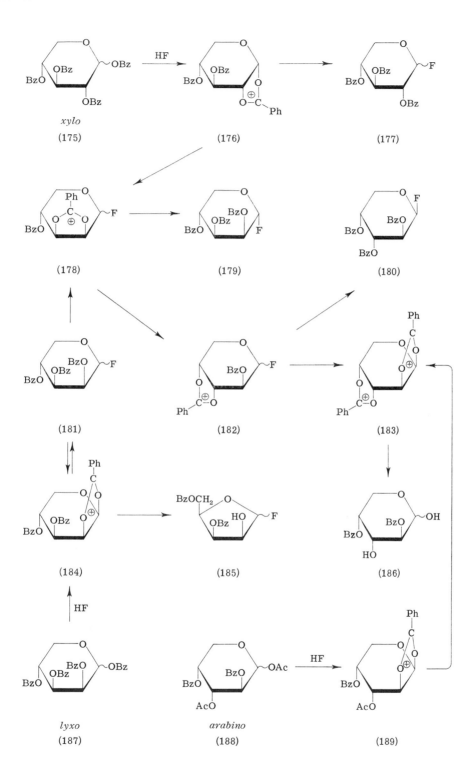

xylo
(175)

(176)

(177)

(178)

(179)

(180)

(181)

(182)

(183)

(184)

(185)

(186)

HF

lyxo
(187)

arabino
(188)

(189)

change is observed in the n.m.r. spectrum when the solution is kept for longer periods. Hydrolysis of the solution gives 2,4-di-O-benzoyl-D-arabinopyranose (186), isolated in 47% yield.[76]

Formation of the di-cation 183 results[75] from acyloxonium rearrangement of the *xylo* ion 176, by way of the *lyxo* ion 178, to the *arabino* ion 182; the latter splits out fluoride anion, to give the di-cation 183. To obtain evidence regarding this mechanism, products that would arise from the intermediates 178 and 182 were sought in hydrolyzates of the solution. After separation of 2,4-di-O-benzoyl-D-arabinopyranose (186), which is the main product of hydrolysis, the remaining mixture was benzoylated, and from the resulting mixture of products were isolated[76] 14.5% of tri-O-benzoyl-α-D-lyxopyranosyl fluoride (179) and 4.5% of the β-D-*arabino* analog 180. Compound 179 must have arisen from the corresponding dibenzoate formed on hydrolysis of the D-*lyxo* ion (178), and similarly, compound 180 must have been formed through the ion 182; these results provide convincing evidence for the existence of the intermediate ions 178 and 182.

As in the *xylo* series, tetra-O-acetyl-D-lyxopyranose reacts rapidly in hydrogen fluoride to give tri-O-acetyl-α,β-D-lyxopyranosyl fluoride, whereas tetra-O-benzoyl-α-D-lyxopyranose (187) gives at once the benzoxonium ion (184), and, by rapid processing of the solution, the corresponding tri-O-benzoyl-α,β-D-lyxopyranosyl fluoride (181) can be isolated.[75] Probably, an equilibrium between the ion 184 and the fluoride 181 exists.

When a solution of 187 in hydrogen fluoride is kept for a longer period (29 hours), the *cis*-dibenzoyloxy groups in the resulting fluoride 181 react to give the ion 178 which, as in the *xylo* series, rearranges, by way of the ion 182, into the di-cation 183; again, this very stable, principal product is readily recognizable from its n.m.r. spectrum.[75] Hydrolysis of the solution gives 2,4-di-O-benzoyl-D-arab-inose (186) in 16% yield.[77] In comparison, however, the formation of the cation 178 from 181 by a *cis* displacement is a slower reaction than the neighboring-group-assisted formation of 178, in the *xylo* series, from 176. In the *lyxo* series, there occurs, to a significant extent, a concurrent reaction that leads by ring-contraction to 3,5-di-O-benzoyl-α-β-D-lyxofuranosyl fluoride (185), probably by way of the ion 184. The presence of the furanosyl derivative 185 can be presumed, because of the isolation of methyl α-D-lyxofuranoside in 8.5% yield after treatment of the mixture with methanol and sub-

(77) C. Pedersen, *Acta Chem. Scand.*, 18, 60 (1964).

sequent saponification.[77] To this extent, the course of the reaction in the *lyxo* series is more complex than in the *xylo* series.

The di-cation **183**, evidently a strongly favored species, can also be obtained as the principal product after treating the *arabino* ester **188** with hydrogen fluoride for 24 hours.[72] The reaction proceeds by way of the ion **189**, from which the di-cation **183** is formed by a *cis*-diacyloxy reaction in which the loss of acetic acid, rather than benzoic acid, is favored.[72] Hydrolysis of the solution gives[72] the anticipated dibenzoate **186**.

The action of hydrogen fluoride has also been examined with pentofuranose derivatives. Upon dissolution in hydrogen fluoride, β-D-ribofuranose tetraacetate (**190a**) and tetrabenzoate (**190b**) react[73] at once to give the acyloxonium derivatives **191a** and **191b**, respectively. Rapid isolation of the products gives[73] the corresponding

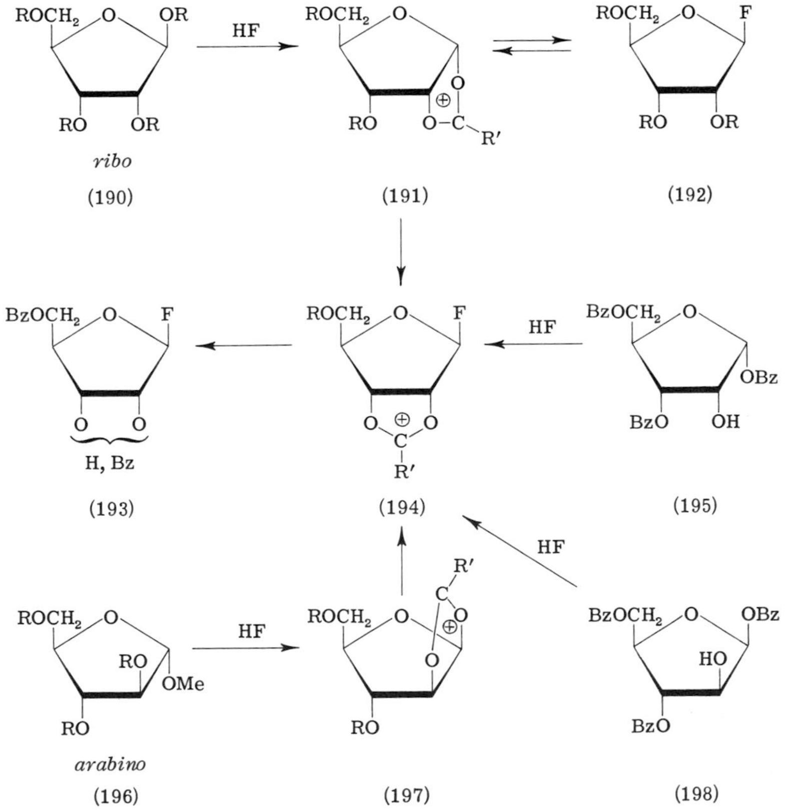

ribo

(190)

(191)

(192)

(193)

(194)

(195)

arabino

(196)

(197)

(198)

(a) R = Ac ; R′ = Me
(b) R = Bz ; R′ = Ph

β-D fluorides, **192a** and **192b**. However, both fluorides are unstable in hydrogen fluoride, and, upon dissolution, the ions **191a** and **191b** are regenerated. Evidently, the dissociation of the glycosyl fluoride takes place particularly readily in the ribofuranose system. In the β-D-ribopyranose system, the tetrabenzoate (**172b**) (see p. 178) gives an acyloxonium ion directly with hydrogen fluoride, in contrast to the tetraacetate (**172a**), which gives instead the β-D fluoride **174a**.

The ions **191a** and **192b** are rearranged after 1–3 days at 20° in hydrogen fluoride into the fluorine-containing ions **194a** and **194b**, respectively. Hydrolysis of the benzoyl derivative **194b** permits the isolation[73] in 44% yield of the anticipated mixture of the 2(and 3),5-di-O-benzoyl-β-D-ribofuranosyl fluorides (**193**). 1,3,5-Tri-O-benzoyl-α-D-ribofuranose (**195**) likewise reacts in hydrogen fluoride during 20 hours at 20° to give the benzoxonium ion **194b**. This reaction proceeds by way of 3,5-di-O-benzoyl-β-D-ribofuranosyl fluoride (**193**, OBz on C-3), which can be detected in the reaction mixture at short reaction-times.[73]

Surprisingly, methyl α-D-arabinofuranoside triacetate (**196a**) and tribenzoate (**196b**), upon dissolution in hydrogen fluoride, also undergo conversion[73] at once into the corresponding ions (**197a** or **197b**). These reactions provide particularly clear evidence that a *trans*-vicinal diol system at C-1 and C-2 in 2-O-acyl monosaccharides is converted into acyloxonium ions with the utmost ease by the action of hydrogen fluoride. After 24 hours in hydrogen fluoride, the ions **197a** and **197b** are rearranged into the ions **194a** and **194b**, respectively, because the n.m.r. spectra of the solutions are identical with those of the final products obtained from the corresponding β-D-ribofuranose esters **190a** and **190b**. Similarly, 1,3,5-tri-O-benzoyl-β-D-arabinofuranose (**198**) in hydrogen fluoride eventually gives[73] the ion **194b**.

Treatment of penta-O-acetyl-β-D-glucopyranose with hydrogen fluoride for 20 hours gives tetra-O-acetyl-α-D-glucopyranosyl fluoride, but only in very low yield.[78] To elucidate the composition of the product mixture, it was treated with sodium methoxide in methanol, and 28% of methyl β-D-mannopyranoside and 11% of 1,6-anhydro-β-D-altropyranose were isolated; also detected were methyl α-D-altropyranoside, methyl β-D-glucopyranoside, and 1,6-anhydro-β-D-glucopyranose.[78] Evidently, successive rearrangements from the D-*gluco* to the D-*manno* and then to the D-*altro* configuration occur, and the sequence is probably analogous to the acyloxonium rearrange-

---

(78) C. Pedersen, *Acta Chem. Scand.*, **16**, 1831 (1962).

ments in the D-*xylo* series through the corresponding ions **176, 178,** and **182**.

Likewise, a solution of penta-*O*-acetyl-α-D-mannopyranose in hydrogen fluoride was kept for 20 hours, and the reaction product was treated with sodium methoxide. From the resultant mixture were obtained 30% of methyl α-D-mannopyranoside, 7% of methyl α-D-altropyranoside, 12% of 1,6-anhydro-β-D-altropyranose, and 1.5% of methyl α-D-idopyranoside.[79] An accurate, n.m.r.-spectral study of the progress of this reaction in the D-*gluco* and D-*manno* series has not yet been reported.

2. Rearrangement of Selectively Protected Pentoses and Hexoses

Selectively methylated monosaccharide derivatives are particularly well suited for studies on acyloxonium rearrangements in hydrogen fluoride, because the methyl ether group is stable to hydrogen fluoride, so that the rearrangement can be blocked at predetermined positions. 2-*O*-Methyl-α,β-D-arabinopyranose tribenzoate (**199**) and the β-D-xylopyranose analog (**204**) cannot give a vicinal, 1,2-cyclic benzoxonium ion in the initial step. The D-*arabino* derivative (**199**) in hydrogen fluoride is converted[75] after 5 minutes into 3,4-di-*O*-benzoyl-2-*O*-methyl-α,β-D-arabinopyranosyl fluoride (**200**), and, during the following 24 hours, the *cis*-diol ester groups at C-3 and C-4 of **200** react to give the benzoxonium ion **201**. The products[75] of hydrolysis of **201** are 3-*O*-benzoyl-2-*O*-methyl-α,β-D-arabinopyranosyl fluoride (**202a**) and the 4-*O*-benzoyl analog (**202b**). Products of ring contraction were not observed in this example.

1,3,4-Tri-*O*-benzoyl-2-*O*-methyl-α,β-D-xylose (**204**) in hydrogen fluoride is converted[75] in 5 minutes into 3,4-di-*O*-benzoyl-2-*O*-methyl-α,β-D-xylosyl fluoride (**205**). This product, in contrast to **200**, has no *cis*-diol ester grouping, but it nevertheless undergoes further reactions. After 24 hours, the principal product, as detected by n.m.r. spectroscopy, is the furanoid ion **206**. Hydrolysis of the solution gave[75] 3-*O*-benzoyl-2-*O*-methyl-α,β-D-xylofuranosyl fluoride (**203a**) and the 5-*O*-benzoyl analog (**203b**). The exact cause of the ring-contraction reaction from **205** to **206** has not yet been ascertained.

On brief treatment with hydrogen fluoride, 1,3,4,6-tetra-*O*-benzoyl-2-*O*-methyl-β-D-glucopyranose reacts to give 3,4,6-tri-*O*-benzoyl-2-*O*-methyl-α-D-glucopyranosyl fluoride in 81% yield.[80] This product is relatively stable, and is still obtainable in 42% yield after a 24-

(79) C. Pedersen, *Acta Chem. Scand.*, **17**, 673 (1963).
(80) C. Pedersen, *Acta Chem. Scand.*, **20**, 963 (1966).

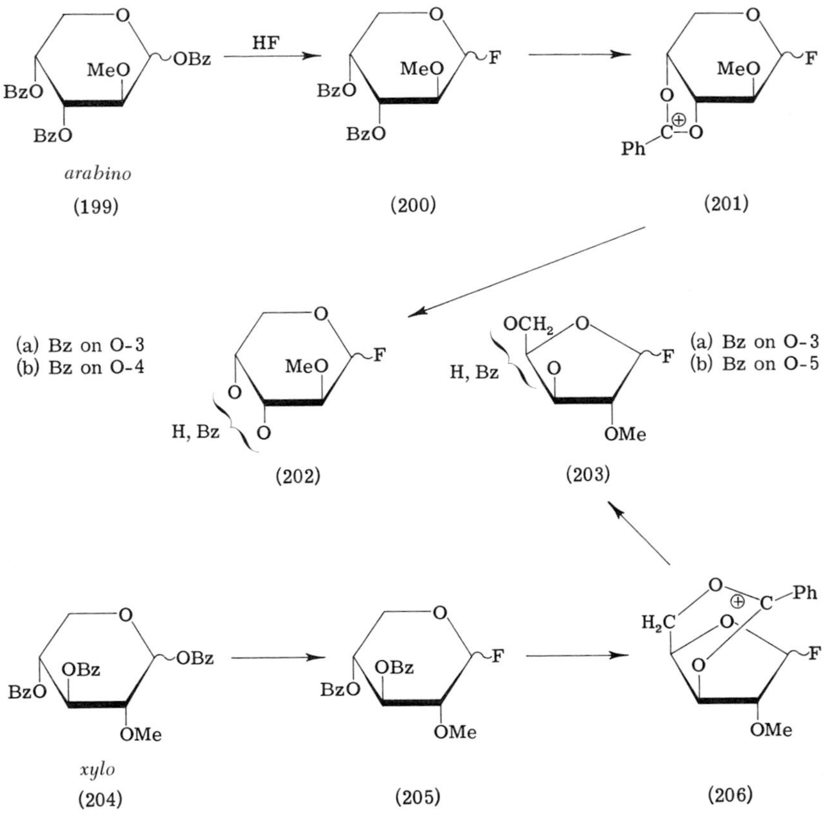

arabino
(199) → (200) → (201)

(a) Bz on O-3
(b) Bz on O-4

(202)   (203)

(a) Bz on O-3
(b) Bz on O-5

xylo
(204) → (205) → (206)

hour reaction-period. After this period, hydrolysis of the mixture gave 3,5- and 3,6-di-O-benzoyl-2-O-methyl-D-glucofuranosyl fluoride as side products.[80] Thus, the *gluco* derivative resembles the *xylo* derivative in following a pyranose → furanose rearrangement analogous to the conversion **205** → **206**.

Treatment of 1,3,4,6-tetra-O-acetyl-2-O-methyl-α-D-glucopyranose with hydrogen fluoride for 48 hours results in quantitative conversion into a furanose derivative, as observed by n.m.r. spectroscopy. When the hydrogen fluoride solution is processed and the product acetylated, 3,5,6-tri-O-acetyl-2-O-methyl-α,β-D-glucofuranosyl fluoride is isolated in 83% yield.[75]

The tetraacetates of 2- and 3-O-methyl-D-mannopyranose, and of 3-deoxy-D-*ribo*-hexopyranose are all completely converted into furanose derivatives when treated with hydrogen fluoride for 1–3 days.[75]

On dissolution in hydrogen fluoride, the tetraacetate (**207a**) of

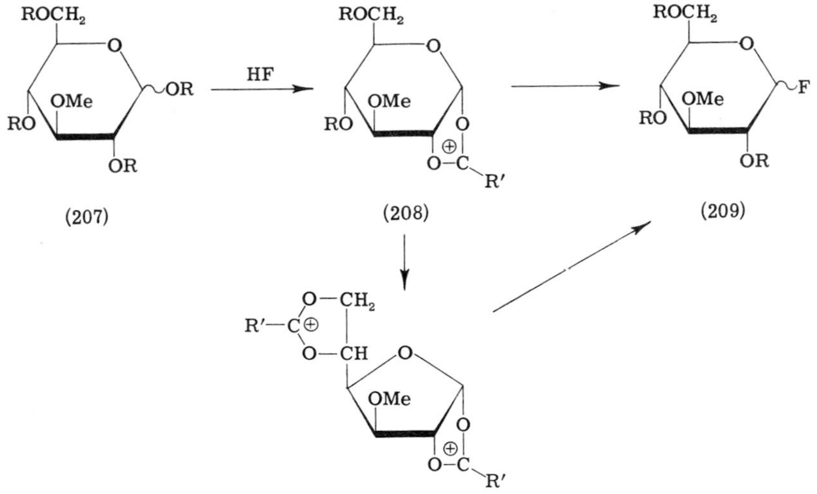

ROCH$_2$ ROCH$_2$ ROCH$_2$

OMe   OR $\xrightarrow{\text{HF}}$ OMe   OMe   F

RO   RO   RO

OR   O–C$^{\oplus}$ R′   OR

(207)   (208)   (209)

R′—C$^{\oplus}$ O—CH$_2$

O—CH   OMe

O–C$^{\oplus}$ R′

(210)

(a) R = Ac ; R′ = Me
(b) R = Bz ; R′ = Ph

3-O-methyl-D-glucose shows[75] at once the n.m.r. spectrum of the glycosyl fluoride **209a**. In contrast, the tetrabenzoate analog (**207b**), similarly treated, gives[75] the benzoxonium ion **208b** directly, and only upon removal of the solvent is the fluoro derivative **209b** formed, presumably because the original ion-pairs are brough together.[81] A neighboring-group reaction in the ion **208** is no longer possible, as the 3-methyl ether group cannot react, and so the ion **208** is relatively stable. Nevertheless, after the action of hydrogen fluoride for 24 hours on the acetate **207a** or benzoate **207b**, the n.m.r. spectra indicated that the furanoid bis(dioxolanylium) di-cation **210a** or **210b** is formed quantitatively. Processing of the hydrogen fluoride solution, which contains **210a**, gives a 50% yield of the fully acetylated pyranosyl fluoride **209a**. This is a strange reaction that has not yet been explained. Apparently, the di-cation **210a** in hydrogen fluoride solution forms a complex with acetic acid, and this complex rearranges rapidly to the glycosyl fluoride **209a** when the hydrogen fluoride solution is processed. The bis(dioxolanylium) di-cation **210b**, formed from **207b**, behaves normally. When the hydrogen fluoride solution is processed, there is obtained a mixture of the corresponding partially benzoylated 3-O-methyl-D-glucofuranosyl fluorides.[75]

1,3,4,6-Tetra-O-benzoyl-2-deoxy-β-D-*arabino*-hexopyranose (**211b**),

(81) I. Lundt, C. Pedersen, and B. Tronier, *Acta Chem. Scand.*, **18**. 1917 (1964).

a compound intrinsically incapable of forming an acyloxonium ion, is surprisingly unstable in hydrogen fluoride at 20°, and, in 3 hours, gives a black tar.[82] In 20 minutes at −70°, 3,4,6-tri-O-benzoyl-2-deoxy-α-D-*arabino*-hexopyranosyl fluoride (212b) is obtained.[72,73,83] After treating 211b or 212b with hydrogen fluoride for one hour at −17°, 2,6-di-O-benzoyl-2-deoxy-β-D-*xylo*-hexopyranosyl fluoride (213) can be isolated in 62% yield.[72] The n.m.r. spectrum of the solution in hydrogen fluoride is concordant with the presence of the ion 216b. The initial assumption that the ion 216b is formed by rearrangement of a 1,3-benzoxonium ion[72] could not be substantiated.[75] Compound 216b evidently arises by addition of hydrogen fluoride to the unsaturated ion 215b. The latter is formed from 211b by an elimination reaction, as was inferred from the observation[75] that, if compound

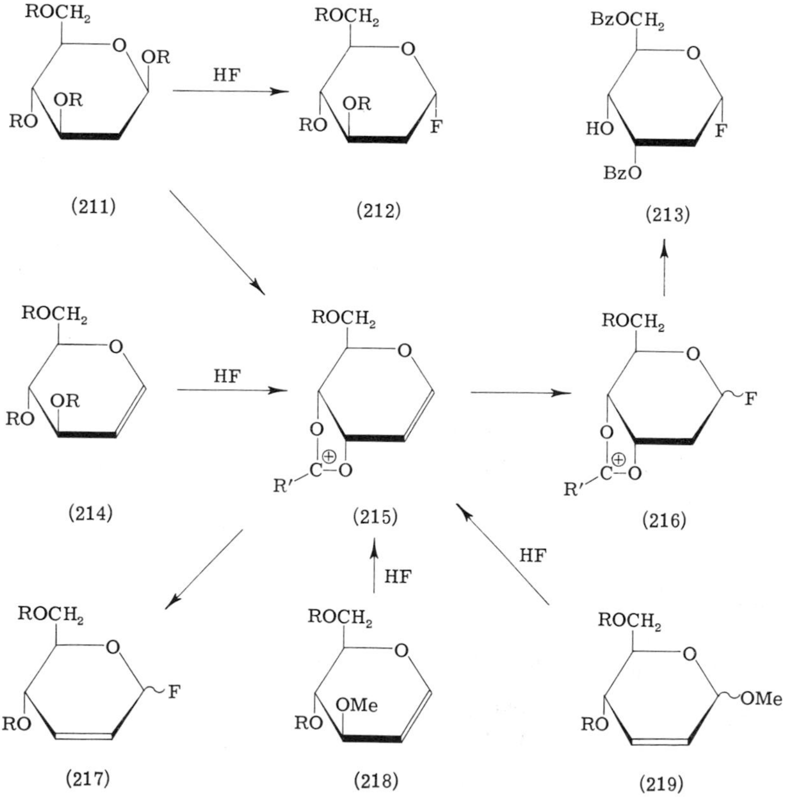

(a) R = Ac ; R′ = Me
(b) R = Bz ; R′ = Ph

(82) I. Lundt and C. Pedersen, *Acta Chem. Scand.*, 21, 1239 (1967).
(83) L. D. Hall and J. F. Manville, *Can. J. Chem.*, 45, 1299 (1967).

211b is treated with liquid deuterium fluoride, a deuterium atom is selectively introduced at C-2 into the isolated product (**213**); this incorporation evidently takes place during the addition step **215b** → **216b**.

In the formation of **215b** from **211b**, the unsaturated derivative **214b** may be considered to be an intermediate. The acetates or benzoates of olefinic sugars, such as **214**, **218**, and **219**, react[84] exceedingly readily with hydrogen fluoride, in 5 minutes at −70°, to give the respective ions **215a** or **215b**. If these reaction mixtures are hydrolyzed at once, the rearranged olefinic sugars **217a** or **217b** can be isolated.[84] After 20 hours in hydrogen fluoride at −70°, compounds **215a** or **215b** add hydrogen fluoride to give the ions **216a** or **216b**, which are evidently the thermodynamically favored products.

The acetate **211a** reacts[82] in 5 minutes at −70° to give 3,4,6-tri-*O*-acetyl-2-deoxy-α-D-*arabino*-hexopyranosyl fluoride (**212a**), and after 24 hours at −70°, likewise by way of the unsaturated ion **215a**, to the ion **216a**. The high stability of the ions of type **216** is also evident in the pentose series. Thus, compounds **220**, **221**, **223**, and

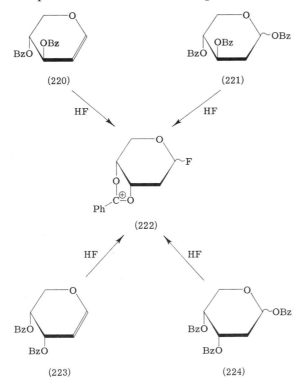

(84) I. Lundt and C. Pedersen, *Acta Chem. Scand.*, **24**, 240 (1970).

224 are all transformed[75] in hydrogen fluoride into the same benz-oxonium ion **222**. Ring-contraction reactions to give furanose derivatives have not been observed with the 2-deoxyaldose derivatives.[82]

As early as 1926, Brauns[85] found that extended treatment of cellobiose octaacetate with hydrogen fluoride led to a rearrangement. A crystalline rearrangement-product, 3,6-di-O-acetyl-4-O-(2,3,4,6-tetra-O-acetyl-β-D-glucopyranosyl)-α-D-mannopyranosyl fluoride, was isolated.[85] The rearrangement of a D-*gluco* precursor to a D-*manno* product is analogous to the course followed with the corresponding monosaccharide derivative, but, inasmuch as C-4 on the reducing end is blocked, no further rearrangement products are formed after the conversion into the D-*manno* derivative.

VI. Rearrangement of Cyclitols with Acetic Acid–Sulfuric Acid

It has been found by Angyal[63,64] that the mixture acetic acid (95%)–sulfuric acid (1.5%) serves as a reagent that is effective in the cyclitol series for causing stereospecific isomerizations. Either acetylated or non-acetylated cyclitols can be used, since acetylation and deacetylation take place concurrently in the reaction mixture. Because of the high temperature (117°) and long times (up to 12 days) of the reaction, it is limited to the cyclitols, which are much more stable under these conditions than, for example, the monosaccharides.

The stereospecificity of the isomerization with acetic acid–sulfuric acid corresponds to that of isomerizations with hydrogen fluoride. Only systems having a *cis-trans*-1,2,3-triol grouping of the type **131** (see p. 170) are rearranged, by a process whereby the configuration of the central hydroxyl group is inverted to give the system **132**; the mechanism is also similar. From the *cis*-1,2-diol diacetate or monoacetate, an acetoxonium ion (**133**) is formed[64] that is rearranged by a neighboring-group reaction to give **134**. The presumed rearrangement-pathway is shown in formulas **131–134**. In the formula schemes that follow, the carbon atom at which inversion takes place is indicated, as in Section IV,1 (see p. 169), by an asterisk.

Starting from either (1,2/3,5)-cyclohexanetetrol (**225**) or (1,2,5/3)-cyclohexanetetrol (**226**) in acetic acid–sulfuric acid for 7 hours at 117°, an equilibrium mixture is obtained that contains[64] 50% of **225** and 29% of **226**. Isomerization of the tetrol **227** leads[64] in 6 days to an equilibrium mixture containing 25% of **227** and 58% of **228**. The

(85) D. H. Brauns, *J. Amer. Chem. Soc.*, **48**, 2776 (1926).

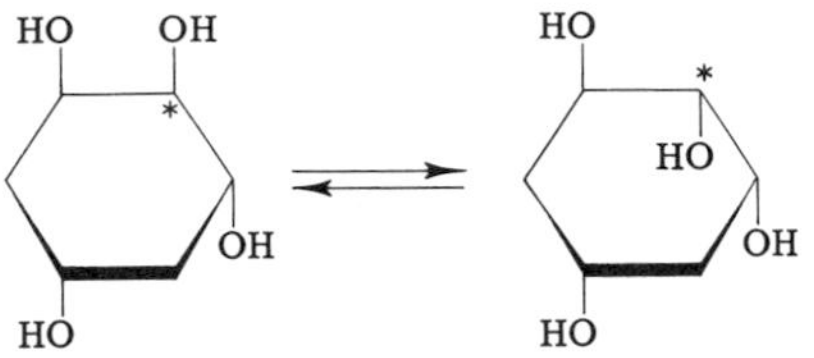

50% *aeee* (225) 29% *eeaa* (226) 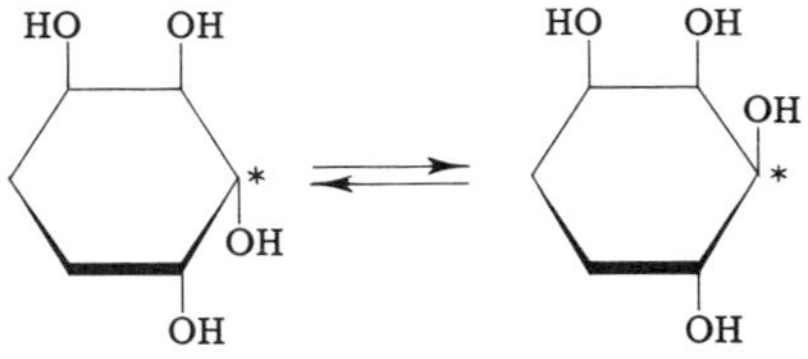 25% *eaea* (227) 58% *eaee* (228)

balance of the mixture consists of side products, but these do not affect the relative proportions of the two tetrols. The reaction mixtures were analyzed periodically by gas–liquid chromatography,[86] and the results showed that the equilibrium position in the isomerization is dependent on the relative conformational free-energies of the participating species. The tetrol **225**, favored to the extent of 50% in the mixture, has only a singly hydroxyl group oriented axially (*aeee* arrangement), whereas the less-favored (29%) tetrol **226** is conformationally less stable, because it is obliged to have two hydroxyl groups attached axially (*eeaa* arrangement).[64] Similarly, in the equilibrium between **227** and **228**, the more-favored (58%) component **228** has only one axially attached substituent (*eaee* arrangement), whereas the less-favored (25%) component **227** has two such axial groups (*eaea* arrangement).

(+)-*proto*-Quercitol (**229**) and (−)-*vibo*-quercitol (**230**) can each be isomerized to give an equilibrium mixture containing the two pentols in the ratio of 39:54, and the optical activity of the compounds is not lost in the process.[64] Likewise, the cyclohexanepentols **231** and **232** are mutually isomerized during 24 hours to give a 3:8 mixture of the

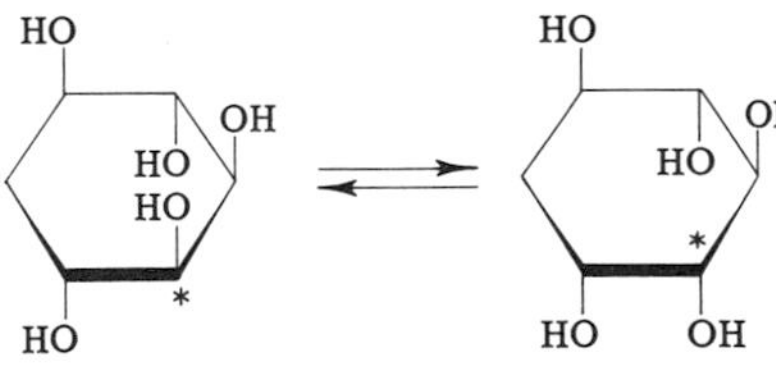

(229) (230) 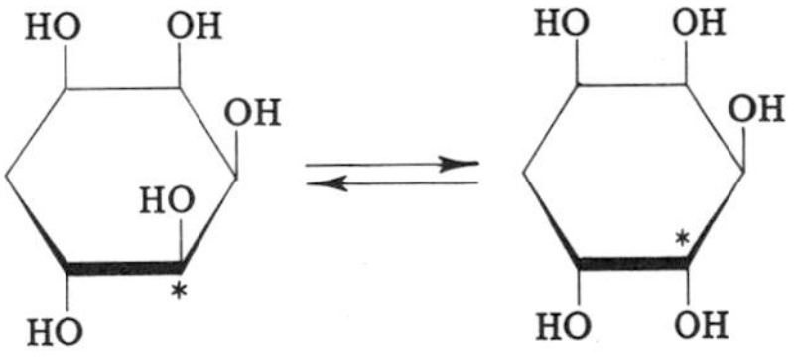 (231) (232)

two.[64] Frequently, isomerization of cyclohexanepentols leads to the formation of additional isomers in small proportion. Treatment of (1,2,3,5/4)-cyclohexanepentol with the acetic acid–sulfuric acid regent gives a reaction mixture that is quite complex.[64]

(86) Z. S. Krzeminski and S. J. Angyal, *J. Chem. Soc.*, 3251 (1962).

Inositols are isomerized during 12 days at 117° to an equilibrium mixture, and the reaction proceeds with high stereospecificity. Thus, *myo*-inositol (**233**) is isomerized[64] in two steps by way of DL-*chiro*-inositol (**234**) to give *muco*-inositol (**235**), and the equilibrium pro-

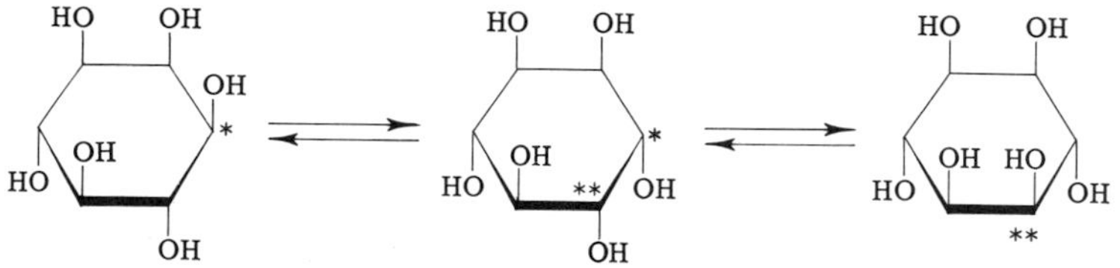

54% *myo*   41% DL-*chiro*   5% *muco*

(233)   (234)   (235)

portions for the process **233** ⇌ **234** ⇌ **235** are 54:41:5. Starting from either *neo*-inositol (**236**), *allo*-inositol (**237**), or *epi*-inositol (**238**), the

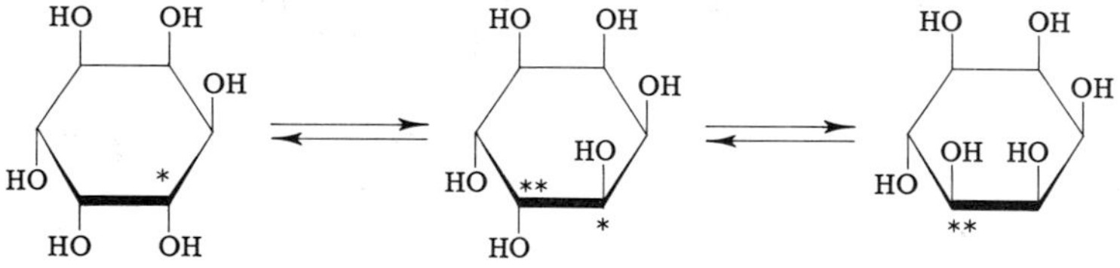

58% *neo*   21% *allo*   15% *epi*

(236)   (237)   (238)

acetic acid–sulfuric acid reagent allows attainment of the second such equilibration process possible in the cyclitol series, and the equilibrium proportions of 58:21:15 were found[64] for the interconversions **236** ⇌ **237** ⇌ **238**. The reaction can be utilized as a preparative route for *neo*-inositol (**236**) from *epi*-inositol (**238**), and **236** can be isolated in 40% yield.[64] *scyllo*-Inositol and *cis*-inositol cannot be isomerized under these conditions, because they lack the requisite *cis-trans*-1,2,3-triol grouping.[64] From the equilibrium concentrations found in the isomerized mixtures, the free-energy differences between the individual inositols can be determined. These differences can likewise be estimated from the intramolecular, steric interactions between the substituents on the cyclohexane ring, and a satisfactory agreement with the experimentally determined values was observed.[63,64]

Cyclopentanepentols that contain a *cis-trans*-1,2,3-triol grouping are isomerized by acetic acid–sulfuric acid; for example, (1,2,3/4,5)-cyclopentanepentol (**239**) is rapidly rearranged into the (1,2,3,4/5) isomer (**240**), and the latter is converted more slowly into the

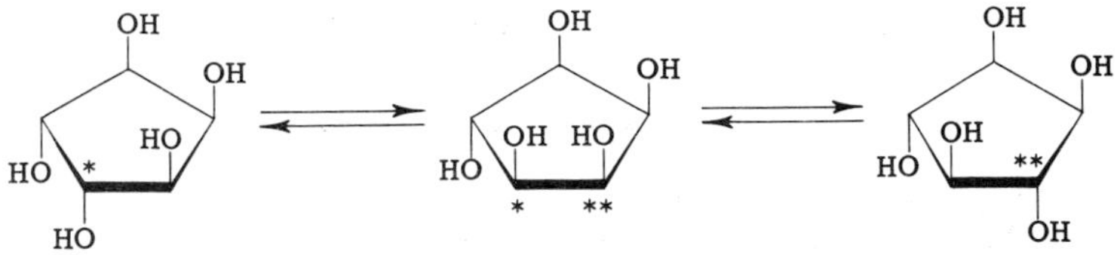

17.5% (1, 2, 3/4, 5)   10.5% (1, 2, 3, 4/5)   72% (1, 2, 4/3, 5)
      (239)   (240)   (241)

(1,2,4/3,5) isomer (**241**). After 21 days of reaction, an equilibrium is established[87] in which the relative proportions in the mixture **241** ⇄ **239** ⇄ **240** are 144:35:21 (±5%). The same equilibrium mixture is obtained by starting with the pentol **241**. The favored isomer (**241**) has the lowest free-energy content as a result of having only one *cis*-1,2-diol interaction, whereas **239** and **240,** having three such interactions, have higher free-energies. It appears that *cis*-1,3-diol interactions also play a role in these cyclopentanepentol systems.[87]

Methyl ethers of inositols can also be isomerized. Quebrachitol (**242**) is converted[64] without racemization into bornesitol (**243**), and the latter is further converted *via* **244** into 1-O-methyl-D-*chiro*-inositol

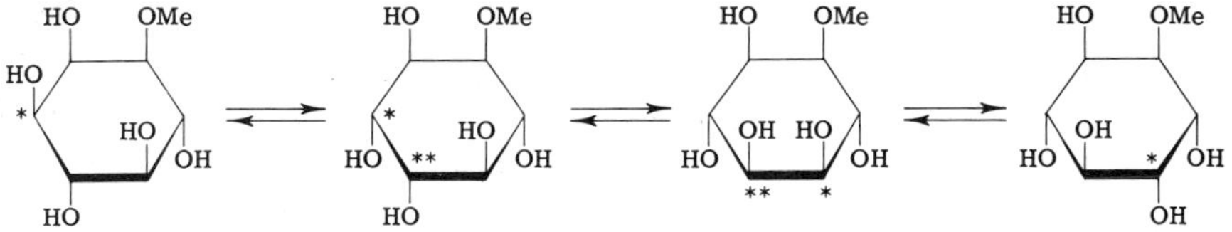

      (242)   (243)   (244)   (245)

(**245**). No inversion occurs at the carbon atoms adjacent to the methyl ether group, in accordance with mechanistic considerations.

Quinic acid (**246**) can be transformed[68] with acetic acid–sulfuric acid into the rearranged lactone **247**. By the same principle, the C-(hydroxymethyl)cyclohexanetetrol **248** is isomerized by acetic acid–sulfuric acid into the isomer **249**, which can be isolated in 14% yield from the reaction mixture.[88]

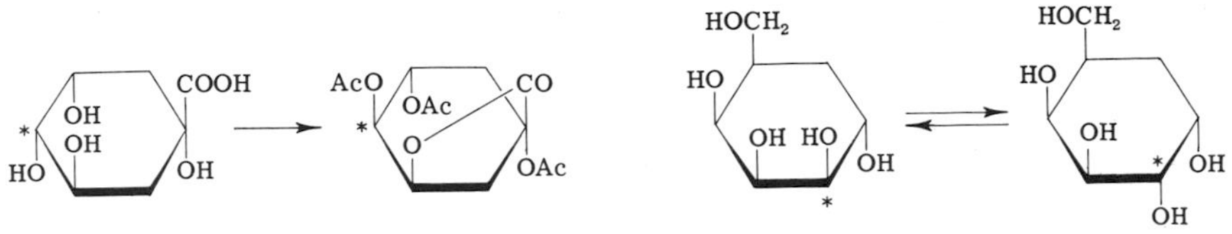

      (246)   (247)   (248)   (249)

(87) S. J. Angyal and B. M. Lutrell, *Aust. J. Chem.*, **23**, 1831 (1970).
(88) G. E. McCasland, S. Furuta, and L. J. Durham, *J. Org. Chem.*, **33**, 2841 (1968).

VII. Rearrangement of Saccharides with Other Lewis Acids

1. Rearrangement with Zinc Chloride–Acetic Anhydride

Treatment of 2,3,4,5-tetra-O-acetyl-6-deoxy-6-iodo-*aldehydo*-D-galactose (205a) or 2,3,4,5-tetra-O-acetyl-6-O-p-tolylsulfonyl-*aldehydo*-D-galactose (250b) with zinc chloride in acetic anhydride for 12 hours at 100° gives optically inactive *aldehydo*-DL-galactose aldehydrol heptaacetate.[89,90] Detailed studies by Micheel and Böhm[91-93] have indicated that this reaction does not involve interchange of the ends of the D-galactose chain, but takes place by total isomerization at all of the carbon atoms along the length of the sugar chain.[91,93] Thus, the sulfonate 250b is first acetylated to give 251a, which is subsequently converted[93] into the chloride 251b. The isomerization reaction observed with 250b is likewise observed with the acetylated sulfonate 251a or chloride 251b. The 6-chloro compound 251b loses chloride ion by the action of the Lewis acid (zinc chloride), and the developing cation at C-6 is stabilized by a neighboring-group reaction, to give the acetoxonium ion 252.

Starting from the ion 252, it is possible for acetoxonium rearrangements to proceed freely along the length of the chain, so that configurational inversion at all of the asymmetric centers is possible. Two such rearrangement steps, from 252 to 256 and from 256 to 255, are shown as illustrative examples. A further configurational inversion can then take place if, as may be assumed, the acetoxonium ring undergoes *trans*-opening by the action of acetic anhydride.[94] Two sugar derivatives can result from each such ring-opening; thus, 252 would give the L-*altro* (253) and D-*galacto* (251c) derivatives, 256 the L-*ido* (257) and D-*galacto* (251c) derivatives, and 255 the L-*gulo* (254) and L-*altro* (253) derivatives. Understandably, by this mechanism, only a few such rearrangement steps are needed to give a number of isomeric sugar derivatives, and, upon further reaction, all sixteen of the hexose derivatives possible will be formed.[93]

On the assumption that the sugar chain adopts a zigzag conformation, the formation of 1,3-dioxolanylium and 1,3-dioxanylium rings can be expected.[94] All rearrangement steps need not occur with

(89) F. Micheel, H. Ruhkopf, and F. Suckfüll, *Chem. Ber.*, **68**, 1523 (1935).
(90) F. Micheel and H. Ruhkopf, *Chem. Ber.*, **70**, 850 (1937).
(91) F. Micheel and R. Böhm, *Tetrahedron Lett.*, 107 (1962).
(92) F. Micheel and R. Böhm, *Chem. Ber.*, **98**, 1655 (1965).
(93) F. Micheel and R. Böhm, *Chem. Ber.*, **98**, 1659 (1965).
(94) F. Micheel, H. Pfetzing, and G. Pirke, *Carbohyd. Res.*, **3**, 283 (1967).

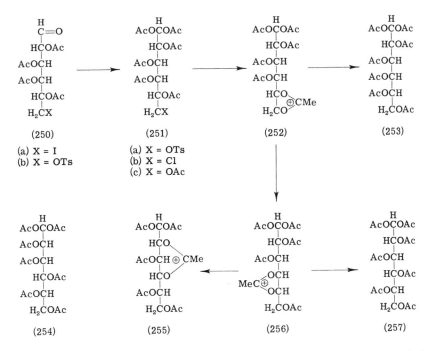

(250)   (251)   (252)   (253)

(a) X = I   (a) X = OTs
(b) X = OTs   (b) X = Cl
          (c) X = OAc

(254)   (255)   (256)   (257)

equal probability; for example, those steps for which twisting of the chain is not required are, presumably, more probable than those wherein such twisting is necessary. Likewise, a selectivity exists in the direction of the *trans*-opening reactions of the acetoxonium ions. Corresponding differences result, in consequence, in the quantitative distribution of the resultant, isomeric sugar derivatives.[94]

In the reaction of 6 grams of **251a** in 180 ml of acetic anhydride with 1.5 grams of zinc chloride for 7 hours at 100°, the principal products obtained were 30% of L-galactose, 17% of D-galactose, and 17% of 6-chloro-6-deoxy-D-galactose. Also formed were D-mannose (11%), D-altrose (7%), L-mannose, and L-altrose, together with the D and L forms of talose, glucose, and allose, in 1–3% yields.[95] 2,3,4,5-Tetra-O-acetyl-6-O-p-tolylsulfonyl-*aldehydo*-D-glucose and 2,3,4,5-tetra-O-acetyl-6-O-p-tolylsulfonyl-*aldehydo*-D-mannose were isomerized by acetic anhydride–zinc chloride to the same products as those formed[93] from **251a**. Chromatographic separation of the products gave DL- and L-galactose, DL-glucose together with allose, DL-mannose together with gulose, and DL-altrose with talose and idose.[93] Methyl 2,4-di-O-acetyl-3,6-anhydro-D-galactopyranoside can be iso-

(95) F. Micheel and E. Matzke, *Carbohyd. Res.*, **4**, 249 (1967).

merized, albeit with difficulty.[91] No reaction is observed with the 6-O-trityl derivative **251** (X = trityl), nor with any of the acetates of D-galactopyranose.[91,93]

Corresponding, acyclic heptose derivatives can be isomerized similarly. 2,3,4,5,6-Penta-O-acetyl-7-O-p-tolylsulfonyl-*aldehydo*-D-*glycero*-D-*gulo*-heptose and 1,2,3,4,5,6-hexa-O-acetyl-7-O-p-tolylsulfonyl-*aldehydo*-D-*glycero*-L-*manno*-heptose aldehydrol were subjected to total isomerization in acetic anhydride–zinc chloride.[94] In the case of the first heptose, there was obtained 40% of a heptose anhydride that most probably has the structure 1,6-anhydro-L-*glycero*-β-D-*gulo*-heptopyranose.[94] Because of the longer chain in these compounds, the possibilities for rearrangement become more complex than for the hexose derivatives, and the number of isomerization products anticipated is correspondingly higher. For both of the heptoses examined, the mixture of isomers formed contained numerous stereoisomers. The anticipated product-composition was estimated by simulation of the reaction by means of a computer; the values calculated were in agreement with those found experimentally.[94]

2. Rearrangement with Aluminum Chloride–Phosphorus Pentachloride

In the reaction of lactose octaacetate (**258**) with phosphorus pentachloride and aluminum trichloride, Hudson and Kunz[96] found, in addition to the anticipated hepta-O-acetyl-lactosyl chloride, a new disaccharide derivative having the structure[97,98] of the heptaacetate of 4-O-β-D-galactopyranosyl-α-D-altropyranosyl chloride (**259**). The yield of **259** was raised to 35–40% when 50 grams of octa-O-acetyl-lactose (**258**) was heated for 20 minutes with 100 grams of aluminum

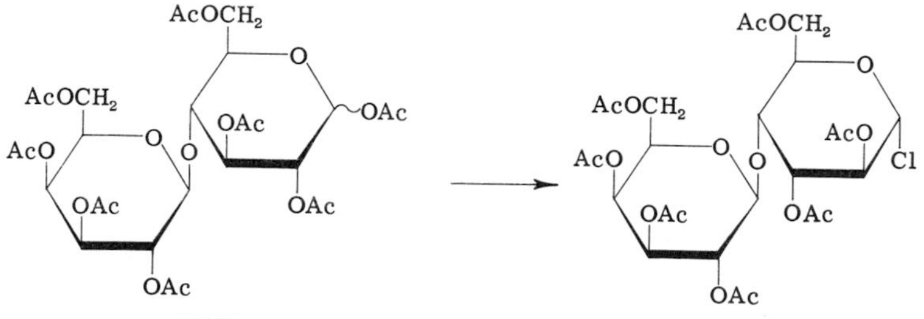

(258)   (259)

(96) C. S. Hudson and A. Kunz, *J. Amer. Chem. Soc.*, **47**, 2052 (1925).
(97) A. Kunz and C. S. Hudson, *J. Amer. Chem. Soc.*, **48**, 1978 (1926).
(98) N. K. Richtmyer and C. S. Hudson, *J. Amer. Chem. Soc.*, **57**, 1716 (1935).

trichloride and 50 grams of phosphorus pentachloride in chloroform.[98] Under these conditions, the two acylated O-glycosylglycosyl halides were formed in the ratio of ~1:1. Difficulties are frequently encountered in reproducing this reaction.[99] In the reaction, the reducing D-glucose moiety of the lactose undergoes configurational inversion at C-2 and C-3, but the mechanism of the rearrangement is as yet unknown. By the mechanistic principles of acetoxonium rearrangements already advanced, it is difficult to explain the inversion at C-3. Because a group is substituted on O-4 of the glucose residue, inversion at C-3 cannot occur through rearside attack by a substituent at C-4.

   Octa-O-acetylcellobiose can, likewise, be rearranged by the action of corresponding excesses of aluminum trichloride and phosphorus pentachloride,[100,101] and thereby gives the heptaacetate of 4-O-β-D-glucopyranosyl-α-D-altropyranosyl chloride[101] in 40–45% yield. Again, inversion at C-2 and C-3 of the reducing residue is observed. The formation of D-mannose and D-altrose derivatives is observed when penta-O-acetyl-D-glucopyranose is treated with aluminum trichloride–phosphorus pentachloride.[102]

   (99) G. Zemplén, Fortschr. Chem. Org. Naturstoffe, 2, 202 (1939).
   (100) C. S. Hudson, J. Amer. Chem. Soc., 48, 2002 (1926).
   (101) N. K. Richtmyer, J. Amer. Chem. Soc., 58, 2534 (1936).
   (102) N. K. Richtmyer, Advan. Carbohyd. Chem., 1, 46 (1946).

CYCLIC ACETALS OF KETOSES[*]

By Robert F. Brady, Jr.

Institute for Materials Research,
National Bureau of Standards, Washington, D. C.

I. Introduction

Cyclic acetals of ketoses are important and useful compounds. Among other applications, they can be used as intermediates in the synthesis of numerous, useful sugar derivatives and as substrates for studies of conformational principles in fused-ring, heterocyclic

[*] A personal contribution by the author.

systems. The name acetal applies to compounds formed when two hydroxyl groups condense with a molecule of an aldehyde or ketone; in order to obtain a cyclic acetal, these two hydroxyl groups must be in the same molecule. Acetals are readily prepared in good yield from aldehydes or ketones, and are unaffected by a wide variety of commonly used reagents. An acetal group can be removed by mild, acid-catalyzed hydrolysis; benzylidene acetal groups can also be removed by hydrogenolysis.

General reviews on cyclic acetals of carbohydrates have appeared.[1,2] The cyclic acetals of the aldoses and aldosides have been treated in this Series by de Belder,[1] but inclusion of cyclic acetals of ketoses was beyond the scope of his Chapter. Other articles, by Barker and Bourne,[3] Mills,[4] and Ferrier and Overend,[5] have been concerned with the stereochemistry and conformation of cyclic acetals of the carbohydrate group. The purpose of the present article is to supplement de Belder's Chapter[1] with a description of the pertinent original work, optimal laboratory preparations, properties, and applications of the cyclic acetals of ketoses, and to provide a summary of the known theoretical aspects of their formation, rearrangement, and hydrolysis.

Emil Fischer first described[6] the condensation of D-fructose with acetone in 1895, and most of the early work on cyclic acetals of ketoses was performed with D-fructose. In 1934, Reichstein and Grüssner published their classic synthesis of L-ascorbic acid (vitamin C), in which L-sorbose was converted into 2,3:4,6-di-O-isopropylidene-α-L-sorbofuranose or other di-alkylidene acetals.[7] The emphasis of research activity then shifted to L-sorbose, and to the elucidation of an optimal procedure for preparing such diacetals. At about the same time, Levene and Tipson[8,9] used isopropylidene acetals as derivatives for the purification of L-*erythro*-pentulose (as the di-isopropylidene acetal) and D-*threo*-pentulose (as the monoisopropylidene acetal). Soon thereafter, Reichstein and coworkers used diisopropylidene acetals of D-[10] and L-psicose,[11] and D-tagatose,[12] to purify the respective sugars.

(1) A. N. de Belder, *Advan. Carbohyd. Chem.*, **20**, 219 (1965).

(2) C. B. Purves, *J. Washington Acad. Sci.*, **36**, 65 (1946).

(3) S. A. Barker and E. J. Bourne, *Advan. Carbohyd. Chem.*, **7**, 137 (1952).

(4) J. A. Mills, *Advan. Carbohyd. Chem.*, **10**, 1 (1955).

(5) R. J. Ferrier and W. G. Overend, *Quart. Rev.* (London), **13**, 265 (1959).

(6) E. Fischer, *Ber.*, **28**, 1145 (1895).

(7) T. Reichstein and A. Grüssner, *Helv. Chim. Acta*, **17**, 311 (1934).

(8) P. A. Levene and R. S. Tipson, *J. Biol. Chem.*, **106**, 603 (1934).

(9) P. A. Levene and R. S. Tipson, *J. Biol. Chem.*, **115**, 731 (1936).

(10) M. Steiger and T. Reichstein, *Helv. Chim. Acta*, **19**, 184 (1936).

(11) M. Steiger and T. Reichstein, *Helv. Chim. Acta*, **18**, 790 (1935).

Interest in these compounds and in their many applications has grown steadily. The impact of modern chemical and physical-chemical techniques, and of the principles of conformational theory, is increasingly felt. Knowledge of the chemistry of these compounds may be expected to expand rapidly in the near future.

II. Preparation and Properties

1. Preparation

Cyclic acetals of ketoses are prepared most commonly from acetone or benzaldehyde; formaldehyde, acetaldehyde, butanone, and cyclohexanone have been used occasionally. These carbonyl reagents are frequently used directly, although such derivatives as 2,2-dimethoxy-[13] or 2,2-diethoxy[14]-propane (acetone dialkyl acetals), or 1,1-dimethoxyethane[15] (acetaldehyde diethyl acetal), are often employed in experiments in which intermediate acetals are of interest,[13,15,16] or in which the presence of water in the reaction mixture adversely affects the yield of products.[14] A polymeric form of an aldehyde is the reagent to be preferred whenever the monomer is volatile; for example, acetaldehyde is often used in the form of a trimer, paraldehyde, and formaldehyde is employed as "formalin" solution, as paraformaldehyde, or as "polyoxymethylene." An excess of the carbonyl reagent is generally used as the solvent, and the condensation is usually effected at room temperature.

The factor most influential in determining the outcome of a condensation reaction is the catalyst. The catalysts most commonly employed are mineral acids, although copper(II) sulfate,[17,18] phosphorus pentaoxide,[19] ethyl metaphosphate,[20] cation-exchange resins in the acid form,[21-23] and zinc chloride (alone[24,25] or in combination

(12) T. Reichstein and W. Bosshard, *Helv. Chim. Acta*, **17**, 753 (1934).
(13) T. Maeda, M. Kiyokawa, and K. Tokuyama, *Bull. Chem. Soc. Jap.*, **38**, 332 (1965).
(14) K. Tokuyama and E. Honda, *Bull. Chem. Soc. Jap.*, **37**, 591 (1964).
(15) T. Maeda, M. Kiyokawa, and K. Tokuyama, *Bull. Chem. Soc. Jap.*, **42**, 492 (1969).
(16) T. Maeda, *Bull. Chem. Soc. Jap.*, **40**, 2122 (1967).
(17) H. Ohle and I. Koller, *Ber.*, **57**, 1566 (1924).
(18) J. R. Patil and J. L. Bose, *Indian J. Chem.*, **5**, 598 (1967).
(19) L. Smith and J. Lindberg, *Ber.*, **64**, 505 (1931).
(20) J. Pacák and M. Černý, *Collect. Czech. Chem. Commun.*, **23**, 490 (1958).
(21) K. Erne, *Acta Chem. Scand.*, **9**, 893 (1955).
(22) J. E. Cadotte, F. Smith, and D. Spriestersbach, *J. Amer. Chem. Soc.*, **74**, 1501 (1952).
(23) D. F. Hinkley and R. H. Beutel, French Pat. 1,541,849; *Chem. Abstr.*, **71**, 102,189 (1969).
(24) H. O. L. Fischer and C. Taube, *Ber.*, **60**, 485 (1927).
(25) W. L. Glen, G. S. Myers, and G. A. Grant, *J. Chem. Soc.*, 2568 (1951).

with phosphorus pentaoxide and phosphoric acid[25-27]) have also been used. For exchange reactions, 0.3–1.0% of *p*-toluenesulfonic acid is used with dialkyl acetals of acetone[14,16] or acetaldehyde.[15]

The concentrations of mineral acids generally used are 0.2–1.5% of hydrogen chloride[7,28,29] or 0.25–5.0% of concentrated sulfuric acid,[17,30,31] although the use of sulfuric acid at concentrations ranging up to 50% has been reported.[7,32,33] The use of 75% of phosphoric acid as catalyst in the formation of methylene acetals has also been described.[32]

The concentration of catalyst and the time for reaction is especially important when isomeric products can be expected. For example, several workers[17,30,31,34] have shown that the proportions of 1,2:4,5-di-*O*-isopropylidene-β-D-fructopyranose (**1**) and 2,3:4,5-di-*O*-isopropylidene-β-D-fructopyranose (**2**), in the product of the reaction of D-

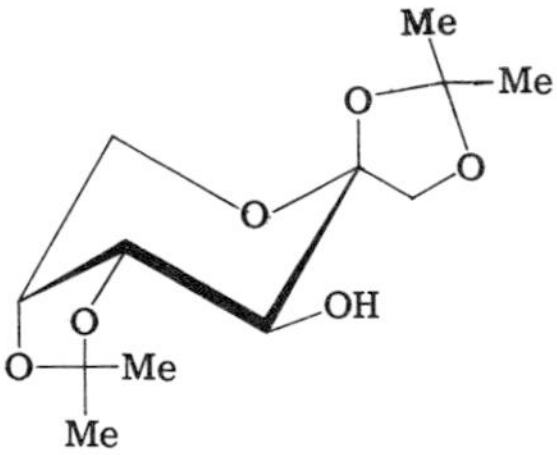

(**1**) 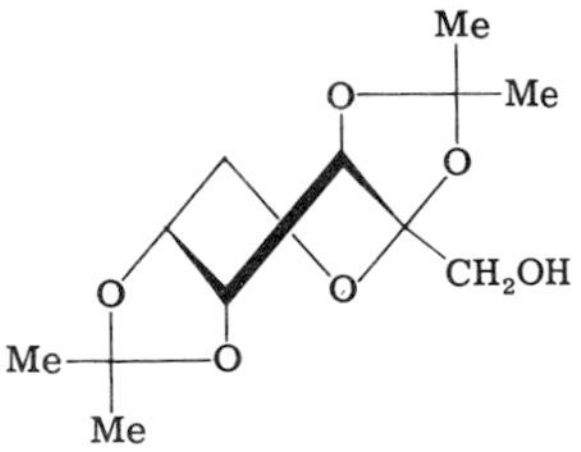   (**2**)

fructose with acetone in the presence of sulfuric acid as the catalyst, are dependent upon the concentration of the catalyst; a low concentration of sulfuric acid favors the formation of **1**, whereas **2** is the preponderant product at high concentrations of catalyst. Long reaction-times favor the formation of **2** when low or moderate concen-

(26) H. van Grunenberg, C. Bredt, and W. Freudenberg, *J. Amer. Chem. Soc.*, **60**, 1507 (1938).
(27) W. L. Glen, G. S. Myers, R. J. Barber, and G. A. Grant, U. S. Pat. 2,715,121; *Chem. Abstr.*, **50**, 8719 (1956).
(28) T. N. Montgomery, *J. Amer. Chem. Soc.*, **56**, 419 (1934).
(29) T. Maeda, M. Kimoto, S. Wakahara, and K. Tokuyama, *Bull. Chem. Soc. Jap.*, **42**, 1668 (1969).
(30) R. F. Brady, Jr., *Carbohyd. Res.*, **15**, 35 (1971).
(31) D. J. Bell, *J. Chem. Soc.*, 1461 (1947).
(32) C. A. Lobry de Bruyn and W. Alberda van Ekenstein, *Rec. Trav. Chim.*, **22**, 159 (1903).
(33) H. Paulsen, I. Sangster, and K. Heyns, *Chem. Ber.*, **100**, 802 (1967).
(34) Y. M. Slobodin and A. N. Klimov, *Zh. Obshch. Khim.*, **15**, 921 (1945); *Chem. Abstr.*, **40**, 6425 (1946).

trations of sulfuric acid are employed as the catalyst.[30] The main product when zinc chloride is used as the catalyst is **1**, and a large proportion of **2** is not formed, even with long times of reaction.[30]

Condensations may be conducted under very mild conditions by using anhydrous copper(II) sulfate alone,[17,18,35] or combined with a low concentration (0.1%) of sulfuric acid.[9,36] At room temperature, in the absence of acid, the reaction is allowed to proceed for periods of from 2 days[9,36] up to 7 days.[18]

The condensation reactions are normally performed at room temperature, because the use of higher temperatures may lead to undesirable side-products resulting from hydrolysis of acid-labile substituents, inversion, or formation of anhydrides.[1] The condensation of L-sorbose with acetone, catalyzed by 4% of sulfuric acid, is reported to be temperature-dependent. Essentially pure 2,3:4,6-di-O-isopropylidene-α-L-sorbofuranose (**3**) is formed[37] below 5°, whereas a mixture composed of equal parts of 3 and 2,3-O-isopropylidene-α-L-sorbofuranose (**4**) is formed[38] at 20°, and only the monoacetal

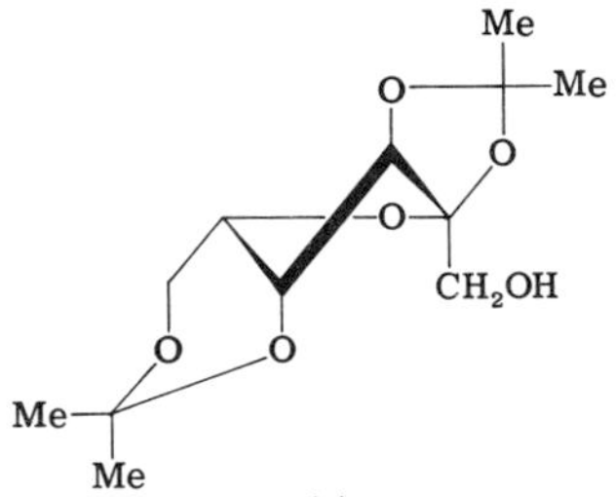

(3) 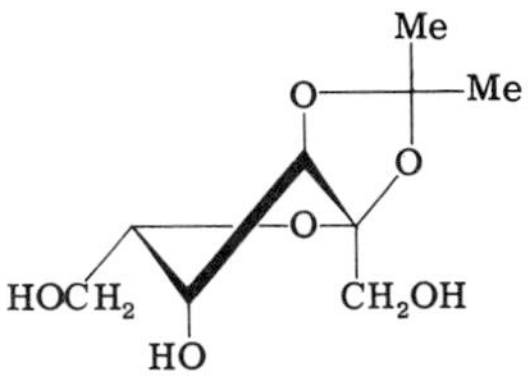   (4)

4 is formed[39] at 30—50°. This effect is reported to be reversible.[38] The yield of **3** was 97% at −8°; upon warming, and keeping the solution at 44° for longer than 20 minutes, 27% of **3** was obtained, but cooling to −8° raised the yield of **3** to 93%. The same results were obtained[38] by starting with "monoacetone sorbose (m. p. 85°)" (presumably **4**, m. p. 91°). It has been stated[38] that D-fructose behaves similarly to L-sorbose, giving a 90% yield of "diacetone fructose" below 0°; however, storage of a solution of D-fructose in acetone

(35) H. Ohle, *Ber.*, **71**, 562 (1938).

(36) R. S. Tipson and R. F. Brady, Jr., *Carbohyd. Res.*, **10**, 549 (1969).

(37) R. G. Krystallenskaya, *Tr. Vsef. Nauch. Issled. Vitamin. Inst.*, 3, 78 (1941); *Chem. Abstr.*, **36**, 3007 (1942).

(38) Y. M. Slobodin, *Zh. Obshch. Khim.*, **17**, 485 (1947); *Chem. Abstr.*, **42**, 871 (1948).

(39) R. G. Krystallenskaya, *Tr. Vsef. Nauch. Issled. Vitamin. Inst.*, 3, 85 (1941); *Chem. Abstr.*, **36**, 3008 (1942).

containing 0.3% of sulfuric acid for 2 hours at $-10°$ did not raise the yield of 1,2:4,5-di-O-isopropylidene-β-D-fructopyranose (1) above that (61%) obtained at room temperature.[30]

Condensations catalyzed by mineral acids are usually allowed to proceed for periods of 6 to 26 hours. Generally, the reaction mixture does not reach equilibrium at the point when the sugar has become completely dissolved, but processing at this stage may yield products resulting from kinetic control of the reaction. In addition, short times of reaction generally favor the formation of monoacetals. The condensation of L-sorbose with acetone, catalyzed by 0.1% of sulfuric acid and "supersonic waves (500 kHz)," is reported[40,41] to give a high yield of 2,3:4,6-di-O-isopropylidene-α-L-sorbofuranose (3) in 1 hour.

2. Stability

The m-dioxane and 1,3-dioxolane rings in the respective cyclic acetals of ketoses are generally unaffected by the routine derivatization procedures that are catalyzed by a base, such as acetylation,[17,42] benzoylation,[17,42] and p-toluenesulfonylation.[17,43,44] In addition, the acetal groups are normally unaffected by the cold, dilute, aqueous acids used in the processing of such reactions. Cyclic acetal groups are not cleaved by such common oxidizing agents as potassium permanganate,[45] methyl sulfoxide–acetic anhydride,[46–48] or ruthenium tetraoxide,[46,49] nor by such reducing agents as sodium borohydride,[46–49] lithium aluminum hydride,[47,49] sodium amalgam,[49] or hydrogen over platinum[49] or palladium[50] catalysts. However, acetal groups are attacked by strong Lewis acids, such as boron trichloride[51,52] or aluminum chloride dihydride (formed in $situ$ in a 1:1 mixture of lithium aluminum hydride and aluminum chloride).[53]

(40) M. Hosokawa, H. Yagi, and K. Naito, U. S. Pat. 2,849,355; Chem. Abstr., 53, 3084 (1959).
(41) M. Hosokawa, Jap. Pat. 7473 (1954); Chem. Abstr., 50, 4204 (1956).
(42) E. Fischer and H. Noth, Ber., 51, 321 (1918).
(43) P. A. Levene and R. S. Tipson, J. Biol. Chem., 120, 607 (1937).
(44) K. Freudenberg, O. Burkhart, and E. Braun, Ber., 59, 714 (1926).
(45) P. Karrer and O. Hurwitz, Helv. Chim. Acta, 4, 728 (1921).
(46) R. S. Tipson, R. F. Brady, Jr., and B. F. West, Carbohyd. Res., 16, 383 (1971).
(47) K. James, A. R. Tatchell, and P. K. Ray, J. Chem. Soc. (C), 2681 (1967).
(48) E. J. McDonald, Carbohyd. Res., 5, 106 (1967).
(49) G. M. Cree and A. S. Perlin, Can. J. Biochem., 46, 765 (1968).
(50) R. F. Brady, Jr., Carbohyd. Res., 20, 170 (1971).
(51) S. Allen, T. G. Bonner, E. J. Bourne, and N. Saville, Chem. Ind. (London), 630 (1958).
(52) T. G. Bonner, E. J. Bourne, and S. McNally, J. Chem. Soc., 2929 (1960).
(53) S. S. Bhattacharjee and P. A. J. Gorin, Can. J. Chem., 47, 1195 (1969).

De Belder[1] has reviewed the stability of cyclic acetals of the aldoses and aldosides to heat, light, bases, and oxidants and reductants. His Chapter gives many examples of the kinds of reactions that may be expected to occur when cyclic acetals of ketoses are subjected to such conditions.

3. Hydrolysis

The acid-catalyzed hydrolysis of cyclic acetals of ketoses is best effected by use of 0.1 M aqueous oxalic acid[9,36,54] or 50–80% aqueous acetic acid[10,11,15,16,29,55–57] as the hydrolyst, because the use of such strong mineral acids as hydrochloric acid or sulfuric acid may lead to considerable decomposition of the ketose liberated.[54] Cation-exchange resins have also been employed for effecting the hydrolysis of acetals.[25] For the hydrolysis of acetal groups, a mild, rapid method that does not attack most other functional groups, involves use of 9:1 trifluoroacetic acid–water at ambient temperature as the hydrolyst.[58] After the cleavage reaction has proceeded to completion, the reagent is removed by evaporation under diminished pressure below 50°.

Selective hydrolysis of diacetals has been accomplished under carefully controlled conditions. Treatment of 1,2:4,5-di-O-isopropyli-dene-β-D-fructopyranose (1) with 0.1% hydrochloric acid for 4.5 hours at 30° gave 1,2-O-isopropylidene-β-D-fructopyranose (5) in 96% yield.[59] Also, 1,2:4,5-di-O-isopropylidene-β-D-psicopyranose (6) was selectively hydrolyzed to 1,2-O-isopropylidene-β-D-psicopyranose (7)

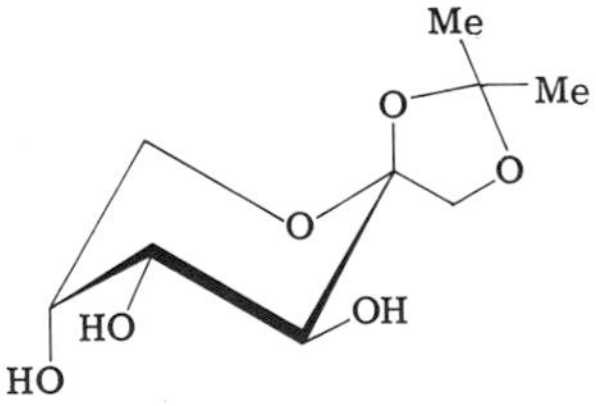

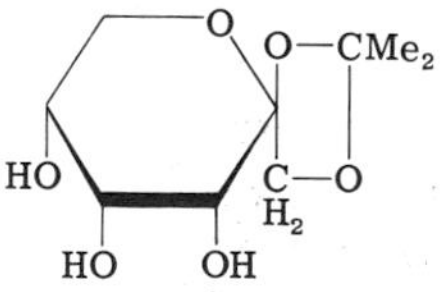

in 73% yield by treatment with 80% aqueous acetic acid for 3.5 hours at room temperature.[49]

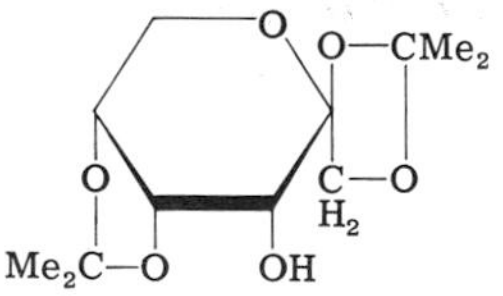

When the difference in the rates of hydrolysis of the two acetal

(54) R. S. Tipson, B. F. West, and R. F. Brady, Jr., Carbohyd. Res., 10, 181 (1969).
(55) K. Tokuyama, M. Kiyokawa, and N. Hōki, Bull. Chem. Soc. Jap., 36, 1392 (1963).
(56) T. Maeda, K. Tori, S. Satoh, and K. Tokuyama, Bull. Chem. Soc. Jap., 42, 2635 (1969).
(57) W. R. Sullivan, J. Amer. Chem. Soc., 67, 837 (1945).
(58) J. E. Christensen and L. Goodman, Carbohyd. Res., 7, 510 (1968).
(59) J. C. Irvine and C. S. Garrett, J. Chem. Soc., 97, 1277 (1910).

groups is not great, selective hydrolysis cannot be accomplished, but interruption of the hydrolysis at the optimal time may lead to a moderate yield of one monoacetal. Thus, partial hydrolysis of 1,2:3,4-di-*O*-isopropylidene-β-D-*erythro*-pentulofuranose (**8**) with 80% aqueous acetic acid for 2 hours at 65° gave 29% of unchanged **8**, 39% of 3,4-*O*-isopropylidene-D-*erythro*-pentulose (**9**), and 32% of D-*erythro*-pentulose.[36] Likewise, 2,3:4,5-di-*O*-isopropylidene-β-D-fructopyranose (**2**) yielded 54% of unchanged **2**, 21% of 2,3-*O*-isopropylidene-β-D-fructopyranose (**10**), and 18% of D-fructose, when treated with 0.5 *M* sulfuric acid for 13 hours at room temperature.[60]

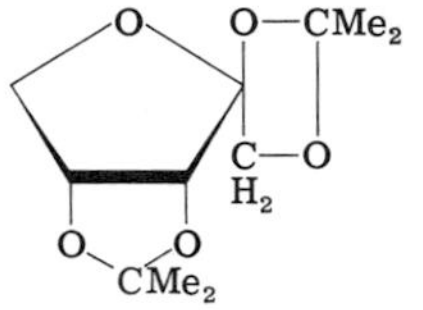

(8) 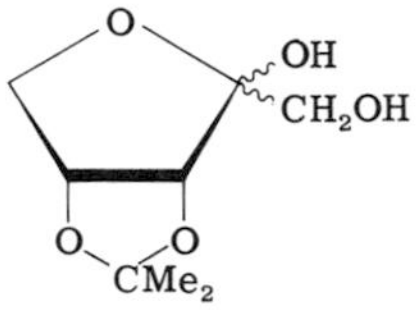   (9)

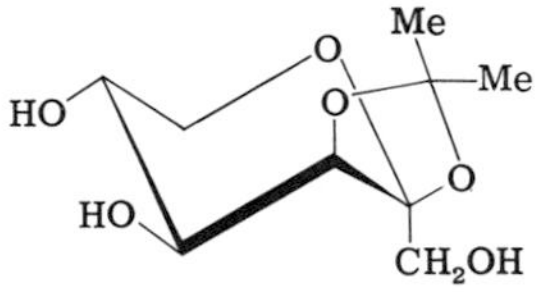

(10)

The rates of hydrolysis at 65° in 0.1 *M* oxalic acid of a variety of isopropylidene acetals of pentuloses and hexuloses have been measured, and found to be related to the structure of the acetals.[54] A *spiro*-fusion at C-2 of a 1,3-dioxolane ring, involving C-1 and C-2 of the ketose and the pyranoid or furanoid ring of the sugar, is revealed by fast hydrolysis (2–3 hours) of the 1,3-dioxolane ring. When this ring is *spiro*-fused at C-2 to a furanoid ring, the hydrolysis is very fast (1.25 hours). In contrast, when this *spiro* system is absent, long times (8 hours) are required for hydrolysis.

In the few examples studied thus far,[36,49,54,59,60] the relative order of stability to hydrolysis of a 2,2-dimethyl-1,3-dioxolane ring is: *cis*-fused to a furanoid ring > *spiro*-fused to a pyranoid ring > *cis*-fused to

(60) M. L. Wolfrom, W. L. Shilling, and W. W. Binkley, *J. Amer. Chem. Soc.*, **72**, 4544 (1950).

a pyranoid ring > *spiro*-fused to a furanoid ring. In general, acetal groups attached to the anomeric carbon atom are hydrolyzed less readily than those attached at other positions, in part because of the special polar environment at the anomeric carbon atom.[1] This effect is not sufficient to confer high relative stability on a 1,3-dioxolane ring *spiro*-fused to a furanoid ring at the anomeric carbon atom,[36] but it is probably responsible for the order of the second and third entries in the series just given. The only examples of a *spiro*-fusion to a pyranoid ring thus far studied are those[49,59] in which the junction is at the anomeric carbon atom. This type of fusion may well prove to be more labile than *cis*-fusion to a pyranoid ring when this stabilization is lacking, and the point deserves further investigation.

III. Mechanistic Aspects

1. The Acetalation of L-Sorbose

Maeda, Tokuyama, and coworkers have published extensively on the synthesis, and mechanism of formation, of isopropylidene, ethylidene, and benzylidene acetals of L-sorbose. In the course of this work, they have isolated and characterized many new mono- and di-acetals that are present in the reaction mixtures in only trace amounts, and they have studied the acid-catalyzed rearrangements of these compounds in the absence of acetalating reagents. Patil and Bose have also investigated the mechanism of the isopropylidenation of L-sorbose. Much exploratory work was necessary before the precise mechanism of acetalation was defined.

a. Early Experiments.—Tokuyama and coworkers[61] first investigated the mechanism of the acid-catalyzed acetonation of 1,2-*O*-isopropylidene-α-L-sorbopyranose (**11**). This monoacetal is formed by

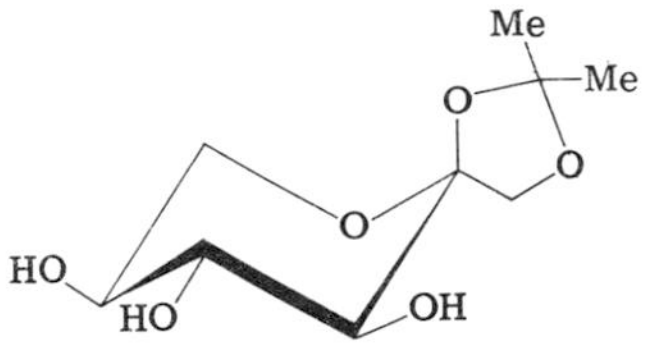

(11)

the reaction of L-sorbose (**14**) with acetone in the absence of acid, and was first reported by Ohle[35] in 1938. The results of titration, with the

(61) K. Tokuyama, E. Honda, and N. Hōki, *J. Org. Chem.*, **29**, 133 (1964).

Karl Fischer reagent, of the water formed during the reaction suggested that the reaction may be second-order with respect to **11**, with a rate constant proportional to the concentration of sulfuric acid. The Japanese workers[61] postulated, for the acetonation, a cyclic pathway (Scheme I) in which a dimer of **11** would be formed during the rate-

$$\textbf{11} \longrightarrow (\textbf{11})_2 \longrightarrow \textbf{14} + \textbf{3} \rightleftharpoons \textbf{4}$$

Scheme I

controlling step; collapse of this intermediate would yield one molecule of 2,3:4,6-di-*O*-isopropylidene-α-L-sorbofuranose (**3**) and one molecule of L-sorbose (**14**), which would be re-acetonated, forming **11**. A second monoacetal observed in the reaction mixture, namely, 2,3-*O*-isopropylidene-α-L-sorbofuranose (**4**), was considered to be a hydrolysis product of the diacetal **3**, and not an intermediate in the formation of **3** from **14**.

Similar results were reported two years later by Patil and Bose.[62] They investigated the acetonation of L-sorbose in acetone containing 4% of concentrated sulfuric acid, and showed by thin-layer chromatography that the 1,2-monoacetal **11** and the diacetal **3** are formed early in the reaction, whereas the 2,3-monoacetal **4** appears much later. Treatment of pure **3** or pure **4** under these acetonating conditions was reported to produce an equilibrium mixture of the two, but no **11** was detected in either experiment. They concluded[62] that the conversion of **11** into **3** is irreversible, and that **4** is formed by subsequent hydrolysis of the more labile 1,3-dioxolane ring involving C-4 and C-6 of **3**.

Hoping to increase the yield of **3** by ensuring anhydrous conditions, Tokuyama and Honda[14] turned to the reaction of L-sorbose with 2,2-diethoxypropane (acetone diethyl acetal), catalyzed by *p*-toluenesulfonic acid. Three main products were isolated: 1-*O*-(2-ethoxy-2-propyl)-2,3:4,6-di-*O*-isopropylidene-α-L-sorbofuranose (**12**), the 2,3:4,6-diacetal **3**, and 1,2:4,6-di-*O*-isopropylidene-α-L-sor-

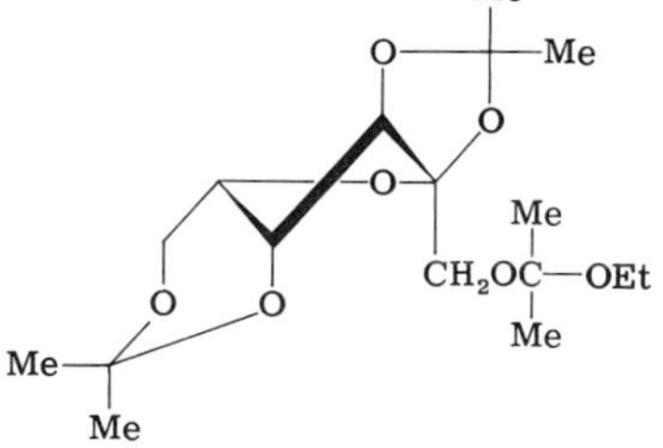

(12)

(62) J. R. Patil and J. L. Bose, *J. Indian Chem. Soc.*, **43**, 161 (1966).

bofuranose (**17**); small proportions of **11**, **14**, and **4** were also detected. Tokuyama and Honda reported that **3** was formed by similar treatment of the 2,3-monoacetal **4**, but not of **11**. They concluded that the mechanism of acetonation by acetal exchange differed from that of direct, acid-catalyzed acetonation, and proposed the pathway shown in Scheme II for this class of acetonation reaction.

$$14 \longrightarrow 4 \longrightarrow 3 \longrightarrow 12$$

Scheme II

In a preliminary communication, Maeda and coworkers proposed[13] a more detailed mechanism (Scheme III) for the acetonation of

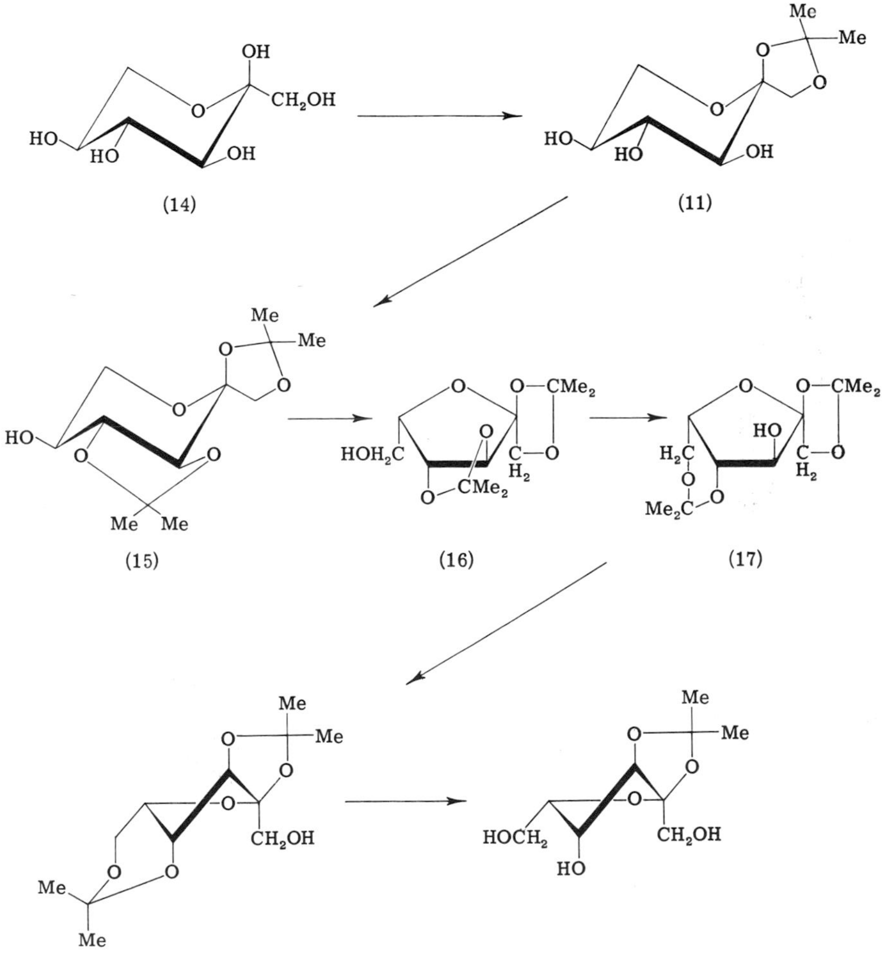

Scheme III

L-sorbose by acetal exchange with 2,2-dimethoxypropane, as catalyzed by *p*-toluenesulfonic acid. The diacetal **16**, namely, 1,2:3,4-di-*O*-isopropylidene-α-L-sorbofuranose, is, however, a most improbable intermediate, as it would have a strained, *trans*-fused system of five-membered rings.

Patil and Bose[18] isolated a monoacetal from the acetonation of L-sorbose catalyzed by anhydrous copper(II) sulfate, and showed it to be 1,2-*O*-isopropylidene-α-L-sorbofuranose (**13**). They proposed

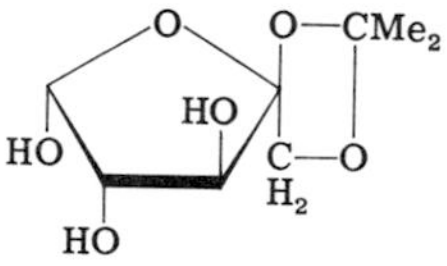

(13)

that the monoacetal **11** would rearrange to **13** under the conditions of the reaction, but it is difficult to envisage a driving force for this step. These workers[18] also isolated the diacetal **17** from the same reaction; the structure depicted was confirmed by mass spectrometry.[63]

In 1967, Maeda published a detailed paper[16] on the reaction of L-sorbose with 2,2-dimethoxypropane. Using thin-layer chromatography for isolation, and infrared and proton magnetic resonance spectroscopy for characterization, he reported four major products [**3**, **17**, 1,3:4,6-di-*O*-isopropylidene-β-L-sorbofuranose (**18**), and 1,2:-4,6-di-*O*-isopropylidene-β-L-sorbofuranose (**20**)] and six minor products (**15**, **19**, **21**, **22–24**) from the reaction, after 4 hours at 45° with *p*-toluenesulfonic acid as the catalyst.

Two years later, the Japanese group reported[15] the results of the ethylidenation of L-sorbose with 1,1-dimethoxyethane, catalyzed by

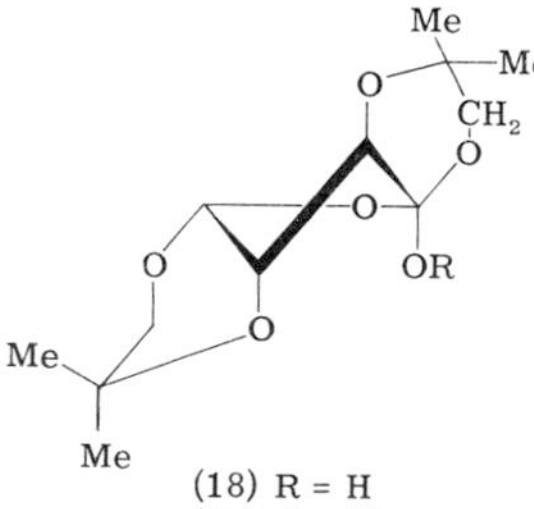

(18) R = H
(19) R = Me

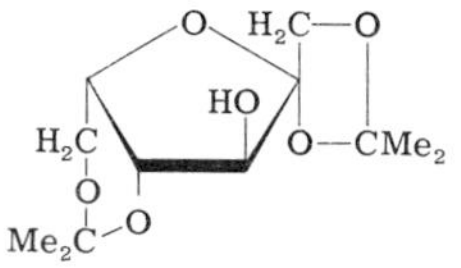

(20)

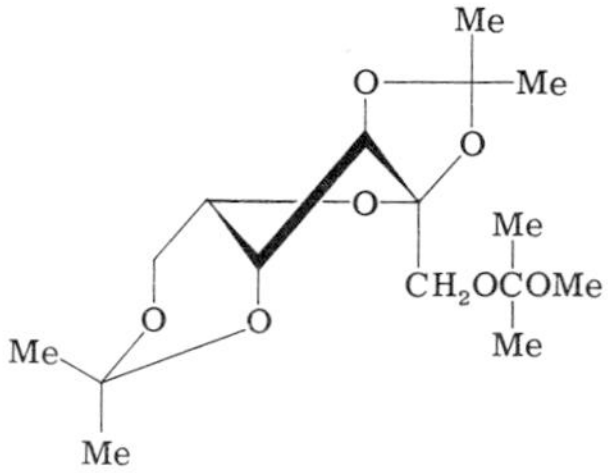

(21)

(63) J. R. Patil, K. G. Das, and J. L. Bose, *Indian J. Chem.*, **5**, 535 (1967).

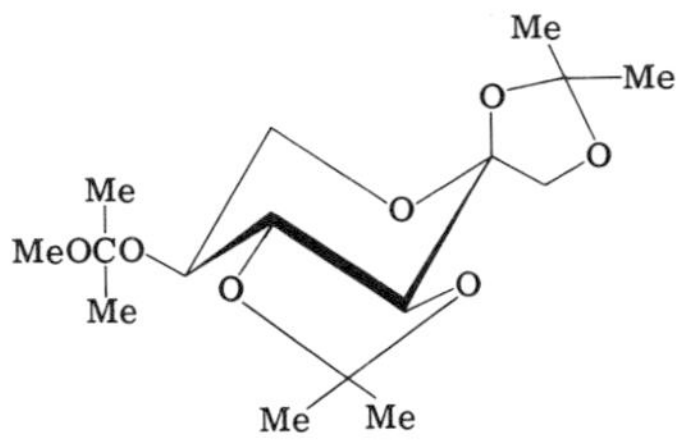

(22)

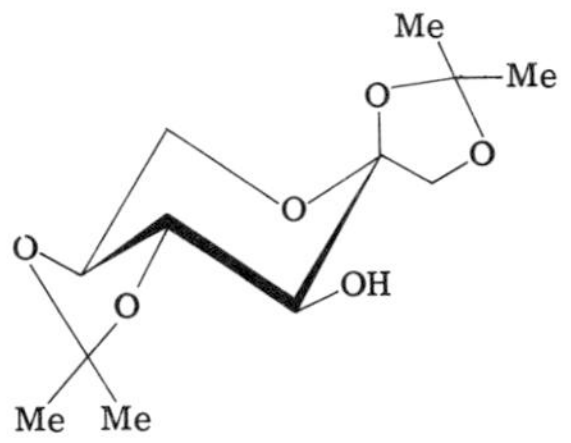

(23)

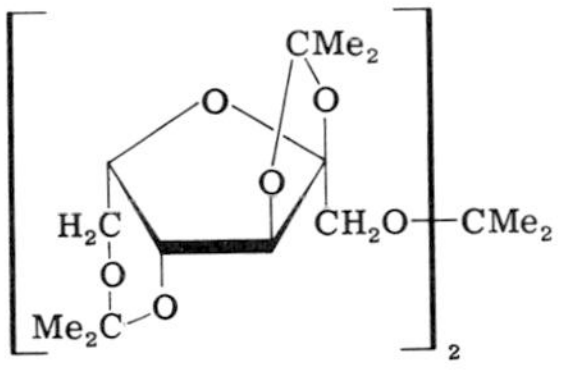

(24)

p-toluenesulfonic acid. Two isomeric monoacetals, 1,2-O-ethylidene-α-L-sorbopyranose (**25**) and 1,3-O-ethylidene-α-L-sorbopyranose (**26**)

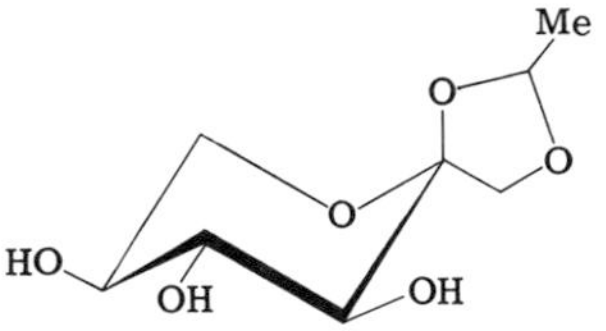

(25)

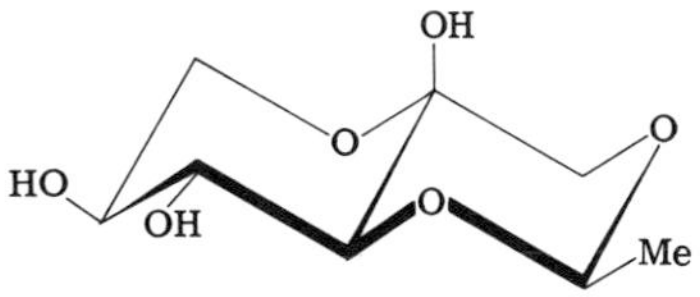

(26)

were obtained when the mixture was processed after a short time had been allowed for reaction, whereas 1,3:4,6-di-O-ethylidene-β-L-sorbofuranose (**27**) was obtained when the reaction was allowed to proceed for longer periods of time. The diacetal **27** is the main product under mildly acidic conditions, but, under highly acidic conditions, it is isomerized to a mixture of two 2,3:4,6-di-O-ethylidene-α-L-sorbofuranoses (**28** and **29**), diastereomeric at the acetalic

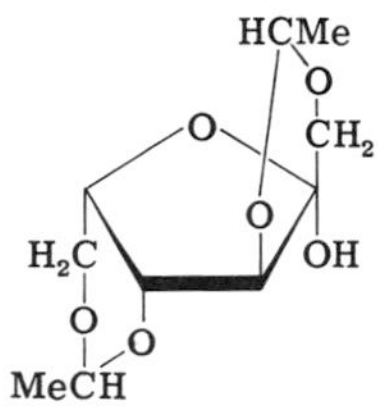

(27)

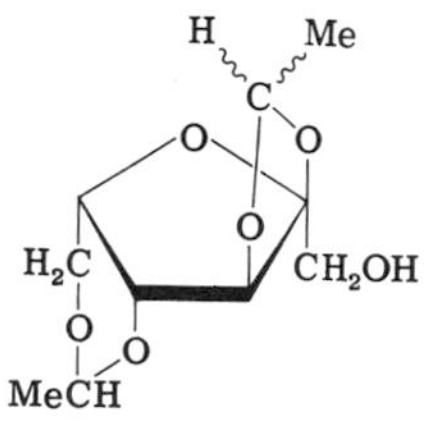

(28, 29)

carbon atom in the 1,3-dioxolane ring. The mechanism shown in Scheme IV was tentatively proposed[15] in order to explain the ethylidenation.

$$14 \longrightarrow [25 \rightleftharpoons 26] \longrightarrow 27 \longrightarrow 28 + 29$$

Scheme IV

The diacetals **26** and **27** represent two new classes of diacetal not detected in the acetonation reactions. Detection of these compounds gives new information about the general mechanism of acetalation.[15] That no compound analogous to **26** has been isolated from the acetonation of L-sorbose can be explained by the Brown–Brewster–Shechter rule,[64] which states that the formation of a m-dioxane ring from acetone and a 1,3-diol is not favored, because one of the two methyl groups is necessarily axially attached in any conformation of the m-dioxane part. In the ethylidenation reactions, no conformational instability is engendered, because of the lack of an axial methyl group on the m-dioxane ring of **26**; thus, the ethylidene acetal is stable and can be isolated. For similar reasons, the stability of the diacetal **27** can be attributed to the lack of an axial methyl group on the m-dioxane ring fused to C-1 and C-3.

Maeda and coworkers[29] subsequently investigated the benzylidenation of L-sorbose. Two isomers of 2,3:4,6-di-O-benzylidene-α-L-sorbofuranose (**30** and **31**), diastereomeric at the acetal carbon in the

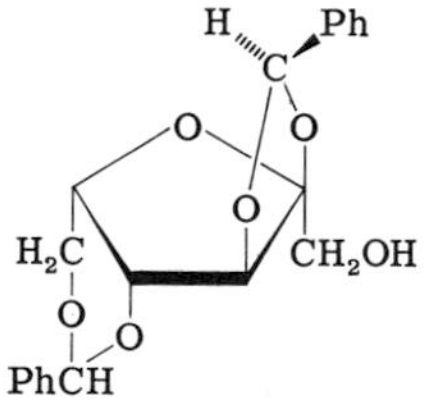

(30) 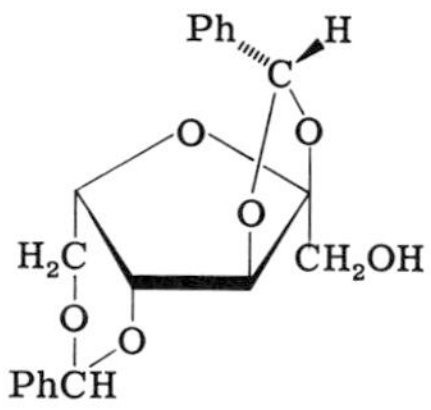   (31) 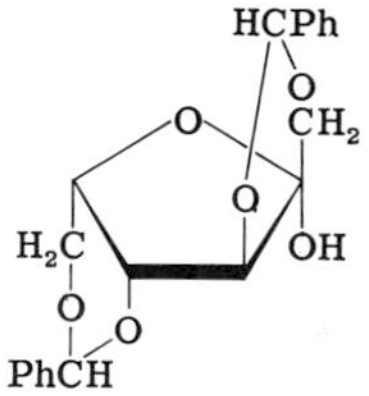   (32)

1,3-dioxolane ring, were isolated when the reaction was catalyzed by dry hydrogen chloride, but the use of a trace of p-toluenesulfonic acid as the catalyst afforded mainly another dibenzylidene acetal, namely, 1,3:4,6-di-O-benzylidene-β-L-sorbofuranose (**32**). When the last reaction was effected in methyl sulfoxide, seven additional products (**33–39**) were isolated. The reaction, catalyzed by p-toluenesulfonic acid, was monitored by proton magnetic resonance spectroscopy with the benzylic-proton singlets as the key signals, and

(64) H. C. Brown, J. H. Brewster, and H. Shechter, *J. Amer. Chem. Soc.*, **76**, 467 (1954).

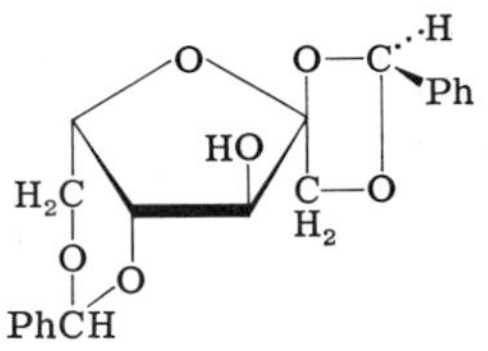

(33) 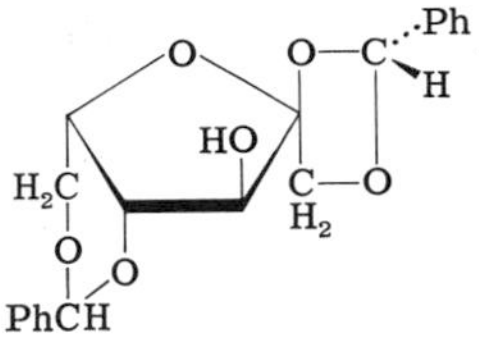   (34) 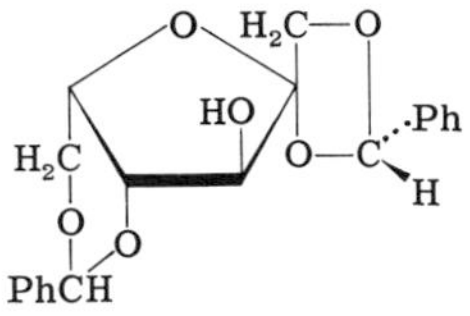   (35)

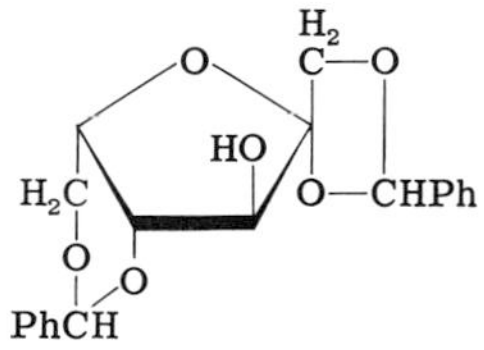

(36) 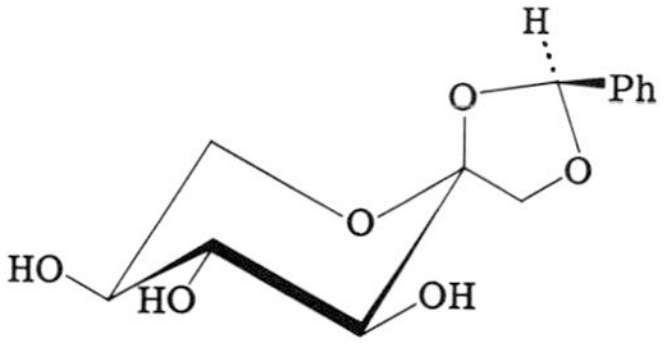   (37)

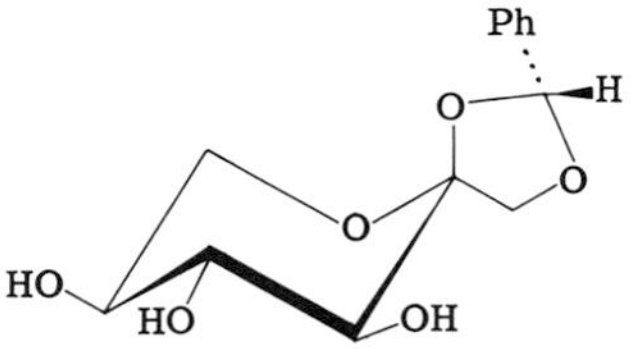

(38) 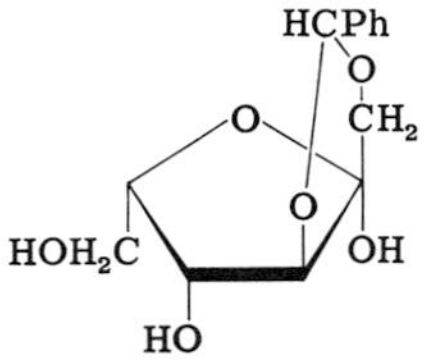   (39)

the reaction was shown to proceed from **32** to **31** to **30**. The authors[29] concluded that the H-*exo* isomer **31** is produced from **32** in a kinetically controlled step, but that it subsequently isomerizes, yielding **30** at equilibrium. Thus, the ring-contraction reaction occurs with retention of the diastereoisomeric configuration at the acetal center.

$$14 \longrightarrow [37, 38 \rightleftharpoons 39] \longrightarrow 32 \longrightarrow 31 \longrightarrow 30$$

Scheme V

A mechanism (see Scheme V) for the benzylidenation was proposed.[64] Subsequently, further work on the tautomeric forms of **39** was reported.[65]

b. The Mechanism of Acetalation.—Maeda and Tokuyama reported[66,67] briefly in 1968 a general mechanism for the acetalation

(65) T. Maeda, M. Kimoto, S. Wakahara, and K. Tokuyama, *Bull. Chem. Soc. Jap.*, **42**, 2021 (1969).

(66) T. Maeda and K. Tokuyama, *Tetrahedron Lett.*, 3079 (1968).

(67) K. Tokuyama and T. Maeda, *Abstr. Papers Amer. Chem. Soc. Meeting*, **156**, Carb 5 (1968).

reactions, and they discussed this mechanism thoroughly in a subsequent full paper.[68] The latter paper contains a discussion of the general pathway for acetalation reactions, but lacks mechanistic details for some of the steps. The complete mechanistic rationale is given in Scheme VI.

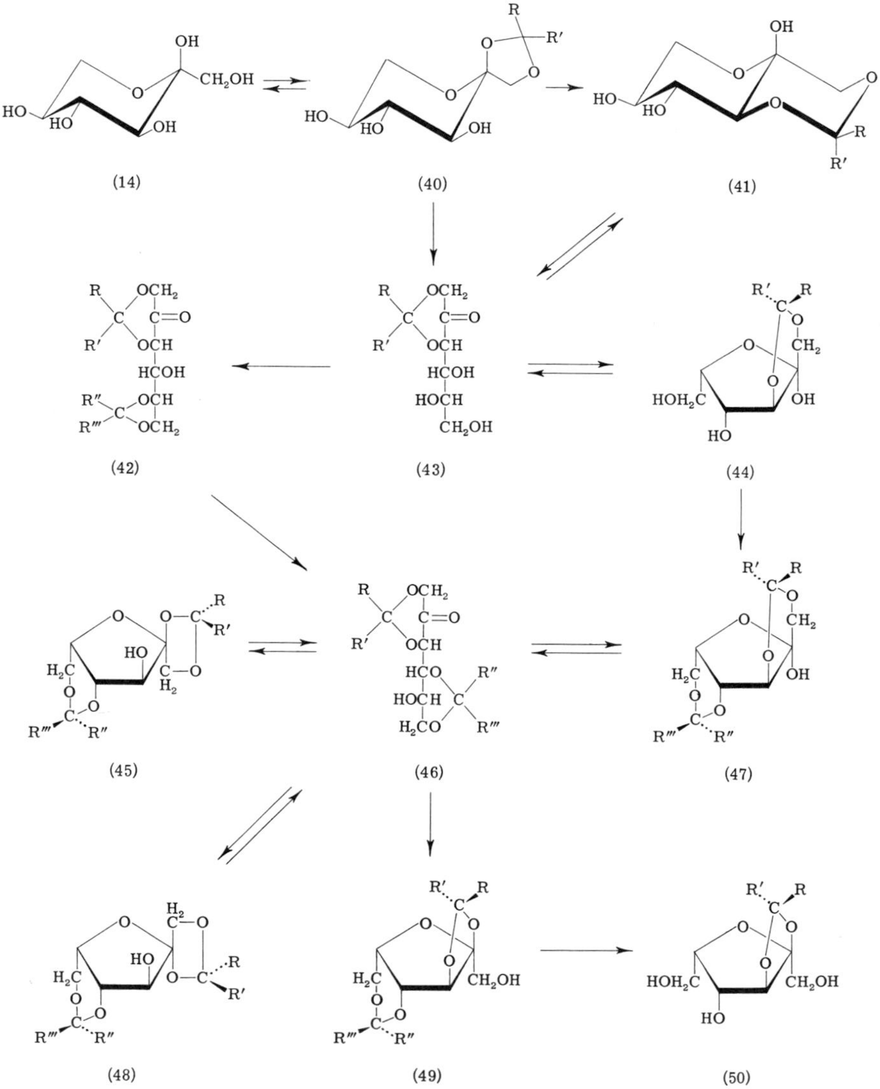

Scheme VI

(68) T. Maeda, Y. Miichi, and K. Tokuyama, *Bull. Chem. Soc. Jap.*, **42**, 2648 (1969).

X-Ray diffraction data[69] have established that L-sorbose crystallizes as the α-L anomer in the $1C(L)$ conformation. Proton magnetic resonance data[70] and optical rotatory dispersion results[71] have shown that the same conformation is favored in aqueous solution. The primary hydroxyl group at C-1 of L-sorbose is the most reactive, and can be expected to combine with the conjugate acid of the carbonyl reagent (or an analogous alkoxy mono- or di-alkyl carbonium ion arising from an acetal of the carbonyl reagent by protonation followed by the loss of a molecule of the alcohol). This species forms a cyclic acetal with the anomeric hydroxyl group, instead of the hydroxyl group at C-3, in a kinetically controlled step in which formation of a five-membered ring is favored, and thus, a 1,2-O-alkylidene-β-L-sorbopyranose (**40**) is formed.

Further acetalation of the *trans*-disposed hydroxyl groups on the pyranoid ring is not sterically favored,[72] but, nonetheless, 1,2:3,4-di-O-isopropylidene-β-L-sorbopyranose (**15**) and 1,2:4,5-di-O-isopropylidene-β-L-sorbopyranose (**23**) have been isolated in small proportions.[13,16,56] These diacetals are not intermediates in the direct formation of **3**, but appear to be the products of secondary pathways.

The *spiro*-fused, five-membered, 1,3-dioxolane ring in **40** is thermodynamically less favored than a *trans*-fused system of six-membered rings and thus **40** rearranges to **41**, as shown in Scheme VII. (For a

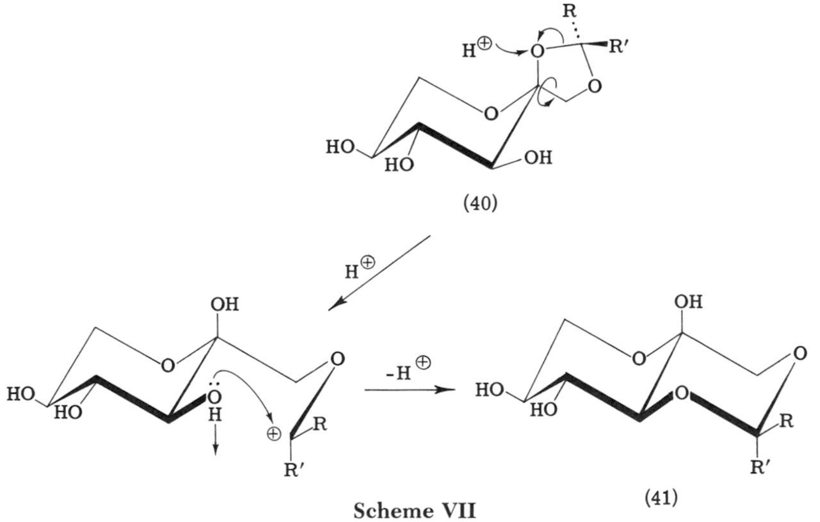

Scheme VII

(69) S. Takagi and R. D. Rosenstein, *Carbohyd. Res.*, **11**, 156 (1969).
(70) J. C. Jochims, G. Taigel, A. Seeliger, P. Lutz, and H. E. Driesen, *Tetrahedron Lett.*, 4363 (1967).
(71) I. Listowsky, S. England, and G. Avigad, *Carbohyd. Res.*, **2**, 261 (1966).
(72) S. J. Angyal and C. G. Macdonald, *J. Chem. Soc.*, 686 (1957).

discussion of the conformational features that preclude rearrangement
to give a 2,3-O-alkylidene ring, see Section III, 2; p. 220.) A m-di-
oxane ring is normally flatter than a cyclohexane ring, and the magnitudes of the 1,3-diaxial interactions are, presumably, diminished.
In **41**, however, this flexibility is restricted in the m-dioxane ring,
and this part may assume a conformation close to the normal geometry
of cyclohexane. Thus, gauche interactions between an axial sub-
situent at C-2 and carbon atoms 4 and 6 of the L-sorbose chain im-
part instability to **41**, and the pyranoid ring may open, to form the
tautomer **43**. Compound **41** has been isolated in examples where
R is phenyl[29] (**39**) or ethyl[15] (**26**) and R' is a hydrogen atom; however, for isopropylidene acetals, an axial methyl group at C-2 of
the m-dioxane ring is unavoidable, and compound **41** (R = R' = CH$_3$)
has not been isolated. In this instance, it is probable that, in the
transition state, the pyranoid ring opens to give the acyclic tautomer
before the m-dioxane ring is formed. Thus, in acetonation reactions,
compound **43**, which is less strained that **40** or **41**, is formed
irreversibly.

The primary hydroxyl group at C-6 of **43** presumably reacts in the
manner described for L-sorbose (**14**), forming the diacetal **42**. This
product has not been detected in any of the acetalation reactions,
but its existence is quite probable, as it has been shown, for acid-
catalyzed acetalations of glycerol, that 1,3-dioxolane compounds are
formed rapidly in the initial, kinetic phase of the reaction, but that
m-dioxanes preponderate at equilibrium.[73] By this process, the
1,3-dioxolane ring would rearrange to give a m-dioxane ring, forming the observed diacetal **46** from the intermediate **42**.

The diacetal **46** may cyclize reversibly to give a tautomer (**47**). In
one conformation (**46a**) of **46**, the carbonyl group at C-2 and the
hydroxyl group at C-5 of the L-sorbose chain can readily achieve

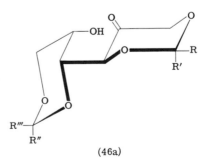

(46a)

(73) N. Baggett, M. Duxbury, A. B. Foster, and J. M. Webber, *Carbohyd. Res.*, **2**, 216
(1966).

the proximity required for this cyclization. Since a six-membered ring is destabilized by an sp^2 hybridized atom,[64] the elimination of the carbonyl group by this step favors closure to form the furanoid ring.

In ethylidenation reactions (R' = R''' = H; R = R'' = Me), compound **47** preponderates over compound **49**, whereas in acetonation reactions (R–R''' = Me), the reverse is true. This fact indicates that the *m*-dioxane ring is more stable in the diethylidene acetal than in the diisopropylidene acetal, a fact that can be attributed[15] to the absence of an axial methyl group in an ethylidene acetal having a *m*-dioxane moiety in the chair conformation.

In the benzylidenation reactions, it has been reported[65] that the monoacetal **43** exists as an equilibrium mixture of a pyranoid form (**41**), the *keto* form (**43**), and a furanoid form (**44**). Acetylation of **43** (R = phenyl, R' = H) in pyridine at low temperature afforded the acetate of **44** (R = phenyl, R' = H). Thus, an alternative, mechanistic route may be available for the benzylidenation reactions, namely **43** → **44** → **47** → **46**.

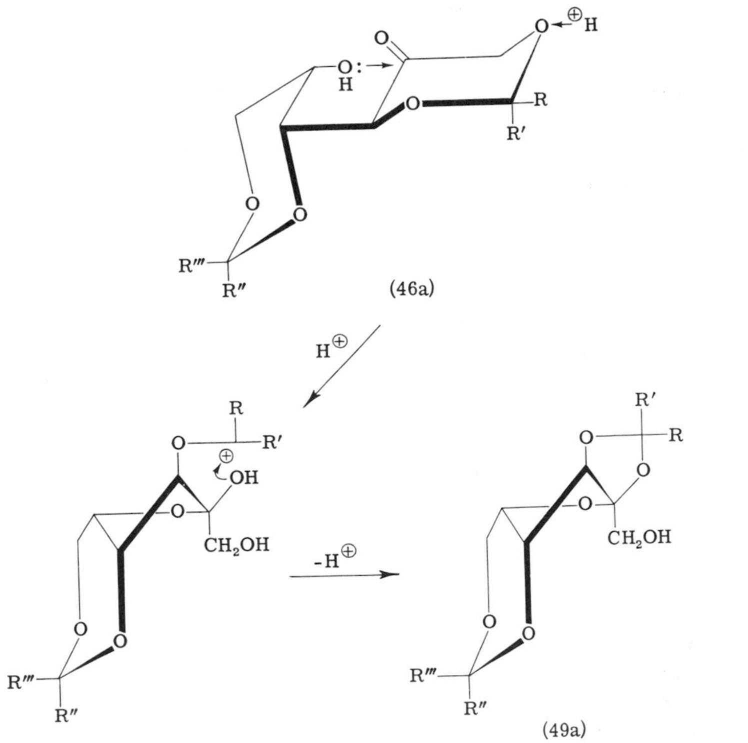

Scheme VIII

The diacetal **46** may also isomerize to diacetals **45** and **48**. Alternatively, **46** may rearrange under acid catalysis to **49** by way of the mechanism shown in Scheme VIII. This step has been shown[29] to proceed with retention of the diastereoisomeric configuration at the acetal center when R = R″ = phenyl, and R′ = R‴ = H. The *endo*-phenyl diacetal (**31**) that is produced is unstable with respect to its *exo*-phenyl isomer (**30**), and **30** is isolated at equilibrium (see p. 210).

Proton magnetic resonance studies[74] have shown that the furanose ring in **49** assumes that conformation (**49a**) in which C-3 is *exo*[75] and C-4 is *endo* to the plane defined by C-2, C-5, and O-5. This conformation is stabilized by the acetal groups, each of which may assume its most favorable orientation at the same time. When R–R‴ = Me (diacetal **3**; p. 207), this rearrangement is favored, as a 2,3-isopropylidene acetal is more stable[74] than a 1,2- or 1,3-isopropylidene acetal of 4,6-*O*-isopropylidene-L-sorbofuranose. This driving force is absent in the ethylidenation reactions, and thus, in the latter, product **49** is formed only with difficulty.

2. The Acetalation of D-Fructose

The mechanism of the acetalation of D-fructose has not been studied so comprehensively as that of L-sorbose. The former reaction appears to be much less complex, but there are many features common to the two reactions. Comparative studies of the acetalation of D-fructose by various aldehydes and ketones have not been reported, although such researches would be a fruitful source of needed and valuable information.

At mutarotational equilibrium in water, D-fructose (**51**) exists preponderantly as the β-D-pyranose anomer in the *1C*(D) conformation.[76,77] A 1,2-alkylidene acetal (**52**) is formed in the same way as for L-sorbose, but this monoacetal has *cis*-disposed hydroxyl groups at C-4 and C-5 that react readily, forming a 1,2:4,5-di-*O*-alkylidene-β-D-fructopyranose (**53**). No evidence is available to indicate that the 1,2-alkylidene acetal might rearrange to a 1,3-alkylidene acetal, and it is to be expected that the activation energy for this isomerization would exceed that for formation of an acetal at O-4 and O-5.

(74) T. Maeda, K. Tori, S. Satoh, and K. Tokuyama, *Bull. Chem. Soc. Jap.*, **41**, 2495 (1968).
(75) The designator *endo* denotes an atom on the same side of the plane defined by the C-2–O–C-5 ring atoms as C-6, whereas *exo* denotes an atom on the opposite side of this plane.
(76) H. Ohle, *Ber.*, **60**, 1168 (1927).
(77) C. P. Barry and J. Honeyman, *Advan. Carbohyd. Chem.*, **7**, 53 (1952).

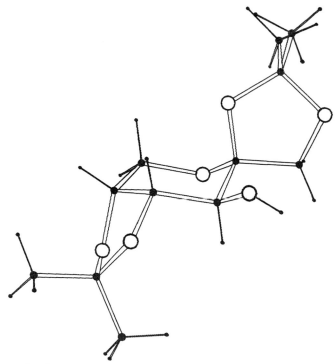

FIG. 1.—Molecular Structure of 1,2:4,5-Di-O-isopropylidene-β-D fructopyranose Drawn by a Computer using the X-Ray Single-crystal Data of Takagi and Rosenstein.[78]

The introduction of a 1,3-dioxolane ring at C-4 and C-5 distorts the pyranoid ring. Takagi and Rosenstein have found[78] that, in crystalline 1,2:4,5-di-O-isopropylidene-β-D-fructopyranose (**1**, see Fig. 1), the conformation of the molecule is intermediate between the $1C(\text{D})$ and $H_0^2(\text{D})$ conformations.[79a] The distance of C-5 from the least-squares plane defined by C-3, C-4, C-6 and O-6 is 44 pm, as compared with 67 pm for D-fructopyranose, and C-2 is 73 pm from this plane, as compared with 67 pm for D-fructopyranose. The portion of the molecule containing O-6, C-2, and C-3 is slightly more puckered, and the portion containing C-4, C-5, and C-6 is relatively less puckered, than in D-fructopyranose; thus, the C-4, C-5, and C-6 portion of the molecule has undergone a conformational distortion toward a skew form. In this process, the dihedral angle subtended

(78) S. Takagi and R. D. Rosenstein, *Acta Crystallogr.*, **A25** [S3], s197 (1969).

(79) (a) H. S. Isbell and R. S. Tipson, *J. Res. Nat. Bur. Stand.*, **64A**, 171 (1960); (b) L.D. Hall, *Chem. Ind.* (London), 950 (1963).

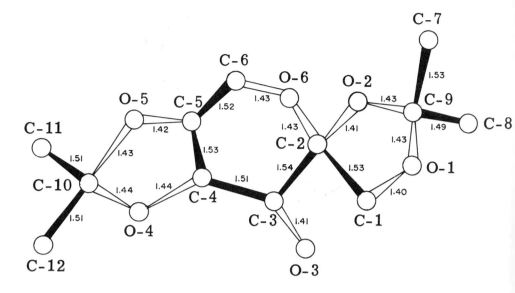

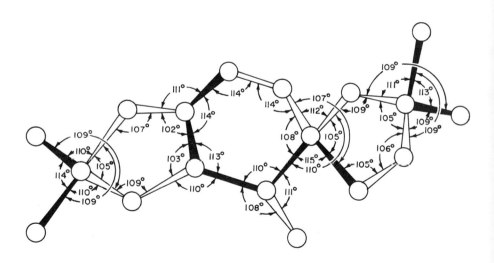

FIG. 2.—Bond Distances (in Å) and Bond Angles Found[78] by X-Ray Single-crystal Analysis for 1,2:4,5-Di-O-isopropylidene-β-D-fructopyranose.

by O-2–C-2–C-3–O-3 is decreased to a magnitude sufficient to accommodate a 1,3-dioxolane ring (see Fig. 2). Under conditions permitting rearrangement, the strained[54] *spiro*-fused ring at C-1 and C-2 is rapidly transformed[30] to give a stable, *cis*-fused ring at C-2 and C-3, forming the diacetal **55**. The ratio of **55** to **53** at equilibrium is[30] 47:3.

2,3:4,6-Di-O-isopropylidene-β-D-fructopyranose (**2**), an example of the general structure **54,** appears from proton magnetic resonance data[56] to adopt the S_4^6 conformation.[79a] This conformation is in accord with those proposed for other diacetals in which the parent sugar has

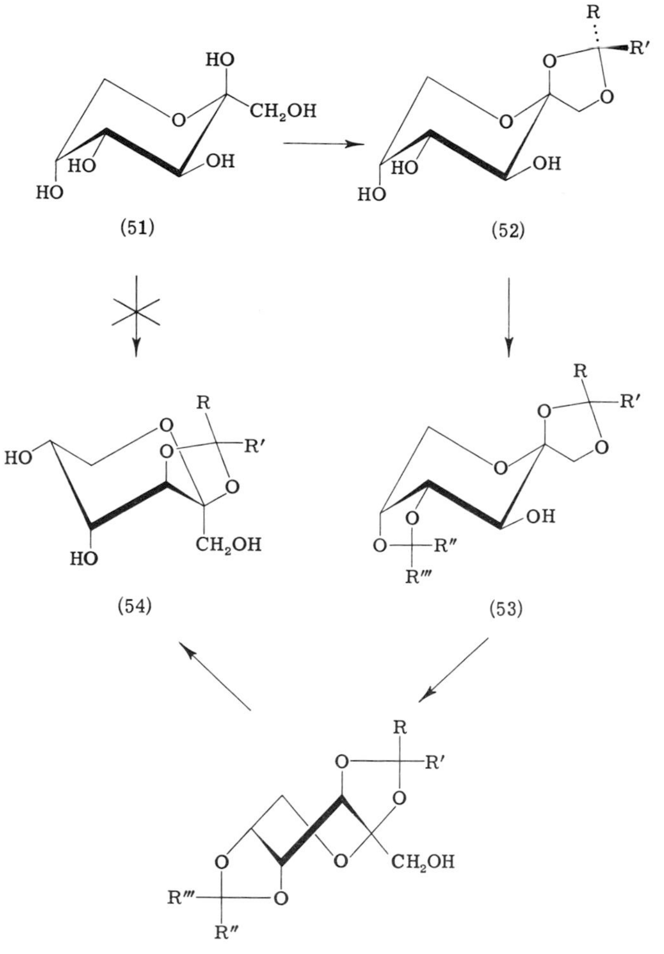

Scheme IX

the *β-arabino* stereochemistry.[80] The diacetal structure **55** is thermodynamically more stable than the isomeric form **53**, and **55** is the end product in acid-catalyzed acetalation reactions of D-fructose.[30]

A second type of monoacetal is known for D-fructopyranose, namely, the 2,3-O-alkylidene-β-D-fructopyranose structure (**54**). A distorted *C1*(D) conformation is indicated for the isopropylidene acetal **5** (**54**, R = R' = Me), from proton magnetic resonance data.[56] Undoubtedly, this conformation is not favored for β-D-fructopyranose itself, because of unfavorable steric and anomeric effects (see This Volume, Chapter 2). It is unreasonable to expect that monoacetal **54** will be formed directly from β-D-fructopyranose, because a much more favored pathway, leading to monoacetal **52**, is available. This consideration precludes a second, possible avenue to diacetal **55**, namely, alkylidenation of O-4 and O-5 of **54**. Monoacetals of type **54** can be prepared by partial hydrolysis of diacetals of type **55**.

The mechanism of acetalation of β-D-fructose is summarized in Scheme IX.

3. Rearrangements of Acetals

Several examples are known in which a hexulose diacetal rearranges, in acetone containing sulfuric acid, to an isomeric acetal. The rearrangements, in acidic acetone, of several diacetals of L-sorbose to 2,3:4,6-di-O-alkylidene-α-L-sorbofuranose (**49**), and of 1,2:4,5-di-O-isopropylidene-β-D-fructopyranose (**1**) to 2,3:4,5-di-O-isopropylidene-β-D-fructopyranose (**2**), were discussed earlier in this Section (see p. 211). Under the same conditions, 1,2:4,5-di-O-isopropylidene-β-D-psicopyranose (**6**) rearranges to 1,2:3,4-di-O-isopropylidene-β-D-psicofuranose (**56**).

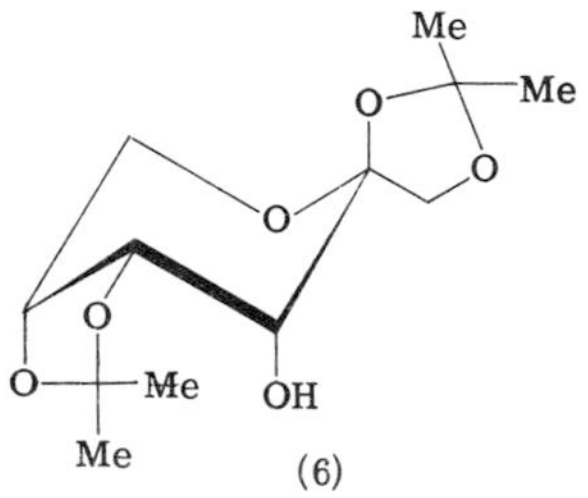

(6) 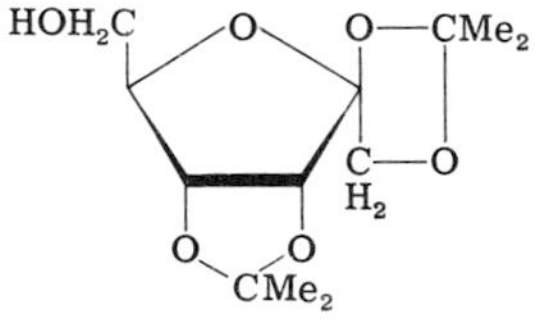   (56)

For **1**, the 1,3-dioxolane ring more resistant to aqueous-acidic hydrolysis undergoes rearrangement, whereas **6**, a diacetal that is hydrolyzed in 3 hours, rearranges to an isomer that is hydrolyzed in

(80) C. Cone and L. Hough, *Carbohyd. Res.*, 1, 1 (1965).

less than 1 hour.[54] Clearly, the order of stability to hydrolysis by aqueous acid is not relevant, *in the absence of water*, to rearrangements in dry acidic acetone. This behavior has also been observed for acetals of alditols.[3,81,82] The rearrangements of the acetals of L-sorbose and D-fructose have been rationalized in the foregoing discussion, but no explanation for the rearrangement of **6** to **56** is evident. This rearrangement may be favored by more-efficient hydrogen-bonding in diacetal **56**.

IV. SPECTROMETRY AND SPECTROSCOPY

1. Proton Magnetic Resonance Spectroscopy

a. Configuration.—The configurations of six compounds constituting three anomeric pairs of derivatives of D-*erythro*-pentulose have been determined by using proton magnetic resonance (p.m.r.) spectroscopy. In each instance, the signals from the two protons on C-1 furnished the information needed.

In the p.m.r. spectra of the α anomer of methyl 1,3,4-tri-O-acetyl-D-*erythro*-pentulofuranoside (**57**) and of the corresponding tribenzoate (**58**), a two-proton singlet was observed for the C-1 methylene group, whereas, in those of the β anomers, two one-proton doublets

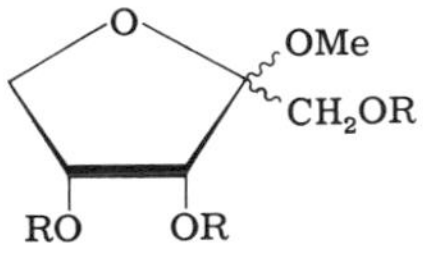

(57) R = Ac
(58) R = Bz

were observed for this group.[36] The multiplicity observed with the β anomers is a consequence of magnetic non-equivalence of the two H-1 protons, arising either from asymmetric shielding or from restricted rotation of the C-1 methylene group when it is *cis* to the 3-acyloxy group. The signal for the methoxyl protons at the anomeric position appears at lower field when the methoxyl group is *cis* to the acyloxy group at C-3, providing complementary evidence for the assignment of anomeric configuration.

The configurations of the anomers of 1,2:3,4-di-O-isopropylidene-D-*erythro*-pentulofuranose were determined[83] by a study of the

(81) S. A. Barker and E. J. Bourne, *J. Chem. Soc.*, 905 (1952).
(82) S. A. Barker, E. J. Bourne, and D. H. Whiffen, *J. Chem. Soc.*, 3865 (1952).
(83) P. M. Collins and P. Gupta, *Chem. Commun.*, 1288 (1969).

chemical shifts of the two protons on C-1. The two C-1 protons of one isomer, and one C-1 proton of the other, resonate in the region τ 5.93–6.04, whereas the other C-1 proton resonates at τ 5.71. Models show that, of the four C-1 protons in the two anomers, only one is situated close enough to an unshared pair of electrons (on O-3) to deshield it, and this occurs in the β anomer (**8**).

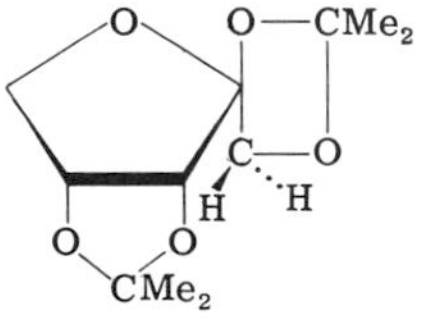

(8) 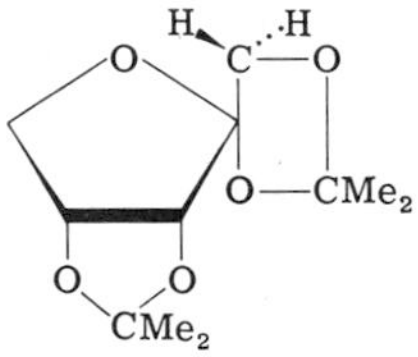   (59)

The relative, specific optical rotations of each pair of anomers support the configurations assigned by proton magnetic resonance spectroscopy.

b. Conformation.—The proton magnetic resonance spectra of several mono- and di-O-isopropylidene-L-sorbofuranoses have been recorded,[74] and these data have been used[74] to differentiate between two[84] possible twist[79b] conformations for the furanoid rings of the acetals, namely, a "C-3-*endo*–C-4-*exo*" (T_3^4) conformation and a "C-3-*exo*–C-4-*endo*" (T_4^3) conformation.[75] Because the ketofuranoses necessarily lack a proton on the anomeric carbon atom, the only coupling constants applicable to study of their conformations are $J_{3,4}$ and $J_{4,5}$. In this example, the dihedral angle between H-4 and H-5 was estimated to be in the range of 10–30° in each conformer, and the coupling constants observed differed by only 0.2 Hz. Therefore, it is surprising that the authors[74] differentiated the conformers on the basis of the $J_{3,4}$ couplings alone.

The furanoid part of 2-O-acetyl-1,3:4,6-di-O-isopropylidene-β-L-sorbofuranose (**60**) was considered[74] to adopt a T_4^3 conformation. The furanoid part of three monoisopropylidene acetals, namely, methyl 4,6-di-O-acetyl-1,3-O-isopropylidene-β-L-sorbofuranoside (**61**), 1,4,6-

(84) The twist conformations were chosen on the basis of reports[85,86] that stated that tetrahydrofuran is thought to exist in a twist conformation having the oxygen atom located on the axis of symmetry, instead of in an envelope conformation. Thus, alternative twist conformations and envelope conformations for the furanoid ring were not rigorously excluded by this study.

(85) K. S. Pitzer and W. E. Donath, *J. Amer. Chem. Soc.*, **81**, 3213 (1959).

(86) J. E. Kilpatrick, K. S. Pitzer, and R. Spitzer, *J. Amer. Chem. Soc.*, **69**, 2483 (1947).

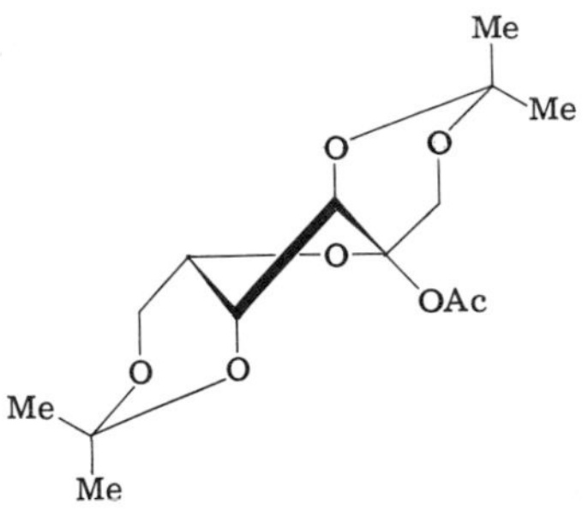

(60)

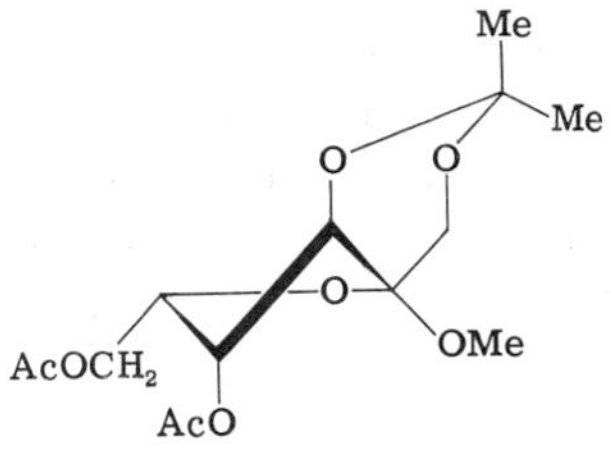

(61)

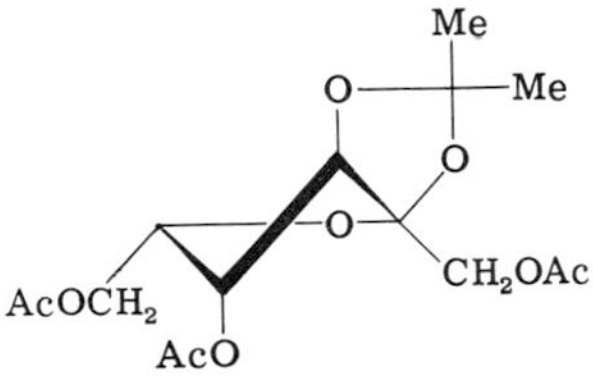

(62)

tri-O-acetyl-2,3-O-isopropylidene-α-L-sorbofuranose (**62**), and methyl 4,6-O-isopropylidene-α-L-sorbofuranoside (**63**) was also said to adopt a T_4^3 conformation.[74] The furanoid parts of other 4,6-O-isopropylidene-L-sorbofuranoses were considered to adopt similar conformations; for such α anomers as **62** and **63**, the 1,3-dioxolane and

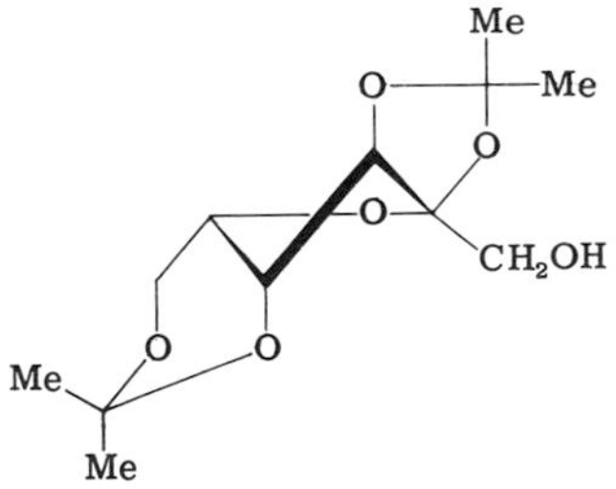

(3)

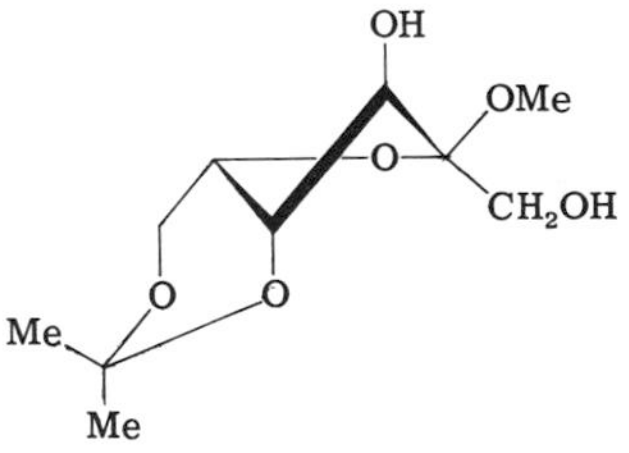

(63)

m-dioxane rings hold the furanoid part in a conformation that is, presumably, not favored by the anomeric effect.[74]

The furanoid part of 2,3:4,6-di-O-isopropylidene-α-L-sorbofuranose (**3**) was also assigned[74] a T_4^3 conformation: this compound was considered to be particularly stable because the conformation assigned to the furanoid part is stabilized by both the 1,3-dioxolane and the m-dioxane rings. More-detailed conformational data are needed for these diacetals, and X-ray crystallographic studies should provide valuable information.

Proton magnetic resonance spectroscopy has also been used to study the conformations of several pyranoid derivatives of D-fructose and L-sorbose.[56] Once more, data obtained in this study are limited, because of the lack of a proton at the anomeric center; the only coupling constants useful in conformational studies of acetals of ketopyranoses are $J_{3,4}$, $J_{4,5}$, $J_{5,6}$, and $J_{5,6'}$, and the conclusions that were reported should not be considered as being definitive.

Among the monoacetals, those containing a 1,2-O-isopropylidene ring, namely, 3,4,5-tri-O-acetyl-1,2-O-isopropylidene-α-L-sorbopyranose (64), 3,4,5-tri-O-acetyl-1,2-O-isopropylidene-β-D-fructopyranose (65), and 4,5-di-O-acetyl-1,2-O-isopropylidene-3-O-methyl-β-D-fructopyranose (66), have been assigned the $1C(\text{D})$ conformations depicted.[56]

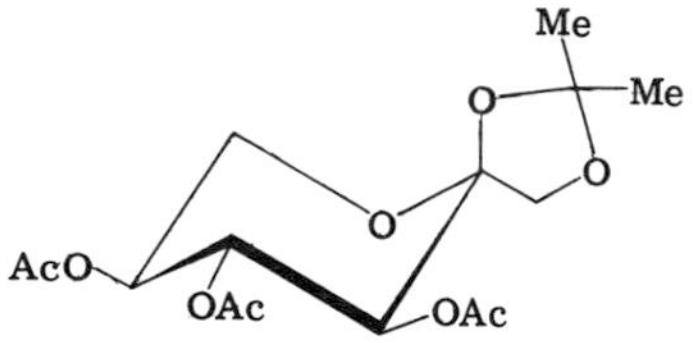

(64) 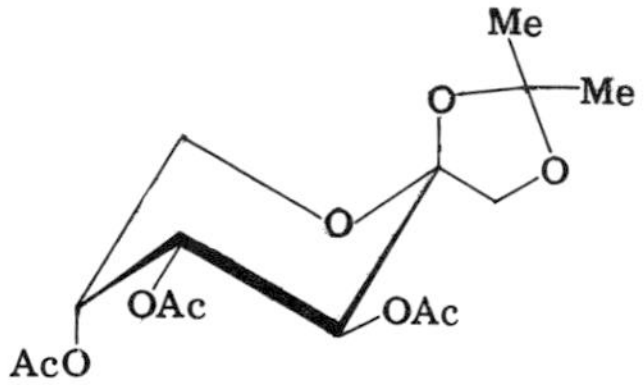   (65)

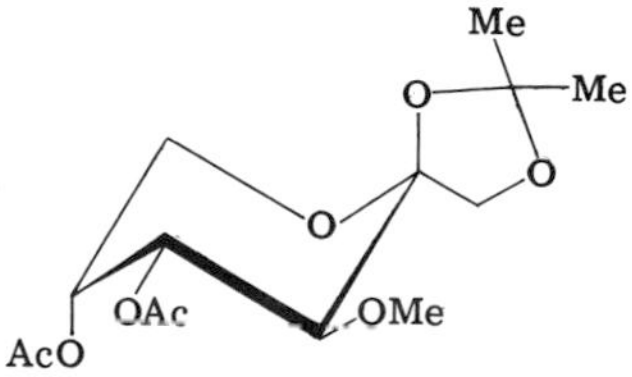

(66)

Ethyl 4,5-di-O-acetyl-1,3-O-isopropylidene-α-L-sorbopyranoside (67) was shown to adopt the expected conformation resembling that of

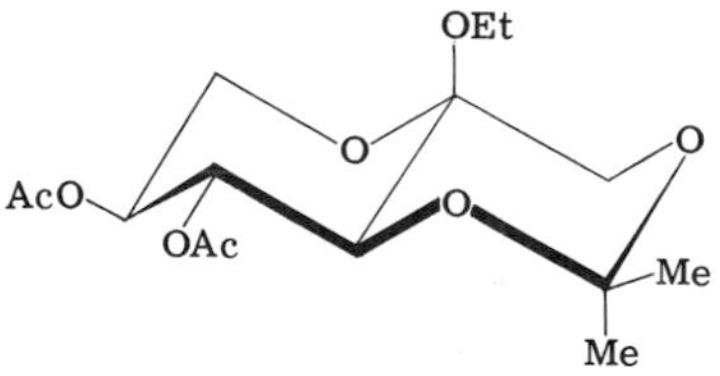

(67)

trans-decalin, in which the pyranoid part assumes the $1C(D)$ conformation. The pyranoid part of 2,3-*O*-isopropylidene-β-D-fructopyranose (**10**) was found to exist in a slightly distorted $C1(D)$ confor-

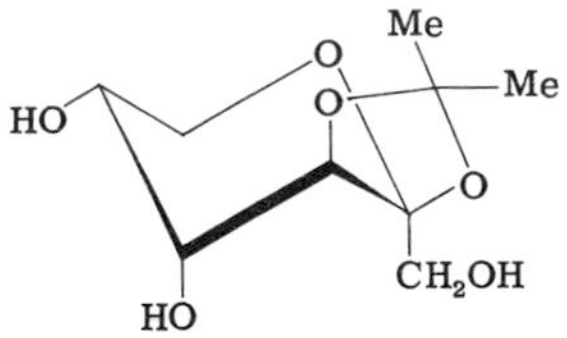

(10)

mation. The observation of a long-range coupling between H-4 and H-6 provides support for assigning this conformation.

The pyranoid moieties in the diacetals containing *trans*-fused 1,3-dioxolane rings, namely, 1,2:3,4-di-*O*-isopropylidene-β-L-sorbopyranose (**15**), its 3-acetate (**68**), and 3-*O*-acetyl-1,2:4,5-di-*O*-isopropylidene-β-L-sorbopyranose (**69**), were considered to exist in slightly distorted, $1C(L)$ conformations.[56] In contrast, the pyranoid parts in diacetals containing *cis*-fused, 1,3-dioxolane rings were found to

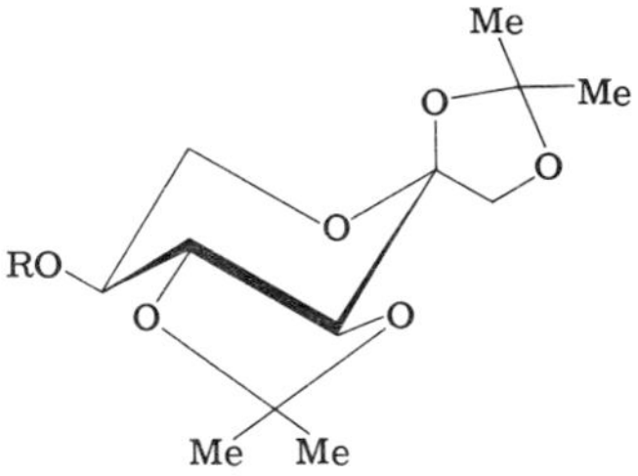

(15) R = H
(68) R = Ac

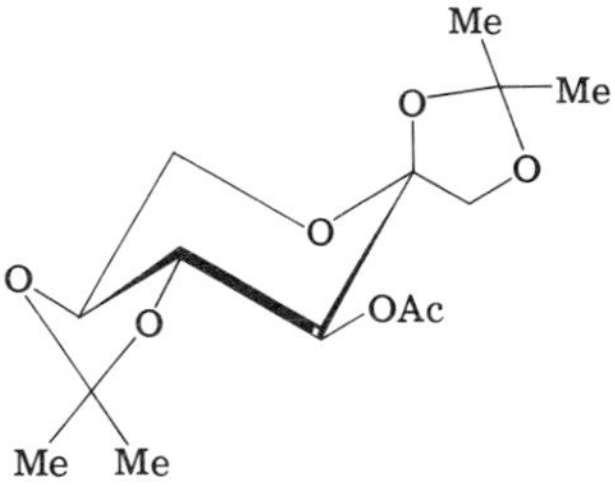

(69)

deviate from ideal chair conformations.[56] The conformations of 1,2:4,5-di-*O*-isopropylidene-β-D-fructopyranose (**1**) and its 3-acetate (**70**), although shown not to have ideal $1C(D)$ shapes, could not

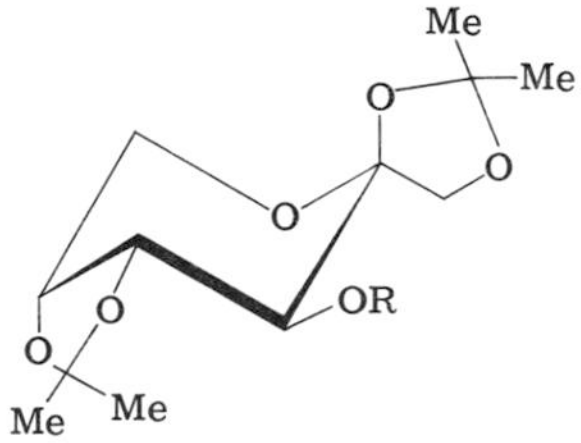

(1) R = H
(70) R = Ac

be rigorously defined on the basis of proton magnetic resonance data alone, as the protons on the pyranoid ring give rise to complex, second-order patterns in chloroform-d, benzene-d_6, and pyridine-d_5, and only incomplete data were obtained.[56] (The precise conformation of **1** in the solid state has been elucidated by the X-ray diffraction technique;[78] see Section III,2; p. 218.) However, 2,3:4,5-di-O-isopropylidene-β-D-fructopyranose (**2**) and its 1-acetate (**71**) were indicated[56]

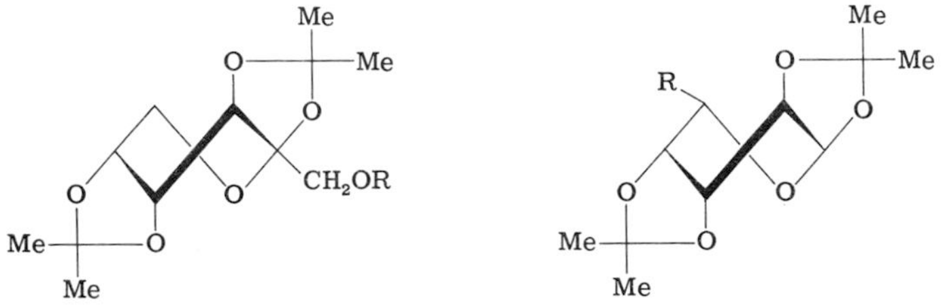

(2) R = H
(71) R = Ac

(72) R = H
(73) R = CH₂OH

by the p.m.r. data to exist in the S_4^6 conformation depicted. This conformation is quite close to those found[80] for two other diacetals having the *cis-anti-cis* disposition of hydroxyl groups, namely, 1,2:3,4-di-O-isopropylidene-β-L-arabinopyranose (**72**) and 1,2:3,4-di-O-isopropylidene-α-D-galactopyranose (**73**).

   c. **Other Applications.**—General applications of proton magnetic resonance spectrometry to carbohydrate chemistry have been reviewed.[87,88] The technique has also been used as a routine aid in determining the structures of the products of acetalation of L-sorbose,[15,16,29,66] and in monitoring the tautomerization of 1,3-O-benzylidene-L-sorbose[65] and of the rearrangement[29] illustrated: **32** → **31** → **30**.

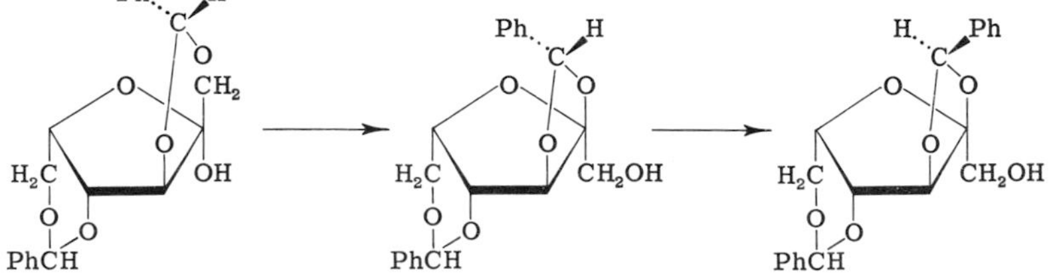

(32)   (31)   (30)

(87) L. D. Hall, *Advan. Carbohyd. Chem.*, **19**, 51 (1964).
(88) T. D. Inch, *Ann. Rev. Nucl. Mag. Res. Spectros.*, **2**, 35 (1969).

The types of free hydroxyl groups present were determined for several acetals of sedoheptulosan by measuring the proton magnetic resonance spectrum of the compound in dry methyl sulfoxide-d_6 and observing the multiplicity of the signal caused by the hydroxyl protons.[89] The p.m.r. spectrum of 2,7-anhydro-4,5-O-isopropylidene-β-D-*altro*-heptulopyranose (**74**) ("isopropylidene-sedoheptulosan") shows a doublet ($J = 7$ Hz) at τ 4.90 and a triplet ($J = 6.5$ Hz) at τ 5.33;

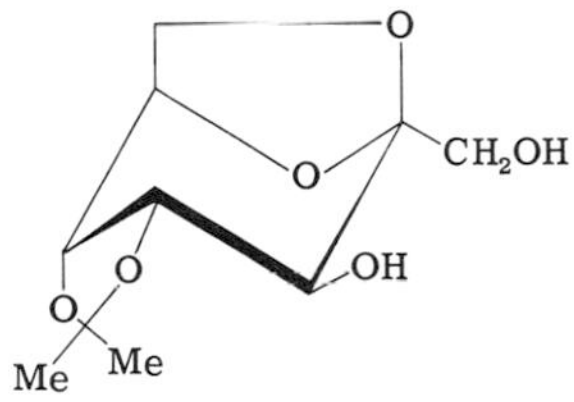

(74)

both signals disappeared after deuterium oxide was added to the solution. Thus, **74** was shown to possess one primary and one secondary hydroxyl group.

2. Infrared Spectroscopy

The infrared spectra of carbohydrates in general have been reviewed by Tipson,[90] and those of cyclic acetals of sugars by S. A. Barker and coworkers,[91] and by Tipson and coworkers.[92]

Barker and coworkers[91] found that sugars bearing a 1,3-dioxolane ring absorb infrared radiation in the regions of 1173–1151, 1151–1132, 1105–1077, and 1053–1038 cm^{-1}. However, Tipson and coworkers studied the infrared spectra of some cyclic acetals of sugars, and concluded that unequivocal detection of a 1,3-dioxolane ring fused to a pyranoid or furanoid ring was not very feasible.[92] Certain strong bands were observed, but they were not readily recognized from an inspection of one or two spectra, and the bands were not characteristic of 1,3-dioxolane rings only.

Barker and Stephens[93] stated that acetals of furanoid sugars absorb infrared radiation at 924 ± 13, 879 ± 7, 858 ± 3, and 799 ± 17 cm^{-1}.

(89) E. Zissis and N. K. Richtmyer, *J. Org. Chem.*, **30**, 462 (1965).
(90) R. S. Tipson, *Nat. Bur. Stand. Monograph* **110** (1968).
(91) S. A. Barker, E. J. Bourne, and D. H. Whiffen, *Methods Biochem. Anal.*, **3**, 213 (1956).
(92) R. S. Tipson, H. S. Isbell, and J. E. Stewart, *J. Res. Nat. Bur. Stand.*, **62**, 257 (1959).
(93) S. A. Barker and R. Stephens, *J. Chem. Soc.*, 4550 (1954).

The first of these absorptions was attributed to the symmetrical stretch ("ring-breathing" mode) of the furanoid ring; the second and third absorptions were thought to arise from vibration modes involving either the stretching of a bond between a substituent atom and a ring atom, or, where present, the rocking vibration of a methylene group. The fourth absorption was assigned to deformation of the C–H bonds of carbon atoms in the ring.[93] Tokuyama and coworkers[55] studied the infrared spectra of twelve mono- and di-O-isopropylidene-L-sorbofuranoses and concluded that, in the spectra of these compounds, the four bands are shifted to 958 ± 24, 895 ± 7, 864 ± 6, and 823 ± 7 cm^{-1}. It is however, apparent, that these modes of absorption are possible for substituted 1,3-dioxolane rings also, and that compounds containing these rings will also show absorption bands at similar frequencies. It is, therefore, not possible to determine, by infrared spectroscopy alone, whether or not a sugar acetal has a furanoid ring. As an example, Maeda[16] studied 21 mono- and di-O-isopropylidene-L-sorbofuranoses and -sorbopyranoses, and reported that four of six pyranose derivatives showed three of the four bands, whereas one showed all four, and one showed only two. In contrast, each of the 15 acetals containing a furanoid ring showed one or more bands in each of the four spectral regions.

Infrared spectroscopy has given useful results in applications of limited scope. Studies[29] of hydrogen bonding in dilute solutions of the acetals 33–36 in carbon tetrachloride substantiated the assignments of anomeric configuration already made by proton magnetic

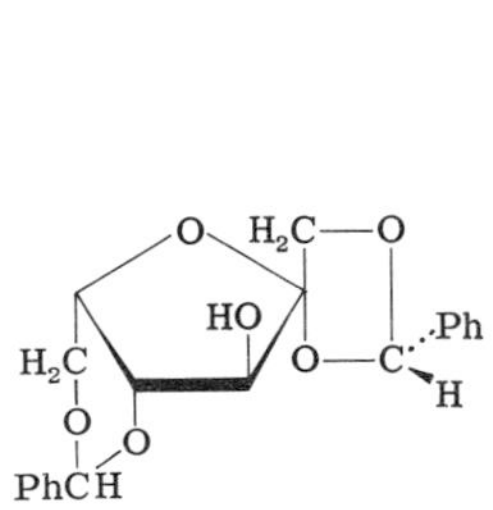

(33) 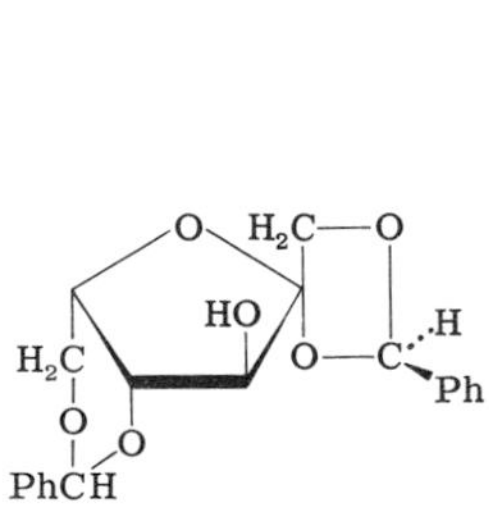   (34)

(35)   (36)

resonance spectroscopy and by relative, specific optical rotations. A characteristic shift of the O–H stretching band due to intramolecular hydrogen-bonding between the 3-OH group and O-2 was observed in the spectra of the α anomers (**33**, 3558 cm^{-1}; **34**, 3556 cm^{-1}), but not in the spectra of the β anomers (**35**, 3630 cm^{-1}; **36**, 3629 cm^{-1}).

Patil and Bose[94] claimed that furanoid and pyranoid structures of hexulose mono- and di-acetals may be differentiated on the basis of the position of the absorption band due to the O–H out-of-plane deformation of the 1-OH group. In i.r. spectra of acetals of hexulofuranoses, the band was observed at 686 ± 4 cm^{-1}, but, for the one hexulopyranose derivative studied, the band was observed at 714 cm^{-1}. Because of the small amount of data provided, their conclusions may not be reliable.

3. Mass Spectrometry

The applications of mass spectrometry to carbohydrate chemistry have been reviewed.[95–97] In general, stereoisomeric acetates or methyl ethers of carbohydrates are not well differentiated by mass spectrometry,[98–100] but when the mode of derivatization depends on the stereochemistry of the parent sugar and leads to new ring-systems, as in acetal formation, structural isomers are produced that are more apt to show significant mass-spectral differences.

Among the acetals of sugars, only the isopropylidene acetals have received intensive mass-spectral investigation.[101] For these compounds, no molecular-ion peak was observed, but loss of a methyl radical from a 2,2-dimethyl-1,3-dioxolane ring forms an intense $(M - 15)^+$ fragment-ion; this facilitates the determination of molecular weight. This and other fragmentation patterns have been confirmed by deuterium labeling and by high-resolution mass measurements.[101]

Some of the fragment-ions identified in the mass spectrum[101] of

(94) J. R. Patil and J. L. Bose, *Carbohyd. Res.*, **7**, 405 (1968).
(95) N. K. Kochetkov and O. S. Chizov, *Advan. Carbohyd. Chem.*, **21**, 39 (1966).
(96) K. Heyns, H. F. Grützmacher, H. Scharmann, and D. Müller, *Fortschr. Chem. Forsch.*, **5**, 448 (1966).
(97) H. Budzikiewicz, C. Djerassi, and D. H. Williams, "Structural Elucidation of Natural Products by Mass Spectrometry," Holden–Day, Inc., San Francisco, California, 1964, Vol. II.
(98) K. Biemann, D. C. DeJongh, and H. K. Schnoes, *J. Amer. Chem. Soc.*, **85**, 1763 (1963).
(99) D. C. DeJongh and K. Biemann, *J. Amer. Chem. Soc.*, **85**, 2289 (1963).
(100) D. C. DeJongh, *J. Org. Chem.*, **30**, 453 (1965).
(101) D. C. DeJongh and K. Biemann, *J. Amer. Chem. Soc.*, **86**, 67 (1964).

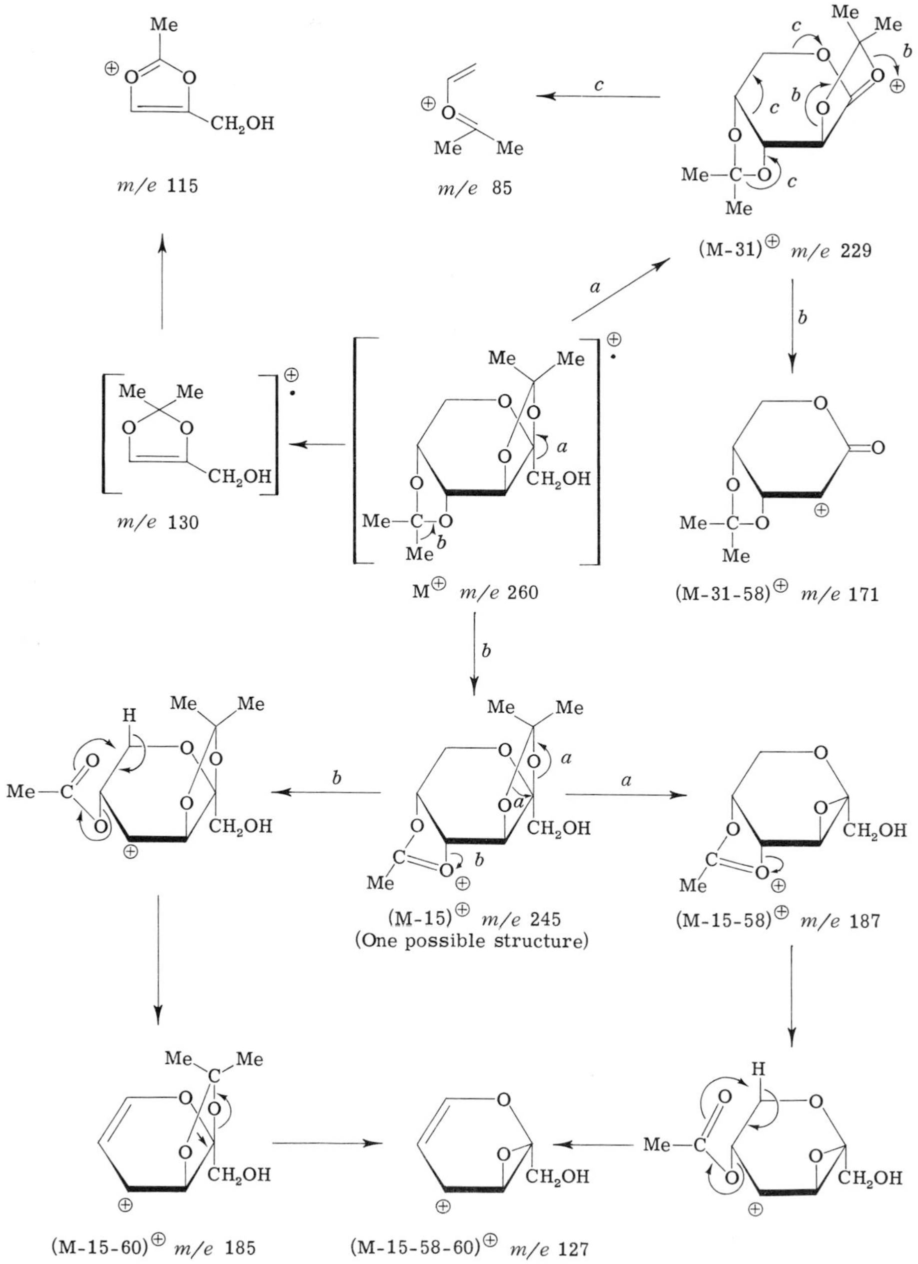
Me
CH₂OH
m/e 115
Me Me
m/e 85
c
Me Me
b
c
c b
Me—C—O
Me
c
(M-31)⊕ m/e 229
a
b
Me Me
Me Me
CH₂OH
m/e 130
a
CH₂OH
Me—C—O
b
Me
M⊕ m/e 260
Me—C—O
Me
(M-31-58)⊕ m/e 171
b
H
Me Me
Me—C
CH₂OH
b
Me Me
a
a
CH₂OH
b
Me
(M-15)⊕ m/e 245
(One possible structure)
a
CH₂OH
Me
(M-15-58)⊕ m/e 187
Me C Me
CH₂OH
(M-15-60)⊕ m/e 185
CH₂OH
(M-15-58-60)⊕ m/e 127
H
Me—C
CH₂OH

Scheme X

2,3:4,5-di-O-isopropylidene-β-D-fructopyranose (2) are depicted in
Scheme X. Loss of a methyl radical from the molecular ion (m/e 260)
leads to an oxonium ion (m/e 245), which undergoes further decomposition to ions resulting from the loss of acetone (m/e 187), or of
acetic acid (m/e 185), and, ultimately, of both fragments (m/e 127).
The loss of the hydroxymethyl group (C-1) gives rise to an appreciable
(M − 31)⁺ ion having m/e 229; this decomposes further by loss of
acetone, giving a peak at m/e 171. Two fragment-ions, corresponding
to C-1 and the 1,3-dioxolane ring involving C-2 and C-3, can be
observed; they have m/e 130 and 115, respectively.

The formation of an ion having m/e 113 was not accounted for in
the original paper.[101] However, since O-4 can readily approach the
ring-oxygen atom, the mechanism shown in Scheme XI is a plausible
route to this fragment-ion.

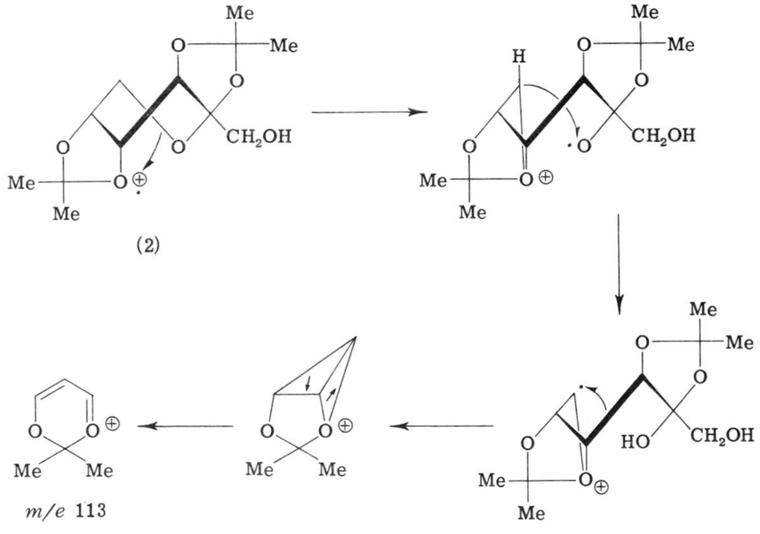

Scheme XI

The mass spectrum of 1,2:4,5-di-O-isopropylidene-β-D-fructopyra-
nose (1) shows the fragment-ions associated with 2,2-dimethyl-1,3-
dioxolane rings (m/e 245, 187, 185, and 127), but differs from that of
its isomer 2, in that the fragment-ions due to a hydroxymethyl group
(m/e 229, 171, 130, and 115) are necessarily absent. New fragment-
ions appear at m/e 117 and 72, and have been assigned to ions containing the 1,3-dioxolane ring at C-1 and C-2. The rearrangements
postulated as leading to these fragment-ions are depicted in Scheme

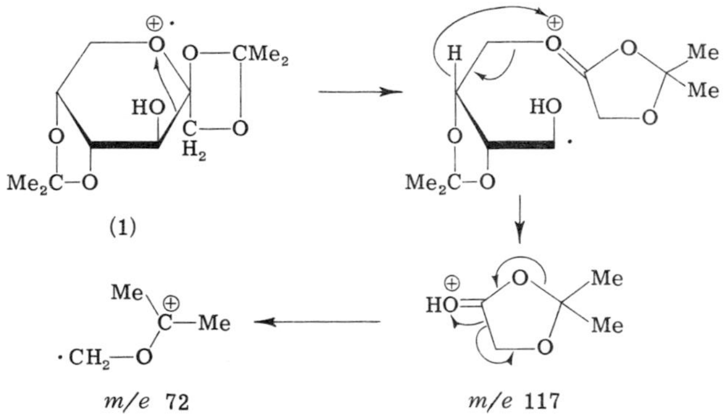

(1)

m/e 72   m/e 117

Scheme XII

XII. There is also a peak at m/e 144, resulting from the loss of two molecules of acetone from the molecular ion.

The mass spectrum[101] of 2,3:4,6-di-O-isopropylidene-α-L-sorbofuranose (3) shows fragment-ions characteristic of a quaternary hydroxymethyl group (m/e 229, 171, 130, and 115), and also those associated with 2,2-dimethyl-1,3-dioxolane rings (m/e 245, 187, 185, and 127). New fragment-ions at m/e 113 and 159 are also observed, and the mechanisms by which they are presumed to be formed are shown in Scheme XIII. With the exception of the peak at m/e 159, the spectrum

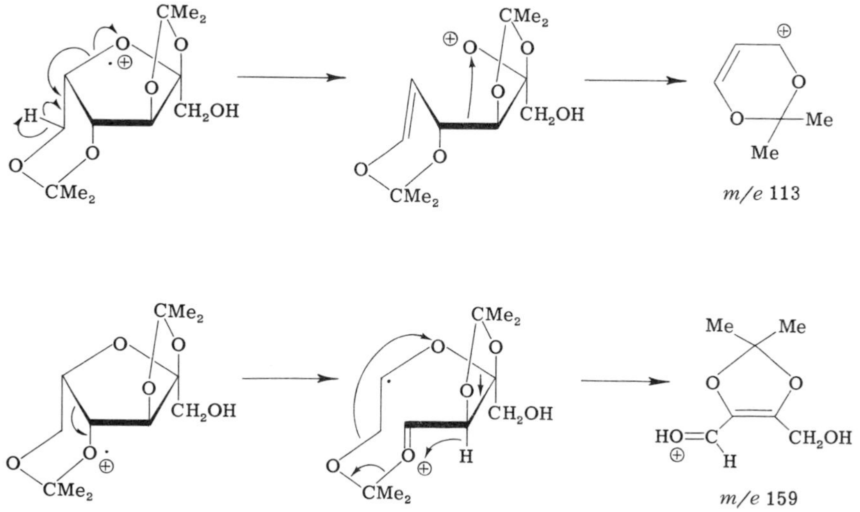

m/e 113

m/e 159

Scheme XIII

of 3 is qualitatively similar to that of 2, differing only in the relative intensities of the peaks. Thus, these two spectra, considered together, provide an example of the limit to the information on such compounds that may be obtained by conventional mass spectrometry.

A diacetal isolated by the acetonation of L-sorbose under catalysis by copper(II) sulfate has been identified, on the basis of its mass spectrum, as 1,2:4,6-di-O-isopropylidene-α-L-sorbofuranose[63] (17). The fragmentation patterns discussed already were used to differentiate between the isomeric structures possible.

V. Chromatography

1. Preparative, Column Chromatography

Wolfrom and coworkers[60] employed a column of Florex XXX–Celite (5:1 by weight) for the preparative fractionation of the mixture of acetals obtained by partial hydrolysis of 2,3:4,5-di-O-isopropylidene-β-D-fructopyranose (2). Elution with 98% isopropyl alcohol gave the starting material and 2,3-O-isopropylidene-β-D-fructopyranose (10); D-fructose was obtained when the column was then washed with 80% ethanol.

2. Paper Chromatography

A method for the separation of acetals from the parent hydroxylated compounds has been described by Barnett and Kent.[102] Paper made of cellulose acetate was used with 3:2 (v/v) methanol–water for separation, and zones were detected either with (2,4-dinitrophenyl)hydrazine or with acidified potassium permanganate. The limit of detection (of 1,2:5,6-di-O-isopropylidene-α-D-glucofuranose) is about 150 μg. Later, Bird[103] introduced the combination of ceric ammonium nitrate and silver nitrate sprays as detection reagents specific for acetals of carbohydrates. By this technique, the limit of detection (of 1,2:5,6-di-O-isopropylidene-α-D-glucofuranose) is 8–12 μg.

3. Thin-layer Chromatography

For the separation and identification of isomeric diisopropylidene, dibenzylidene, or diethylidene acetals of L-sorbose on layers of silica gel, such developing solvents as chloroform–acetone (9:1 v/v[16,29,65,68,74] or 1:1 v/v[29]), hexane–ether (various proportions from 2:1 to 1:5 v/v),

(102) J. E. G. Barnett and P. W. Kent, Nature, 192, 556 (1961).
(103) J. W. Bird, Can. J. Chem., 40, 1716 (1962).

[15,16,29,65,74,104] benzene–ether (2:1 v/v[65] or 3:1 v/v[29]), and hexane–acetone (various proportions from 3:1 to 1:4 v/v)[18,62,63,105] have been used. Many acetals of L-sorbose have been separated on a preparative scale by using these systems.[15,16,18,29,65,68,74]

The mono- and di-acetal products of a reaction of L-sorbose with acetone have been examined separately, without preliminary fractionation.[18] When the mixture of products was developed with 3:1 (v/v) hexane–acetone, the diacetals were dispersed, but the mono-acetals were practically immobile. Alternatively, irrigation of the mixture of products with 2:3 (v/v) hexane–acetone effected a good separation of the isomeric monoacetals, whereas the diacetals gave rise to a single, fast-moving spot.

The oxidation of 1,2:4,5-di-O-isopropylidene-β-D-fructopyranose (1) to 1,2:4,5-di-O-isopropylidene-β-D-*erythro*-2,3-hexodiulo-2,6-pyranose (75) has been monitored by thin-layer chromatography on silica

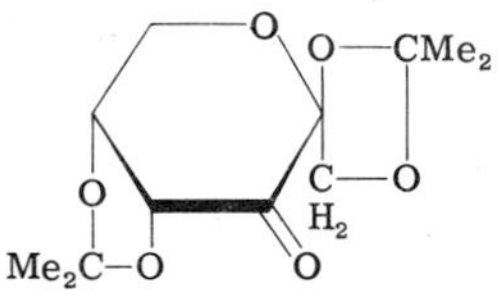

(75)

gel[47] and on Silica Gel G[46,48,49] by using pentane–ethyl acetate (1:3 v/v[46,48] or 3:1 v/v[46]), and benzene–methanol (19:1 v/v[49] or 24:1 v/v[46,47]). The subsequent reduction of 75 to 1,2:4,5-di-O-isopropylidene-β-D-psicopyranose (6) has been monitored by using ethyl acetate,[49] in addition to the irrigants already noted.

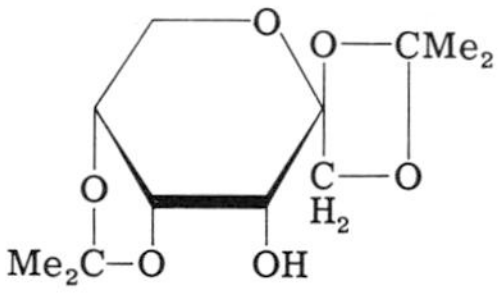

(6)

A solution of concentrated sulfuric acid in methanol has proved to be a most useful reagent for detecting zones on developed chromatoplates.[18,46,48,49,63] Other reagents that have been employed are iodine vapor,[15,16,29,68,74] resorcinol–phosphoric acid,[16,68] naphthoresorcinol–

(104) A. T. Ness, R. M. Hann, and C. S. Hudson, *J. Amer. Chem. Soc.*, **70**, 765 (1948).
(105) B. D. Modi, J. R. Patil, and J. L. Bose, *Indian J. Chem.*, **2**, 32 (1964).

phosphoric acid,[47] the Seliwanoff reagent,[29] and naphthoresorcinol–sulfuric acid.[105] With the exception of iodine vapor, heating for 10–15 minutes at 110° (after spraying) is necessary for development of the color.

An apparatus for descending, thin-layer chromatography of iso-propylidene acetals of L-sorbose has been described.[105]

4. Gas–Liquid Chromatography

Jones and coworkers[106] were the first to report a separation of carbohydrate acetals by gas–liquid chromatography. They employed a column (117 × 0.5 cm inside diameter) packed with 40 cm of an intimate 1:1 (v/v) mixture of 20% (wt./wt.) Apiezon M grease on Chromosorb W and 20% (wt./wt.) of butanediol succinate polyester on Chromosorb W, on top of 77 cm of 1% (wt./wt.) of SE-30 medium on glass beads. The temperature of the column was kept at 206°. With

TABLE I

Retention Times of Per(trimethylsilyl) Ethers and Other Derivatives of Some Acetals of D-Fructose and D-Psicose

Compound	Liquid phase		
	15% EGS[47]		3% SE-30[46]
	Temperature (degrees)		
	132	157	170
	Retention time[a]		
1,2:3,4-Di-O-Isopropylidene-3-O-(trimethylsilyl)-β-D-fructopyranose	40 m	22.6 m	100 s
1,2:3,4-Di-O-isopropylidene-3-O-(trimethylsilyl)-β-D-psicopyranose	68.8 m	36.8 m	111 s
2,3:4,5-Di-O-isopropylidene-1-O-(trimethylsilyl)-β-D-fructopyranose	62.4 m		114 s
1,2:3,4-Di-O-isopropylidene-5-O-(trimethylsilyl)-β-D-psicofuranose	24.8 m[b]	18.4 m	
3-O-Acetyl-1,2:4,5-di-O-isopropylidene-β-D-fructopyranose			15 s
1,2:4,5-Di-O-isopropylidene-3-O-[(methylthio)methyl]-β-D-fructopyranose			215 s
1,2:4,5-Di-O-isopropylidene-β-D-erythro-2,3-hexodiulo-2,6-pyranose			68 s

[a] Abbreviations: m, minutes; s, seconds.

[b] Measurement made at 141°.

(106) H. G. Jones, J. K. N. Jones, and M. B. Perry, *Can. J. Chem.*, **40**, 1559 (1962).

a flow rate of argon of 150 ml/min, a mixture of isopropylidene acetals of D-fructose was separated and eluted within 12 minutes.

The oxidation of **1** to **75** and the reduction of the latter to **6**, have been monitored by two groups[46,47] by use of gas–liquid chromatography. In each case, the compounds were chromatographed as their per(trimethylsilyl) ethers. Long retention-times (see Table I) were encountered when a polar liquid-phase (15% of ethylene glycol succinate on silanized Chromosorb W) was employed[47] for the separation, whereas a rapid separation was achieved[46] in six-foot glass columns containing a non-polar liquid phase [3% of SE-30 medium on VarAport 30 (a proprietary acid-washed, silanized Chromosorb W)]. The formation and isomerization of isopropylidene acetals of D-fructose was also studied by gas–liquid chromatography with such a column.[30]

VI. Cyclic Acetals of Specific Ketoses

1. *erythro*-Pentulose ("Ribulose")

Reichstein[107] was the first to prepare 1,2:3,4-di-O-isopropylidene-β-L-*erythro*-pentulofuranose (**76**); this derivative was used to purify

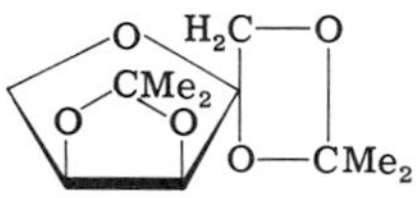

(76)

the ketose ("adonose") that had been prepared by fermentation of ribitol. A more practical preparation of L-*erythro*-pentulose, from L-arabinose, was described by Levene and Tipson[9] two years later; they also purified the ketose by using the diacetal **76**. A simple procedure for the preparation of D-*erythro*-pentulose was published by Tipson and Brady[36] over 30 years later; the β-D anomer of the 1,2:3,4-diisopropylidene acetal (**8**) was used for purification of the ketose. These workers also reported[36] the preparation of crystalline 3,4-O-isopropylidene-α-D-*erythro*-pentulofuranose (**77**) by partial hydrolysis, with concomitant anomerization, of **8**.

The anomer of **8**, 1,2:3,4-di-O-isopropylidene-α-D-*erythro*-pentulofuranose (**59**), was reported shortly thereafter.[83] Anomers **8** and **59**

(107) T. Reichstein, *Helv. Chim. Acta*, **17**, 996 (1934).

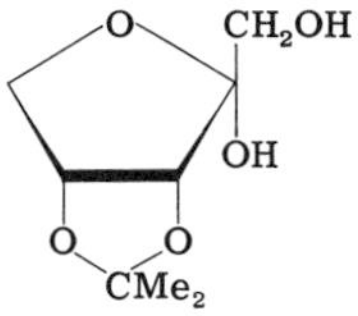

(77)

in the ratio of 8:5 were prepared by photochemical decarbonylation of the diulose **75** (see Scheme XIV).

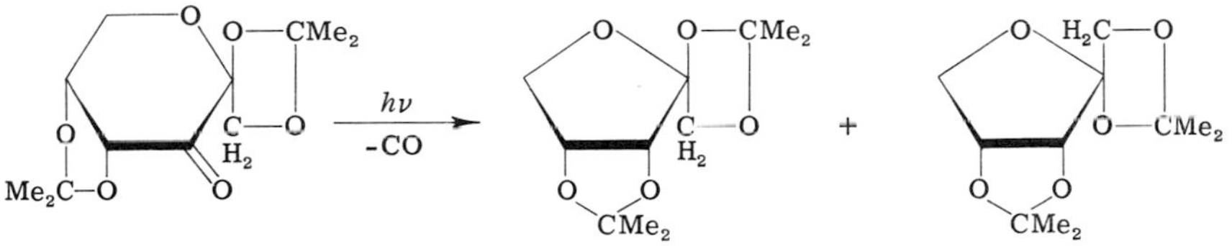

(75)   (8)   (59)

Scheme XIV

A di-ether of an acetal of *keto*-D-*erythro*-pentulose, namely, 3,4-*O*-isopropylidene-1,5-di-*O*-trityl-*keto*-D-*erythro*-pentulose (**78**), has been

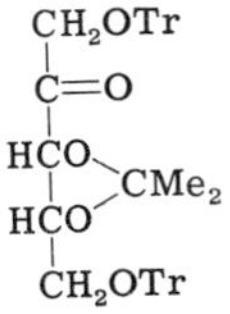

(78)

prepared[108] by oxidation of the corresponding derivative of D-arabinitol with chromic acid.

2. D-*threo*-Pentulose ("D-Xylulose")

Levene and Tipson[8,9] reported the preparation of 2,3-*O*-isopropylidene-β-D-*threo*-pentulofuranose (**79**) in 1936; this derivative was

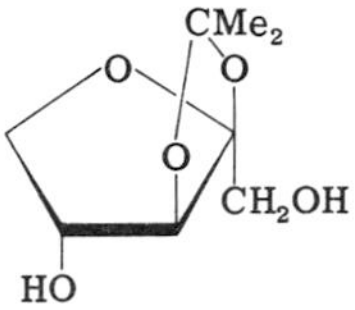

(79)

(108) D. H. Rammler and D. L. MacDonald, *Arch. Biochem. Biophys.*, **78**, 359 (1958).

employed to purify the free ketose, which had been prepared by the isomerization of D-xylose in pyridine. Proof of the structure of this acetal was described in a separate paper[43] published a year later. Tipson and Brady[36] published much later a simplified preparation of the free ketose, and also used acetal **79** to purify the sugar.

An acetal of *keto*-D-*threo*-pentulose, namely 1,5-di-*O*-benzoyl-3,4-

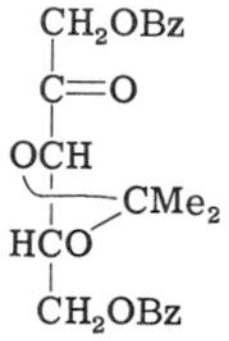

$$
\begin{array}{c}
CH_2OBz \\
| \\
C{=}O \\
| \\
OCH \\
\quad\diagdown CMe_2 \\
HCO\diagup \\
| \\
CH_2OBz
\end{array}
$$

(80)

O-isopropylidene-*keto*-D-*threo*-pentulose (**80**), has been prepared[109] by oxidation of the corresponding derivative of D-arabinitol with chromic acid.

3. D-Fructose

Two crystalline diisopropylidene acetals of D-fructose were prepared[6] by E. Fischer in 1895 by the action of acetone containing 0.2% of hydrogen chloride on the sugar at room temperature. The acetal isolated first was provisionally termed "α" until structural determinations had been made; it was obtained in 50% yield. A small proportion of a second diacetal, arbitrarily termed "β," was on one occasion isolated from the mother liquors. Fischer was unable to repeat the experiment in which he had obtained the "β" form.

Irvine and Hynd[110,111] methylated "α"-di-*O*-isopropylidene-D-fructose and obtained a crystalline monomethyl ether, which they hydrolyzed to yield a mono-*O*-methyl-D-fructose,[112] obtained crystalline. This ether was oxidized with bromine water to yield a syrupy acid which, on the basis of titration with barium hydroxide and of combustion data, was formulated as a dihydroxymethoxybutanoic acid ($C_5H_{10}O_5$), not a trihydroxymethoxypentanoic acid ($C_6H_{12}O_6$). After prolonged heating at 100° in a vacuum, the analytical figures for the compound remained unaltered, and titration gave the results

(109) D. H. Rammler and C. A. Dekker, *J. Org. Chem.*, **26**, 4615 (1961).
(110) J. C. Irvine and A. Hynd, *Proc. Chem. Soc.*, **25**, 176 (1909).
(111) J. C. Irvine and A. Hynd, *J. Chem. Soc.*, **95**, 1220 (1909).
(112) J. C. Irvine and G. Robertson, *J. Chem. Soc.*, **109**, 1305 (1916).

calculated for the free acid, leading Irvine and Hynd to conclude that the compound was incapable of forming a lactone. This result indicated to them that the oxidation product was 4-O-methyl-D-erythronic acid, and thus the parent monomethyl-D-fructose was concluded to be 6-O-methyl-D-fructose. They were then compelled to formulate "α"-di-O-isopropylidene-D-fructose as 1,2:3,4-di-O-isopropylidene-D-fructofuranose (81).

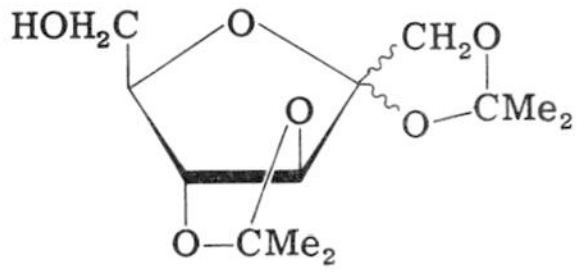

(81) 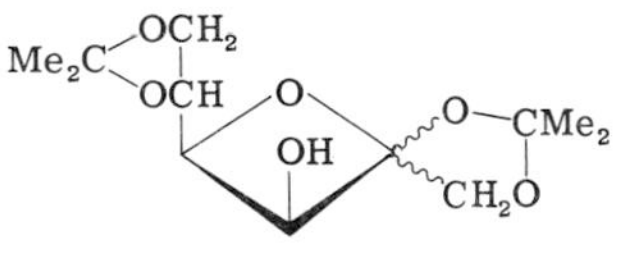   (82)

Early, incorrect structures proposed for
"α"-di-O-isopropylidene-D-fructose

Irvine and Garrett[59,113] noted a difference in the rates of acid hydrolysis of "α"- and "β"-di-O-isopropylidene-D-fructose, and reported that one of the acetone residues in the former diacetal was hydrolyzed more readily to give, in high yield, "α"-mono-O-isopropylidene-D-fructose. In contrast, "β"-di-O-isopropylidene-D-fructose required hydrolytic conditions considerably more severe than those used for its "α" isomer, and no appreciable difference was noted in the rates of hydrolysis of its two acetone residues. No "β"-mono-O-isopropylidene-D-fructose was obtained. Irvine and Garrett also concluded[59] that only a small proportion of "β"-di-O-isopropylidene-D-fructose could be produced by the action of acetone containing 0.2% of hydrogen chloride, but they improved the yield of this diacetal by treating the syrupy residue left from the preparation of "α"-di-O-isopropylidene-D-fructose with more acetone containing 0.2% of hydrogen chloride.

Irvine and Scott[114] found that the mono-O-methyl-D-fructose obtained by methylation of "α"-di-O-isopropylidene-D-fructose and subsequent hydrolytic removal of the acetone groups gave the same mono-O-methyl-D-hexulose phenylosazone as that obtained by methylation of 1,2:5,6-di-O-isopropylidene-α-D-glucofuranose, hydrolytic cleavage of the acetone residues, and formation of the phenylosazone. Further evidence for the structures of these two diacetals was obtained when Karrer and Hurwitz[45] showed that each was unaffected

(113) J. C. Irvine and C. S. Garrett, Proc. Chem. Soc., 26, 143 (1910).
(114) J. C. Irvine and J. P. Scott, J. Chem. Soc., 103, 564 (1913).

by treatment with warm, alkaline potassium permanganate solution, whereas 1,2-O-isopropylideneglycerol rapidly decolorized the reagent. They concluded that the compounds contained no unsubstituted, primary hydroxyl groups, and postulated structure **82** (1,2:5,6-di-O-isopropylidene-D-fructo"tetranose") for "α"-di-O-isopropylidene-D-fructose and structure **83** (1,2:5,6-di-O-isopropylidene-

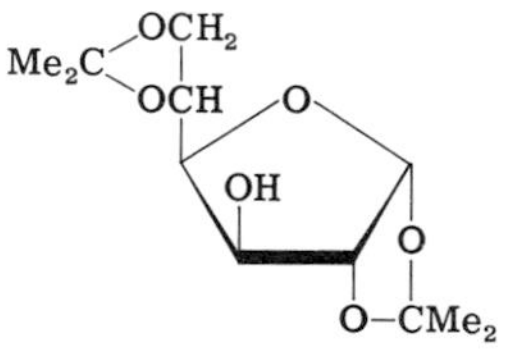

(83)

D-glucofuranose) for the di-O-isopropylidene-D-glucose (the latter structure had first been proposed by J. L. A. Macdonald[115] in 1913).

These structures, and the arguments in their favor, were vigorously rebutted by Irvine and Patterson.[116] They based their argument on earlier work by Irvine and his students that had "established" that, in both the mono-O-methyl-D-glucose[117] and the mono-O-methyl-D-fructose,[110,111] the methoxyl group was situated on C-6; the identity of the osazones prepared from each compound was cited as evidence to support this view. Irvine and Patterson[116] strongly reasserted their view that "α"-di-O-isopropylidene-D-fructose may be formulated as **81** and di-O-isopropylidene-D-glucose may be formulated as 1,2:4,5-di-O-isopropylidene-D-gluco"tetranose" (**84**).

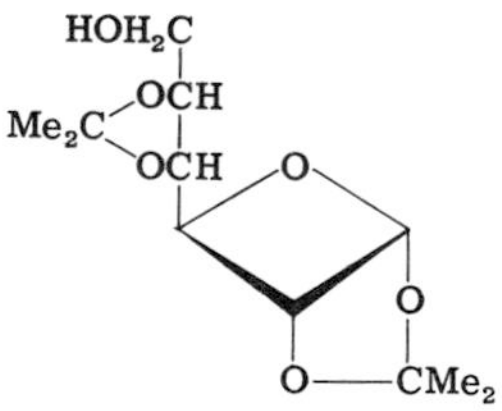

(84)

Irvine and Patterson's proposed[116]
structure for di- O -isopropylidene-
D-glucose

In 1923, Freudenberg and Doser[118] treated, in separate experiments, the p-toluenesulfonic esters of "α"-di-O-isopropylidene-D-

(115) J. L. A. Macdonald, *J. Chem. Soc.*, **103**, 1896 (1913).
(116) J. C. Irvine and J. Patterson, *J. Chem. Soc.*, **121**, 2146 (1922).
(117) J. C. Irvine and T. P. Hogg, *J. Chem. Soc.*, **105**, 1386 (1914).
(118) K. Freudenberg and A. Doser, *Ber.*, **56**, 1243 (1923).

fructose and di-O-isopropylidene-D-glucose with hydrazine, thus preparing a hydrazino derivative of each compound. Upon treatment with cold, concentrated hydrochloric acid, each was converted into 3-(1,2,3-trihydroxy-D-*erythro*-propyl)pyrazole hydrochloride (85), which was oxidized to the known pyrazole-3-carboxylic acid (86).

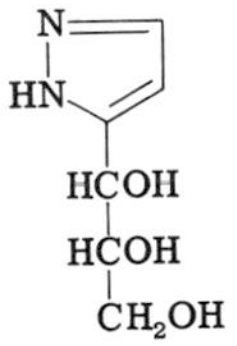

(85) 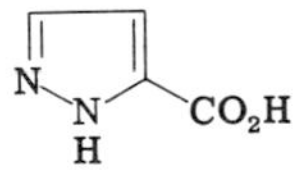   (86)

This sequence established that the free hydroxyl group in each compound was situated at C-3. Although the structure postulated by Karrer and Hurwitz possessed a free hydroxyl group at C-3, Freudenberg and Doser postulated a structure containing a pyranoid ring for "α"-di-O-isopropylidene-D-fructose, and formulated the structure as 87, in view of the strain presumed present in the alternative formulation, namely, 1,2:4,6-di-O-isopropylidene-D-fructofuranose (88),

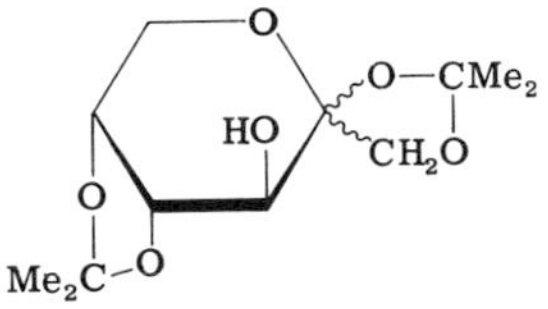

(87) 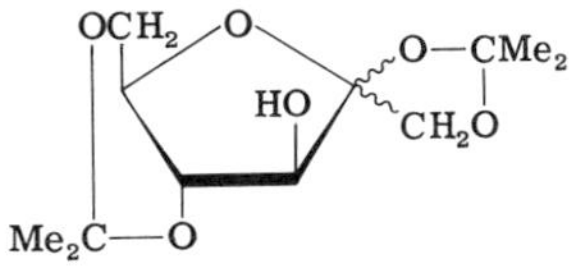   (88)

which contains a *m*-dioxane ring *trans*-fused to a furanoid ring. Ohle and Just[119] tentatively assigned the β-configuration to diacetal 87 on the basis of specific optical rotatory relationships with derivatives of L-sorbose. The β-D configuration and the $1C(\text{D})$ conformation of compound 87 in the crystalline state (see Figs. 1 and 2, pp. 216 and 217) were established[69,78] by X-ray diffraction in 1969. The mono-O-methyl-D-fructose obtained by methylation and hydrolysis of 87 was subsequently shown[120] to be 3-O-methyl-D-fructose. Later, the work of Irvine and Hynd[112] was repeated,[121] and found to be erroneous.

In 1925, Ohle[122] determined the structure of "β"-di-O-isopropyli-

(119) H. Ohle and F. Just, *Ber.*, **68**, 601 (1935).
(120) C. F. Allpress, *J. Chem. Soc.*, 1720 (1926).
(121) C. G. Anderson, W. Charlton, W. N. Haworth, and V. S. Nicholson, *J. Chem. Soc.*, 1337 (1929).
(122) H. Ohle, *Ber.*, **58**, 2577 (1925).

dene-D-fructose. He methylated the diacetal, removed the acetone residues by treatment with dilute acid, and obtained a mono-O-methyl-D-fructose that was not identical with either 3- or 6-O-methyl-D-fructose. The new methyl ether failed to form an osazone. He also oxidized the diacetal with two equivalents of potassium permanganate, and obtained the crystalline, potassium salt of a di-O-isopropylidenehexulonic acid. This was hydrolyzed to the free acid, which was identified as α-D-*arabino*-hexulopyranosonic acid by formation of the calcium, cadmium, and brucine salts. Ohle and Koller had concluded earlier[17] that the non-"glycosidic" acetone group is bound to two secondary hydroxyl groups, and thus they formulated "β"-di-O-isopropylidene-D-fructose as 2,3:4,5-di-O-isopropylidene-β-D-fructopyranose (2).

Later, Wolfrom, Shilling, and Binkley[60] treated "β"-di-O-isopro-

(2)

pylidene-D-fructose (2) with 0.5 M sulfuric acid for 13 hours at room temperature, conditions that do not effect complete hydrolysis. The mixture of reactant and products was neutralized, de-ionized, and resolved chromatographically into starting material (2), D-fructose, and a syrupy monoacetal. This monoacetal, purified through its crystalline triacetate (89), consumed one molecular equivalent of

(10) R = H
(89) R = Ac

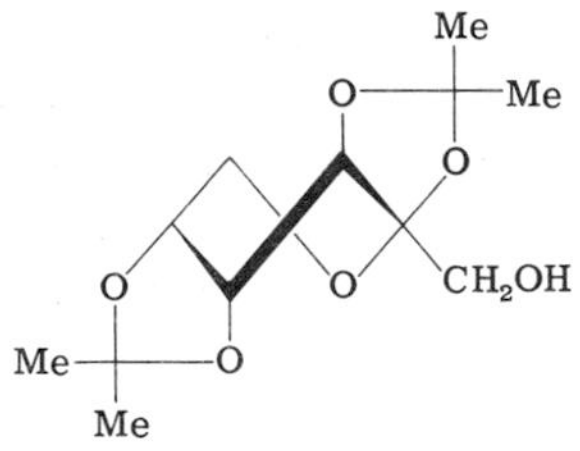

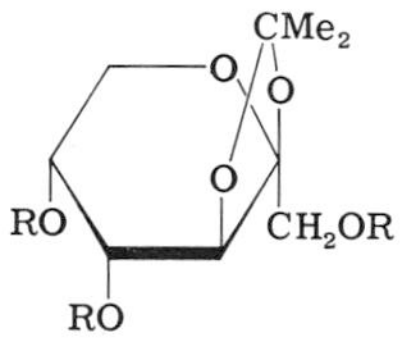

periodate. Thus, this monoacetal is 2,3-O-isopropylidene-β-D-fructopyranose (10), and, as hydrolysis by aqueous acid does not cause migration of an O-isopropylidene group,[60] the structure of "β"-di-O-isopropylidene-D-fructose was verified as being 2,3:4,5-di-O-isopropylidene-β-D-fructopyranose (2).

Ohle and Koller,[17] Slobodin and Klimov,[34] Bell,[31] and Brady[30] showed that low concentrations (≤0.5%) of sulfuric acid as catalyst for 1–3 days at room temperature lead to the formation of 1. Slightly higher concentrations of catalyst and shorter reaction times (3–24 hours) favor an increase in the proportion of 2, and lead to the isolation of mixtures of diacetals 1 and 2. However, at high concentrations of acid (≥4%), or after prolonged reaction, compound 2 was found to be the preponderant product. Evidently (see also Section III,2; p. 218), the diacetal 1 is the product of kinetically controlled condensation of D-fructose with acetone, but it rearranges, at a rate proportional to the concentration of the sulfuric acid catalyst, to the diacetal 2, which is thermodynamically more stable than 1, and thus preponderates at equilibrium.[30] The use of zinc chloride as the catalyst is known[24,30] to favor the formation of 1.

The diacetal 1 is hydrolyzed much faster than diacetal 2 by aqueous acids.[6,17,31,54,59,113] Hydrolysis of diacetal 1 is complete in 2 hours at 65° when 0.1 M oxalic acid is used as the hydrolyst, whereas eight hours is required for complete hydrolysis of diacetal 2 under these conditions.[54] Diacetal 2 has been prepared on a kilogram scale[123] by acetonation of sucrose followed by removal of 1,2:5,6-di-O-isopropylidene-α-D-glucofuranose (83) and diacetal 1 by hydrolysis with 25 mM sulfuric acid and appropriate treatment.

Zervas and Sessler[124] synthesized an isomeric mono-O-isopropylidene-D-fructose by acetonating 1,6-di-O-benzoyl-D-fructose, prepared by treatment of D-fructose cyanohydrin with two molar proportions of benzoyl chloride in the presence of pyridine. The cyclic acetal obtained after debenzoylation was named "2,3-acetone-α-D-fructo-furanose" by the authors, although, quite evidently, it is 2,3-O-isopropylidene-β-D-fructofuranose (90).

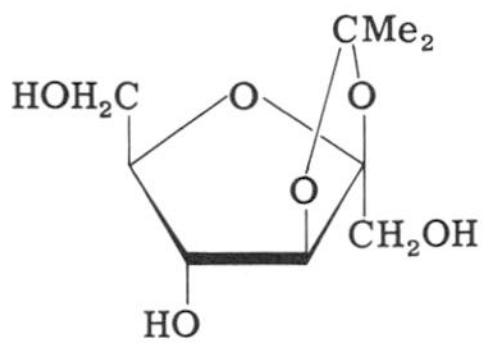

(90)

(123) H. Ohle and R. Wolter, Ber., 63, 843 (1930).
(124) L. Zervas and P. Sessler, Ber., 66, 1698 (1933).

Brigl and Schinle[125] condensed D-fructose with benzaldehyde containing zinc chloride, and obtained a crystalline di-O-benzylidene-D-fructose. The structure of this diacetal was established when Brigl and Widmaier[126] methylated the compound, hydrolyzed the benzylidene acetal groups, and converted the resulting amorphous mono-O-methyl-D-fructose into 2,3:4,6-di-O-isopropylidene-1-O-

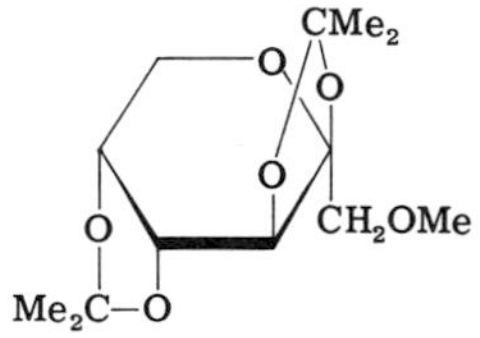

(91)

methyl-β-D-fructopyranose (91). Furthermore, the dibenzylidene acetal was hydrolyzed to a mono-O-benzylidene-D-fructose that did not reduce Fehling solution, and consumed one molecular equivalent of lead tetraacetate. This compound must, therefore, be 2,3-O-benzylidene-β-D-fructopyranose (92), and thus the structure of the diacetal is that of 2,3:4,5-di-O-benzylidene-β-D-fructopyranose (93). The iso-

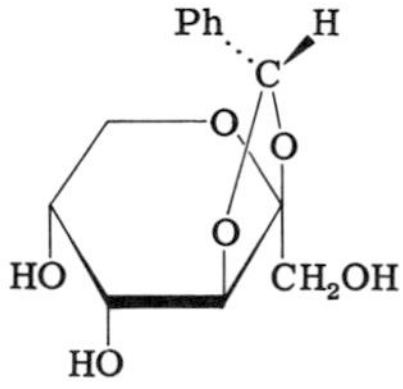

(92)

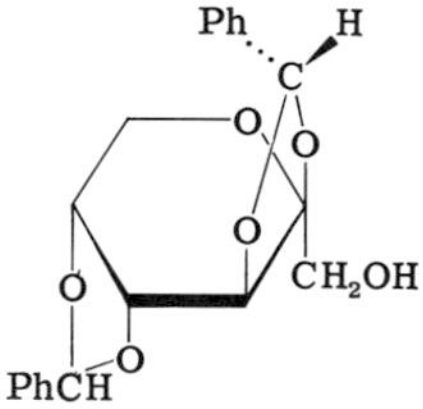

(93)

meric 2,4:3,5 diacetal is excluded by virtue of the structure of the monoacetal.

A di-O-cyclohexylidene-D-fructose was prepared[127] by treating a D-gluco-D-fructan (shown to contain one D-glucose and about thirteen D-fructose residues) present in the tuber of the Hawaiian Ti plant with cyclohexanone containing 1.6% of sulfuric acid. The structure of the diacetal was not rigorously determined. It melted at 145° and showed $[\alpha]_D$ −123° in acetone, and, on this basis, it was assigned the structure of 1,2:4,5-di-O-cyclohexylidene-D-fructopyranose (94).

(125) P. Brigl and R. Schinle, Ber., 66, 325 (1933).
(126) P. Brigl and O. Widmaier, Ber., 69, 1219 (1936).
(127) L. A. Boggs and F. Smith, J. Amer. Chem. Soc., 78, 1880 (1956).

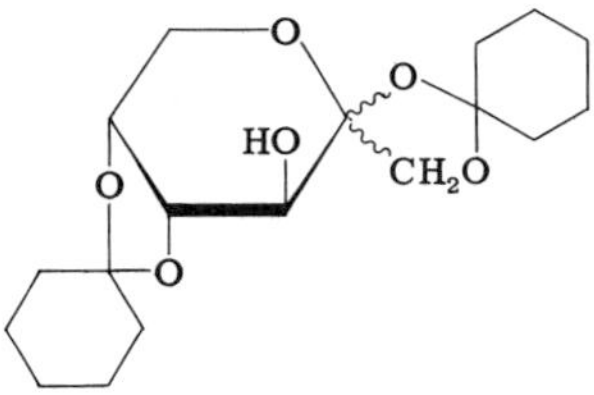

(94)

Mićović and Stojiljković[128] synthesized a di-O-cyclohexylidene-D-fructose, m.p. 142°, which, on the basis of its lack of reaction with alkaline potassium permanganate, was assumed to be 1,2:4,5-di-O-cyclohexylidene-D-fructopyranose (94). They also prepared[128] a syrupy di-O-cyclopentylidene-D-fructose, which was characterized as a crystalline p-toluenesulfonic ester; it, too, did not react with alkaline potassium permanganate.

Lobry de Bruyn and Alberda van Ekenstein obtained a di-O-methylene-D-fructose by treating a mixture of D-fructose and paraformaldehyde with either 50% sulfuric acid or 75% phosphoric acid.[32] The diacetal, which melted at 92° and had $[\alpha]_D$ −34.9° in water, did not reduce Fehling solution nor react with phenylhydrazine, but it did form a monoacetate, from which the parent diacetal could be recovered by saponification. By analogy with 2,3:4,5-di-O-isopropylidene-β-D-fructopyranose (2) ($[\alpha]_D$ −33.7°), the dimethylene acetal may be formulated as 2,3:4,5-di-O-methylene-β-D-fructopyranose (95).

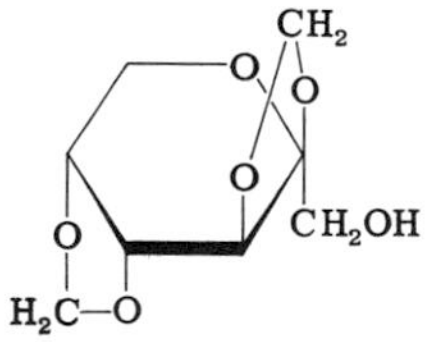

(95)

4. L-Sorbose

The first study of the condensation of L-sorbose with acetone was reported by Reichstein[7] in 1933. A suspension of the sugar in acetone containing 4% of concentrated sulfuric acid was stirred for 20 hours; processing followed by distillation of the mixture of products afforded a 50% yield of 2,3:4,6-di-O-isopropylidene-α-L-sorbofuranose (3) together with some 2,3-O-isopropylidene-α-L-sorbofuranose (4).

(128) V. M. Mićović and A. Stojiljković, *Tetrahedron*, 4, 186 (1958).

Processing of the reaction mixture after three hours yielded a much greater proportion of **4**. Later, Reichstein reported[129] the synthesis of 2,3:4,6-di-O-isopropylidene-α-D-sorbofuranose (**96**).

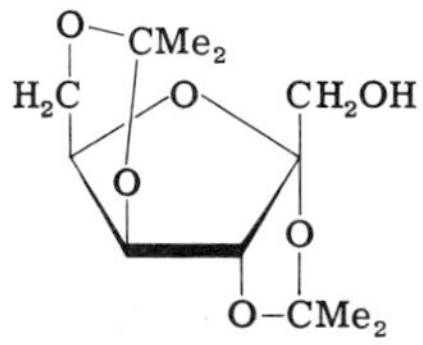

(96)

Ohle[35] isolated an isomeric mono-O-isopropylidene-L-sorbose from a similar reaction. After removing diacetal **3**, he acetylated the remaining mixture containing monoacetals. Fractional recrystallization of the distilled mixture of triacetates was effected from methanol–water, and, on saponification, one fraction of the crystalline material gave 1,2-O-isopropylidene-α-L-sorbopyranose (**11**) in 3% yield from L-sorbose. Periodate oxidation of **11** showed consumption of two moles of oxidant per mole, with formation of one mole of formic acid.[62] The monoacetal **11** was also isolated in 16% yield[35] when a suspension of L-sorbose and copper(II) sulfate in "impure" acetone was shaken for seven days. Other products were **3** and an isomeric di-O-isopropylidene-L-sorbose which was obtained in 1% yield and regarded by Ohle[35] as slightly impure. Its properties (m.p. 155–157°, $[\alpha]_D$ +44.9° in acetone) closely match those of 1,3:4,6-di-O-isopropylidene-β-L-sorbofuranose (**18**) (m.p. 159–160°, $[\alpha]_D$ +43.4° in acetone), identified by Maeda[16] in 1967. Patil and Bose[18] later isolated, but did not identify, a diacetal which was probably **18** (m.p. 155–157°, $[\alpha]_D$ +4.49° in acetone).

Ohle and Just[119] synthesized 1,2-O-isopropylidene-4-O-methyl-β-D-sorbopyranose (**99**) from 1,2-O-isopropylidene-3-O-p-tolylsulfonyl-β-D-fructopyranose (**97**) by successive Walden inversions involving 3,4-anhydro-1,2-O-isopropylidene-β-D-psicopyranose (**98**) as the intermediate (see Scheme XV).

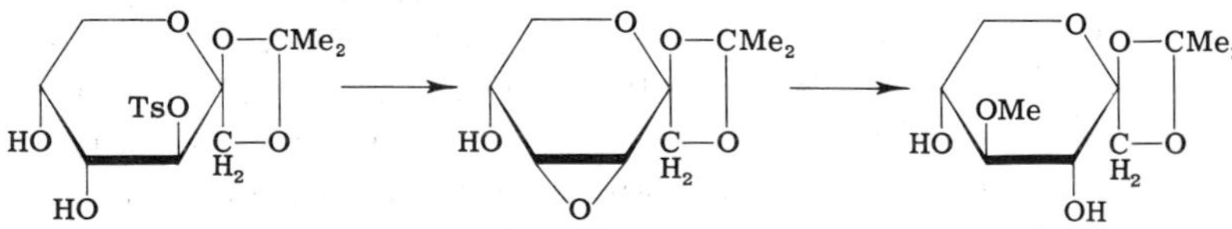

(97)   (98)   (99)

Scheme XV

(129) K. Gätzi and T. Reichstein, *Helv. Chim. Acta*, **21**, 456 (1938).

Many other isopropylidene acetals of L-sorbose have been isolated[13,14,16,18,62,63,66] in small, variable proportions, and identified, during researches on the mechanism of condensation of L-sorbose with acetone. These compounds are not of value for synthetic work, but they have been discussed in detail in Section III,1 (see p. 205).

Treatment of L-sorbose with 1,1-dimethoxyethane and a trace of *p*-toluenesulfonic acid yields a wide variety of mono- and di-ethylidene acetals.[15] The main products in the early stages of the reaction are **25** and **26**, whereas **27** may be obtained after longer intervals.

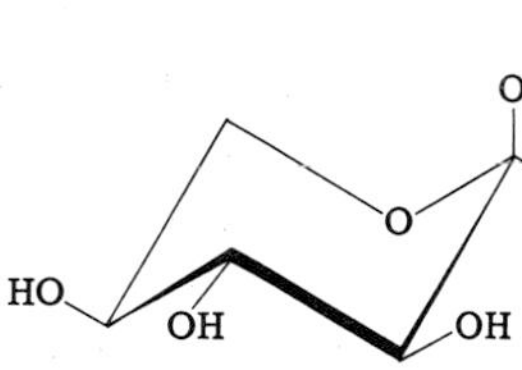

(25) 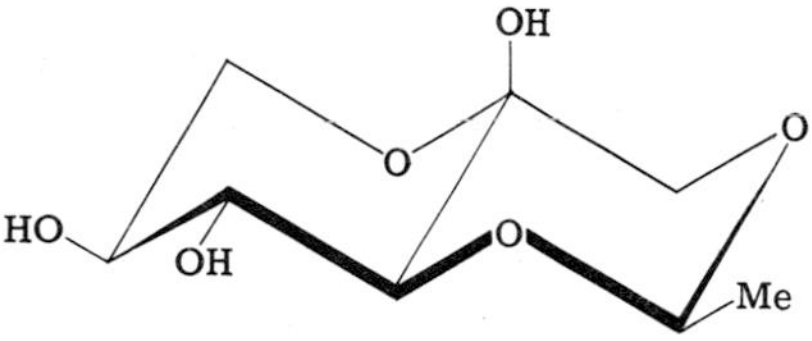   (26)

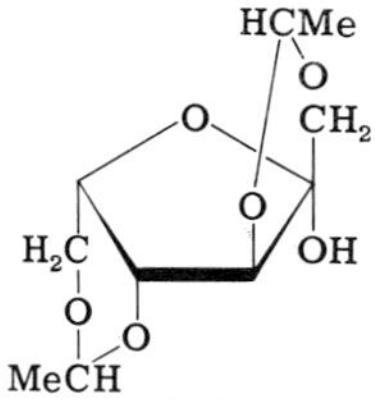

(27)

Strongly acidic conditions are needed in order to transform **27** into a mixture of the diastereoisomers **28** and **29**.

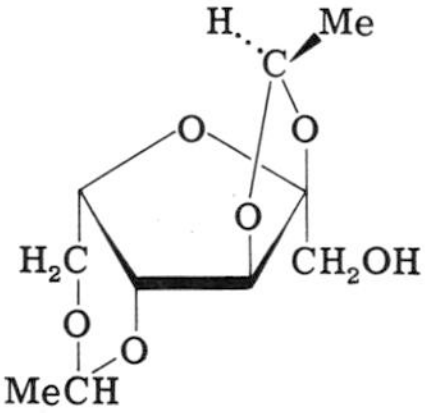

(28) 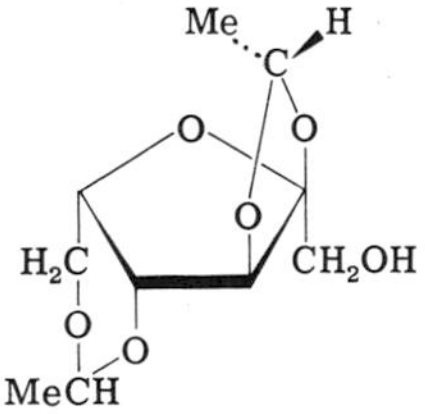   (29)

A diacetal of *keto*-L-sorbose, namely, 1-O-benzoyl-3,5:4,6-di-O-ethylidene-L-sorbose (**100**), has been prepared[57] by the oxidation of 6-O-benzoyl-1,3:2,4-di-O-ethylidene-D-glucitol with chromic acid.

Lobry de Bruyn and Alberda van Ekenstein[32] were the first to synthesize a dimethylene acetal of both D- and L-sorbose. By treating the separate enantiomorphs with trioxymethylene in 50% sul-

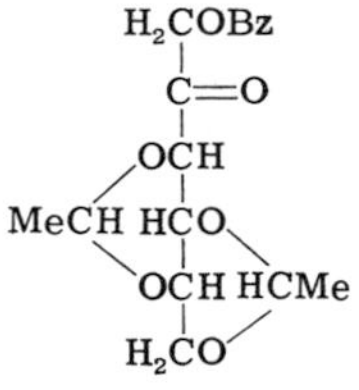

H₂COBz
|
C=O
|
OCH
|
MeCH HCO
            OCH HCMe
|
H₂CO

(100)

furic acid, they obtained a di-*O*-methylene-D-sorbose that melted at
54° and had [α]_D +25° in water, and a di-*O*-methylene-L-sorbose that
also melted at 54° but showed [α]_D −25° in water. They did not
determine the structure of either diacetal. Over 30 years later, Reichstein and Grüssner[7] prepared a di-*O*-methylene-L-sorbose (**101**) that
melted at 77–78° and had [α]_D −45.7° in water. They found that
oxidation of **101** with permanganate, and hydrolysis of the product,
gave α-D-*arabino*-hexulopyranosonic acid, the same compound that
was obtained by similar treatment of **3**. Thus, the structure of **101**
was shown to be 2,3:4,6-di-*O*-methylene-α-L-sorbofuranose.

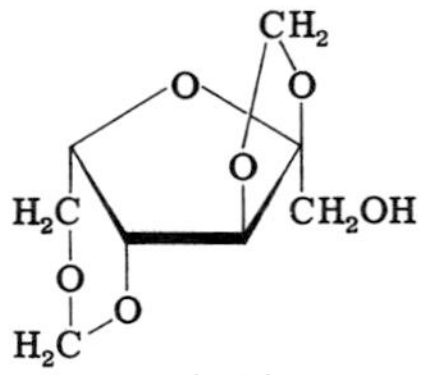

(101)

Reichstein and Grüssner[7] also prepared di-*O*-benzylidene and di-*O*-
(2-butylidene) acetals of L-sorbose, and found them to be useful intermediates in synthesis of L-ascorbic acid. By analogy, therefore, their
structures are 2,3:4,6-di-*O*-benzylidene-α-L-sorbofuranose (**30**) and
2,3:4,6-di-*O*-(2-butylidene)-α-L-sorbofuranose (**102**).

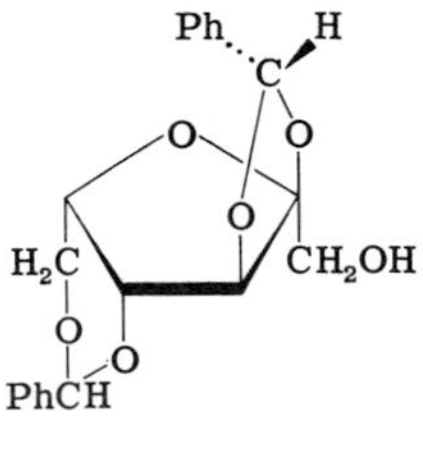

(30) 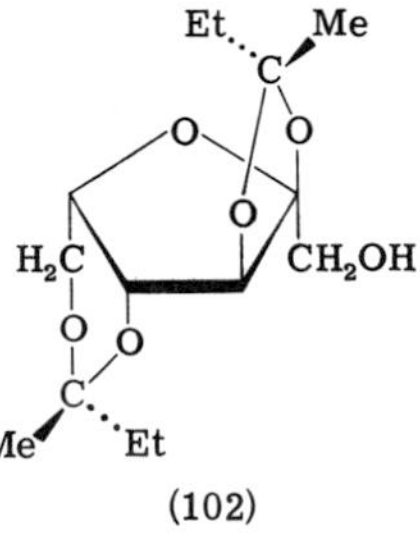   (102)

Maeda and coworkers[29] repeated the synthesis of **30** by this
method,[7] and isolated two isomers of 2,3:4,6-di-*O*-benzylidene-α-L-

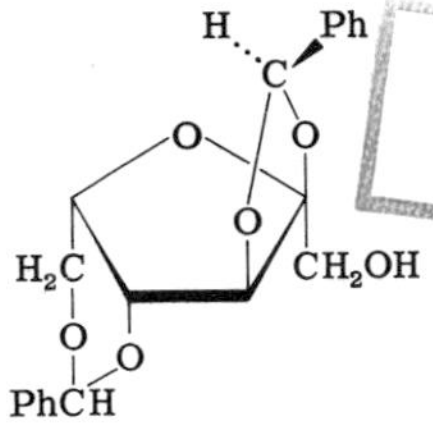

(31)

sorbofuranose (**30** and **31**), diastereoisomeric at the acetalic carbon atom linking O-2 and O-3. When slightly acidic conditions were used, however, small proportions of eight additional products (**32–39**) were formed and isolated (see p. 210–211).

A di-O-cyclohexylidene-L-sorbose has been prepared[130] in moderate yield by treating the sugar with cyclohexanone containing 15% of sulfuric acid. The structure of the diacetal, reported[131] in a separate paper, was determined as follows. It was oxidized with permanganate, and the product hydrolyzed, to give α-D-*arabino*-hexulopyranosonic acid; the diacetal also formed a monotrityl ether. Thus, the diacetal may be formulated as 2,3:4,6-di-O-cyclohexylidene-α-L-sorbofuranose (**103**).

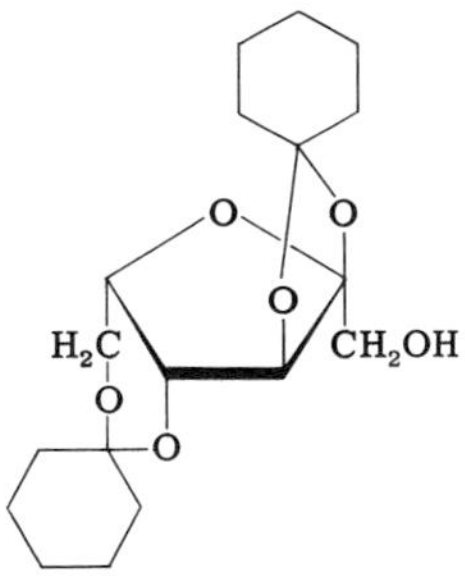

(103)

5. Psicose

The synthesis of 1,2:3,4-di-O-isopropylidene-L-psicofuranose (**104**) was reported[11] in 1935 by Steiger and Reichstein. This derivative was

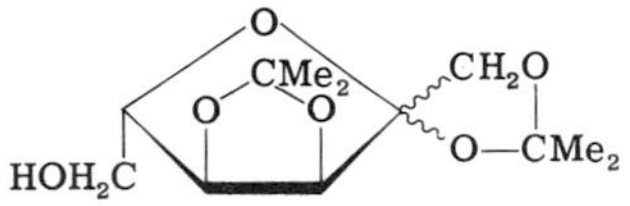

(104)

(130) V. F. Kazimirova, *J. Gen. Chem. USSR (Engl. Transl.)*, **24**, 637 (1954).
(131) V. F. Kazimirova, *J. Gen. Chem. USSR (Engl. Transl.)*, **25**, 1559 (1955).

used to purify L-psicose prepared from allitol by microbiological oxidation. The following year, Steiger and Reichstein reported[10] the preparation of the enantiomorph, 1,2:3,4-di-O-isopropylidene-D-psicofuranose (56), which was used to purify D-psicose prepared by

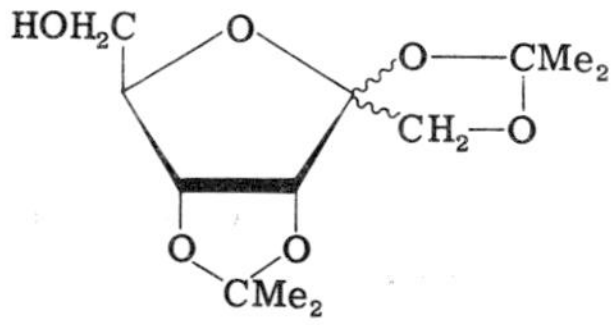

(56)

the isomerization of D-allose in pyridine.

An isomeric diacetal, 1,2:4,5-di-O-isopropylidene-β-D-psicopyranose (6), has been prepared[46-49] (Scheme XVI) by reduction of 1,2:4,5-di-O-isopropylidene-β-D-erythro-2,3-hexodiulo-2,6-pyranose (75), obtainable by oxidation of 1,2:4,5-di-O-isopropylidene-β-D-fructopyranose (1) with methyl sulfoxide–acetic anhydride[46-48] or ruthenium tetraoxide.[46,49] This reduction has been found to be completely stereospecific,[46] in contrast to reports in earlier work[47-49]; but to ensure that

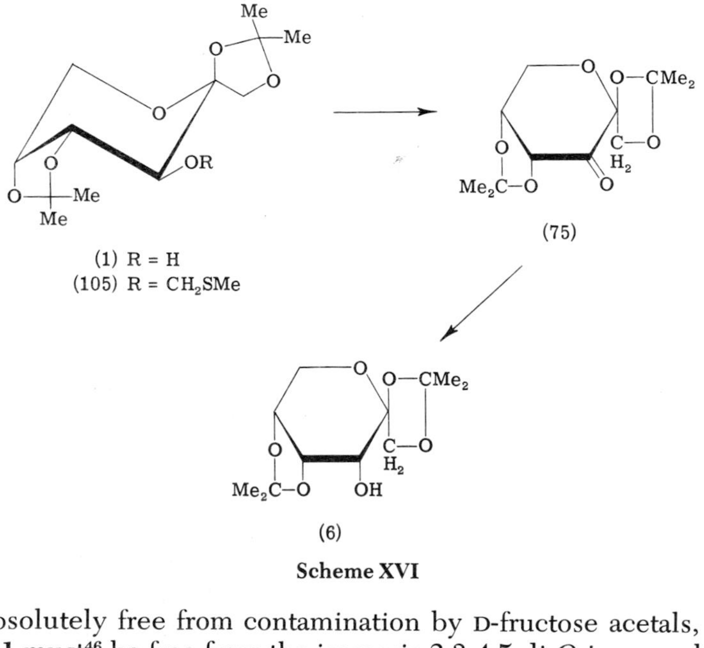

(1) R = H
(105) R = CH₂SMe

(75)

(6)

Scheme XVI

6 is absolutely free from contamination by D-fructose acetals, the diacetal 1 must[46] be free from the isomeric 2,3:4,5-di-O-isopropylidene-

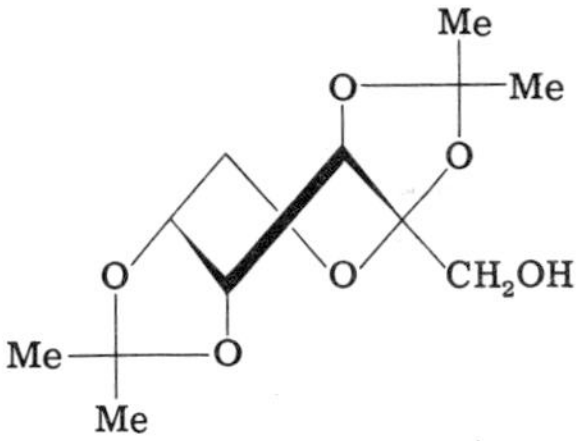

(2)

β-D-fructopyranose (2), and the diulose 75, when prepared by oxidation of 1 with methyl sulfoxide–acetic anhydride, must be free
from 1,2:4,5-di-O-isopropylidene-3-O-[(methylthio)methyl]-β-D-fruc-
topyranose[46,49] (105).

The monoacetal, 1,2-O-isopropylidene-β-D-psicopyranose (7), is

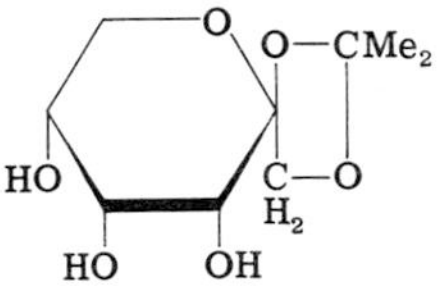

(7)

obtainable by partial hydrolysis of 6 with 80% acetic acid at room
temperature.[49]

Diacetal 56 is obtainable in high yield by treatment of diacetal 6
with anhydrous acetone containing 0.5% of sulfuric acid.[47,49,50]

The diacetal 1,2:4,5-di-O-cyclohexylidene-β-D-psicopyranose (107)
has been prepared[47] by reduction of 1,2:4,5-di-O-cyclohexylidene-
β-D-erythro-2,3-hexodiulo-2,6-pyranose (106).

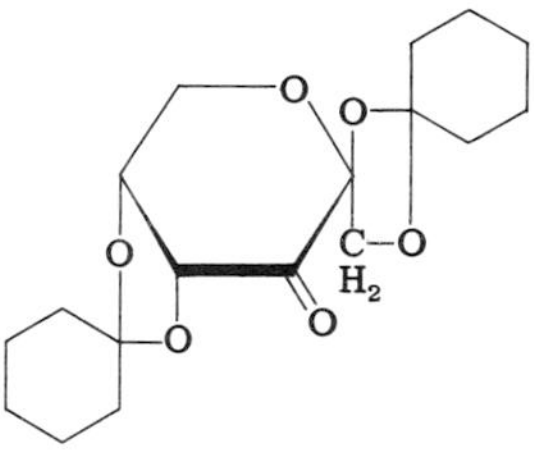

(106)

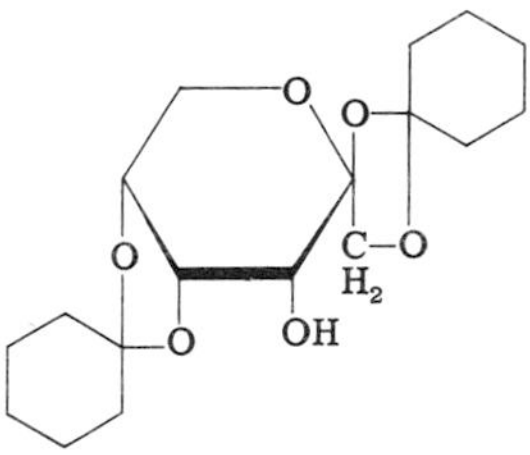

(107)

6. D-Tagatose

Reichstein and Bosshard[12] synthesized 1,2:3,4-di-O-isopropyli-
dene-D-tagatofuranose (108) in 40% yield by treating crystalline

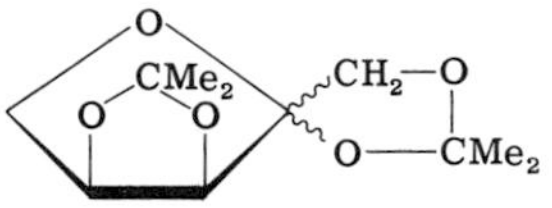

(108)

D-tagatose with acetone containing 4% of concentrated sulfuric acid for 16 hours at room temperature.

7. Heptuloses

Crystalline isopropylidene (**74**) and dimethylene (**109**) acetals of

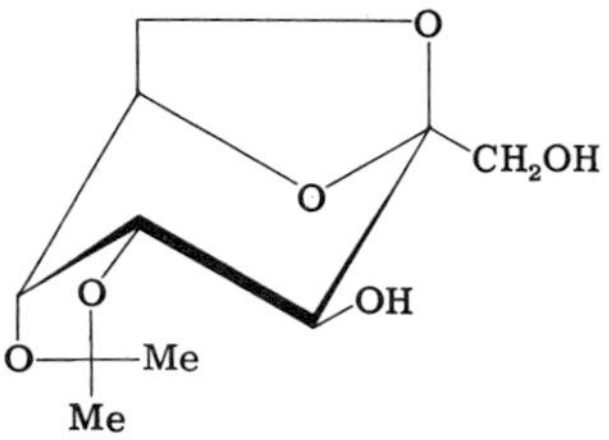

(74) 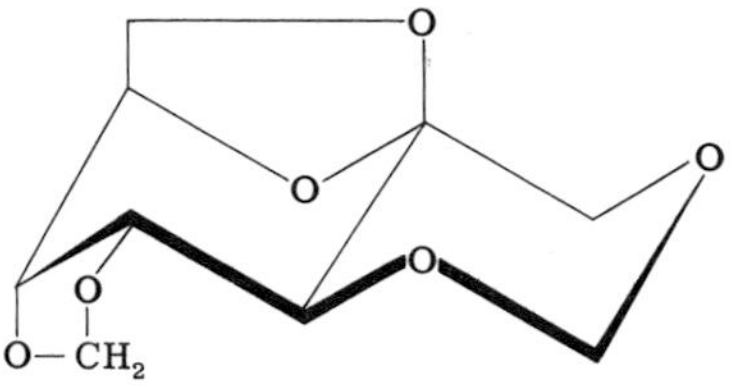   (109)

2,7-anhydro-β-D-*altro*-heptulose (sedoheptulosan) were prepared[132] as part of the investigations made to elucidate the structure of the anhydride. After the structure had been determined, a proton magnetic resonance study[89] of **74** (see Section IV,1; p. 227) confirmed that the structure of the acetal was that of 2,7-anhydro-4,5-O-isopropylidene-β-D-*altro*-heptulopyranose. The structure of **109** has not been explicitly defined, but is most probably that of 2,7-anhydro-1,3:4,5-di-O-methylene-β-D-*altro*-heptulopyranose.

A crystalline isopropylidene acetal of 2,7-anhydro-β-D-*manno*-

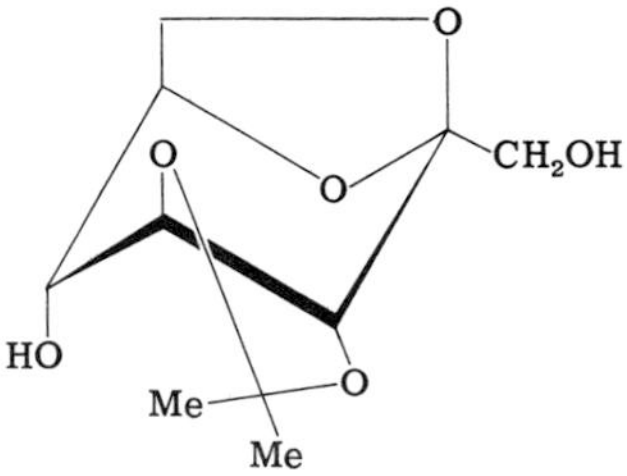

(110)

(132) W. T. Haskins, R. M. Hann, and C. S. Hudson, *J. Amer. Chem. Soc.*, **74**, 2198 (1952).

heptulopyranose has been prepared[133] and shown to be 2,7-anhydro-3,4-O-isopropylidene-β-D-*manno*-heptulopyranose (**110**).

Crystalline 5,7-O-ethylidene acetals of *keto*-D-*manno*-heptulose (**111**) and D-*altro*-heptulopyranose (**112**) have been prepared[134] by

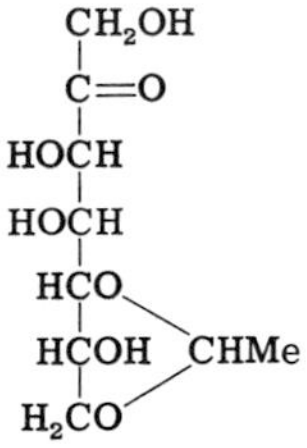

(111) 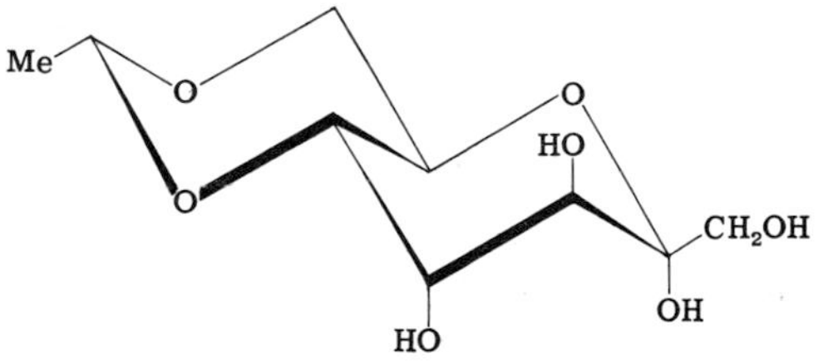   (112)

the reaction of 2,4-O-ethylidene-D-erythrose with 1,3-dihydroxy-2-propanone. No mutarotation of **111** was observed in water, whereas the specific rotation of **112** in methanol diminished with time; for this reason, **112** was formulated as 5,7-O-ethylidene-α-D-*altro*-heptulopyranose.

Mutarotation of **112** may relieve the gauche interactions between the axially attached hydroxyl groups on C–2 and C–4 and the carbon atoms β to them. The lack of mutarotation in **111** suggests that this acetal exists either as the stable α anomer **113** or as the *keto* tautomer **111**.

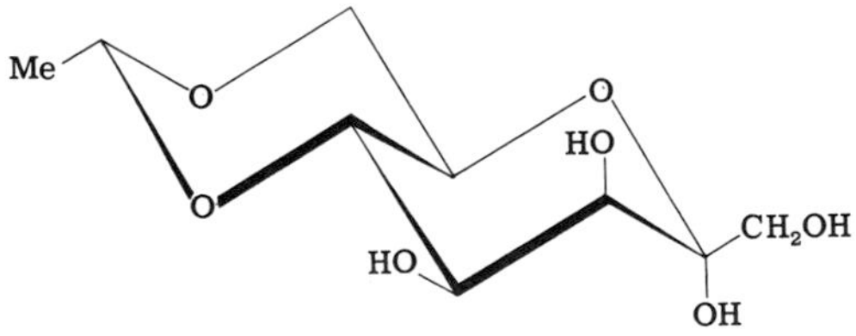

(113)

8. 3-Pentuloses and 3-Hexuloses

Chromium trioxide in acetic acid has been used[135] to oxidize 1,5-di-O-benzoyl-2,4-O-methylenexylitol. The product, 1,5-di-O-benzoyl-2,4-O-methylene-*keto*-*erythro*-3-pentulose (**114**), was subsequently

(133) E. Zissis, L. C. Stewart, and N. K. Richtmyer, *J. Amer. Chem. Soc.*, **79**, 2593 (1957).
(134) R. Schaffer and H. S. Isbell, *J. Org. Chem.*, **27**, 3268 (1962).
(135) A. Sera, *Bull. Chem. Soc. Jap.*, **35**, 2033 (1962).

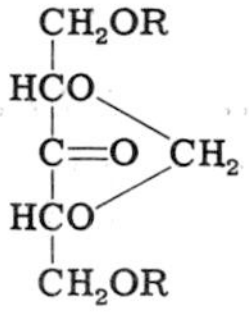

CH₂OR
|
HCO
   \
    C=O CH₂
|
HCO
|
CH₂OR

(114) R = Bz
(115) R = H

debenzoylated, to yield 2,4-O-methylene-*keto-erythro*-3-pentulose
(**115**). Because of the method of preparation, the new ketose was
given the trivial name "3-xylulose"; the use of this poor choice would
undoubtedly cause great confusion.

A novel diacetal of L-*xylo*-3-hexulose has been isolated in the
course of studies on a general method for the synthesis of 3-hexuloses
by oxidation of fully acetylated, hexitol monoacetals with chromium
trioxide.[136] Oxidation of 1,2,5,6-tetra-O-acetyl-3,4-O-methylene-D-
glucitol with chromium trioxide in acetic acid yielded principally
1,2,5,6-tetra-O-acetyl-4-O-formyl-D-*ribo*-3-hexulose. The material remaining in the mother liquors was saponified with sodium methoxide
in methanol and the product treated with acetone containing concentrated sulfuric acid; a di-O-isopropylidene derivative, presumably
1,2:3,4-di-O-isopropylidene-β-L-*xylo*-3-hexulofuranose (**116**), was
obtained.

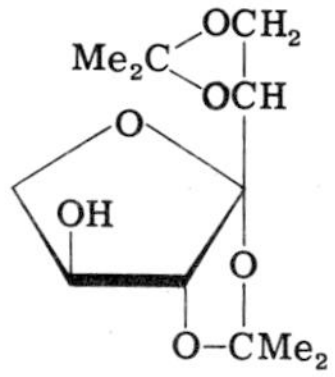

OCH₂
|
Me₂C   OCH
   \   /
    O
   /   \
  OH   \
         O
         |
         O—CMe₂

(116)

Likewise, oxidation of 3-O-benzoyl-1,2:5,6-di-O-isopropylidene-D-
mannitol with chromium trioxide in pyridine yielded crystalline 4-O-
benzoyl-1,2:5,6-di-O-isopropylidene-*keto*-D-*arabino*-3-hexulose[137]
(**117**). Oxidation of 6-O-benzoyl-1,2:4,5-di-O-isopropylidene-DL-ga-
lactitol with chromium trioxide in pyridine gave crystalline 6-O-
benzoyl-1,2:4,5-di-O-isopropylidene-*keto*-DL-*xylo*-3-hexulose[138] (**118**).

(136) S. J. Angyal and K. James, *Chem. Commun.*, 320 (1970).
(137) J. M. Sugihara and G. U. Yuen, *J. Amer. Chem. Soc.*, **79**, 5780 (1957).
(138) G. U. Yuen and J. M. Sugihara, *J. Org. Chem.*, **26**, 1598 (1961).

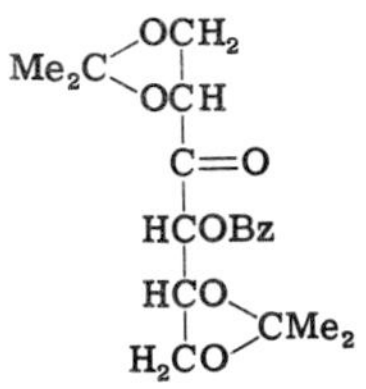

$$
\begin{array}{l}
Me_2C{\overset{\displaystyle OCH_2}{\underset{\displaystyle OCH}{\big\backslash}}}\,| \\
\quad\;\; C{=}O \\
\quad\;\; HCOBz \\
\quad\;\; HCO{\diagdown}_{CMe_2} \\
\quad\;\; H_2CO{\diagup}
\end{array}
\qquad\qquad
\begin{array}{l}
{\overset{\displaystyle H_2CO}{\underset{\displaystyle HCO}{\diagdown}}}CMe_2 \\
\quad\; C{=}O \\
\quad\; OCH{\diagdown}_{CMe_2} \\
\quad\; HCO{\diagup} \\
\quad\; H_2COBz
\end{array}
$$

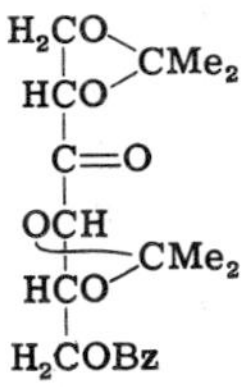

                                              D-Isomer

(117)   (118)

VII. TABLES OF PROPERTIES OF CYCLIC ACETALS OF KETOSES

The following Tables do not constitute a comprehensive list of all of the known derivatives of acetals of ketoses. In order to keep the Tables within manageable proportions, only the more common esters and ethers of the acetals have been included. Halogen-containing compounds and thioethers, and most of the deoxy and anhydro sugars, phosphorus-containing compounds, and esters of inorganic acids have been omitted. Some of these compounds may be found in the Tables of Barry and Honeyman.[77]

For the first time in this Series, references to proton magnetic resonance (p.m.r.), infrared (i.r.), mass (m.s.), and optical rotatory dispersion (o.r.d.) spectra are included in the Tables. This has been done in recognition of the increasing role played by spectroscopy and spectrometry in the identification of derivatives of sugars.

The solvents used for measuring specific rotation are abbreviated as follows: A, acetone; B, benzene; Bu, butanone; C, chloroform; Cy, cyclohexane; E, ethanol; M, methanol; Mf, NN-dimethylformamide; Ms, methyl sulfoxide; P, pyridine; Pe, petroleum ether; T, 1,1,2,2-tetrachloroethane; and W, water.

The letter (d) denotes melting with decomposition.

TABLE II

Derivatives of D- and L-*erythro*-Pentulose ("D- and L-Ribulose")

Compound	M.p. (°C.)	B.p. (°C./torr)	$[\alpha]_D$, degrees (solvent)	Spectra	References
α-D-*erythro*-Pentulofuranose					
1,2:3,4-di-O-isopropylidene-	87–89		−5.7 (C)	p.m.r.	83
3,4-O-isopropylidene-	70–71		−5.4 (A)	p.m.r., i.r.	36
β-D-*erythro*-Pentulofuranose					
1,2:3,4-di-O-isopropylidene-		61–72/0.5	−104.4 (A)	p.m.r., i.r.	36
			−111 (A)	p.m.r.	83
D-*erythro*-Pentulofuranoside,					
methyl 3,4-O-isopropylidene-		88–90/0.3	+113.2 (M)		107
D-*erythro*-Pentulose					
3,4-O-isopropylidene-					
1,5-di-O-trityl-	187–188		+32.6 (C)		108
β-L-*erythro*-Pentulofuranose					
1,2:3,4-di-O-isopropylidene-	5	55/0.1	+105.5 (A)		9
	5	76–78/0.3	+107 (A)		107

TABLE III

Derivatives of D-*threo*-Pentulose ("D-Xylulose")

Compound	M.p. (°C.)	B.p. (°C./torr)	$[\alpha]_D$, degrees (solvent)	Spectra	References
β-D-*threo*-Pentulofuranose					
2,3-O-isopropylidene-	70–71	110–118/0.1	+1.6 (A)		9
	73		+1.5 (A)	p.m.r., i.r.	36
				i.r.	92
1,4-di-O-methyl-		47/0.1	−12.6 (A)		43
1,4-di-O-*p*-tolylsulfonyl-	71–73		+6.3 (A)		43
D-*threo*-Pentulose					
1,5-di-O-benzoyl-					
3,4-O-isopropylidene-	78		−21.9 (C)		109

TABLE IV

Derivatives of D-Fructose

Compound	M.p. (°C.)	B.p. (°C./torr)	[α], degrees (solvent)	Spectra	References
β-D-Fructofuranose					
2,3-O-isopropylidene-	80		+18.9 (W)		124
			+14.1 (A)		124
			+10 (E)		124
1,6-di-O-benzoyl-	80–81				139
	118		+13.5 (E)		124
1,6-di-O-p-tolylsulfonyl-	132–133		+15.0 (E)		139,140,177
	131		+22.1 (C)		141
4-O-acetyl-	80–82		+15.8 (E)	p.m.r.	177
4-O-methyl-	112–113		+23 (E)		140
1,6-di-O-trityl-	155		−5.2 (P)		124
4-O-methyl-		150/0.05	+6.5 (E)		140
1,4,6-tri-O-acetyl-	55		−8 (E)		124
1,4,6-tri-O-benzoyl-	137		−9.1 (A)		124
1,4,6-tri-O-p-tolylsulfonyl-	130		+22 (E)		142
β-D-Fructopyranose					
2,3-O-benzylidene-	181		−183.3 (P)		126
1,4,5-tri-O-acetyl-	112		−158.9 (C)		126
1,4,5-tri-O-benzoyl-			−327.1 (C)		126
1,2-O-cyclohexylidene-	186–187				143
2,3:4,5-di-O-benzylidene-	160		−22.9 (C)		125,126
1-O-acetyl-	160			i.r., p.m.r., m.s. 178	
	145–146		−37.5 (C)	i.r., p.m.r., m.s. 178	125
	145				
1-O-benzoyl-	147		−18.1 (C)		125
1-O-methyl-	113–114		−30.5 (C)		126
1-O-p-tolylsulfonyl-	171 (d)		−34.9 (C)		126
1,2:4,5-di-O-cyclohexylidene-	145–146		−133.5 (C)	p.m.r	47
	145		−133.7 (A)		47
	142		−123 (A)		127
					128

TABLE IV (continued)

Compound	M.p. (°C.)	B.p. (°C./torr)	[α]D, degrees (solvent)	Spectra	References
β-D-Fructopyranose					
3-O-methyl-	61–62		−132.6 (C)		47
1,2:4,5-di-O-isopropylidene-	119		−154.8 (A)		30,46
			−145.0 (C)		30,46
			−158.0 (E)		30,46
	119–120		−161.3 (W)		6
	119		−147.3 (C)		31
			−161.0 (W)		31
	118–119		−146.6 (C)		144
3-O-acetyl-				i.r.	92
	76–77		−175.9 (E)	p.m.r.	56
	76–77		−176.3 (E)	i.r.	46
3-O-benzoyl-	107–108		−161.4 (E)		42
				p.m.r.	56
3-O-benzyl-	100–101	145/0.05	−95.0 (E)		145
3-O-(ethylsulfonyl)-	115		−163.6 (T)		44
3-O-methyl-			−136.4 (M)		110,111
			−149.4 (B)		111
			−135.3 (A)		111
			−140 (C)		47
3-O-(methylsulfonyl)-	115–116		−161.4 (C)		146
	104–105		+26.1 (C)		46
3-O-[(methylthio)methyl]-	82.5–83.0		+12.7 (A)	i.r.	46
			+26.8 (C)		47
3-O-nitro-	81.0–82.5		−124.3 (W)	p.m.r.	28
	61–62				
3-O-sulfo-, potassium salt	165 (d)		−161 (M)		147
3-O-p-tolylsulfonyl-	97–98		−159.5 (T)		140
	97				44

(continued)

TABLE IV (continued)

Compound	M.p. (°C.)	B.p. (°C./torr)	$[\alpha]_D$, degrees (solvent)	Spectra	References
β-D-Fructopyranose					
3-O-(trimethylsilyl)-	65–66		−122.9 (C)		148
2,3;4,5-di-O-isopropylidene-	97		−38.1 (A)		30,46
			−24.7 (C)		30
			−33.6 (W)		30
	97		−33.7		6
	95–96		−33.1 (W)		31
			−24.8 (C)		31
	97	111/0.2	−29.0 (B)		17
			−36.7 (E)		17
				p.m.r., i.r.	56
				i.r.	92
1-O-acetyl-	66		−36.0 (E)		17
1-O-benzoyl-	81			p.m.r.	56
	82–83		−21.8 (E)		17
1-O-tert-butyl-			−22 (E)		149
1-O-methyl-	48–49		−23.7 (Pe)		150
			−38.3 (E)		122
			−29.5 (C)		122
1-O-(methylsulfonyl)-	125–126		−29.3 (C)		146
1-O-p-tolylsulfonyl-	83		−27.1 (E)		17
	82		−26 (M)		140
	81–82		−27.2 (E)		151
1-O-(trimethylsilyl)-	92	106/0.06	−24.9 (Cy)		148
2,3;4,5-di-O-methylene-			−34.9 (W)		32
1-O-acetyl-			−46 (E)		32
1,2-O-isopropylidene-	120–121		−158.9 (W)		59,113

TABLE IV (*continued*)

Compound	M.p. (°C.)	B.p. (°C./torr)	[α]$_D$, degrees (solvent)	Spectra	References
β-D-Fructopyranose					
3-O-acetyl-	154–155		−179.6 (E)		42
	149		−153 (W)		152
	152–153				153
4,5-di-O-benzoyl-	108–109		−269.4 (E)		42
4,5-di-O-p-tolylsulfonyl-	127–128		−119.5 (C)		153
3-O-benzoyl-	202–204		−151.9 (E)		42
4,5-di-O-acetyl-	77–78		−132.5 (E)		42
4,5-di-O-p-tolylsulfonyl-	164–165		−175.0 (C)		153
3,4-di-O-acetyl-	81				154
4,5-di-O-acetyl-					
3-O-methyl-	101–102		−130.8 (A)	p.m.r.	56
	97–98		−119 (E)		155
3-O-(methylsulfonyl)-	84–86				146
3-O-p-tolylsulfonyl-	97		−133.0 (C)		119
4,5-di-O-benzoyl-					
3-O-p-tolylsulfonyl-	144		−316.2 (C)		119
4,5-di-O-methyl-	64–65		−169 (M)		140
3-O-p-tolylsulfonyl-	84–85		−121 (M)		140
3-O-methyl-			−126 (E)		155
3-O-(methylsulfonyl)-	133 (d)		−138.0 (A)		146
	142				154
3-O-nitro-	151–152				28
3-O-sulfo-, potassium salt	150		−112 (W)		147
3-O-p-tolylsulfonyl-	124–125		−112 (C)		140
			−128 (M)		140
	124.5		−113.0 (C)		119
3,4,5-tri-O-acetyl-	99–101		−135.6 (E)		42
	98.5–99.5		−135.7 (E)		153
				p.m.r.	56

(*continued*)

TABLE IV (continued)

Compound	M.p. (°C.)	B.p. (°C./torr)	$[\alpha]_D$, degrees (solvent)	Spectra	References
β-D-Fructopyranose					
3,4,5-tri-O-methyl-		111/0.08	−158.7 (W)		119
3,4,5-tri-O-(methylsulfonyl)-	128–130		−115.5 (C)		146
3,4,5-tri-O-p-tolylsulfonyl-	125.5		−121.4 (C)		119
3,4,5-tri-O-(trimethylsilyl)-		90/10⁻⁴	−26.2 (C)		148
2,3-O-isopropylidene-			+17.5 (W)		60
			+28.2 (E)		60
			+29 (E)		140
			+27.6 (E)		17
1,4,5-tri-O-acetyl-	55.5–56.0		+18 (E)		60
	55–56		+18 (E)		140
				p.m.r.	56
1,4,5-tri-O-methyl-		100/0.09	+35 (E)		140
1,4,5-tri-O-p-tolylsulfonyl-	130		+22 (E)		142
β-D-Fructopyranoside,					
methyl 1,3-O-benzylidene-	136–137		−73 (C)		156
4,5-di-O-(methylsulfonyl)-	145–146		−107 (C)		156
4-O-methylsulfonyl-	134–135		−83 (C)		156
5-O-benzoyl-	146–148		−167 (C)		156
D-Fructose					
3,4:5,6-di-O-isopropylidene-1-deoxy-		86/0.2	0 (M)	i.r.	157

TABLE V

Derivatives of D- and L-Sorbose

Compound	M.p. (°C.)	B.p. (°C./torr)	$[\alpha]_D$, degrees (solvent)	Spectra	References
α-D-Sorbofuranose					
2,3,4,6-di-O-isopropylidene-		109–110/0.1			129
β-D-Sorbopyranose					
1,2-O-isopropylidene-					
4-O-methyl-	112–113		−81.6 (C)		119
3,5-di-O-acetyl-	75		−92.8 (C)		119
3,4,5-tri-O-methyl-		105–106/0.3	−59.6 (W)		119
α-L-Sorbofuranose					
(R)-2,3-O-benzylidene-					
1,4,6-tri-O-acetyl-			−1.6 (C)	p.m.r.	29
(S)-2,3-O-benzylidene-					
1,4,6-tri-O-acetyl-			−15.6 (C)	p.m.r.	29
(R)-4,6-O-benzylidene-			−8.0 (A)	p.m.r.	29
(R)-1,2:(R)-4,6-di-O-benzylidene-	207–208		+34.7 (C)	i.r.	66
	122–125		+34.7 (C)	p.m.r., i.r.	29
3-O-acetyl-	121–125		−28.6 (C)	p.m.r.	29
	145–146			p.m.r.	66
(S)-1,2:(R)-4,6-di-O-benzylidene-	114–115		−31.0 (C)	i.r.	66
	114–115		−31.0 (C)	p.m.r., i.r.	29
3-O-acetyl-			−38.6 (C)	p.m.r.	29
				p.m.r.	66
(R)-2,3:(R)-4,6-di-O-benzylidene-	129–131		−8.7 (C)	p.m.r.	29,66
1-O-acetyl-			−17.9 (C)	p.m.r.	29
1-O-trityl-	166–168		−8.6 (C)	p.m.r.	29
(S)-1,2:(R)-4,6-di-O-benzylidene-			−10.0 (C)	p.m.r.	66
			−10.6 (C)	p.m.r.	29
1-O-acetyl-			−12.1 (C)	p.m.r.	29

(continued)

TABLE V (continued)

Compound	M.p. (°C.)	B.p. (°C./torr)	$[\alpha]_D$, degrees (solvent)	Spectra	References
1-O-trityl-	195–197		−20.4 (C)		29
2,3:4,6-di-O-sec-butylidene-	96–99	140/0.6	−16.6 (Bu)		7
	105			i.r.	94
2,3:4,6-di-O-cyclohexylidene-	118–119		−12.6 (B)		130,131
	124		−23 (M)		158
1-O-acetyl-	63–65		−19.5 (B)		131
1-O-benzoyl-	101–102		−29.0 (B)		131
1-O-trityl-	122–123				131
(R)-2,3:(R)-4,6-di-O-ethylidene-[a]	82–83		−9.5 (C)	p.m.r.	15,66
1-O-acetyl-	71–72		−21.5 (C)	p.m.r.	15
1-O-p-tolylsulfonyl-	70–72 (d)		−8.2 (C)	p.m.r.	15
1-O trityl-	130–133		−32.8 (C)	p.m.r.	15
(S)-2,3:(R)-4,6-di-O-ethylidene-[a]	58–59		−8.5 (C)	p.m.r.	15,66
1-O-acetyl-	53–54		−24.0 (C)	p.m.r.	15
1-O-p-tolylsulfonyl-	72–73		−7.0 (C)	p.m.r.	15
1-O-trityl-	162–163		−35.5 (C)	p.m.r.	15
"1,2:3,4-di-O-isopropylidene-"[b]					
1,2:4,6-di-O-isopropylidene-	78–79		−36.7 (A)	m.s.	18,63
	72–73		−23.9 (A)	i.r.	14
					63
				i.r.	16
3-O-acetyl-	44–46		−61.4 (A)	p.m.r.	74
				i.r., p.m.r.	16
3-O-(2-methoxy-2-propyl)-			+17.5 (A)	p.m.r.	74
2,3:4,6-di-O-isopropylidene-	77–78	135/0.3	−18.1 (A)		13
			−4.9 (W)		7,18,61,159
					7
				i.r.	55,92
				i.r., p.m.r.	16
				p.m.r.	74

TABLE V (continued)

Compound	M.p. (°C.)	B.p. (°C./torr)	$[\alpha]_D$, degrees (solvent)	Spectra	References
1-O-acetyl-	62		-15.4 (A)		33
	52	168/6	-16.4 (A)	i.r.	55
				i.r., p.m.r.	16
				p.m.r.	118
1-O-benzoyl-	133.5		-3.4 (M)		57
1-O-(diphenylphosphono)-			-11.7 (C)		160
1-O-(2-ethoxy-2-propyl)-	99-100		+1.3 (A)		14
					16
1-O-(methylsulfonyl)-	115		-4.7 (C)	i.r.	161
1-O-(2-methoxy-2-propyl)-	116-117				146
			-10.9 (A)		13
1-O-methyl-	53	111/0.2	-10.9 (A)	i.r., p.m.r.	16
	54-55		-11.1 (A)		162
	102-103		-11 (A)		163
1-O-p-tolylsulfonyl-	101-102		+4.2 (C)		57
	65-74		+2.8 (C)		164
	182		+3.0 (C)		165
1-O-trityl-	77-78	132/0.2	-45.7 (W)		166
2,3;4,6-di-O-methylene-	91-92		-75.1 (A)		7
1,2-O-isopropylidene-			-67.3 (A)	p.m.r.	18
3,4,6-tri-O-acetyl-	93		+7.0 (W)		74
2,3-O-isopropylidene-	90-91		+7.0 (E)		7
	91		+7.0 (W)		33
					61
1-O-acetyl-		205/11	+7.3 (A)	i.r.	16,55
4,6-di-O-p-tolylsulfonyl-			-3.8 (A)	i.r.	55
6-O-p-tolylsulfonyl-	120-122		-12.7 (A)		55
			+4.2 (A)		33
4-O-acetyl-	100		0 (W)		55
			+23.0 (C)		166
					166

(continued)

TABLE V (*continued*)

Compound	M.p. (°C.)	B.p. (°C./torr)	[α]ᴅ, degrees (solvent)	Spectra	References
1,6-di-O-trityl-	224–225		+23.0 (C)		35
1,4-anhydro-	92		−34.5 (C)		167
6-O-p-tolylsulfonyl-	115		−11.6 (A)	i.r.	55
4,6-anhydro-	102–103		−40.6 (C)		168
1-O-benzoyl-	55–57		−21.7 (C)		165
1-O-methyl-		100/25	−26.8 (C)		165
1-O-(methylsulfonyl)-	109–110		−29.0 (C)		168
1-O-(p-nitrobenzoyl)-	116–117		−19.3 (C)		168
1-O-p-tolylsulfonyl-	115–116		−12.8 (C)		168
			−11.6 (A)		168
	114–115		−11.2 (A)		165
4-O-phenylsulfonyl-	163–164		+17.6 (C)		165
6-O-p-tolylsulfonyl-					
6-O-phenylsulfonyl-					
1-O-p-tolylsulfonyl-	143–145		+15.5 (C)		165
1-O-benzoyl-	91–92		+8.0 (C)		168
	92		+7.2 (C)		57
6-O-p-tolylsulfonyl-	155–156 (d)		0 (C)		168
1,6-di-O-(diphenylphosphono)-	100–101		−4.2 (C)		160
4,6-di-O-(methylsulfonyl)-					
1-O-p-tolylsulfonyl-	122–123		+33.8 (C)		168
1,6-di-O-p-tolylsulfonyl-	131		+8.9 (E)		169
			+8.4 (A)	i.r.	55
4,6-di-O-p-tolylsulfonyl-	125		+33.3 (C)	i.r.	162
1-O-methyl-		140/2			68
		149/2			
6-O-p-tolylsulfonyl-	37		+21.4 (C)		162
	40–50		+25.3 (C)		165
1-O-(methylsulfonyl)-	93–94		+5.5 (W)		170
6-O-p-tolylsulfonyl-	132–133		+3.3 (C)		168
	132–134		+3.2 (C)		170

TABLE V (*continued*)

Compound	M.p. (°C.)	B.p. (°C./torr)	[α]$_D$, degrees (solvent)	Spectra	References
4-O-(methylsulfonyl)-					
1,6-di-O-p-tolylsulfonyl-	121–122		+24.9 (C)		170
6-O-(methylsulfonyl)-					
1-O-p-tolylsulfonyl-	159–160 (d)		+21.9 (C)		168
4-O-(tetrahydropyran-2-yl)-					
1-O-p-tolylsulfonyl-	A:^{c} 126–127		−10.9 (Mf)		170
	B:^{c} 96–97		+48.3 (Mf)		170
1-O-p-tolylsulfonyl-	120–121		+12.6 (M)		171
	120		+15.7 (C)		57
	106–108		+16.3 (C)		165
6-O-p-tolylsulfonyl-	105		+15.1 (A)	i.r.	55
	103–104		+14.0 (A)		168
			+12.2 (M)		168
	90–92		+5.2 (E)		33
	91–92		+10.3 (A)		165
1,4,6-tri-O-acetyl-			+2.6 (C)	i.r., p.m.r.	16
		136/11		p.m.r.	74
1,4,6-tri-O-methyl-			+29.6 (C)		172
			+32.2 (M)		172
1,4,6-tri-O-(methylsulfonyl)-	125–126		+8.5 (P)		170
β-L-Sorbofuranose					
(R)-1,3-O-benzylidene-	183–185		+30.5 (M)		29
			+20.2 (P)		29
			+20.4 (Ms)		29
6-O-acetyl-	153–155		+16.5 (C)	p.m.r., i.r., o.r.d.	65
4,6-di-O-acetyl-			−3.3 (C)	p.m.r.	29
2,4,6-tri-O-acetyl-	75–77		−10.5 (C)	p.m.r.	29

(*continued*)

TABLE V (continued)

Compound	M.p. (°C.)	B.p. (C./torr)	$[\alpha]_D$, degrees (solvent)	Spectra	References
(R)-1,2:(R)-4,6-di-O-benzylidene-	186–194		+33.8 (A)	i.r.	66
3-O-acetyl-	186–196		+22.7 (C)	p.m.r., i.r.	29
			+34.4 (C)	p.m.r.	29
(S)-1,2:(R)-4,6-di-O-benzylidene-	148–154		+175.4 (A)	i.r.	66
	148–154		+166.6 (C)	p.m.r., i.r.	29
3-O-acetyl-	137–139		+147.1 (C)	p.m.r.	29
				p.m.r.	66
(R)-1,3:(R)-4,6-di-O-benzylidene-	164–166		+48.8 (C)	p.m.r.	66
	165–167		+48.8 (C)	p.m.r.	29
(R)-1,3:(R)-4,6-di-O-ethylidene-			+81.3 (C)	p.m.r.	66
1,2:4,6-di-O-isopropylidene-			+58.6 (A)	i.r.	16
3-O-acetyl-			+58.0 (A)	p.m.r., i.r.	16
				p.m.r.	74
1,3:4,6-di-O-isopropylidene-[a]	159–160		+43.4 (A)	p.m.r., i.r.	16
				p.m.r.	15,74
2-O-acetyl-	67–69		+1.3 (A)	p.m.r., i.r.	16
				p.m.r.	74
1,2-O-isopropylidene-3,4,6-tri-O-acetyl-α-L-Sorbofuranoside			+45.6 (A)	p.m.r.	74
methyl 4,6-O-isopropylidene-1,3-di-O-acetyl-β-L-Sorbofuranoside	108–109		−65.0 (A)	p.m.r.	74
	104–105		−63.9 (A)		74
methyl (R)-1,3-O-benzylidene-4,6-di-O-acetyl-	87.5		+23.5 (C)	p.m.r.	29
methyl (R)-1,3:(R)-4,6-di-O-benzylidene-	114–116		+56.3 (C)	p.m.r.	29

TABLE V (*continued*)

Compound	M.p. (°C.)	B.p. (°C./torr)	$[\alpha]_D$, degrees (solvent)	Spectra	References
methyl (*R*)-1,3:(*R*)-4,6-di-*O*-ethylidene-			+68.1 (C)		15
methyl 1,3:4,6-di-*O*-isopropylidene-	46–47		+25.4 (A)	p.m.r., i.r.	16
methyl (*R*)-1,3-*O*-ethylidene-4,6-di-*O*-acetyl-	61–62		+53.7 (C)	p.m.r.	15
methyl (*R*)-4,6-*O*-ethylidene-1,3-di-*O*-acetyl-			−31.2 (C)	p.m.r.	15
methyl 1,3-*O*-isopropylidene-4,6-di-*O*-acetyl-			+34.9 (A)	p.m.r.	74
α-L-Sorbopyranose					
1,2-*O*-benzylidene-3,4,5-tri-*O*-acetyl-	A:[e] 128–130		−125.2 (C)	p.m.r.	29,66
	B:[e] 111–115		−89.2 (C)	p.m.r.	29,66
				p.m.r.	56
(*R*)-1,3-*O*-benzylidene-	150–158		−24.1 (M)	p.m.r.	65
4,5-di-*O*-acetyl-	181–183		−47.6 (C)	p.m.r.	29
2,4,5-tri-*O*-acetyl-	203–205		−38.7 (C)	p.m.r.	65
				p.m.r.	56
1,2:3,4-di-*O*-isopropylidene-	103–105		−91.2 (A)	p.m.r.	13,16
				i.r.	16,56
				i.r.	16
5-*O*-acetyl-	129–130		−55.6 (A)	i.r.	16
5-*O*-(2-methoxy-2-propyl)-	133–135		−82.2 (C)	p.m.r.	16,56
				p.m.r., i.r.	13,16
1,2:4,5-di-*O*-isopropylidene-	80–83		−69.2 (C)	i.r.	16
				p.m.r.	56
3-*O*-acetyl-			−78.5 (A)	p.m.r., i.r.	16

(*continued*)

TABLE V (continued)

Compound	M.p. (°C.)	B.p. (°C./torr)	[α]$_D$, degrees (solvent)	Spectra	References
1,2-O-ethylidene-			-45.7 (C)	p.m.r.	15
3,4,5-tri-O-acetyl-			-45.5 (C)	p.m.r.	66
				p.m.r.	56
(R)-1,3-O-ethylidene-	140–142		-24.7 (C)	p.m.r.	15
4,5-di-O-acetyl-	110–112		-13 (C)	p.m.r., i.r.	66
1,2-O-isopropylidene-	142		-85.2 (W)	i.r.	35,61
	142		-88.7 (W)	i.r.	62
				i.r.	16
3-O-methyl-	121–122		-66.3 (E)	p.m.r., i.r.	153
3,4,5-tri-O-acetyl-	88–89		-72.8 (C)	p.m.r.	35
	77–78		-72.0 (C)	p.m.r.	16
				p.m.r.	56
β-L-Sorbopyranose					
(R)-1,3-O-benzylidene-					
2,4,5-tri-O-acetyl-			+14.7 (C)	o.r.d.	65
α-L-Sorbopyranoside					
ethyl 1,3-O-isopropylidene-	178–179		-60.4 (A)	p.m.r.	56
4,5-di-O-acetyl-	183–184		-48.6 (A)	p.m.r.	56
methyl (R)-1,3-O-benzylidene-	185–186		-54.6 (C)	p.m.r.	156
4,5-di-O-acetyl-			-74.0 (C)	p.m.r.	29
				p.m.r.	56
4,5-di-O-(methylsulfonyl)-	168–169		-63.9 (C)	p.m.r.	156
methyl (R)-1,3-O-ethylidene-	93–94		-58.4 (C)	p.m.r.	15
4,5-di-O-acetyl-				p.m.r.	56

TABLE V (*continued*)

Compound	M.p. (°C.)	B.p. (°C./torr)	$[\alpha]_D$, degrees (solvent)	Spectra	References
L-Sorbose					
3,5-O-benzylidene-					
1,6-di-O-methyl-	89–93		−6.0 (A)		68
4-O-acetyl-	98–100		−46.9 (A)	p.m.r.	68
3,5:4,6-di-O-ethylidene-					
1-O-benzoyl-	105		−110.5 (C)		57
1-O-p-tolylsulfonyl-	141–142		−76.3 (C)		57
3,4:5,6-di-O-isopropylidene-					66,68
1-O-methyl-			+10.5 (A)	i.r.	66

a The physical constants for these two compounds are interchanged in Refs. 15 and 66. Ref. 15 is later and is, presumably, correct; entries in this Table correspond to Ref. 15. *b* A compound assigned[13] this structure was later shown[16] to be 1,2:4,5-di-O-isopropylidene-α-L-sorbopyranose. *c* These two derivatives are stereoisomeric at C-2 of the tetrahydropyran ring. The absolute configuration of the two compounds has not yet been determined. *d* Ohle[35] and Patil and Bose[18] both isolated a di-O-isopropylidene-L-sorbose having very similar properties. These diacetals were not identified, but they seem to be identical with this compound. *e* These two derivatives are stereoisomeric at the carbon atom linking O-1 and O-2. The absolute configuration of the two compounds has not yet been determined.

TABLE VI

Derivatives of D- and L-Psicose

Compound	M.p. (°C.)	B.p. (°C./torr)	$[\alpha]_D$, degrees (solvent)	Spectra	References
β-D-Psicofuranose					
1,2:3,4-di-O-isopropylidene-	57–58.5	104/0.3	−98.2 (A)		10
	57–57.5		−89.6 (C)		47
6-O-benzoyl-	56–58		−94.6 (C)		173
6-O-p-tolylsulfonyl-	70–71		−59.6 (C)		173
	98–99		−42.4 (C)		173
β-D-Psicopyranose					
1,2:4,5-di-O-cyclohexylidene-	112–114		−106.4 (C)	p.m.r.	47
			−119.2 (A)		47
1,2:4,5-di-O-isopropylidene-	68–69[a]		−116.4 (A)	i.r.	46,50
			−120.7 (C)		46
			−118.9 (E)		46
				p.m.r.	47
				p.m.r., o.r.d.	49
3-O-acetyl-	64–66		−125.5 (C)		173
3-O-benzoyl-	143–144	160/0.1	−116 (E)	p.m.r., o.r.d.	49
3-O-p-tolylsulfonyl-	108–109		−141.7 (C)		173
1,2-O-isopropylidene-	175–176		−103.2 (C)		173
			−114 (E)	p.m.r., o.r.d.	49
3,4-anhydro-	92		−47.6 (C)		119
5-O-acetyl-	81		+5.9 (C)		119
5-O-benzoyl-	111.5		+24 (C)		119
5-O-methyl-		90–92/0.01	−2.8 (C)		119
5-O-p-tolylsulfonyl-	95		+4.7 (C)		119
β-L-Psicofuranose					
1,2:3,4-di-O-isopropylidene-	56.5–57	105/0.2	+99 (A)		11
3,4-O-isopropylidene-2,6-anhydro-	137		+97.2 (A)		11

[a] The melting point of freshly prepared material is 68–69° (Ref. 46). After storage at room temperature during several months, the melting point changes to 56–58° (Refs. 46 and 47). This change is attributable to a change in the crystal modification[46] not to isomerization or decomposition.

TABLE VII

Derivatives of D-Tagatose

Compound	M.p. (°C.)	[α]D, degrees (solvent)	Spectra	References
D-Tagatofuranose				
1,2:3,4-di-O-isopropylidene-	65–66	+81.5 (A)		12
		+71.8 (W)		12
		+64 (C)		174
6-O-p-tolylsulfonyl-	99–100	+33.9 (A)		175
1,2-O-isopropylidene-				
3,4-anhydro-	81–82	−80.7 (C)		153
		−60.0 (W)		153
5-O-acetyl-	80–81	−28.6 (C)		153
5-O-p-tolylsulfonyl-	117–118	−27.0 (C)		153
2,3-O-isopropylidene-		+15.4 (C)		177
1,6-di-O-tolylsulfonyl-		+ 7.8 (E)	p.m.r., m.s.	177
1,4,6-tri-O-acetyl-		− 3.9 (C)	p.m.r.	177

TABLE VIII

Derivatives of 3-Pentuloses and 3-Hexuloses

Compound	M.p. (°C.)	[α]$_D$, degrees (solvent)	Spectra	References
D-*arabino*-3-Hexulose				
1,2:5,6-di-O-isopropylidene-				
4-O-benzoyl-	94–95	+9.2 (C)		137
phenylhydrazone	131–132	+293 (C)		137
DL-*xylo*-3-Hexulose				
1,2:4,5-di-O-isopropylidene-				
6-O-benzoyl-	68			138
phenylhydrazone	111			138
β-L-*xylo*-3-Hexulofuranose				
1,2:3,4-di-O-isopropylidene-	99–100	−15.4 (C)		136
erythro-3-Pentulose				
2,4-O-methylene-	96–97		i.r.	135
1,5-di-O-benzoyl-	116		i.r.	135

TABLE IX

Derivatives of Heptuloses

Compound	M.p. (°C.)	[α]b, degrees (solvent)	Spectra	References
α-D-*altro*-Heptulopyranose				
5,7-O-ethylidene-	149–151	+55 (W)		134
		+40 (M)		134
β-D-*altro*-Heptulopyranose				
2,7-anhydro-				
1,3:4,5-di-O-methylene	183–184	−131 (W)		132
4,5-O-isopropylidene-	226–227	−124 (W)		132
			p.m.r.	89
1,3-di-O-benzoyl-	131–132	−79.0 (C)	p.m.r.	89
1,3-di-O-p-tolylsulfonyl-	102–104	−78.1 (C)		89
1-O-p-tolylsulfonyl-			p.m.r.	89
3-O-p-tolylsulfonyl-				
1-O-trityl-	177–179	−7.3 (C)		176
D-*manno*-Heptulose				
5,7-O-ethylidene-	200–201	+4.4 (W)		134
2,7-anhydro-3,4-O-				
isopropylidene-	106–107	−21.8 (W)		133

VIII. Further Developments

A new acetal of D-tagatofuranose has been prepared (see Scheme XVII) by oxidation of 2,3-O-isopropylidene-1,5-di-O-p-tolylsulfonyl-

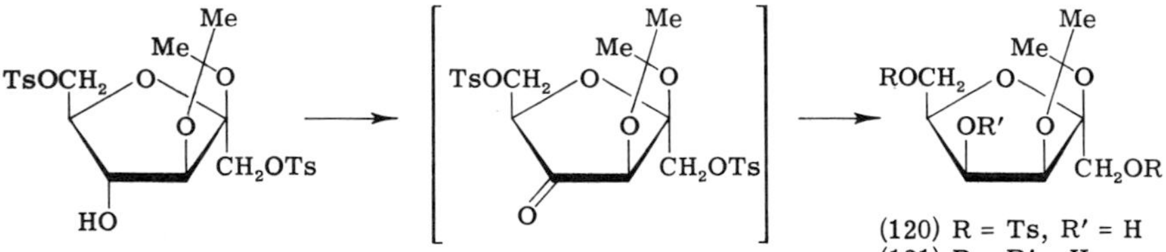

(120) R = Ts, R' = H
(121) R = R' = H
(122) R = R' = Ac

Scheme XVII

β-D-fructopyranose (119) with methyl sulfoxide–acetic anhydride, followed by reduction of the intermediate diulose (which was not isolated) with sodium borohydride.[177] Following hydrolysis of the di-p-toluenesulfonic ester (120), the product, 2,3-O-isopropylidene-β-D-tagatofuranose (121), was isolated as a colorless syrup and characterized as the 1,4,6-triacetate (122).

An attempt (see Scheme XVIII) to prepare a diacetal of D-tagatopyranose, namely, 1,2:4,5-di-O-isopropylidene-β-D-tagatopyranose, was

(139) W. T. J. Morgan and T. Reichstein, *Helv. Chim. Acta*, **21**, 1023 (1938).
(140) E. L. Hirst, W. E. A. Mitchell, E. E. Percival, and E. G. V. Percival, *J. Chem. Soc.*, 3170 (1953).
(141) M. S. Feather and R. L. Whistler, *J. Org. Chem.*, **28**, 1567 (1963).
(142) J. K. N. Jones and J. L. Thompson, *Can. J. Chem.*, **35**, 955 (1957).
(143) N. P. Klyushnik, *Ukr. Khim. Zh.*, **33**, 67 (1967); *Chem. Abstr.*, **66**, 115,900 (1967).
(144) D. H. Brauns and H. L. Frush, *J. Res. Nat. Bur. Stand.*, **6**, 449 (1931).
(145) P. A. J. Gorin and A. S. Perlin, *Can. J. Chem.*, **36**, 480 (1958).
(146) B. Helferich and H. Jochinke, *Ber.*, **73**, 1049 (1940).
(147) H. Ohle and G. Coutsicos, *Ber.*, **63**, 2912 (1930).
(148) V. Prey and K.-H. Gump, *Ann.*, **682**, 228 (1965).
(149) H. Bredereck and W. Protzen, *Chem. Ber.*, **87**, 1873 (1954).
(150) V. Prey and F. Grundschober, *Chem. Ber.*, **95**, 1845 (1962).
(151) S. B. Baker, *Can. J. Chem.*, **33**, 1459 (1955).
(152) P. A. J. Gorin, L. Hough, and J. K. N. Jones, *J. Chem. Soc.*, 2699 (1955).
(153) H. Ohle and C. A. Schultz, *Ber.*, **71**, 2302 (1938).
(154) J. K. N. Jones and W. A. Nicholson, *J. Chem. Soc.*, 3050 (1955).
(155) G. N. Richards, *J. Chem. Soc.*, 3222 (1957).
(156) D. Murphy, *J. Chem. Soc. (C)*, 1732 (1967).
(157) J. S. Brimacombe, J. G. H. Bryan, A. Husain, M. Stacey, and M. S. Tolley, *Carbohyd. Res.*, **3**, 318 (1967).
(158) F. Hoffmann-La Roche and Co., Akt.-Ges., Ger. Pat. 703,227; *Chem. Abstr.*, **36**, 1619 (1942).

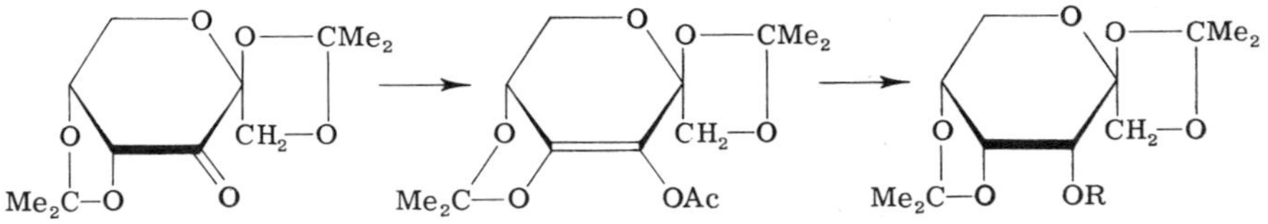

Scheme XVIII

unsuccessful.[50] Treatment of diulose **75** with an acetic anhydride–triethylamine reagent produced 3-*O*-acetyl-1,2:4,5-di-*O*-isopropylidene-β-D-*glycero*-hex-3-enulopyranose (**123**). Reduction of enediol acetate **123** with hydrogen in the presence of a palladium-on-carbon catalyst afforded 3-*O*-acetyl-1,2:4,5-di-*O*-isopropylidene-β-D-psicopyranose (**124**), which was characterized as the known diacetal **6** obtained on deacetylation with barium methoxide in methanol.

Koerner, Younathan, and Wander[178] have provided additional evidence for the β configuration of 2,3:4,5-di-*O*-benzylidene-β-D-fructopyranose (**93**). The p.m.r. spectrum of **9.3** in carbon tetrachloride solution was measured at 100 MHz and 35° before and after serial additions of a saturated solution of tris[2,2,6,6-tetramethyl-3,5-heptanedionato]europium (III) in carbon tetrachloride. Apparently, this reagent coordinates selectively with O-1, because a two-proton

(159) S. Maruyama, *Sci. Papers Inst. Phys. Chem. Res.* (Tokyo), **27**, 56 (1935).
(160) K. H. Mann and H. A. Lardy, *J. Biol. Chem.*, **187**, 339 (1950).
(161) K. Tokuyama, M. Kiyokawa, and M. Katsuhara, *J. Org. Chem.*, **30**, 4057 (1965).
(162) K. Tokuyama and M. Katsuhara, *Bull. Chem. Soc. Jap.*, **39**, 2728 (1966).
(163) J. C. Sowden and I. I-ling Mao, *J. Org. Chem.*, **25**, 1461 (1960).
(164) T. S. Gardner and J. Lee, *J. Org. Chem.*, **12**, 733 (1947).
(165) L. Hough and B. A. Otter, *Carbohyd. Res.*, **4**, 126 (1967).
(166) F. Valentin, *Chem. Zvesti*, **1**, 2 (1947); *Chem. Abstr.*, **43**, 7431 (1949).
(167) K. Tokuyama, *Bull. Chem. Soc. Jap.*, **37**, 1133 (1964).
(168) Ö. Fehér and L. Vargha, *Acta Chim. Acad. Sci. Hung.*, **50**, 371 (1966).
(169) H. Müller and T. Reichstein, *Helv. Chim. Acta*, **21**, 263 (1938).
(170) Ö. Fehér and L. Vargha, *Acta Chim. Acad. Sci. Hung.*, **37**, 443 (1963).
(171) K. Tokuyama and M. Kiyokawa, *J. Org. Chem.*, **29**, 1475 (1964).
(172) H. H. Schlubach and P. Olters, *Ann.*, **550**, 140 (1942).
(173) M. Haga, M. Takano, and S. Tejima, *Carbohyd. Res.*, **14**, 237 (1970).
(174) P. A. J. Gorin, J. K. N. Jones, and W. W. Reid, *Can. J. Chem.*, **33**, 1116 (1955).
(175) J. Barnett and T. Reichstein, *Helv. Chim. Acta*, **20**, 1529 (1937).
(176) E. Zissis, *J. Org. Chem.*, **33**, 2844 (1968).
(177) A. A. H. Al-Jobore, R. D. Guthrie, and R. D. Wells, *Carbohyd. Res.*, **16**, 474 (1971).
(178) T. A. W. Koerner, Jr., E. S. Younathan, and J. D. Wander, *Carbohyd. Res.*, in press (1972).

signal from H-1 and H-1′ was displaced downfield by the greatest amount. The H-3 doublet exhibited the next-greatest downfield displacement. Thus, H-3 is *cis* to the paramagnetic center, and it follows that the configuration of **93** is β. This work is the first application of a lanthanide shift-reagent to the study of cyclic acetals of ketoses by p.m.r. spectroscopy.

TABLES OF THE PROPERTIES OF DEOXY SUGARS AND THEIR SIMPLE DERIVATIVES

By Roger F. Butterworth and Stephen Hanessian

*Department of Chemistry, University of Montreal,
Montreal, Quebec, Canada*

The following five Tables are complementary to the Chapter entitled "Deoxy Sugars" by Stephen Hanessian, *Advan. Carbohyd. Chem.*, **21**, 143 (1966).

Where a deoxy sugar or its derivative has been characterized in more than one publication, the reference number printed in bold type indicates the publication from which the physical constants were taken. The specific optical rotations recorded in the Tables are equilibrium values for those compounds for which mutarotation values have been reported. The abbreviations used to indicate the rotation solvent are as follows: A, acetone; C, chloroform; D, *p*-dioxane; E, ethanol; M, methanol; P, pyridine; W, water.

The Tables contain the melting point and optical rotation of the deoxyaldoses and deoxyketoses. An immediate derivative is included for most of the sugars listed, in order to provide an additional criterion for identification. Whenever possible, the same derivative has been chosen for various deoxy sugars within a series, in order to make available a further basis for differentiation. Ring forms of the deoxy sugars are not included, but α and β anomers are designated whenever they have been reported. Because of the voluminous literature on deoxy sugars, only those references are included wherein the free deoxy sugar has been obtained and characterized.

The literature was surveyed up to July, 1971.

TABLE I

Deoxytetroses

Tetrose[a]	M.p., °C	$[\alpha]_D$, degrees	Rotation solvent	References[b]
2-Deoxy-				
D-glycero-	67–69	−4	W	1
(p-nitrophenyl)hydrazone	136–138			1
L-glycero-	67–69	+3.6	W	1
4-Deoxy-				
L-erythrose				
phenylosazone	145–146	+50	P	2

[a] 2-Tetrulose in the case of the osazone derivative.
[b] All references appear in a separate list starting on page 291.

TABLE II

Deoxypentoses

Pentose[a]	M.p., °C	$[\alpha]_D$, degrees	Rotation solvent	References
2-Deoxy-				
D-*erythro*-	96–98	−58	W	3–15, **16**
2-benzyl-2-phenylhydrazone	125.5–126.5	−15.25	P	9, 13, 17, 18, **19**, 20, 21
L-*erythro*-	92–95	+59	W	12, **16**, 20, 22
aniline derivative	174–175	−58	P	22, **23**
D-*threo*-	92–96	−2	W	**20**, 24, 24a, 25, 26, 27
2-benzyl-2-phenylhydrazone	115–116	+13.5	P	**20**, 24, 25, 28, 29
L-*threo*-	91			17, **30**
2-benzyl-2-phenylhydrazone	122–124	−12	P	17
3-Deoxy-				
D-*erythro*-[b]	syrup	−8.2	W	31, **32**, 33
(p-nitrophenyl)osazone	259–260	+45	D	3, 31, 32, **34**, 35
L-*erythro*-	syrup	+8.7	W	36
(p-nitrophenyl)osazone	254–256			36
D-*threo*-	syrup			24
2-benzyl-2-phenylhydrazone	86–86.5	+16.2	C	24
4-Deoxy-				
L-*erythro*-	syrup			36a
2-benzyl-2-phenylhydrazone	102–103	+23.1	W	36a
5-Deoxy-				
D-arabinose	syrup	+7	W	37
phenylosazone	172–174	−65	P–E	38
2-benzyl-2-phenylhydrazone	99	+10.2	M	37
L-arabinose	syrup	−6.9	W	38a, **39**, 40, 41
phenylosazone	172–174	+66	P–E	39, 40, 41, 42, **43**

(*continued*)

TABLE II (continued)

Pentose[a]	M.p., °C	$[\alpha]_D$[b], degrees	Rotation solvent	References
5-Deoxy-				
D-lyxose	syrup	+32.4	W	44, 45
(p-bromophenyl)osazone	143–144			44, **45**
L-lyxose	68–69	−31.5	W	46, **47**, 48
2-benzyl-2-phenylhydrazone	100.5–101	−36.4	P	47, 48, **49**
D-ribose	syrup	+20	W	50, 51, **52**
phenylosazone	175–177	−61	P-E	43, **51**, 53
D-xylose	syrup	+13.26	W	54, **55**
phenylosazone	178–180	+66.6	P-E	**55**, 56
L-xylose	syrup	−13	W	56a
2,3-Dideoxy-				
L-*glycero*-	syrup	+22.7	W	57
dibenzyl dithioacetal	74	+20.4	C	57
2,5-Dideoxy-				
D-*erythro*-	syrup	+16.6	P	58
dimethyl dithioacetal	97–98	−26.2	C	58
3,5-Dideoxy-				
D-*erythro*-	syrup			59

[a] 2-Pentulose in the case of osazone derivatives.
[b] Tosylhydrazone derivative, m.p. 143–144°; P. Szabó and L. Szabó, *J. Chem. Soc.*, 2944 (1965).

Table III

Deoxyhexoses

Hexose[a]	M.p., °C	$[\alpha]_D$, degrees	Rotation solvent	References
2-Deoxy-				
D-*arabino*-(α anomer)	128–129	+47	W	63
(β anomer)	146	+46	W	60, 61, 62, **64**, 65
2-benzyl-2-phenylhydrazone	158–159	+7.6	M	**63**, 66
L-*arabino*-				
aniline derivative	176–178	+41	W	67
		–34	E	67
D-*lyxo*-	120–121	+60.5	W	26, 68, 69, 70, 71, **72**, 73
p-toluidine derivative	142–143	–115	P	70
D-*ribo*-	140–142	+57.5	W	74, **74a**, 75, 75a
(p-nitrophenyl)hydrazone	61–62	–54.6	M	75a
D-*xylo*-	syrup	+12	W	67, **76**, 77
phenylhydrazone	124–126	–8.1	M	77
p-toluidine derivative	74–76	–44	E	77
p-toluidine derivative (two forms)	141–142	–67	E	67, **77**
3-Deoxy-				
D-*arabino*-	141–142	53.1	W	**78**, 79
dimethyl dithioacetal	82	–38.6	M	**78**, 80
phenylhydrazone	128–129	–46.7	M	79
D-*lyxo*-	syrup	–10	W	32, **81**
(2,5-dichlorophenyl)osazone	225–226	–204	P	**32**, 82
L-*lyxo*-				
dimethyl dithioacetal	46.5–48	48	C	83
D-*ribo*- (α anomer)	108–111	+29	W	80, **84**, 85, 86, 87
(β anomer)	137	+30.4	W	86
(p-nitrophenyl)osazone	248			88
D-*xylo*-	syrup	+6.94	W	24a, 25, **81**, 89
(2,4-dinitrophenyl)hydrazone	129–130	–15.1	M	89

(continued)

TABLE III (continued)

Hexose[a]	M.p., °C	$[\alpha]_D$, degrees	Rotation solvent	References
4-Deoxy-				
D-*arabino*-				
dibenzyl dithioacetal	103–105	+146	E	**90**, 91
D-*lyxo*-				
(*p*-nitrophenylsulfonyl)-hydrazone	syrup	+3	W	92
	113–120	−53	P	92
D-*ribo*-	89–93	−50	W	92
(*p*-nitrophenylsulfonyl)-hydrazone	139–141	−75	P	92
D-*xylo*-	136–140	+58	W	84, 89, **90**, 92a, 93, 94
dibenzyl dithioacetal	106–108	+88	E	**90**, 94
5-Deoxy-				
D-*ribo*-	syrup	+19.4	W	95, **96**
phenylosazone	137–139	+15.1	M	95, **96**
D-*xylo*-	syrup	+40	W	96, 97, **98**, 99
phenylosazone	156–159	−40.7	M	96, **97**, 98, 99, 100
6-Deoxy-				
D-allose	151–152	+1.2	W	43, 101, 102, 103, **104**, 105
phenylosazone	182–183	−72.3	P–E	**43**, 101, 106
phenylhydrazone	163–163.5	+20	W	103
D-altrose	syrup	+16.2	W	**103**, 107, 108
(*p*-bromophenyl)osazone	177–177.5	+8	C	103
L-altrose	syrup	−17.3	W	109
(*p*-bromophenyl)osazone	132	−1	P	109
D-galactose (D-Fucose)	139–142	+75	W	44, 93, 110, 111, **112**, 113
2,2-diphenylhydrazone	187–188			114

TABLE III (*continued*)

Hexose[a]	M.p., °C	$[\alpha]_D$, degrees	Rotation solvent	References
6-Deoxy-				
L-galactose (L-Fucose)	140–141	−76	W	115, **116**, 117, 118
phenylosazone	178	−70	P–E	109
D-glucose	146	+27.9	W	**119**, 120, 121
phenylosazone	189–191	−77	P–E	93, **109**, 122, 123
L-glucose	143–145	−29.9	W	**124**, 125
diethyl dithioacetal	97–98			124
D-gulose	130–131	−38	W	**55**, 110, 126
phenylosazone	182–183			55
L-gulose	syrup	+40.8	W	56a
phenylosazone	183–184			56a
L-idose	107–108	−23.6	W	**127**, 128, 129
phenylosazone	184–185	−33.5	P–E	**129**, 130, 131, 132
D-mannose (D-Rhamnose)				133, 134, **135**
monohydrate	90–91	−8.2	W	136
(*p*-bromophenyl)hydrazone	167			134
L-mannose (L-Rhamnose)				137, 138, **139**
α anomer, monohydrate	93–94	+9.1	W	**140**, 140a, 141
β anomer	123.5–124.5	+8.9	W	109, 142, **144**
phenylosazone	190			143
(2,4-dinitrophenyl)hydrazone	164–165			133
D-talose	129–131	+20.6	W	44
2-methyl-2-phenylhydrazone	136			**145**, 146, 147, 148, 149
L-talose	126–127	−20.5	W	149
2-methyl-2-phenylhydrazone	137			
2,3-Dideoxy-				
D-*erythro*-				
ethyl α-pyranoside	72–72.5	+138	W	150, **151**
D-*threo*-				
ethyl β-pyranoside	56–57	+129	C	152

(*continued*)

TABLE III (*continued*)

Hexose[a]	M.p., °C	$[\alpha]_D$, degrees	Rotation solvent	References
2,4-Dideoxy-				
D-*threo*-	syrup	+9.6	W	152a
(p-tolylsulfonyl)hydrazone	169–170	−20.8	P	152a
2,6-Dideoxy-				
D-*arabino*-	100–103	+20	W	153, **154**, 155
(2,4-dinitrophenyl)hydrazone	132–132.5			156
L-*arabino*-	93–94	−18.2	W	**157**, 158
L-*lyxo*-	103–106	−61.6	W	158
(2,4-dinitrophenyl)hydrazone	214		A	158
diethyl dithioacetal	101–102	−12.6		158
D-*ribo*-	110–112	+50.2	W	43, 75a, 158a, 159, **160**, 161
diethyl dithioacetal	40–41			162
D-*xylo*-	103	+3.8	W	163, **164**
L-*xylo*-	syrup	−6	W	129
3,6-Dideoxy-				
D-*arabino*-	97–99	+23	W	32, **165**
(p-nitrobiphenyl)sulfonyl-hydrazone	143–144			166
L-*arabino*-	97–99	−25	W	165, **167**
(p-tolylsulfonyl)hydrazone	124			168
D-*lyxo*-	syrup	+13	W	**32**, 169
(p-nitrobiphenyl)sulfonyl-hydrazone	240			169
L-*lyxo*-	syrup	−20	M	165
(p-nitrobiphenyl)sulfonyl hydrazone	143–145			165

TABLE III (*continued*)

Hexose[a]	M.p., °C	$[\alpha]_\text{D}$, degrees	Rotation solvent	References
3,6-Dideoxy-				
D-*ribo*-				
(*p*-nitrobiphenyl)sulfonyl-hydrazone	syrup	+10	W	**32**, 165
D-*xylo*-				170
(*p*-nitrobiphenyl)sulfonyl-hydrazone	syrup	−12.2	W	171, **172**, 172a
	146			173, 174
L-*xylo*-				
hydrazone	syrup	+4	W	169, **174**
4,6-Dideoxy-				
D-*xylo*-	137–138	+33	W	92a, 175, 175a
5,6-Dideoxy-				
D-*arabino*-				
diethyl dithioacetal	108–109			176
L-*arabino*-				
diethyl dithioacetal	110			177
D-*xylo*-	syrup	+2	M	**178**, 179
phenylosazone	169–170	+14	P–E	**178**, 179, 179a
2,3,6-Trideoxy-				
D-*erythro*-				
(2,4-dinitrophenyl)hydrazone	syrup	+28.6	W	**180**, 183
D-*threo*-	154–154.5	−9.8	P	180, **181**, 182, 183
(2,4-dinitrophenyl)hydrazone	syrup	−0.2	W	183
L-*threo*-	121–122	13.7	P	183
(2,4-dinitrophenyl)hydrazone	121–122	−14.9	P	183

[a] 2-Hexulose in the case of osazone derivatives.

TABLE IV

Deoxyheptoses and Deoxyoctoses

Heptoses[a] and octoses[a]	M.p., °C	$[\alpha]_D$, degrees	Rotation Solvent	References
2-Deoxy-				
D-galacto-heptose	syrup	−43.2	W	184
2-methyl-2-phenylhydrazone	171–172	−21	M	184
D-gluco-heptose	174.5–175.5	+11.1	W	186
D-manno-heptose	syrup	+52	W	186
2-methyl-2-phenylhydrazone	132	+10	M	186
7-Deoxy-				
L-glycero-L-galacto-heptose	186.5–187.5	−62.5	W	62, 185
pentaacetate (α anomer)	93.5–94.5	−135.6	C	187
(β anomer)	108–108.5	−30.7	C	187
(DL)-glycero-L-manno-heptose	180–181	−61.4	W	188, 189
phenylosazone	200			189
L-glycero-D-manno-heptose	syrup	+17	W	189a
phenylhydrazone	206.5			189a
2,7-Dideoxy-				
L-manno-heptose	syrup	−47.1	W	190
8-Deoxy-				
D-erythro-D-galacto-octose	118–122	+57.4	W	190a
(DL)-erythro(threo)-L-manno-octose	amorph.	+8.4	W	189
phenylosazone	200			189

[a] The corresponding 2-glyculoses in the case of osazone derivatives.

TABLE V

Deoxy-2-ketoses

Deoxy-2-ketose	M.p., °C	$[\alpha]_D$, degrees	Rotation solvent	References
5-Deoxy-				
D-*threo*-pentulose	syrup	−15	M	179, **191**
phenylosazone	171–173	+7	P–E	179
1-Deoxy-				
D-*arabino*-hexulose	amorph.	−90	W	192, 192a
phenylosazone	137–140	−28	P	192, 192a
L-*arabino*-hexulose	syrup			
phenylosazone	137–140	+29	P	192
D-*lyxo*-hexulose	121–123	−14	W	193
D-*ribo*-hexulose	97–98	−0.1	W	**194**, 195
phenylosazone	128–130	+75.7	P	195
3-Deoxy-				
D-*erythro*-hexulose	112–114	−43.4	W	196
(*p*-nitrophenyl)osazone	246–247	+455	P	196
4-Deoxy-				
D-*erythro*-hexulose	syrup	+5	E	197
phenylosazone	167–168	−34	P–E	197
5-Deoxy-				
D-*threo*-hexulose	110	−67.0	W	100
phenylosazone	153			99, **100**
6-Deoxy-				
D-*arabino*-hexulose	syrup	−13	W	**198**, 199, 200
(*o*-nitrophenyl)osazone	136–137	+40	E	200

(*continued*)

TABLE V (continued)

Deoxy-2-ketose	M.p., °C	$[\alpha]_D$, degrees	Rotation solvent	References
6-Deoxy-				
D-*ribo*-hexulose	syrup			200a
phenylosazone	179–180	−70.4	P–E	**127**, 200a
L-*ribo*-hexulose	amorph	+8.2	W	127
phenylosazone	179–180	+66.3	P–E	127
D-*lyxo*-hexulose	syrup	−2	W	201
(*o*-nitrophenyl)osazone	160–168	+72.5	M	201
L-*lyxo*-hexulose	68–69	+3.4	W	201
(*p*-nitrophenyl)osazone	161–162	−69.5	M	201
L-*xylo*-hexulose	84–85	−29.8	W	**127**, 130, 132, 198
phenylosazone	184–185			132, 201a
1,3-Dideoxy-				
D-*erythro*-hexulose	syrup			202
dimethyl dithioacetal	syrup	−37.8	C	202
5,6-Dideoxy-				
D-*threo*-hexulose	syrup	−13	M	179, **203**
phenylosazone	164–165	+5	P–E	**179**, 203
1-Deoxy-				
D-*galacto*-heptulose	syrup	+65	W	204
(2,5-dichlorophenyl)hydrazone	186.5	−13.1	P	204
D-*manno*-heptulose	90–91	+24.2	W	205
phenylosazone	176–178			205

References

(1) H. Venner, *Chem. Ber.*, **90**, 121 (1957).

(2) J. Fried, D. E. Walz, and O. Wintersteiner, *J. Amer. Chem. Soc.*, **68**, 2746 (1946).

(3) L. Vargha and J. Kuszmann, *Chem. Ber.*, **96**, 411 (1963).

(4) H. W. Diehl and H. G. Fletcher, Jr., *Biochem. Prepn.*, **8**, 49 (1961).

(5) H. W. Diehl and H. G. Fletcher, Jr., *Arch. Biochem. Biophys.*, **78**, 386 (1958).

(6) E. Hardegger, M. Schellenbaum, R. Huwyler, and A. Züst, *Helv. Chim. Acta*, **40**, 1815 (1957).

(7) J. Kenner and G. N. Richards, *J. Chem. Soc.*, 3019 (1957).

(8) G. N. Richards, *J. Chem. Soc.*, 3638 (1954).

(9) J. C. Sowden, *J. Amer. Chem. Soc.*, **76**, 3541 (1954).

(10) P. A. J. Gorin and J. K. N. Jones, *Nature*, **172**, 1051 (1953).

(11) L. Hough, *J. Chem. Soc.*, 3066 (1953).

(12) R. E. Deriaz, W. G. Overend, M. Stacey, E. G. Teece, and L. F. Wiggins, *J. Chem. Soc.*, 1879 (1949).

(13) J. C. Sowden, *J. Amer. Chem. Soc.*, **71**, 1897 (1949).

(14) W. G. Overend, M. Stacey, and L. F. Wiggins, *J. Chem. Soc.*, 1358 (1949).

(15) P. A. Levene, L. A. Mikeska, and T. Mori, *J. Biol. Chem.*, **85**, 785 (1930).

(16) J. Meisenheimer and H. Jung, *Ber.*, **60**, 1462 (1927).

(17) H. Kuzuhara and II. G. Fletcher, Jr., *J. Org. Chem.*, **33**, 1816 (1968).

(18) Y. Matsushima and Y. Imanaga, *Bull. Chem. Soc. Jap.*, **26**, 506 (1953).

(19) Y. Matsushima and Y. Imanaga, *Nature*, **171**, 475 (1953).

(20) P. A. Levene and T. Mori, *J. Biol. Chem.*, **83**, 803 (1929).

(21) H. Kiliani and H. Naegell, *Ber.*, **35**, 3528 (1902).

(22) G. Nakaminami, M. Kakagawa, S. Shioi, Y. Sugiyama, S. Isemura, and M. Shibuya, *Tetrahedron Lett.*, **40**, 3983 (1967).

(23) K. Butler, S. Laland, W. G. Overend, and M. Stacey, *J. Chem. Soc.*, 1433 (1950).

(24) G. Casini and L. Goodman, *J. Amer. Chem. Soc.*, **86**, 1427 (1964).

(24a) H. Zinner, G. Wulf, and R. Heinatz, *Chem. Ber.*, **97**, 3536 (1964).

(25) F. Weygand and H. Wolz, *Chem. Ber.*, **85**, 256 (1952).

(26) W. G. Overend, F. Shafizadeh, and M. Stacey, *J. Chem. Soc.*, 1027 (1950).

(27) A. M. Gakhokidze, *Zh. Obshch. Khim.*, **15**, 539 (1945); *Chem. Abstr.*, **40**, 4675 (1946).

(28) G. Rembarz, *Chem. Ber.*, **95**, 1565 (1962).

(29) J. C. Sowden, M. G. Blair, and D. J. Kuenne, *J. Amer. Chem. Soc.*, **79**, 6450 (1957).

(30) A. M. Gakhokidze, *Zh. Obshch. Khim.*, **10**, 507 (1940); *Chem. Abstr.*, **34**, 7858 (1940).

(31) P. W. Kent, M. Stacey, and L. F. Wiggins, *J. Chem. Soc.*, 1232 (1949).

(32) R. J. Ferrier and G. H. Sankey, *J. Chem. Soc.* (C), 2339 (1966).

(33) P. W. Kent, M. Stacey, and L. F. Wiggins, *Nature*, **161**, 21 (1948).

(34) H. Kato, *Agr. Biol. Chem.* (Tokyo), **26**, 187 (1962).

(35) C. D. Anderson, L. Goodman, and B. R. Baker, *J. Amer. Chem. Soc.*, **81**, 898 (1959).

(36) S. Mukherjee and A. R. Todd, *J. Chem. Soc.*, 969 (1947).

(36a) P. W. Kent and P. F. V. Ward, *J. Chem. Soc.*, 416 (1953).

(37) H. Zinner, K. Wessely, and H. Kristen, *Chem. Ber.*, **92**, 1618 (1959).

(38) F. Micheel, *Ber.*, **63**, 359 (1930).

(38a) R. Kuhn, W. Bister, and W. Dafeldecker, *Ann.*, **617**, 115 (1958).

(39) L. Hough and T. J. Taylor, *J. Chem. Soc.*, 3544 (1955).

(40) O. Ruff, *Ber.*, **35**, 2360 (1902).

(41) E. Fischer, *Ber.*, **29**, 1377 (1896).

(42) M. Schulz and H. Boeden, *Tetrahedron Lett.*, **25**, 2843 (1966).

(43) F. Micheel, *Ber.*, **63**, 347 (1930).

(44) E. Votoček and F. Valentin, *Collect. Czech. Chem. Commun.*, **2**, 36 (1930).

(45) E. Votoček, *Ber.*, **56**, 35 (1917).

(46) R. K. Hulyalkar and M. B. Perry, *Can. J. Chem.*, **43**, 3241 (1965).

(47) R. Kuhn, W. Bister, and W. Dafeldecker, *Ann.*, **628**, 186 (1959).

(48) Y. Wang, W. Lin, W. Yi, and T. Ku, *Hua Hsueh Hsueh Pao*, **25**, 265 (1959); *Chem. Abstracts*, **54**, 18370h (1960).

(49) O. T. Schmidt, W. Mayer, and A. Distelmaier, *Ann.*, **555**, 26 (1944).

(50) K. A. Folkers and C. H. Shunk, U. S. Pat. 2,847,413 (1958); *Chem. Abstr.*, **53**, 3252d (1959).

(51) C. H. Shunk, J. B. Lavigne, and K. A. Folkers, *J. Amer. Chem. Soc.*, **77**, 2210 (1955).

(52) K. Iwadare, *Bull. Chem. Soc. Jap.*, **17**, 90 (1942).

(53) K. J. Ryan, H. Arzoumanian, E. M. Acton, and L. Goodman, *J. Amer. Chem. Soc.*, **86**, 2497 (1964).

(54) P. Karrer and A. Boettcher, *Helv. Chim. Acta*, **36**, 837 (1953).

(55) P. A. Levene and J. Compton, *J. Biol. Chem.*, **111**, 335 (1935).

(56) P. A. J. Gorin, L. Hough, and J. K. N. Jones, *J. Chem. Soc.*, 2140 (1953).

(56a) H. Müller and T. Reichstein, *Helv. Chim. Acta*, **21**, 251 (1938).

(57) R. Allerton, W. G. Overend, and M. Stacey, *J. Chem. Soc.*, 255 (1952).

(58) H. Zinner and H. Wigert, *Chem. Ber.*, **92**, 2893 (1959).

(59) L. Goodman, *J. Amer. Chem. Soc.*, **86**, 4167 (1964).

(60) H. J. Bestmann and J. Angerer, *Tetrahedron Lett.*, **41**, 3665 (1969).

(61) J. Adamson and A. B. Foster, *Carbohyd. Res.*, **10**, 517 (1969).

(62) Yu. A. Zhdanov, V. I. Kornilov, and Lun Dink Chung, *Zh. Obshch. Khim.*, **39**, 2360 (1969); *Chem. Abst.*, **72**, 55803f (1970).

(63) J. C. Sowden and H. O. L. Fischer, *J. Amer. Chem. Soc.*, **69**, 1048 (1947).

(64) W. G. Overend, M. Stacey, and J. Staněk, *J. Chem. Soc.*, 2841 (1949).

(65) M. Bergmann, H. Schotte, and W. Lechinsky, *Ber.*, **55**, 158 (1922).

(66) M. Bergmann and H. Schotte, *Ber.*, **54**, 440 (1921).

(67) Yu. A. Zhdanov and V. G. Alexeeva, *Carbohyd. Res.*, **10**, 184 (1969).

(68) C. Tamm and T. Reichstein, *Helv. Chim. Acta*, **35**, 61 (1952).

(69) W. G. Overend, F. Shafizadeh, and M. Stacey, *J. Chem. Soc.*, 992 (1951).

(70) A. B. Foster, W. G. Overend, and M. Stacey, *J. Chem. Soc.*, 974 (1951).

(71) C. Tamm and T. Reichstein, *Helv. Chim. Acta*, **31**, 1630 (1948).

(72) H. S. Isbell and W. W. Pigman, *J. Res. Nat. Bur. Stand.*, **22**, 397 (1939).

(73) P. A. Levene and R. S. Tipson, *J. Biol. Chem.*, **93**, 637 (1931).

(74) M. B. Perry and J. Furdova, *Can. J. Chem.*, **46**, 2859 (1968).

(74a) W. W. Zorbach and A. P. Ollapally, *J. Org. Chem.*, **29**, 1790 (1964).

(75) H. R. Bolliger and M. Thürkauf, *Helv. Chim. Acta*, **35**, 1426 (1952).

(75a) M. Gut and D. A. Prins, *Helv. Chim. Acta*, **30**, 1223 (1947).

(76) M. B. Perry and A. C. Webb, *Can. J. Chem.*, **47**, 1245 (1969).

(77) T. Golab and T. Reichstein, *Helv. Chim. Acta*, **44**, 616 (1961).

(78) G. Rembarz, *Chem. Ber.*, **93**, 622 (1960).

(79) H. R. Bolliger and D. A. Prins, *Helv. Chim. Acta*, **29**, 1061 (1946).

(80) H. B. Wood, Jr., and H. G. Fletcher, Jr., *J. Org. Chem.*, **26**, 1969 (1961).

(81) H. Huber and T. Reichstein, *Helv. Chim. Acta*, **31**, 1645 (1948).

(82) E. F. L. J. Anet, *J. Amer. Chem. Soc.*, **82**, 1502 (1960).

(83) J. Němec, Z. Kefurtová, K. Kefurt, and J. Jarý, *Collect. Czech. Chem. Commun.*, **33**, 2097 (1968).
(84) E. J. Hedgley, W. G. Overend, and R. A. C. Rennie, *J. Chem. Soc.*, 4701 (1963).
(85) J. W. Pratt and N. K. Richtmyer, *J. Amer. Chem. Soc.*, **79**, 2597 (1957).
(86) E. F. L. J. Anet, *Chem. Ind.* (London), 345 (1960).
(87) M. Černý and J. Pacák, *Collect. Czech. Chem. Commun.*, **21**, 1003 (1956).
(88) R. Kuhn, G. Krüger, H. J. Haas, and A. Seeliger, *Ann.*, **644**, 122 (1961).
(89) M. Dahlgard, B. H. Chastain, and Ru-Jen Lee Han, *J. Org. Chem.*, **27**, 929 (1962).
(90) M. Černý and J. Pacák, *Collect. Czech. Chem. Commun.*, **27**, 94 (1962).
(91) H. W. H. Schmidt and H. Neukom, *Tetrahedron Lett.*, 2063 (1964).
(92) M. Černý, J. Stanek, and J. Pacák, *Collect. Czech. Chem. Commun.*, **34**, 1750 (1969).
(92a) G. Siewert and O. Westphal, *Ann.*, **720**, 161 (1968).
(93) N. K. Kochetkov and A. I. Usov, *Tetrahedron*, **19**, 973 (1963).
(94) M. Cerny, J. Staněk, and J. Pacák, *Chem. Ind.* (London), 945 (1961).
(95) J. A. Montgomery and K. Hewson, *J. Org. Chem.*, **29**, 3436 (1964).
(96) K. J. Ryan, H. Arzoumanian, E. M. Acton, and L. Goodman, *J. Amer. Chem. Soc.*, **86**, 2503 (1964).
(97) E. J. Hedgley, O. Mérész, and W. G. Overend, *J. Chem. Soc.* (C), 888 (1967).
(98) R. E. Gramera, T. R. Ingle, and R. L. Whistler, *J. Org. Chem.*, **29**, 2074 (1964).
(99) M. L. Wolfrom, K. Matsuda, F. Komitsky, Jr., and T. E. Whitely, *J. Org. Chem.*, **28**, 3551 (1963).
(100) P. P. Regna, *J. Amer. Chem. Soc.*, **69**, 246 (1947).
(101) A. Hunger and T. Reichstein, *Helv. Chim. Acta*, **35**, 1073 (1952).
(102) M. Keller and T. Reichstein, *Helv. Chim. Acta*, **32**, 1607 (1949).
(103) K. Iwadare *Bull. Chem. Soc. Jap.*, **17**, 296 (1942).
(104) P. A. Levene and J. Compton, *J. Biol. Chem.*, **116**, 169 (1936).
(105) A. Windaus and G. Schwarte, *Nachr. Ges. Wiss. Göttingen Math. Phys. Klasse*, 1 (1926); *Chem. Abstr.*, **21**, 3618 (1927).
(106) E. J. Reist, L. Goodman, and B. R. Baker, *J. Amer. Chem. Soc.*, **80**, 5775 (1958).
(107) A. D. Albein, H. Koffler, and H. R. Garner, *Biochim. Biophys. Acta*, **56**, 165 (1962).
(108) M. Gut and D. A. Prins, *Helv. Chim. Acta*, **29**, 1555 (1964).
(109) K. Freudenberg and K. Raschig, *Ber.*, **62**, 373 (1929).
(110) J. A. Moore, C. Tamm, and T. Reichstein, *Helv. Chim. Acta*, **37**, 755 (1954).
(111) H. Schmid and P. Karrer, *Helv. Chim. Acta*, **32**, 1371 (1949).
(112) H. H. Schlubach and E. Wagenitz, *Ber.*, **65**, 304 (1932).
(113) K. Freudenberg and K. Raschig, *Ber.*, **60**, 1633 (1927).
(114) L. H. Sternbach, S. Kaiser, and M. W. Goldberg, *J. Amer. Chem. Soc.*, **80**, 1639 (1958).
(115) C. H. Trabert, *Arch. Pharm.*, **293**, 278 (1960).
(116) R. C. Hockett, F. P. Phelps, and C. S. Hudson, *J. Amer. Chem. Soc.*, **61**, 1658 (1939).
(117) J. Minsaas, *Rec. Trav. Chim.*, **50**, 424 (1931).
(118) B. Tollens and F. Rorive, *Ber.*, **42**, 2009 (1909).
(119) J. D. Chanley, R. Ledeen, J. Wax, R. F. Nigrelli, and H. Sobotka, *J. Amer. Chem. Soc.*, **81**, 5180 (1959).
(120) P. Karrer and A. Boettcher, *Helv. Chim. Acta*, **36**, 570 (1953).
(121) E. Fischer and K. Zach, *Ber.*, **45**, 3761 (1912).
(122) E. Votoček, *Ber.*, **44**, 819 (1911).

(123) E. Fischer and C. Liebermann, *Ber.*, **26**, 2415 (1893).
(124) E. Zissis, N. K. Richtmyer, and C. S. Hudson, *J. Amer. Chem. Soc.*, **73**, 4714 (1951).
(125) E. Fischer and H. Herborn, *Ber.*, **29**, 1961 (1896).
(126) D. E. Schittler and T. Reichstein, *Helv. Chim. Acta*, **31**, 688 (1948).
(127) H. Kaufmann and T. Reichstein, *Helv. Chim. Acta*, **50**, 2280 (1967).
(128) G. Charalambous and E. Percival, *J. Chem. Soc.*, 2443 (1954).
(129) A. S. Meyer and T. Reichstein, *Helv. Chim. Acta*, **29**, 139, 152 (1946).
(130) M. L. Wolfrom and S. Hanessian, *J. Org. Chem.*, **27**, 1800 (1962).
(131) A. C. Arcus and N. L. Edson, *Biochem. J.*, **64**, 385 (1956).
(132) H. Müller and T. Reichstein, *Helv. Chim. Acta*, **21**, 263 (1938).
(133) A. Markovitz, *J. Biol. Chem.*, **237**, 1767 (1962).
(134) W. W. Zorbach and C. O. Tio, *J. Org. Chem.*, **26**, 3543 (1961).
(135) W. T. Haskins, R. M. Hann, and C. S. Hudson, *J. Amer. Chem. Soc.*, **68**, 628 (1946).
(136) E. Votoček and F. Valentin, *Chem. Listy*, **21**, 7 (1927); *Chem. Abstr.*, **21**, 1969 (1927).
(137) F. G. Torto, *J. Chem. Soc.*, 5234 (1961).
(138) J. Minsaas, *Kgl. Norske Videnskabs. Forh.* **68**, 177 (1934); *Chem. Abstr.*, **28**, 5047 (1934).
(139) C. S. Hudson and E. Yanovsky, *J. Amer. Chem. Soc.*, **39**, 1032 (1917).
(140) E. L. Jackson and C. S. Hudson, *J. Amer. Chem. Soc.*, **59**, 1076 (1937).
(140a) R. S. Tipson and H. S. Isbell, *J. Res. Nat. Bur. Stand.*, **66A**, 31 (1962).
(141) E. Fischer, *Ber.*, **29**, 324 (1896).
(142) K. Akhtardzhiev and D. Kolev, *Pharm. Zentralhalle*, **100**, 14 (1961); *Chem. Abstr.*, **55**, 15829 (1961); *Compt. Rend. Acad. Bulgare Sci.*, **10**, 387 (1957); *Chem. Abstr.*, **52**, 14203 (1958).
(143) J. A. Dominguez, *J. Amer. Chem. Soc.*, **73**, 849 (1951).
(144) J. P. McKinnell and E. Percival, *J. Chem. Soc.*, 3141 (1962).
(145) P. M. Collins and W. G. Overend, *J. Chem. Soc.*, 1912 (1965).
(146) A. P. MacLennan, *Biochim. Biophys. Acta*, **48**, 600 (1961).
(147) J. K. N. Jones and W. H. Nicholson, *J. Chem. Soc.*, 3050 (1955).
(148) J. Schmutz, *Helv. Chim. Acta*, **31**, 1719 (1948).
(149) E. Votoček and V. Kučerenko, *Collect. Czech. Chem. Commun.*, **2**, 47 (1930).
(150) S. Laland, W. G. Overend, and M. Stacey, *J. Chem. Soc.*, 738 (1950).
(151) M. Bergmann, *Ann.*, **443**, 223 (1923).
(152) A. B. Foster, R. Harrison, J. Lehmann, and J. M. Webber, *J. Chem. Soc.*, 4471 (1963).
(152a) A. F. Cook and W. G. Overend, *Chem. Ind.* (London), 1141 (1966).
(153) W. W. Zorbach and J. P. Ciaudelli, *J. Org. Chem.*, **30**, 451 (1965).
(154) P. Studer, S. K. Pavanaram, C. R. Gavilanes, H. Linde, and K. Meyer, *Helv. Chim. Acta*, **46**, 23 (1963).
(155) M. Miyamoto, Y. Kawamatsu, M. Shinohara, Y. Nakadaira, and K. Nakanishi, *Tetrahedron*, **22**, 2785 (1966).
(156) B. Iselin and T. Reichstein, *Helv. Chim. Acta*, **27**, 1146 (1944).
(157) B. Iselin and T. Reichstein, *Helv. Chim. Acta*, **27**, 1200 (1944).
(158) S. Takahashi, M. Kurabayashi, and E. Ohki, *Chem. Pharm. Bull.* (Tokyo), **15**, 1657 (1967).
(158a) M. Haga, M. Chonan, and S. Tejima, *Carbohyd. Res.*, **16**, 486 (1971).
(159) S. Hanessian and N. R. Plessas, *J. Org. Chem.*, **34**, 2163 (1969).
(160) H. R. Bolliger and P. Ulrich, *Helv. Chim. Acta*, **35**, 93 (1952).

(161) B. Iselin and T. Reichstein, *Helv. Chim. Acta*, **27**, 1203 (1944).

(162) D. C. DeJongh and S. Hanessian, *J. Amer. Chem. Soc.*, **88**, 3114 (1966).

(163) H. R. Bolliger and T. Reichstein, *Helv. Chim. Acta*, **36**, 302 (1953).

(164) O. Schindler and T. Reichstein, *Helv. Chim. Acta*, **35**, 730 (1952).

(165) C. Fouquey, J. Polonsky, and E. Lederer, *Bull. Soc. Chim. Fr.*, 803 (1959).

(166) O. Westphal, O. Lüderitz, I. Fromme, and N. Joseph, *Angew. Chem.*, **65**, 555 (1953).

(167) O. Westphal and O. Lüderitz, *Angew. Chem.*, **72**, 881 (1960).

(168) C. Fouquey, J. Polonsky, and E. Lederer, *Bull. Soc. Chim. Biol.*, **39**, 101 (1957).

(169) C. Fouquey, E. Lederer, O. Lüderitz, J. Polonsky, A. M. Staub, S. Stirm, R. Tinelli, and O. Westphal, *Compt. Rend. (C)*, **246**, 2417 (1958).

(170) D. A. L. Davies, A. M. Staub, I. Fromme, O. Lüderitz, and O. Westphal, *Nature*, **181**, 822 (1958).

(171) G. Siewert and O. Westphal, *Ann.*, **720**, 171 (1968).

(172) K. Čapek, J. Němec, and J. Jarý, *Collect. Czech. Chem. Commun.*, **33**, 1758 (1968).

(172a) H. Zinner, B. Ernst, and F. Kreienbring, *Chem. Ber.*, **95**, 821 (1962).

(173) O. Westphal and S. Stirm, *Ann.*, **620**, 8 (1959).

(174) O. Lüderitz, A. M. Staub, S. Stirm, and O. Westphal, *Biochem. Z.*, **330**, 193 (1958).

(175) J. Hill, L. Hough, and A. C. Richardson, *Carbohyd. Res.*, **8**, 19 (1968).

(175a) B. T. Lawton, W. A. Szarek, and J. K. N. Jones, *Carbohyd. Res.*, **14**, 255 (1970).

(176) R. L. Mann and D. O. Woolf, *J. Amer. Chem. Soc.*, **79**, 120 (1957).

(177) D. J. Ball, A. E. Flood, and J. K. N. Jones, *Can. J. Chem.*, **37**, 1018 (1959).

(178) J. K. N. Jones and J. L. Thompson, *Can. J. Chem.*, **35**, 955 (1957).

(179) P. A. J. Gorin, L. Hough, and J. K. N. Jones, *J. Chem. Soc.*, 2699 (1955).

(179a) L. D. Hall, L. Hough, and R. A. Pritchard, *J. Chem. Soc.*, 1537 (1961).

(180) C. L. Stevens, K. Nagarajan, and T. H. Haskell, *J. Org. Chem.*, **27**, 2991 (1962).

(181) E. L. Albano and D. Horton, *J. Org. Chem.*, **34**, 3519 (1969).

(182) A. H. Haines, *Carbohyd. Res.*, **10**, 466 (1969).

(183) C. L. Stevens, P. Blumbergs, and D. L. Wood, *J. Amer. Chem. Soc.*, **86**, 3592 (1964).

(184) M. B. Perry and A. C. Webb, *Can. J. Chem.*, **46**, 789 (1968).

(185) D. T. Williams and M. B. Perry, *Can. J. Chem.*, **47**, 2763 (1969).

(186) M. B. Perry, *Can. J. Chem.*, **45**, 1295 (1967).

(187) E. L. Jackson and C. S. Hudson, *J. Amer. Chem. Soc.*, **75**, 3000 (1953).

(188) E. Fischer and O. Piloty, *Ber.*, **23**, 3827 (1890).

(189) E. Fischer and O. Piloty, *Ber.*, **23**, 3102 (1890).

(189a) E. Votoček and F. Valentin, *Collect. Czech. Chem. Commun.*, **10**, 77 (1938).

(190) J. Yoshimura, H. Komoto, H. Ando, and T. Nakagawa, *Bull. Chem. Soc. Jap.*, **39**, 1775 (1966).

(190a) D. G. Lance, W. A. Szarek, J. K. N. Jones, and G. B. Howarth, *Can. J. Chem.*, **47**, 2871 (1969).

(191) P. A. J. Gorin, L. Hough, and J. K. N. Jones, *J. Chem. Soc.*, 2140 (1953).

(192) A. Ishizu, B. Lindberg, and O. Theander, *Carbohyd. Res.*, **5**, 329 (1967).

(192a) C. R. Haylock, L. D. Melton, K. N. Slessor, and A. S. Tracy, *Carbohyd. Res.*, **16**, 375 (1971).

(193) M. L. Wolfrom and R. B. Bennett, *J. Org. Chem.*, **30**, 1284 (1965).

(194) J. Šmejkal and J. Farkaš, *Collect. Czech. Chem. Commun.*, **28**, 1345 (1963).

(195) M. L. Wolfrom, A. Thompson, and E. F. Evans, *J. Amer. Chem. Soc.*, **67**, 1793 (1945).

(196) R. Kuhn, H. J. Haas, and A. Seeliger, *Chem. Ber.*, **94**, 2534 (1961).

(197) P. A. J. Gorin, L. Hough, and J. K. N. Jones, *J. Chem. Soc.*, 4700 (1954).
(198) L. Hough and J. K. N. Jones, *J. Chem. Soc.*, 4052 (1952).
(199) L. Anderson and H. A. Lardy, *J. Amer. Chem. Soc.*, **70**, 594 (1948).
(200) W. T. J. Morgan and T. Reichstein, *Helv. Chim. Acta*, **21**, 1023 (1938).
(200a) M. Haga, M. Takano, and S. Tejima, *Carbohyd. Res.*, **14**, 237 (1970).
(201) J. Barnett and T. Reichstein, *Helv. Chim. Acta*, **21**, 913 (1938).
(201a) L. Hough, J. K. N. Jones, and D. L. Mitchell, *Can. J. Chem.*, **37**, 725 (1959).
(202) B. F. West, K. V. Bhat, and W. W. Zorbach, *Carbohyd Res.*, **8**, 253 (1968).
(203) P. A. J. Gorin, L. Hough, and J. K. N. Jones, *J. Chem. Soc.*, 3843 (1955).
(204) B. Coxon and H. G. Fletcher, Jr., *J. Amer. Chem. Soc.*, **86**, 922 (1964).
(205) J. C. Sowden, C. H. Bowers, and K. O. Lloyd, *J. Org. Chem.*, **29**, 130 (1964).

MORPHOLOGY AND BIOGENESIS OF
CELLULOSE AND PLANT CELL-WALLS

By F. Shafizadeh and G. D. McGinnis

Wood Chemistry Laboratory,° *School of Forestry and Department of Chemistry,*
University of Montana, Missoula, Montana

I. Introduction

Adaptation of the physical and chemical disciplines for investigation of biological materials and processes has created a new and exciting frontier in science[1,2] that may be called molecular biology or cellular chemistry. An outstanding result of the integrated or multidis-

° Established through a grant from Hoerner–Waldorf Corporation of Montana.

(1) R. D. Preston, in "Cellular Ultrastructure of Woody Plants," W. A. Côté, Jr., ed., Syracuse University Press, Syracuse, 1965, p. 1.
(2) W. G. Whaley, in "Recent Advances in Phytochemistry," T. J. Mabry, R. E. Alston, and V. C. Runeckles, eds., Meredith Corporation, New York, 1968, Vol. 1, p. 1.

ciplinary approach has been the unravelling of metabolic processes by which natural products, such as the carbohydrates[3] and terpenoids, are formed.[4] For the latter group, understanding of the biosynthetic pathways has led to rationalization of the numerous structural variations[5,6] and to decoding of some genetic or taxonomic relationships that these variations entail.[7,8]

In the field of carbohydrates, biosyntheses of sugars and polysaccharides[3] are being related to the functions of various cells and their components, to provide for the subject a broader outlook that may be called biogenesis. For plant cell-wall polysaccharides, this includes the correlation of biosynthesis with the organelles and functions of the cell, as a programmed sequence of events that is responsible for division, growth, and differentiation processes.[9] This correlation therefore involves not only the production and molecular properties of the polysaccharides, but also the architecture and function of the cell wall,[10,11] in which they serve as highly specific, structural materials. The broader ramification of biogenesis, as compared with biosynthesis, is further realized by considering that the biological processes involved are geared to the production and functioning of the cell wall, rather than merely to the production of polysaccharides as chemical entities.

Plant cell-walls normally contain cellulosic fibrils that are embedded in a matrix of pectins and hemicelluloses. The framework of the mature cells is further encrusted with lignin. The multidisciplinary investigation of the molecular morphology and biogenesis of cellulosic microfibrils provides a fascinating and controversial chapter of molecular biology. The controversy centers not only on the interpretation of physical or biological data, but also on the problem of the extent to which molecular and structural information can be used for understanding the biological processes, and *vice versa*. One of the basic questions is whether production of microfibrils is controlled entirely by the biological functions of the cell or (at least in part) by exocellular chemical and physical forces.

(3) W. Z. Hassid, *Science*, **165**, 137 (1969).
(4) J. B. Pridham and T. Swain, "Biosynthetic Pathways in Higher Plants," Academic Press, New York, 1965.
(5) T. A. Geissman and D. H. G. Crout, "Organic Chemistry of Secondary Plant Metabolism," Freeman, Cooper & Company, San Francisco, 1969.
(6) J. B. Pridham, "Terpenoids in Plants," Academic Press, New York, 1967.
(7) T. Swain, "Comparative Phytochemistry," Academic Press, New York, 1966.
(8) H. G. H. Erdtman, in Ref. 2, p. 13.
(9) D. H. Northcote, *Proc. Roy. Soc.*, Ser. B, **173**, 21 (1969).
(10) K. Mühlethaler, *Proc. Cellulose Conf. 6th, Syracuse 1969*, No. 28, 305.
(11) R. D. Preston, *Endeavour*, **23**, 153 (1964).

It should be noted that many of the problems surrounding the structure and origin of lignin, the encrusting material of the plant cell-walls, were solved with the discovery that lignin is formed as an amorphous and random polymer by oxidative coupling of coniferyl alcohol units.[12] Although the oxidation process occurs enzymically, the polymerization and deposition take place as a chemical reaction. The microfibrillar and crystalline arrangements of cellulosic materials present a structural system far more complicated than the amorphous condition of lignin as an encrusting material. Nevertheless, an understanding of the biogenesis of the microfibrils, and of the cell wall in general, could explain some of the controversy concerning the physical properties and molecular structure of cellulose, the most abundant natural product. Conversely, definitive information on the crystalline structure and arrangement of the cellulose macromolecules within the microfibrils could lead to selection among the most plausible extant theories on biogenesis. The prevailing concepts about the organization and structure of the matrix materials are also in process of modification and development through investigation of how they are formed and how they function. A good example is the discovery of glycoprotein components as possible cross-linking materials involved in the expansion mechanism of the cell wall (see p. 346).

A coherent discussion of the above topic should, therefore, include consideration of: the relevant aspects of composition, structure, and evolution of the plant cell-walls; development of present knowledge concerning the physicochemical properties of cellulosic microfibrils; the biological processes that account for the production and properties of the wall materials; and the cell organelles that are instrumental in the execution of these processes. The present article provides a brief review of these multidisciplinary subjects. The chemical and physical properties of the structural polysaccharides and their biosyntheses,[13-16] have been excellently reviewed in previous Volumes of this Series. Therefore, consideration of these subjects has, so far as possible, been avoided in favor of discussing development of our present knowledge through different approaches and diverse opinions.

It is hoped that this Chapter will present the subject in its proper perspective and help further to expand the interest of carbohydrate chemists in the cognate fields of biological and physical sciences.

(12) K. Freudenberg, *Science*, **148**, 595 (1965).
(13) D. M. Jones, *Advan. Carbohyd. Chem.*, **19**, 219 (1964).
(14) E. F. Neufeld and W. Z. Hassid, *Advan. Carbohyd. Chem.*, **18**, 309 (1963).
(15) T. E. Timell, *Advan. Carbohyd. Chem.*, **19**, 247 (1964); **20**, 409 (1965).
(16) R. H. Marchessault and A. Sarko, *Advan. Carbohyd. Chem.*, **22**, 421 (1967).

II. Plant Cell-walls

1. Chemical Composition

Plant cells are distinguished from animal cells by the presence of true cell-walls containing polysaccharides as the structural material. Cell walls of the higher plants generally contain cellulose which, as a crystalline and rigid material, forms a framework of microfibrils. This framework is embedded in a matrix composed of hemicelluloses, pectins, and some proteins. Mature cell-walls also contain lignin as the incrusting material. There are a few exceptions to this general rule; for example, the wall of endosperm cells in palm seeds contains large proportions of a D-mannan as microfibrils,[17] and very little cellulose. However, in this material, the mannan occurs as a food reserve and disappears upon germination of the seed.[18]

The evolutionary development of the plant kingdom that culminates in the angiosperms (*Anthophyta*) and gymnosperms (*Coniferophyta*) is shown[19] in Fig. 1. In the lower scale of evolution (*Thallophyta*), some of the marine algae and other plant organisms contain a variety of polysaccharides as the crystalline and matrix materials.[11,20–23] A small group of *Chlorophyta* (green algae), including *Valonia*, *Cladophora*, and *Chaetomorpha*, contains large proportions of highly pure, crystalline cellulose I. Other forms of algae, and plant cell-walls in general, contain microfibrils composed of cellulose having various degrees of purity.

A mannan has been detected in some species of *Rhodophyta* (red algae) and *Chlorophyta* as a structural polysaccharide; these species include *Porphyra umbilicalis* and *Bangia fuscopurpurea* (red algae), *Acetabularia calyculus* (of the *Dasycladales* order), and various species of *Codium* (of the *Codiales* order; green algae). X-Ray crystallographic studies have shown that the seaweed mannan is partly crystalline and that the molecular chains lie parallel to each other in the cell wall as for cellulose, but electron-microscope studies show the presence of a granular structure and no readily observable microfibrils. This apparent discrepancy is believed to result from the low

(17) H. Meier, *Biochim. Biophys. Acta*, **28**, 229 (1958).
(18) W. Pigman, ed., "The Carbohydrates," Academic Press, New York, 1957.
(19) A. Cronquist, "Introductory Botany," Harper Bros., New York, 1961, p. 72.
(20) E. Percival and R. H. McDowell, "Chemistry and Enzymology of Marine Algal Polysaccharides," Academic Press, New York, 1967.
(21) R. D. Preston, *Sci. American*, **218**, 102 (1968).
(22) E. Frei and R. D. Preston, *Nature*, **192**, 939 (1961).
(23) E. Frei and R. D. Preston, *Proc. Roy. Soc. Ser. B*, **160**, 293 (1964).

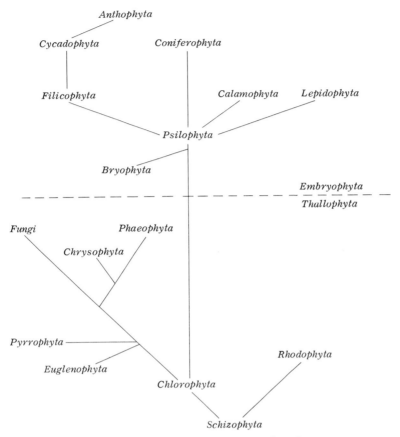

FIG. 1.—Evolution of the Plant Kingdom.[19]

molecular weight of the polysaccharide and the overlapping of the parallel array of short-chain molecules, which gives the appearance of granular structure.

A xylan has been found in *Chlorophyta, Rhodophyta,* and *Phaeophyta* (brown algae). The skeletal polysaccharide of the green algae *Caulerpa, Halimeda,* and *Udotea* is a $(1 \rightarrow 3)$-β-D-xylopyranan, coiled helically to form microfibrils.[24] In contrast to the crystalline lattice of cellulose, which could not be penetrated by water, the helical arrangement in the β-D-$(1 \rightarrow 3)$-linked xylan allows in the lattice considerable hydration and hydrogen bonding, and these contribute to

(24) E. D. T. Atkins, K. D. Parker, and R. D. Preston, *Proc. Roy. Soc., Ser. B,* **173,** 209 (1969).

the crystallinity and mechanical properties of the microfibrils in water.

Two groups of plants are known to combine the structural materials found in the mannan- and xylan-containing algae. *Porphyra,* a red seaweed, contains $(1 \rightarrow 3)$-β-D-xylopyranan in the cell wall and $(1 \rightarrow 4)$-β-D-mannopyranan in the cutiles covering the wall. An interesting correlation between the chromosome complement of the cells and the composition of the wall is noted for *Halicystis* and *Derbesia,* which are two generations of a *Chlorophyta.* During the reproduction cycle, the cell nuclei of the first generation, which is haploid (contains one set of chromosomes), fuse in pairs to form the second generation as a diploid. This plant eventually undergoes cell division to regenerate the original haploid form. *Halicystis* is composed of large, globular cells, and *Derbesia* is filamentous. The former contains crystalline $(1 \rightarrow 3)$-β-D-xylopyranan and some cellulose. The latter contains a mannan, but neither cellulose nor a xylan.

In fungi, the cell wall frequently contains chitin as a replacement or supplement for cellulosic microfibrils,[25,26] and various mannans have been isolated from yeast.[26a,26b,26c]

Interestingly, although variation in the crystalline or fibrillar component of the cell walls is confined to lower plants, variations in the composition of matrix polysaccharides and encrusting materials are effected in higher plants; this is evident from the substantial differences observed in the composition of the hemicelluloses and lignin of gymnosperms (softwoods) and angiosperms (hardwoods).[15,27] There are some notable changes in the composition of the polysaccharides in different layers of the cell wall, described in the following Section. Cellulose is shown to have a random, and lower, molecular weight in the primary wall, and a uniform and much higher molecular weight in the secondary wall (see p. 333). The chemical composition of the hemicelluloses, and their distribution and variation through different layers of the cell wall, are discussed in excellent articles by Timell.[15]

2. Laminar Structure

Fully to appreciate the intricate structure of the plant cell-walls and the molecular morphology of the structural polysaccharides, the

(25) A. J. Michell and G. Scurfield, *Arch. Biochem. Biophys.,* **120,** 628 (1967).

(26) K. M. Rudall, in Ref. 10, p. 83.

(26a) D. H. Northcote and R. W. Horne, *Biochem. J.,* **51,** 232 (1952).

(26b) G. O. Aspinall, E. Percival, D. A. Rees, and M. Rennie, in "Rodd's Chemistry of Carbon Compounds," S. Coffey, ed., Elsevier, Amsterdam, 2nd Edition, 1967, Vol. 1F, p. 688.

(26c) P. A. J. Gorin and J. F. T. Spencer, *Advan. Carbohyd. Chem.,* **23,** 367 (1968).

(27) F. Shafizadeh and W. T. Nearn, *Materials Res. Standards,* **6,** 593 (1966).

reader is referred to several excellent monographs and articles[1,28-31] in which this subject is presented in greater detail. The present discussion will briefly cover some main features relevant to the topic.

The fibrous and woody tissues of plants contain several layers of cell wall surrounded by an amorphous, intercellular substance known as the middle lamella. The multi-layer structure was first observed by means of a polarizing microscope that distinguished the layers in a transverse section by virtue of the differences in the content and orientation of the cellulosic microfibrils. This phenomenon, which will be discussed later (see p. 308), was called micellar orientation. By use of this method, Kerr and Bailey[32] divided the cell-wall layers of a tracheid of a normal gymnosperm into the intercellular layer, the primary wall, and the secondary wall. The latter was further divided into the outer, central, and inner layers. The existence of these layers was confirmed by numerous investigations employing X-ray diffraction and electron microscopy,[1,29,30] which also revealed structural features in much finer detail. Electron microscopy allows direct observation of various surface-layers that are distinguished by the orientation of microfibrils, especially when the encrusting materials are removed by delignification (see p. 307).

The main, structural features of mature, tracheid cell-walls may be seen in Fig. 2, from a schematic representation by Dadswell and Wardrop.[33] This diagram shows the amorphous intercellular or middle lamella (I), containing mostly lignin; the primary wall (P), with a loose, random network of microfibrils embedded in a matrix that, until recently, was considered to consist of amorphous pectins and hemicellulose molecules lacking structural orientations; the outer layer of the secondary wall (S_1) containing an increased number of fibrils, arranged in a cross-hatch pattern; the middle layer of the secondary wall (S_2), which accounts for a major part of the cell-wall volume and is composed of parallel fibrillar units oriented at a slight angle to the cell axis; and, finally, the tertiary layer of the secondary wall (S_3), which contains parallel fibrillar units forming a flat helix in the transverse direction.

(28) A. B. Wardrop, in "The Formation of Wood in Forest Trees," M. H. Zimmermann, ed., Academic Press, New York, 1964, p. 87.

(29) H. Harada, in Ref. 1, p. 215.

(30) R. D. Preston, "The Molecular Architecture of Plant Cell Walls," Chapman and Hall, London, 1952.

(31) A. Frey-Wyssling and K. Mühlethaler, "Ultrastructural Plant Cytology," Elsevier, Amsterdam, 1965.

(32) T. Kerr and I. W. Bailey, *J. Arnold Arboretum*, **15**, 327 (1934).

(33) H. E. Dadswell and A. B. Wardrop, *Proc. World Forestry Congr. 5th, Seattle, 1960*, **2**, 1279.

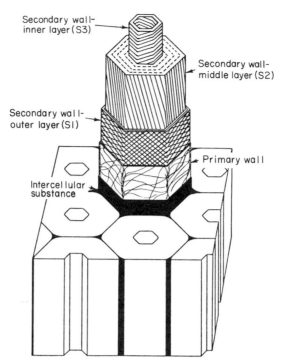

Fig. 2.—Schematic Representation of Cell Walls, According to Dadswell and Wardrop.[33]

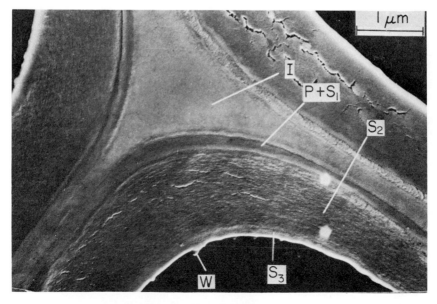

Fig. 3.—Various Layers of the Wall in a Tracheid of *Pinus densiflora*, as Shown in a Micrograph by Harada.[29]

The cells in soft tissues contain only the intercellular layer (cell plate) and the primary wall (cambial layer), which are formed during the periods of cell division and growth. The secondary wall is formed during the period of differentiation and wall thickening (see p. 338).

The thickness of the various layers of an early tracheid (spring wood cell) of *Pinus densiflora* (see Fig. 3) has been measured by electron microscopy[29] and found to be as follows: $P = 0.06$, $S_1 = 0.31$, $S_2 = 1.93$, and $S_3 = 0.17$ μm. The walls of late-wood tracheids (summer-wood cells) are much thicker, especially at the S_2 layer, and the difference in density of the bands that contain these two types of cell creates the annual rings.

The parallel orientation of microfibrils shown in Fig. 4 is greatest in

FIG. 4.—The Replica of a Tracheid of *Picea jezoensis*, Showing the Parallel Orientation of the Microfibrils in S_2 and S_3 Layers.[29]

the S_2 layer and least in the S_3. Furthermore, the S_1, S_2, and S_3 layers are each composed of several lamellae that contain microfibrils of almost the same orientation; however, between the S_1 and S_2, or the S_2 and S_3, there are some intermediate lamellae in which the angle between the orientation of the microfibril and the cell axis changes gradually.[29]

The lamellar structure of the cell walls and the orderly arrangement of the microfibrils reach a high degree of perfection in *Valonia, Chaetomorpha,* and *Cladophora*, which, as already noted, contain a high proportion of unusually pure cellulose. In *Chaetomorpha melagonium* (see Fig. 5) and *Cladophora rupestris*,[11,16,34-36] the wall consists of alternating lamellae that contain parallel-oriented microfibrils

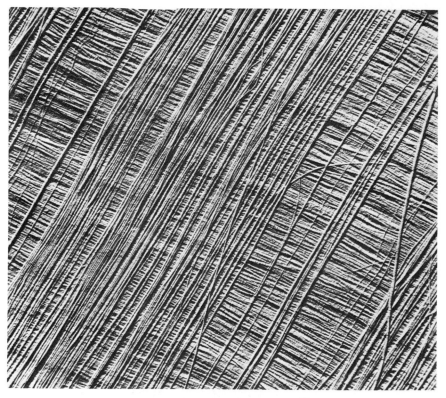

FIG. 5.—Orientation of Microfibrils in the Cell Wall of *Chaetomorpha melangonium*, as Shown in a Micrograph by Frei and Preston.[35]

(34) R. D. Preston, in Ref. 28, p. 169.
(35) E. Frei and R. D. Preston, *Proc. Roy. Soc., Ser. B,* **154,** 70 (1961).
(36) E. Frei and R. D. Preston, *Proc. Roy. Soc., Ser. B,* **155,** 55 (1961).

crossing each other approximately at right angles. One set forms a flat spiral, and the other, a steep spiral; these correspond geometrically to the S_1 and S_2 layers of tracheids. In *Chaetomorpha princeps* and *Cladophora prolifera*, a third type of lamella is present in which the microfibrils have a diagonal orientation.

As will be discussed later (see p. 332), these data provide strong support for the argument that the microfibrils are produced by "on-the-site" synthesis and orientation (apposition) of the cellulosic microfibrils under the guiding influence of the living cell, rather than by a mechanism that proposes synthesis of the microfibrils within the cell and subsequent translocation and crystallization (deposition) of microfibrils on the cell wall by exocellular factors. Further factors relevant to these opposing theories emerge from study of the fine structure of the cellulosic microfibrils, as discussed in the following Section.

III. MICROFIBRILS

1. Original Concepts

The combination of rigidity and stress resistance that is required in higher plants is provided by the structure of the various layers and lamellae of the cell wall that contain crystalline microfibrils, matrix polysaccharides, and amorphous lignin. According to an early interpretation by Freudenberg,[37] the microfibrils in the cell wall act in the same way as the reinforcing rods in prestressed concrete. A better mathematical model is furnished by filament-wound, reinforced-plastic structures,[38] such as pressure vessels.

When the cell-wall materials are treated with dilute acid, alkali, or other suitable chemical compounds in order to remove the amorphous materials, the cellulosic fibrillar structure is exposed and may be examined by electron microscopy.[39–41] Microfibrils may be described as thin threads, the width of which has been variously estimated, according to the materials and methods used. Preston, working with algae, reported that microfibrils vary in width from 8.0 to 30.0 nm and are approximately half as thick as they are wide.[1,11,42] Hodge and

(37) K. Freudenberg, *J. Chem. Educ.*, 9, 1171 (1932).
(38) R. Mark, in Ref. 1, p. 493.
(39) R. D. Preston, E. Nicolai, R. Reed, and A. Millard, *Nature*, 162, 665 (1948).
(40) A. Frey-Wyssling, K. Mühlethaler, and R. W. G. Wyckoff, *Experientia*, 4, 475 (1948).
(41) K. Mühlethaler, *Biochim. Biophys. Acta*, 3, 15 (1949).
(42) R. D. Preston, *Discussions Faraday Soc.*, 11, 165 (1951).

Wardrop[43] found a range of 5.0 to 10.0 nm for wood, and Vogel, [44] of 17.0 to 20.0 nm for ramie microfibrils.

A long series of chemical, physical, and microscopic investigations has provided greater understanding of the composition and structure of the microfibrils. The chemical constitution of cellulose isolated from plant cell-walls has been established as a linear polymer of β-D-$(1 \rightarrow 4)$-linked D-glucopyranose.[45,46] Furthermore, the cellulosic composition of microfibrils within the cell wall has been shown by its characteristic diffraction diagram.[1,30,47] It should be noted that the material left behind after removal of amorphous substances from cell walls is not necessarily the "pure" cellulose of organic chemists; because, on hydrolysis, this material may provide various proportions of D-xylose and other sugars besides D-glucose.[48] These sugars are regarded as an integral part of the microfibrils rather than as residual hemicelluloses from the matrix. Cellulose of high purity is, however, found in the microfibrils of the group of algae represented by *Valonia* (see p. 300). Preston has suggested the name "eucellulose" for this substance. A different terminology is used by technologists, who consider the pure β-D-$(1 \rightarrow 4)$-linked glucan to be an asymptotic entity that represents the alkali-insoluble portion of the pulp isolated from the cell walls; they are accustomed to referring to this entity as α-cellulose. The terms "cellulosic materials"[49] and "cellulosic microfibrils" are used here to describe the natural product having various degrees of purity.

Earlier concepts about the fine structure of microfibrils have been reviewed by several authors[30,50-52] and will be only briefly mentioned here as background for the description of later developments. In 1858, Nägeli[1,50,53] proposed the micellar theory; this assumed the presence of separate, brick-like, crystalline micelles within the cellulose fibers (see Fig. 6) to explain the birefringence observed with a polarized

(43) A. J. Hodge and A. B. Wardrop, *Nature*, **165**, 272 (1950).
(44) A. Vogel, *Makromol. Chem.*, **11**, 111 (1953).
(45) W. N. Haworth, E. L. Hirst, and H. A. Thomas, *J. Chem. Soc.*, 824 (1931).
(46) C. B. Purves, in "Cellulose and Cellulose Derivatives," E. Ott and H. M. Spurlin, eds., Interscience Publishers, New York, 1954, Part 1, p. 29.
(47) R. D. Preston and G. W. Ripley, *Nature*, **174**, 76 (1954).
(48) J. Cronshaw, A. Myers, and R. D. Preston, *Biochim. Biophys. Acta*, **27**, 89 (1958).
(49) F. Shafizadeh, *Advan. Carbohyd. Chem.*, **23**, 419 (1968).
(50) J. A. Howsmon and W. A. Sisson, in Ref. 46, p. 231.
(51) A. Frey-Wyssling, "Submicroscopic Morphology of Protoplasm and Its Derivatives," Elsevier, New York, 1948, p. 45.
(52) P. H. Hermans, *J. Phys. Chem.*, **45**, 827 (1941).
(53) C. Nägeli, "Die Stärkekörner," F. Schulthess, Zürich, 1858.

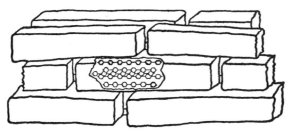

FIG. 6.—The Brick-like Structure of Cellulose Fibers, According to the Original Micellar Theory.[53]

microscope. Many years elapsed before the crystalline structure of cellulose was proved by the X-ray diffraction method.[54-61] The ensuing investigations showed that, with the possible exception of cellulose in the *Halicystis* plant, the cellulosic materials from all sources, including bacteria and animals, have the same crystalline structure, called[54,62] cellulose I. These studies culminated in the description by Meyer and Misch[59,60] of the crystallographic unit-cell for cellulose I, with antiparallel orientation of the neighboring molecular chains.

When the presence of a crystalline structure in the cellulose was confirmed, the micellar theory was revived, and expanded[57,58] to accomodate new observations, including the diffuse, X-ray diagram of cellulose, having broad diffraction lines. Broadening of the diffraction lines is caused by discontinuity resulting from small crystal size or other factors[1,50] (see p. 315). On the assumption that the limited size of the crystallites was the main factor, the breadth of the diffraction lines from the planes parallel and perpendicular to the fiber axis was used for estimating the dimensions of the crystalline micelles; these indicated that the micelles in ramie microfibrils are approximately 5.0–10.0 nm in diameter and 50.0–60.0 nm in length.[63,64] The X-ray scattering, resulting in a diffused diagram, was attributed to the presence of amorphous materials. It was proposed that these materials

(54) R. O. Herzog and W. Jancke, Z. Physik., 3, 196 (1920).
(55) M. Polanyi, Naturwissenschaften, 9, 288 (1921).
(56) O. L. Sponsler, J. Gen. Physiol., 9, 677 (1926).
(57) K. H. Meyer and H. Mark, Ber., 61, 593 (1928).
(58) K. H. Meyer and H. Mark, Z. Physik. Chem., B2, 115 (1929).
(59) K. H. Meyer, Ber., 70, 266 (1937).
(60) K. H. Meyer and L. Misch, Helv. Chim. Acta, 20, 232 (1937).
(61) H. Mark, Chem. Rev., 26, 169 (1940).
(62) W. A. Sisson, Science, 87, 350 (1938).
(63) R. O. Herzog, J. Phys. Chem., 30, 457 (1926).
(64) J. Hengstenberg and H. Mark, Z. Kristallogr., 69, 271 (1928).

act as the cementing compound between the micelles and are responsible for the swelling of the cellulose fibers.[65]

The micellar theory, which assumed that straight chains of cellulose
molecules, 60.0 nm in length, form the crystalline particles (see Fig.
6), was confirmed by determining the degree of polymerization (d.p.)
of the polysaccharide. End-group analysis[66-68] at that time indicated
for different types of cellulose a d.p. of 100–200 that was in close
agreement with a minimum value of about 120 estimated from the
X-ray data. However, it subsequently became quite clear that these
measurements represented low values, and, with further refinement
of end-group analysis and the development of newer physical
methods, the values obtained for the molecular weight or the chain
length of cellulose were substantially higher.[30] Viscosity measurements by Staudinger and Mohr[69] indicated for native cellulose a d.p.
of about 3,000, which corresponds with the chain length of 1.5 μm.
Shortly after that, Gralén and Svedberg[70] obtained, by ultracentrifuga-
tion, d.p. values of about 10,000 for native cellulose; this is close to
the range of values since obtained by other methods. It then became
obvious that, in its early form, the micellar theory was no longer
tenable.

At the same time, a strong argument was presented for the existence
of a continuous structure in which long molecules are arranged parallel, with some interspersed discontinuities[71-74] (see Fig. 7a). However, the presence of amorphous and crystalline regions in cellulose
was supported by the observation that the X-ray diagram of cellulose
remains the same after swelling of the cellulose in water, and it was
argued that this lack of change is because water enters the amorphous
structure between the crystalline micelles.[75-77] These conflicting
ideas were reconciled in the fringed-micellar theory[51-52] postulated by
Kratky[78] and Frey-Wyssling.[79] According to this theory, the micro-

(65) R. O. Herzog, *Kolloid Z.*, **39**, 98 (1926).
(66) M. Bergmann and H. Machemer, *Ber.*, **63**, 316, 2304 (1930).
(67) W. N. Haworth and H. Machemer, *J. Chem. Soc.*, 2270, 2372 (1932).
(68) A. M. Sookne and M. Harris, in Ref. 46, p. 197.
(69) H. Staudinger and R. Mohr, *Ber.*, **70**, 2296 (1937).
(70) N. Gralén and T. Svedberg, *Nature*, **152**, 625 (1943).
(71) F. T. Pierce, *Trans. Faraday Soc.*, **29**, 50 (1933).
(72) S. M. Neale, *Trans. Faraday Soc.*, **29**, 228 (1933).
(73) O. L. Sponsler, *Quart. Rev. Biol.*, **8**, 1 (1933).
(74) W. T. Astbury, *Trans. Faraday Soc.*, **29**, 193 (1933).
(75) J. R. Katz, *Ergeb. Exakt. Naturw.*, **3**, 365 (1923).
(76) J. R. Katz, *Trans. Faraday Soc.*, **29**, 279 (1933).
(77) K. Mühlethaler, in Ref. 1, p. 191.
(78) O. Kratky, *Kolloid Z.*, **70**, 14 (1935).
(79) A. Frey-Wyssling, *Protoplasma*, **25**, 261 (1936).

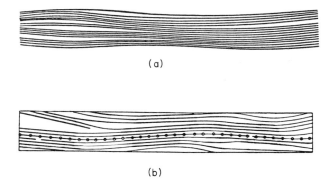

(a)

(b)

FIG. 7.—Arrangement of Cellulose Molecules in: (a) Continuous and (b) Fringed-Micellar Theories.[50]

fibrils shown in Fig. 7b are composed of statistically distributed, crystalline and amorphous regions formed by the transition of the cellulose chain from an orderly arrangement in the direction of the microfibrils in the crystalline regions to a less orderly orientation in the amorphous area. The presence of crystalline and amorphous (or paracrystalline) regions in cellulose has been investigated by a number of physical and chemical methods.[1,30,50,80] The density of cellulose was found to be lower than the theoretical density calculated from the unit-cell, crystallographic data;[30,81,82] this result showed the existence of regions less densely packed, or amorphous. Density measurements were also used for estimating the relative crystallinity by comparing the data with the values expected from crystalline and amorphous materials. Assuming that crystalline regions provide a discrete X-ray diffraction pattern, and that the amorphous parts contribute to the background scattering, an estimate of about 70% relative crystallinity was obtained for native cellulose.[82–85] Chemical methods, such as deuterium exchange,[85,86] periodate oxidation, hydrolysis[87–90] (see p. 312), and sub-

(80) S. A. Rydholm, "Pulping Processes," Interscience Publishers, New York, 1965, p. 109.
(81) F. C. Brenner, V. Frilette, and H. Mark, J. Amer. Chem. Soc., 70, 877 (1948).
(82) P. H. Hermans, "Physics and Chemistry of Cellulose Fibers," Elsevier, Amsterdam, 1949.
(83) P. H. Hermans and A. J. Weidinger, J. Polym. Sci., 4, 317 (1949).
(84) P. H. Hermans, Makromol. Chem., 6, 25 (1951).
(85) J. Mann, Methods Carbohyd. Chem., 3, 114 (1963).
(86) V. J. Frilette, J. Hanle, and H. Mark, J. Amer. Chem. Soc., 70, 1107 (1948).
(87) C. C. Conrad and A. G. Scroggie, Ind. Eng. Chem., 37, 592 (1945).
(88) E. L. Lovell and O. Goldshmid, Ind. Eng. Chem., 38, 811 (1946).
(89) R. F. Nickerson, Ind. Eng. Chem., 34, 1480 (1942).
(90) F. T. Ratliff, Tappi, 32, 357 (1949).

stitution[91] reactions, have also been used; these give crystallinity values ranging from 55 to 95%.

These data often provide a measure of the unavailability or inaccessibility of certain groups, rather than of the crystallinity of the polymer. This inaccessibility includes the hydroxyl groups that are hydrogen bonded in the uniform structure of the crystallites. However, the molecules covering the surface area of the crystalline regions and the hydroxyl groups projecting from them behave as the amorphous regions.[85] As the surface molecules account for a substantial proportion of the cellulose in the fine microfibrils (see p. 315), the validity of these data as proof for the existence of crystalline and amorphous regions is questionable.

2. New Models

The fundamental concept of molecular chains of cellulose extending along the direction of microfibrils, to form crystalline and amorphous or paracrystalline regions, was well accepted and set the trend for interpretation of all relevant data. However, folded-chain conformations have since been advocated as a strong possibility. Also, several other models based on the extended-chain concept have been proposed in order to define better the distribution of crystallites and the disposition of the molecular chain within the microfibrils.

The rate of heterogeneous hydrolysis of cellulose with acid has been used for investigating crystallinity;[87-90] this procedure involves the observation of an initial, high rate of reaction that indicates attack in the amorphous or more accessible component, and then a lower, more constant rate, that represents the hydrolysis of the crystalline material.[89] After extensive heterogeneous hydrolysis and ultrasonic dispersion, the resulting fractions, assumed to be the remaining crystalline micelles, were observed by electron microscopy as small rodlets, about 50.0–60.0 nm long and 5.0–10.0 nm wide.[92-94] These dimensions are in general agreement with the X-ray data and with values obtained by other methods. They were, therefore, accepted as meaningful values, although another study indicated that the length may range from 120.0 to 330.0 nm for the crystallites in different, natural, cellulosic fibers.[95] Furthermore, it was suggested that the micelles

(91) H. Tarkow and A. J. Stamm, *J. Phys. Chem.*, **56**, 262, 266 (1952).
(92) B. G. Rånby and E. Ribi, *Experientia*, **6**, 12 (1950).
(93) B. G. Rånby and B. Grinberg, *Compt. Rend.*, **230**, 1402 (1950).
(94) B. G. Rånby, *Discussions Faraday Soc.*, **11**, 158 (1951).
(95) F. F. Morehead, *Text. Res. J.*, **20**, 549 (1950).

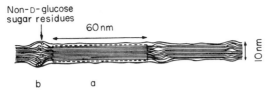

FIG. 8.—Cellulose Microfibril in which Crystallites (*a*) are Linked by Paracrystalline regions, or (*b*) Contain Non-D-glucose Sugar Residues.[93]

are linked together, along the microfibril, by paracrystalline-cellulose chains containing residues of uronic acid and other sugars, as well as D-glucose[96,97] (see Fig. 8).

Another model that differs from the fringed micellar concept assumes a regular arrangement of the crystalline and paracrystalline segments, rather than a statistical distribution. This model, proposed by Hess and coworkers,[98] is based on an electron-microscope examination that revealed a lateral order in cellulose after deposition of iodine or thallium in the disordered region. As shown in Fig. 9, the alternating layers of crystalline and disordered regions are formed by the alignment of the neighboring microfibrils.

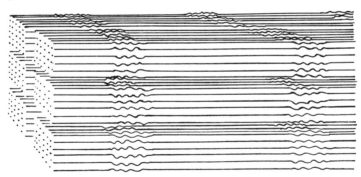

FIG. 9.—Lateral Order of Crystalline and Paracrystalline Regions of Neighboring Microfibrils.[98]

In one of the studies by electron microscopy, it was found that, on disintegration in a blender, ultrathin sections of ramie fibers produce flat filaments or elementary fibrils, having the dimensions 3.0 × 10.0 nm, that aggregate laterally. This led to the Frey-Wyssling model,[99] shown in Fig. 10, which presents the cross section of a microfibril

(96) B. G. Rånby, in "Fundamentals of Papermaking Fibers," F. Bolam, ed., Tech. Sect. Brit. Paper & Board Makers' Assoc., Kenley, Surrey, England, 1958, p. 55.

(97) G. P. Berlyn, *Forest Prods. J.*, **14**, 467 (1964).

(98) K. Hess, H. Mahl, and E. Gütter, *Kolloid Z.*, **155**, 1 (1957).

(99) A. Frey-Wyssling, *Science*, **119**, 80 (1954).

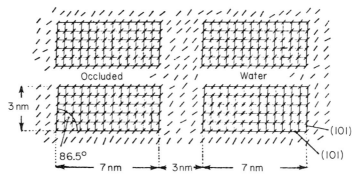

Fig.10.—Cross Section of a Microfibril Composed of Elementary Fibrils, According to Frey-Wyssling.[99]

composed of elementary fibrils having a core of crystalline cellulose chains embedded in a cortex of paracrystalline cellulose. In this model, the plane of the ribbon corresponds with the 101 crystallographic plane of the cellulose lattice.

Another model for microfibrils[1,100] is shown in Fig. 11. According to this model, proposed by Preston, microfibrils contain a single core of

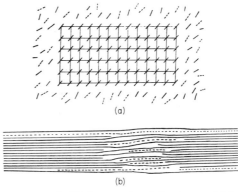

Fig. 11.—(a) Transverse and (b) Longitudinal Sections of Microfibrils in which Molecular Chains Containing Non-D-glucose Residues Surround Cores of Pure, Crystalline Cellulose.[48]

crystalline, pure cellulose ("eucellulose") surrounded by a paracrystalline cortex containing sugar residues other than those of D-glucose. The model is based on the observation that removal of non-cellulosic material from the cell wall leaves crystalline cellulosic microfibrils that, on hydrolysis, normally yield D-glucose as well as other sugars.[48] However, the rodlets that arise from the crystalline core of these

(100) R. D. Preston, in "The Interpretation of Ultrastructure," R. J. C. Harris, ed., Academic Press, New York, 1962, p. 325.

microfibrils provide only D-glucose on hydrolysis. It is further noted that a comparison in size between the crystallites and the microfibrils seems to indicate that the microfibrils are not aggregates of smaller units.

The molecules at the surface may contribute to the accessibility and other properties of the cellulosic microfibrils; such properties had previously been attributed to the amorphous materials. These molecules, which represent structural discontinuity, cause "small-particle scattering," thus increasing the width of X-ray deflections and decreasing the resolution. Both of these simulate, or give the illusion of, the presence of amorphous materials.[101–103] Thus, it is obvious that the width of the microfibrils is a significant factor in estimating the extent of crystallinity. For a microfibril that is 6.0 nm in width, it has been calculated that 37% of the macromolecules are on the surface.[101] Bonart and coworkers[104] have also reported that a considerable portion of the X-ray scattering of cellulose is due to lattice distortion rather than to the amorphous region; consequently, cellulose must be more crystalline than had generally been supposed. This conclusion conforms with data, obtained at the U. S. Forest Products Laboratory,[105] indicating that amorphous polysaccharides undergo rapid, initial hydrolysis to a small extent (10%).

Mühlethaler[10,77,106] investigated the existence of amorphous regions in ramie fibers and other cellulosic materials by means of electron microscopy, using the negative-staining technique.[107] In this method, the substrate is treated with phosphotungstic acid; this penetrates the capillary spaces and loose textures, and gives them a dark contrast on electron-microscope examination. This is a much better method than metal shadowing for studying the internal structure of microfibrils. The results indicated that microfibrils are composed of iso-diametric, elementary fibrils having an average width of 3.5 nm. The elementary fibrils seemed to be crystalline along their entire length, as there was no sign of penetration of the phosphotungstic acid. Were crystalline and paracrystalline sections present, there should have been alternating dark and light segments. This observation also explained the vari-

(101) R. D. Preston, *Polymer*, **3**, 511 (1962).
(102) P. S. Rudman, in "The Encyclopedia of X-Rays and Gamma Rays," G. L. Clark, ed., 1963, p. 551.
(103) N. B. Patil, N. E. Oweltz, and T. Radhakrishnan, *Text. Res. J.*, **32**, (6) 460 (1962).
(104) R. Bonart, R. Hosemann, F. Motzkus, and H. Ruck, *Norelco Reptr.*, **7**, 81 (1960).
(105) M. A. Millett, W. E. Moore, and J. F. Saeman, *Ind. Eng. Chem.*, **46**, 1493 (1954).
(106) K. Mühlethaler, Z. *Schweiz. Forstv.*, **30**, 55 (1960).
(107) S. Brenner and R. W. Horne, *Biochim. Biophys. Acta*, **34**, 103 (1959).

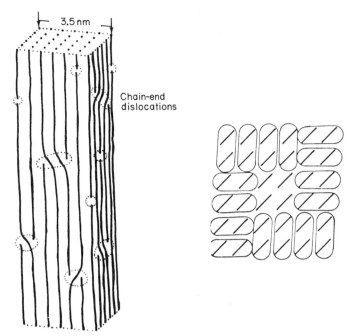

3.5 nm

Chain-end
dislocations

FIG. 12.—The Structural Model of Elementary Fibrils, as Proposed by Mühle-thaler.[10,77]

ation in the width of the microfibrils noted by previous in-vestigators,[44,96,108] because two or more elementary fibrils could fasciculate to form wider threads or microfibrils. The cross section of the elementary fibrils (3.5 × 3.5 nm) should contain about 36 cellulose chains. Assuming that these chains are antiparallel, as in-dicated by the Meyer–Misch unit cell, Mühlethaler proposed the model shown in Fig. 12 to accommodate the new data. Notably, in this model, 16 of the 18 pairs of antiparallel chains are at the surface. Furthermore, the only disorder or disturbance in this structure is caused by chain-end dislocation, which, according to work with syn-thetic polymers, can cause as much scattering of X-rays as can the en-tangled or amorphous regions.[10]

Elementary fibrils have also been observed by Manley,[109,110] who called them protofibrils, Ohad and Danon,[111] and other investigators (see p. 319).

(108) I. Günther, *J. Ultrastructure Res.*, **4**, 304 (1960).
(109) R. S. J. Manley, *Nature*, **204**, 1155 (1964).
(110) R. S. J. Manley, "Pioneering, Research Program, Summary Report," Institute of Paper Chemistry, 1965, p. 88.
(111) I. Ohad and D. Danon, *J. Cell. Biol.*, **22**, 302 (1964).

In Manley's experiments, ultrasonic disintegration and negative staining of ramie and other fibers not only produced the protofibrils of 3.5-nm diameter, but also a periodic structure resembling a string of pearls. To account for this beaded appearance, or periodic variation in the structure, Manley abandoned the extended-chain concept and proposed for the protofibrils an elegant model based on the folded-chain concept. The folded-chain structure, found in several synthetic and natural crystalline polymers, has been discussed by Marchessault and Sarko.[16,112] A folded-chain model for cellulose microfibrils has also been proposed by Dolmetsch and Dolmetsch.[113] According to Manley, the molecular chain of cellulose forms a ribbon 3.5 nm wide by folding in concertina fashion. With the ribbon wound as a tight helix, the straight segments of the molecular chain become parallel to the helical, fibril axis, as shown in Fig. 13a. The protofibrils are arranged within the cross sections of native cellulose fibers in an array that permits a distance between microfibril centers of 4.0 nm. At this proximity, the microfibril surfaces are held laterally by hydrogen bonds, and so the microfibrils are considered to be quasi single-λ crystals that do not contain amorphous regions. The diffuseness of the X-ray diffraction pattern could be accounted for by various types of lattice imperfection.

One of the notable features of the folded-chain conformation is that it readily explains the antiparallel arrangement of the cellulose molecules within the crystalline structure of the microfibrils. This type of arrangement has also been suggested by Asunmaa,[114] Marx-Figini and Schulz[115] (see Fig. 13b), and Bittiger and coworkers.[116] The last group studied crystallization, or the transition from coils to fibrillar solid structures, with some suitable polymers, including mannan and cellulose derivatives. Precipitation of these polymers from solution produced microfibrils of 4.0 to 20.0 nm diameter. Based on this investiga-

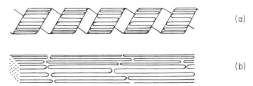

(a)

(b)

FIG. 13.—Folded-chain Structure of Protofibrils: (a) According to Manley[109] and (b) According to Marx-Figini and Schulz.[115]

(112) A. Sarko and R. H. Marchessault, in Ref. 10, p. 317.
(113) H. Dolmetsch and H. Dolmetsch, *Kolloih Z.*, **185**, 106 (1962).
(114) S. K. Asunmaa, *Tappi*, **49**, 7, 319 (1966).
(115) M. Marx-Figini and G. V. Schulz, *Biochim. Biophys. Acta*, **112**, 81 (1966).
(116) H. Bittiger, E. Husemann, and A. Kuppel, in Ref. 10, p. 45.

tion, Bittiger's group proposed a variation of Manley's model that has a folded length of 15.0 to 20.0 nm, instead of 3.5 nm. A significant conclusion inferred from this work is that uniform fibrils can be formed by physical factors alone, without operation of any biological processes. The folded-chain structure also eliminates the need to explain how the antiparallel chains are synthesized or arranged in opposite directions.

In order to decide whether cellulose molecules have an extended or a folded-chain conformation in the native cell-wall, Muggli and co-workers[117,118] (see also, Ref. 10) devised an interesting experiment based on determination of the molecular-weight distribution in ramie microfibrils before and after they had been cut into 2-μm sections. As shown in Fig. 14, should the molecules have an extended conformation, many of the molecular chains would be cut in a random fashion (Gaussian distribution), resulting in substantial diminution in the molecular weight. On the other hand, in the folded conformation, a few macromolecules would be cut, but the rest would retain their

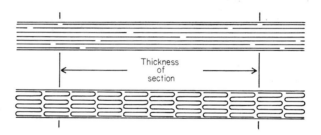

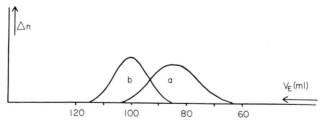

FIG. 14.—Sectioning of the Cellulose Fibers Having an Extended or Folded-chain Conformation and the Resulting Change in Molecular-weight Distribution Determined by Gel-permeation Chromatography. [(a) Before and (b) after sectioning.[10,117]]

(117) R. Muggli, J. Cellulose Chem. Technol., 2, 549 (1968).
(118) R. Muggli, H. G. Elias, and K. Mühlethaler, Makromol. Chem., 121, 290 (1969).

original molecular weight. The experiment showed that the observed degree of polymerization is lowered from 3,900 to 1,600 by the sectioning; this result provides strong evidence that the conformation of cellulose molecules is extended, as originally proposed in the fringed micellar theory that eventually developed into the Mühlethaler model for elementary fibrils.[106,119]

Despite all of these studies and developments, the diameter of the elementary fibrils is still a debatable subject. It has even been suggested that the microfibril is an artifact produced by electron microscopy and that its ultimate diameter is the width of the molecule, as has been shown to be true for extended-chain, polyethylene crystals. There is a good likelihood that cellulose molecules may aggregate in 1.0–6.0-nm fibrils to lessen the surface area.[112] According to Sullivan,[120] wood cellulose contains innumerable structures that are less than 5.0 nm in diameter. These structures, called protofibrils after Manley, are not the substructure of microfibrils or of any larger organization. Brown and coworkers,[121] who investigated scale formation in a *Chrysophyta (Pleurochrysis scherffelii)*, also reported variable dimensions, from 1.2 to 2.5 × 2.5 to 4.0 nm for the ribbon-like, elementary cellulosic fibrils observed in this organism.

IV. BIOSYNTHESIS OF CELLULOSE

1. Biological Systems

The preceding, brief discussion about the complex structure of the cell walls and the fine features of the cellulosic microfibrils clearly indicates that the formation and functioning of this system must involve not only the chemical synthesis of the raw materials, but also other fundamental processes, such as the packing of the cellulose molecules in the microfibrils and the orientation of the microfibrils within the cell walls. These complex processes are so closely related in biological processes that a discussion of one aspect without consideration of others would be rather meaningless. However, for the sake of convenience and simplicity, the biosynthesis of cellulose, or the process of polymerization of the D-glucose residues, will be considered first, irrespective of the physical structure of the final product.

Early experiments on the biosynthesis of cellulose mainly involved

(119) A. Frey-Wyssling and K. Mühlethaler, *Makromol. Chem.*, **62**, 25 (1963).
(120) J. D. Sullivan, *Tappi*, **51**, 501 (1968).
(121) R. M. Brown, Jr., W. W. Franke, H. Kleinig, H. Falk, and P. Sitte, *J. Cell Biol.*, **45**, 246 (1970).

the introduction of D-glucose specifically labeled with ^{14}C and of other metabolites into such biological systems as cultures of *Acetobacter xylinum* and *A. acetigenum*,[122–125] maturing cotton-bolls,[126] and growing wheat-seedlings.[127,128] The resulting products containing radioactive cellulose were subsequently isolated, and investigated by radiochemical methods involving determination of distribution of ^{14}C in the D-glucose residues. These investigations generally indicated that the labeled sugar had been incorporated into the metabolic pool, instead of being directly polymerized to cellulose. In several experiments, Shafizadeh, Wolfrom, and Webber[129–133] introduced D-glucose-*1-^{14}C*, -2-^{14}C, and -6-^{14}C into maturing cotton-bolls at the period of rapid formation of cellulose, by use of the method developed by Greathouse.[126] These experiments provided different samples of ^{14}C-labeled cellulose in radiochemical yields of up to 23.5%. Distribution of the ^{14}C-label within the D-glucose residues of these products was determined by use of the sequence of reactions shown in Fig. 15, and by other methods. The resulting data indicated that about 60% of the ^{14}C-label remained in the original position, and that a major part of the balance, ~22%, had migrated to the corresponding position (C-6) at the opposite end of the D-glucose residues. The distribution pattern, which has also been observed with growing wheat-seedlings,[127,128] is consistent with the assumption that labeled D-glucose has been broken down and resynthesized through the Embden–Meyerhof pathway. In this pathway, D-glucose is converted into D-fructose 1,6-diphosphate, which splits into two C_3 fragments, namely, 1,3-dihydroxy-2-propanone 1-phosphate and D-glyceraldehyde 3-phosphate. Reversible isomerization of these compounds by an

(122) F. W. Minor, G. A. Greathouse, H. G. Shirk, A. M. Schwartz, and M. Harris, *J. Amer. Chem. Soc.*, **76**, 1658, 5052 (1954); G. A. Greathouse, H. G. Shirk, and F. W. Minor, *ibid.*, **76**, 5157 (1954); F. W. Minor, G. A. Greathouse, and H. G. Shirk, *ibid.*, **77**, 1244 (1955).
(123) E. J. Bourne and H. Weigel, *Chem. Ind.* (London), 132 (1954).
(124) M. Stacey and S. A. Barker, "Polysaccharides of Micro-Organisms," Clarendon, Oxford, 1960, p. 77.
(125) K. S. Barclay, E. J. Bourne, M. Stacey, and M. Webb, *J. Chem. Soc.*, 1501 (1954).
(126) G. A. Greathouse, *Science*, **117**, 553 (1953).
(127) S. A. Brown and A. C. Neish, *Can. J. Biochem. Physiol.*, **32**, 170 (1954).
(128) J. Edelman, V. Ginsburg, and W. Z. Hassid, *J. Biol. Chem.*, **213**, 843 (1955).
(129) F. Shafizadeh and M. L. Wolfrom, *J. Amer. Chem. Soc.*, **77**, 5182 (1955).
(130) F. Shafizadeh and M. L. Wolfrom, *J. Amer. Chem. Soc.*, **78**, 2498 (1956).
(131) M. L. Wolfrom, J. M. Webber, and F. Shafizadeh, *J. Amer. Chem. Soc.*, **81**, 1217 (1959).
(132) F. Shafizadeh and M. L. Wolfrom, in Ref. 85, p. 375.
(133) F. Shafizadeh and M. L. Wolfrom, in Ref. 85, p. 377.

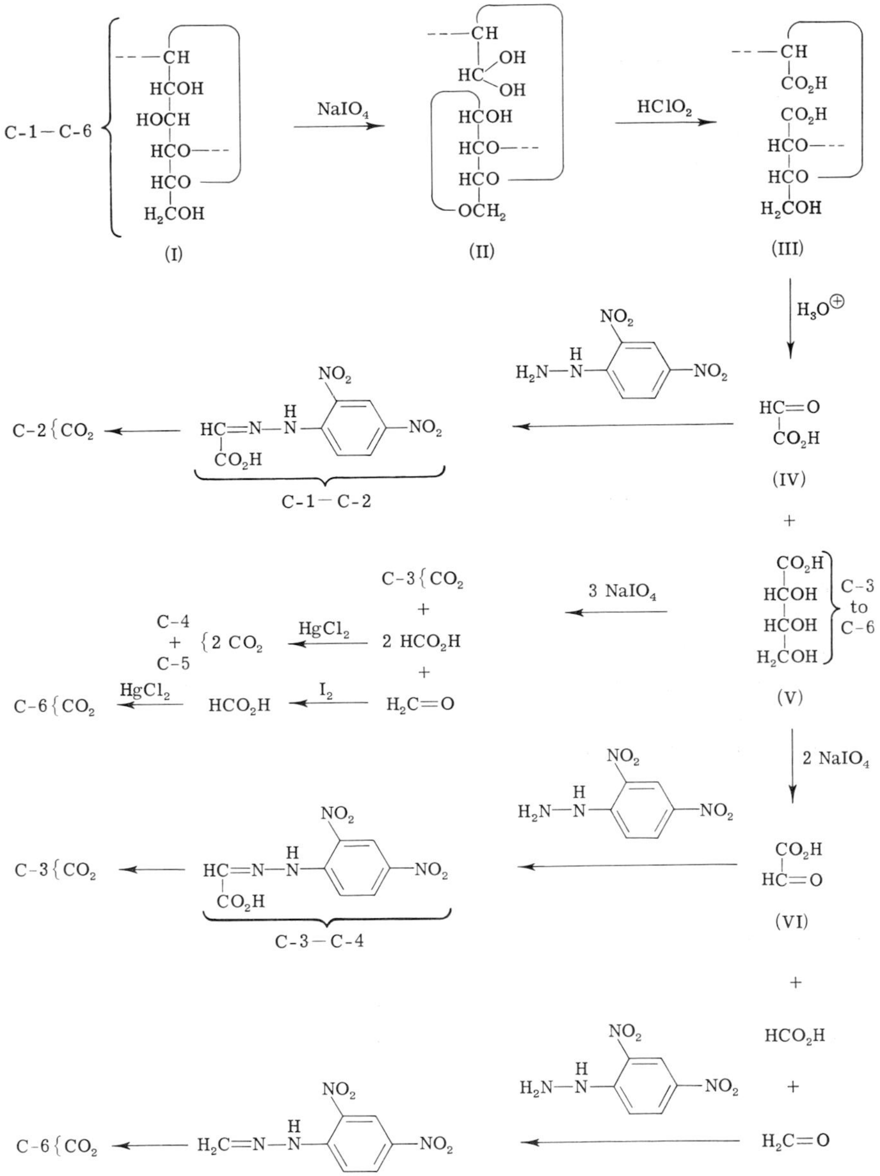

FIG. 15.—Determination of Distribution of ^{14}C-label in Radioactive Cellulose.[131]

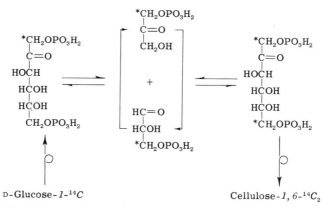

Fig. 16.—Breakdown and Resynthesis of D-Glucose Residues Forming the Cellulose Molecule.[131]

enzyme, triose isomerase, results in the distribution of the label as shown in Fig. 16. Further evidence for the involvement of the C_3 fragments was provided by the fact that some of the specifically labeled D-glucose residues in the plant had been converted into a glycerol component of the seed oil,[130,131] apparently through the "triose" phosphate intermediates.

Because *A. xylinum* and *A. acetigenum* produce bacterial cellulose as an exocellular polysaccharide,[134,135] these organisms have been extensively investigated in relation to the *in vitro* synthesis of cellulose. The methods employed generally involved disruption of the cells in such a way as to leave the cellulose-synthesizing capacity of the system still operational. Hestrin and coworkers[136] reported that D-glucose, D-gluconate, 1,3-dihydroxy-2-propanone, and glycerol in the presence of oxygen are rapidly converted into cellulose by a dry cell-preparation of *A. xylinum*.

Greathouse[137] reported the isolation of a cell-free enzyme-system by homogenizing cultures of *A. xylinum;* this system incorporated D-glucose-1-^{14}C into labeled cellulose, with 96% of the label remaining at C-1 of the D-glucose residues. He also made the significant observation that adenosine 5'-triphosphate (ATP) should be added to the enzyme system in order to provide the energy required for polymerization of the sugar, and that practically no cellulose is produced without

(134) K. Mühlethaler, *Biochim. Biophys. Acta*, **3**, 527 (1949).
(135) I. Ohad, D. Danon and S. Hestrin, *J. Cell Biol.*, **12**, 31 (1962).
(136) M. Schramm, Z. Gromet, and S. Hestrin, *Bull. Res. Council Israel*, **5A**, No. 1, 99 (1955).
(137) G. A. Greathouse, *J. Amer. Chem. Soc.*, **79**, 4503 (1957).

addition of this compound. Colvin[138] also found that ATP stimulates the synthesis of cellulose by the cell homogenate. Experiments with cotton hairs have since shown that uridine 5'-(α-D-glucopyranosyl pyrophosphate) (UDP-D-glucose) serves as an excellent substrate for the synthesis of cellulose, whereas D-glucose is incorporated only to a lesser extent.[139]

2. Nucleotide Precursors

The mechanism of the biosynthesis of polysaccharides has been reviewed in authoritative articles by Hassid and coworkers[14,139a] in this Volume and an earlier Volume of this Series. The present discussion will include only the biosynthesis of cellulose, because of its significance in the composition of plant cell-walls and because of the complexities that have been revealed by recent investigations.

It had been generally known that sugar phosphates are involved in the enzymic formation and cleavage of the glycosidic links in oligo- and poly-saccharides.[3] For instance, sucrose could be synthesized from α-D-glucosyl phosphate and D-fructose in the presence of a bacterial enzyme obtained from *Pseudomonas saccharophila*.[140] It was recognized that the phosphate group "activates" the sugar and provides the energy for the transformation, but the general mechanism involved in the synthesis of sucrose was not clear. Leloir and coworkers later found that the donor of D-glucose for formation of sucrose is the "sugar nucleotide" uridine 5'-(α-D-glucopyranosyl pyrophosphate), rather than α-D-glucosyl phosphate.[141-143] Since then, it has been well established that the oligo- and poly-saccharides are formed from glycosyl esters of nucleoside pyrophosphates by the process of transglycosylation.[3,144,145] Biosynthesis of cellulose through this process was shown by Glaser,[146] who obtained from *Acetobacter xylinum* a cell-free, particulate-enzyme preparation that catalyzed incorporation of the D-glucopyranosyl group from "UDP-D-glucose" into cellulose. As noted before, there was also some evidence for the

(138) J. R. Colvin, *Arch. Biochem. Biophys.*, **70**, 294 (1957).
(139) G. Franz and H. Meier, *Phytochemistry*, **8**, 579 (1969).
(139a) See H. Nikaido and W. Z. Hassid, *This Volume*, p. 351.
(140) W. Z. Hassid, M. Doudoroff, and H. A. Barker, *J. Amer. Chem. Soc.*, **66**, 1416 (1944); W. Z. Hassid and M. Doudoroff, *Advan. Enzymol.*, **10**, 123 (1950).
(141) C. E. Cardini, L. F. Leloir, and J. Chiriboga, *J. Biol. Chem.*, **214**, 149 (1955).
(142) L. F. Leloir and C. E. Cardini, *J. Biol. Chem.*, **214**, 157 (1955).
(143) E. W. Putman and W. Z. Hassid, *J. Biol. Chem.*, **207**, 885 (1954).
(144) L. F. Leloir, *Biochem. J.*, **91**, 1 (1964).
(145) W. Z. Hassid, *Ann. Rev. Plant Physiol.*, **18**, 253 (1967).
(146) L. Glaser, *J. Biol. Chem.*, **232**, 627 (1958).

involvement of ATP in the synthesis of cellulose by similar systems.[137,138]

Later, Hassid and coworkers showed that a particulate, enzyme system prepared from roots of mung bean (*Phaseolus aureus*) seedlings[147,148] and maturing cotton-bolls[149] incorporates ^{14}C-labeled D-glucose from guanosine 5'-(α-D-glucopyranosyl-^{14}C pyrophosphate)- ("GDP-D-glucose-^{14}C") into an alkali-insoluble compound characterized as cellulose. This enzyme system was also capable of synthesizing other polysaccharides from guanosine 5'-(α-D-manno-pyranosyl-^{14}C pyrophosphate) (GDP-D-mannose-^{14}C") or a combination of GDP-D-glucose and GDP-D-mannose. However, it was highly specific for the guanosine derivatives, and was not effective with other glycosyl esters of nucleoside pyrophosphates investigated. This observation was in contrast to the reported activity of the enzyme preparation from *Acetobacter xylinum* with UDP-D-glucose. Brummond and Gibbons also prepared from lupin (*Lupinus albus*) an enzyme system that, with UDP-D-glucose, produced a series of cellodextrins and polysaccharides, including an alkali-insoluble fraction.[150,151]

The controversy has been extensively investigated, and found to be related to the synthesis of callose or similar polysaccharides. Callose is a β-D-(1 → 3)-linked glucan containing ~2% of uronic acid residues;[152,153] it is insoluble in water, but dissolves in hot alkali and is degraded therein by the "peeling" reaction.[154] It plays a significant role in development of sieve plates in phloem tissues[9,155] (see p. 348), and is produced as a result of injury to other tissues of the plant.[156] Feingold and coworkers[157] have shown that a particulate-enzyme system prepared from mung beans, or the soluble enzymes extracted from this preparation with digitonin, could synthesize callose from UDP-D-glucose.

(147) A. D. Elbein, G. A. Barber, and W. Z. Hassid, *J. Amer. Chem. Soc.*, **86**, 309 (1964).
(148) G. A. Barber, A. D. Elbein, and W. Z. Hassid, *J. Biol. Chem.*, **239**, 4056 (1964).
(149) G. A. Barber and W. Z. Hassid, *Biochim. Biophys. Acta*, **86**, 397 (1964).
(150) D. O. Brummond and A. P. Gibbons, *Biochem. Biophys. Res. Commun.*, **17**, 156 (1964).
(151) D. O. Brummond and A. P. Gibbons, *Biochem. Z.*, **342**, 308 (1965).
(152) G. O. Aspinall and G. Kessler, *Chem. Ind.* (London), 1296 (1957).
(153) G. Kessler, *Ber. Schweiz. Botan. Ges.*, **68**, 5 (1958).
(154) R. L. Whistler and J. N. BeMiller, *Advan. Carbohyd. Chem.*, **13**, 289 (1958).
(155) W. Escherich, *Protoplasma*, **47**, 487 (1956).
(156) H. B. Currier, *Amer. J. Bot.*, **44**, 478 (1957).
(157) D. S. Feingold, E. F. Neufeld, and W. Z. Hassid, *J. Biol. Chem.*, **233**, 783 (1958).

Ordin and Hall[158,159] found that particulate preparations from oat coleoptiles could utilize both GDP-D-glucose and UDP-D-glucose as the substrate for polysaccharide formation. Hydrolysis with cellulase of the polysaccharide derived from UDP-D-glucose gave cellobiose and, to a lesser extent, a trisaccharide containing mixed β-D-(1 → 4) and β-D-(1 → 3) D-glucosyl linkages. However, when GDP-D-glucose was used as the substrate, the resulting polysaccharide gave only cellobiose and cellotriose, indicating that it contained only β-D-(1 → 4)-links. Furthermore, it was reported that the type of linkage produced depends on the conditions employed and that, by changing the conditions, cellulose, callose, or a glucan having mixed β-D-(1 → 4)- and β-D-(1 → 3)-links may be obtained.

Villemez and coworkers[160] have also reported that an enzyme preparation from mung beans could produce an alkali-insoluble D-glucan from UDP-D-glucose that contains both β-D-(1 → 3)- and β-D-(1 → 4)-links. However, it was not determined whether the product was a mixture of cellulose and callose or a single polysaccharide having mixed linkages.

Ray and coworkers[161] found that a particulate preparation from peas could synthesize cellulose from both UDP-D-glucose and GDP-D-glucose. It is noteworthy that the synthetase particles consisted of Golgi fragments and membranes bearing vesicles (see p. 341); this aspect of the synthesis of cellulose will be discussed later (see p. 341).

Flowers and coworkers,[162,163] carefully re-examined the enzyme preparations from mung bean and lupin, and found that the β-D-(1 → 3)-glucan is formed from UDP-D-glucose, and the β-D-(1 → 4)-glucan, from GDP-D-glucose. An investigation of the mung-bean enzyme-system by another group[164] also confirms the synthesis of β-D-(1 → 3)-links (laminaran or callose) from UDP-D-glucose, and of β-D-(1 → 4)-links (cellulose) from GDP-D-glucose. As noted before (see p. 324), the synthesis of callose from UDP-D-glucose was first reported by Feingold and coworkers,[157] who obtained an alkali-soluble polysaccharide. In the following experiments,[162-164] it was found that ~ 90% of the polysaccharide was alkali-soluble and ~ 10% alkali-insoluble;

(158) L. Ordin and M. A. Hall, *Plant Physiol.*, **42**, 205 (1967).
(159) L. Ordin and M. A. Hall, *Plant Physiol.*, **43**, 473 (1968).
(160) C. L. Villemez, Jr., G. Franz, and W. Z. Hassid, *Plant Physiol.*, **42**, 1219 (1967).
(161) P. Ray, T. Shiniger, and M. Ray, *Proc. Nat. Acad. Sci. U. S.*, **64**, 605 (1969).
(162) H. M. Flowers, K. K. Batra, J. Kemp, and W. Z. Hassid, *Plant Physiol.*, **43**, 1703 (1968).
(163) H. M. Flowers, K. K. Batra, J. Kemp, and W. Z. Hassid, *J. Biol. Chem.*, **244**, 4969 (1969).
(164) J. Chambers and A. D. Elbein, *Arch. Biochem. Biophys.*, **138**, 620 (1970).

however, both fractions had the same chemical constitution. Furthermore, in addition to partial hydrolysis, Smith degradation[165] was used for identification of the products. With this method, the (1 → 3)-linked D-glucan gave D-glucose, whereas the (1 → 4)-linked polysaccharide gave erythritol.

General consideration of the preceding data leads to the conclusion that, with *P. aureus* and *L. albus*, GDP-D-glucose controls the formation of cellulose. However, it is possible that cellulose, or related glucans containing β-D-(1 → 4)-links, could be synthesized by D-glucosyl esters of other nucleotides. This possibility is indicated by variations in the data that may be attributable to a seasonal factor involved in the production of the enzymes, and by the observation that different polysaccharides may be produced from more than one "sugar nucleotide." For instance, glycogen is normally produced from UDP-D-glucose, but it could also be formed from ADP-D-glucose at a lower rate. In the same way, starch is formed from ADP-D-glucose ten times as fast as from UDP-D-glucose, which requires another enzyme.[3,144]

It has been shown that soluble, enzyme preparations from mung beans, peas, and other plants show pyrophosphorylase activity, and can form GDP-D-glucose from GTP and α-D-glucosyl phosphate.[166] These enzyme preparations also contain various carbohydrate compounds, including small proportions of cellulose, that could act as acceptor. Therefore, it was postulated that cellulose in higher plants is formed through the following reactions.[163]

$$\text{GTP} + \alpha\text{-D-glucopyranosyl phosphate} \underset{}{\overset{\text{pyrophosphorylase}}{\rightleftharpoons}}$$
$$\text{GDP-D-glucose} + \text{pyrophosphate}$$
$$n(\text{GDP-D-glucose}) + \text{acceptor} \longrightarrow$$
$$\text{acceptor-}[(1 \longrightarrow 4)\text{-}\beta\text{-D-glucose}]_n + n(\text{GDP})$$

The presence of an epimerase that converts D-glucose into D-mannose could result in the incorporation of D-mannose into the enzymically produced cellulose.[163] This possibility is consistent with the observation that cellulose in higher plants generally contains small proportions of D-mannose and other sugars.[48]

V. DEVELOPMENT OF FIBRILLAR STRUCTURES

1. Formation of Cellulosic Microfibrils

In the previous Section, the biochemical synthesis of cellulose was discussed without consideration of the formation of microfibrils. In

(165) I. J. Goldstein, G. W. Hay, B. A. Lewis, and F. Smith, in Ref. 85, Vol. 5, p. 361.
(166) G. A. Barber and W. Z. Hassid, *Biochim. Biophys. Acta*, **86**, 397 (1964).

this Section, the formation of cellulosic microfibrils by A. *xylinum* and A. *acetogenum* will be discussed without considering the problems of orientation or incorporation of these microfibrils in the plant cell-walls.

Electron-microscope investigation of cultures of bacteria has shown that the cellulose "elements" are not a physical appendage of the cell, but occur free and are scattered within the medium.[134,135] This discovery has raised questions as to the processes or steps involved in the formation of the microfibrils and the extent to which they are influenced by cell organelles or carried out by exogenous chemical interactions and mechanical forces. Ohad and coworkers[167] considered that the steps involved may be resolved into (a) polymerization of the activated, monomeric precursor to form cellulose molecules of high molecular weight, (b) transport of the molecule from the site of synthesis to that of crystallization, (c) crystallization or fibril formation, and (d) orientation of fibrils during deposition.

It was further suggested that the enzymes involved in the synthesis of cellulose by A. *xylinum* are anchored to the cell membrane, so that the rate of synthesis is not decreased by repeated washing.[168] Furthermore, it was reported[167] that cell-free extracts of A. *xylinum* could incorporate D-glucose-1-^{14}C into water- and alkali-insoluble polysaccharides. The alkali-insoluble polysaccharide consists of aggregates formed from granules of variable size having no well-defined units. This material could be filtered through Millipore filters of 0.45- and 0.8-μm pore diameter. As the microfibrils observed under the electron microscope had a cross section of about 3.0×1.6 nm and a minimal length of 3 to 5 μm, the aggregates represent cellulose molecules in a prefibrillar state.[135] On the basis of these data, it was contended that individual cellulose molecules are released at cell surfaces as an amorphous or "prefibrous" material that diffuses into the medium and crystallizes to form microfibrils without the necessity of extracellular enzymes.[167] Moreover, it was believed that orientation of the microfibrils is a mechanical process. Fibrils produced in a shaken cell-suspension intertwine and form aggregates visible to the naked eye. In undisturbed suspensions, the fibrils form pellicles that float on the surface. Addition of the sodium salt of O-(carboxymethyl)cellulose (CMC), a soluble cellulose derivative, to the medium results in co-crystallization, and delays the aggregation of the fibrils, presumably because of the electrostatic repulsion between the carboxylic groups in the polyelectrolyte. The cellulose pellicles formed under these

(167) G. Ben-Hayyim and I. Ohad, *J. Cell Biol.*, **25**, 191 (1965).
(168) S. Hestrin and M. Schramm, *Biochem. J.*, **58**, 345 (1954).

conditions from unshaken cell-suspensions contain crossed and superimposed layers of parallel-oriented, cellulose fibrils.[167] It was, therefore, proposed that orientation of the fibrils in the lamellae of the plant cell-wall is also catalyzed in a similar way by the presence of charged polysaccharides. If this is so, then, among the aforementioned processes, only the first one (that is, the biosynthesis of cellulose) takes place enzymically.

One of the attractions of this theory, and its earlier version[134] called[169] the "Intermediate High Polymer Hypothesis," is that it could explain the formation of amorphous areas in the microfibrils through the entanglement of the preformed polysaccharide chain. However, as noted before, the concept of crystalline and amorphous regions in microfibrils, as proposed by the fringed micellar theory, is no longer generally accepted. Experiments on the preparation of polyethylene have now shown that, under conditions of heterogeneous catalysis, where simultaneous polymerization and crystallization take place, formation of the extended-chain structure is favored,[112,170] whereas crystallization of cellulose by precipitation from solution gives folded-chain microfibrils. Therefore, this theory is more consistent with the folded-chain structure, and the crystallization step proposed receives indirect support from the investigation of Bittiger and coworkers[116] already described (see p. 317).

The intermediate high-polymer hypothesis has been contested by Roelofsen,[171] Setterfield and Bayley,[172] Colvin,[169] Preston,[11] and Mühlethaler[10] for a variety of reasons. As noted before, natural microfibrils show the crystalline structure of cellulose I, whereas regenerated or recrystallized materials generally adopt the cellulose II structure. Therefore, native fibrils having the unstable, cellulose I structure could not be formed by a crystallization process, although Marx-Figini has claimed that, when cellulose is precipitated from a very dilute cupriethylenediamine hydroxide (Cuene) solution by dialysis, it forms fibrils having the cellulose I structure.[173] The cellulose molecules are highly insoluble in water and, as dispersed particles, are unlikely to form well-defined microfibrils. The microfibrils observed could be formed by *A. xylinum* in stiff gels of *O*-(carboxymethyl)-cellulose which should interfere with the diffusion of the "prefibrous"

(169) J. R. Colvin, in Ref. 28, p. 189.
(170) H. D. Chanzy and R. H. Marchessault. *Macromolecules*, **2**, 108 (1969).
(171) P. A. Roelofsen, in "Encyclopedia of Plant Anatomy," W. Zimmerman and P. G. Ozenda, eds., Gebrüder Borntraeger, Berlin, 1959, Part 4.
(172) G. Setterfield and S. T. Bayley, *Ann. Rev. Plant Physiol.*, **12**, 35 (1961).
(173) M. Marx-Figini, in Ref. 10, p. 57.

cellulose molecules. The intermediate particles (that, it was supposed, consist of prefibrous cellulose) may actually have been insoluble proteins and other artifacts. Under the microscope, it could be seen that the microfibrils grow at their tips, which appear smooth and tapered, whereas, under the conditions postulated by the foregoing hypothesis, they should be splayed and uneven.[174,175]

A logical alternative to the intermediate-polymer theory (crystallization of preformed, cellulose molecules) in the biogenesis of microfibrils would involve the transfer or addition of D-glucose residues to the end of the cellulose molecules in a growing microfibril. This process, proposed by Roelofsen,[171] supposes simultaneous polymerization and crystallization as the insoluble polymer is formed; such a process would be similar to the simultaneous polymerization and crystallization occurring during formation of polyethylene under conditions of heterogeneous catalysis, a process that results in exclusive formation of extended-chain structures.[170] Because, as noted already, the former mechanism proposed should provide a folded-chain structure, it seems that the mechanism of the biogenesis could indicate folded or extended packing of the molecular chain in the microfibrils and *vice versa*.

The idea of endwise synthesis is especially attractive as a possible route for the formation of plant cell-walls, wherein it would result in the apposition of the fibrillar structure rather than deposition of the cellulosic materials. However, even with bacterial cellulose, Colvin and coworkers[169] have strongly favored this process.

To account for the formation of the bacterial microfibrils in the medium external[134,135,174] to the cell, these authors suggested that D-glucose or one of its precursors is activated by the biochemical mechanism leading to UDP-D-glucose and/or other intermediates already discussed (see p. 323). The sugar is then transferred to a lipid carrier, with partial retention of the available free-energy in the D-glucosyl phosphate link. The lipid–D-glucose complex (perhaps with other components) then diffuses through the bacterial cell-wall into the external medium. Once outside the cell membrane, the D-glucosyl group is transferred, by an extracellular trans-D-glucosylase, to the end of a cellulose chain at the tip of the microfibril. Thus, polymerization and crystallization into a lattice are simultaneously achieved.

The lipid carrier was postulated as a mechanism necessary for crossing the cell membrane, and a cellulose precursor isolated from

(174) J. R. Colvin and M. Beer, *Can. J. Microbiol.*, **6**, 631 (1960).
(175) B. Millman and J. R. Colvin, *Can. J. Microbiol.*, **7**, 383 (1961).

active cell-extracts was found to contain D-glucose and a lipid.[169,176–179] In other experiments,[180] on the polysaccharide of the surface O-antigen of *Salmonella newington,* it has been shown that the glycosyl groups are transferred from the "sugar nucleotide" to a lipid–oligosaccharide precursor that then gives rise to the polysaccharide.

Although the theory of endwise synthesis adequately answers the objections raised against the intermediate-polymer or the prefibrous hypothesis, it faces some difficulties of its own.[169] If the activated D-glucosyl group is transferred from "sugar nucleotide" or a lipid derivative to the cellulose molecule in the medium, the question arises as to the fate of the remaining material that is continuously generated. To answer this question, it was assumed that the lipid carrier is readsorbed by the cell membrane, to be reused in the same process. However, glycolipids have not been found as intermediates in the biosynthesis of cellulose[10,148] by plant-enzyme systems. Also, if it is assumed that microfibrils contain extended, antiparallel chains of cellulose (according to the Meyer–Misch crystalline structure), two types of enzyme would be required for the synthesis of the cellulose molecules in the two different directions. Alternatively, if the chains are parallel, the two ends of the microfibrils should show different properties.[169,171] Both of these possibilities present unsolved problems.

Investigations by Dennis and Colvin[181] have now shed considerable light on this question. Electron micrographs obtained in their studies showed that *A. xylinum* has a multi-layered cell-wall and no capsule. The space between the inner and outer layers of the wall can be heavily stained by osmium, indicating a high content of lipid. Immediately within the cell wall is a double-layered, cytoplasmic membrane. In lysed cells, the cytoplasmic membrane flows out of the cell wall and can be precipitated by magnesium ions. Incubation of the lysed-cell preparation with trypsin removes the cytoplasmic membrane and associated materials completely; yet the cell walls remain intact, apparently because of the high content of lipid. The precipitated membrane could still synthesize cellulose, but treatment with trypsin completely destroyed the cellulose-synthesizing capacity of the lysed cells.[181,182] This result suggested that the cytoplasmic membrane is the

(176) J. R. Colvin, *Nature,* **183,** 1135 (1959).
(177) A. W. Khan and J. R. Colvin, *J. Polym. Sci.,* **51,** 1 (1961).
(178) A. W. Khan and J. R. Colvin, *Science,* **133,** 2014 (1961).
(179) J. R. Colvin, *Can. J. Biochem. Physiol.,* **39,** 1921 (1961).
(180) P. W. Robbins, D. Bray, M. Dankert, and A. Wright, *Science,* **158,** 1536 (1967).
(181) D. T. Dennis and J. R. Colvin, in Ref. 1, p. 199.
(182) A. M. Brown and J. A. Gascoigne, *Nature,* **187,** 1010 (1960).

most likely site for the synthesis of the cellulose precursor; according to Colvin and associates, this material then diffuses across the cell wall, where the D-glucosyl group is transferred to the cellulose molecule. This hypothesis is in conformity with available information on the role of plasmalemma in the synthesis of the plant cell-wall (see p. 342). However, the exact mechanism involved in the synthesis of the microfibrils in the bacterial media still remains to be clarified.

2. Mechanism of Orientation in the Cell Walls

The general discussion and controversy as to whether the microfibrils are formed by apposition or deposition of cellulosic materials also applies to the plant cell-wall; but, here, the question assumes much greater significance, especially with respect to the architecture of the cell wall and the precise orientation of the microfibrils within its layers and lamellae. How these structures are formed and to what extent the processes involved are carried out and controlled by the living cell, or by inanimate, physical forces, pose major questions that have been extensively investigated and discussed. Various theories for the passive and active orientation of the microfibrils in growing-plant cell-walls have been reviewed in several botanical articles[51,172,183] with excellent discussions, and will only briefly be mentioned here as background.

It has been suggested that microfibrils are oriented by streaming of the flowing cytoplasm over the cell wall.[184] However, this hypothesis is not feasible, because plasmalemma (see p. 330) separates the cytoplasm from the wall. The orientation of microfibrils has also been attributed to the stress[185] or strain[186] produced by the turgor pressure of the growing cell. In the multi-net growth theory proposed by Roelofsen and Houwink,[187] it was assumed that the microfibrils in the primary wall form a sheaf of nets that are oriented first in the transverse direction and that, as the cell grows and elongates, the nets stretch and the microfibrils orient more and more in the longitudinal direction. These theories have been criticized as being inconsistent with the orientation of newly formed microfibrils in elongating cells.[183,188] The formation of the secondary wall, in particular, is not accompanied by any considerable expansion of the cell; therefore, the

(183) A. Frey-Wyssling, in Ref. 100, p. 307.
(184) G. Van Iterson, *Chem. Weekbl.*, **24**, 165 (1927).
(185) E. S. Castle, *J. Cell. Comp. Physiol.*, **10**, 113 (1937).
(186) P. B. Green and J. C. W. Chen, *Z. Wiss. Mikroskop.*, **64**, 482 (1960).
(187) P. A. Roelofsen and A. L. Houwink, *Acta Botan. Neerl.*, **2**, 218 (1953).
(188) K. Mühlethaler, in Ref. 1, p. 51.

crossed lamellae in this layer having well oriented microfibrils must be preformed, rather than rearranged by physical forces.[183]

As already noted (see p. 327), some experimental evidence has been educed for parallel orientation of the microfibrils in crossed layers under the influence of O-(carboxymethyl)cellulose. It has been proposed that microfibrils in the plant cell-walls are oriented in the same way under the influence of charged polysaccharides[167] (such as pectins) found in the middle lamella and the primary wall (see p. 348).

Prior to the discovery of proteins in cell walls (see p. 346), Frey-Wyssling had suggested[183] that a papilla containing cytoplasmic materials penetrates the wall matrix, to provide the mechanism for a genetically controlled arrangement of microfibrils.

Preston has proposed a different guiding mechanism for the direction of the microfibrils,[11,34-36] based on the following considerations. The precise orientation of microfibrils at right angles, or at right angles plus the diagonal direction in some seaweeds (see p. 306), is more consistent with the idea that they are formed by apposition or by the endwise synthesis rather than by the deposition and lateral aggregation of preformed chains. When the organization of the cell membrane is disturbed, as in plasmolysis (collapse of the cell contents, involving separation of the cytoplasm from the cell wall) and sporulation, the guiding mechanism is lost, but the ability to synthesize cellulose is retained. Consequently, the microfibrils produced lie in a random fashion. The successive lamellae, formed by the crosswise orientation of the microfibrils, can often be separated completely, but sometimes the microfibrils of one lamella weave back and forth between the microfibrils of the adjacent lamella, and turn 90° to become part of another lamella, or penetrate the neighboring lamella to be incorporated with one that is more remote. This theory seems to indicate that the microfibrils lying in the two (or three) directions are synthesized simultaneously, rather than successively, and that the synthesizing and orienting mechanism involves a surface layer that is at least three microfibrils thick.

After plasmolysis, the inner face of the wall carries granular aggregates from which microfibrils pass out. It is, therefore, suggested that the granular aggregates may represent parts of the microfibril-synthesizing system. Such a system, according to Preston, may consist of three layers of granules, each about 50.0 nm in diameter, covering the cell surface with close cubic packing as shown in Fig. 17. Such granules would constitute an enzyme complex that could transfer D-glucosyl groups to the growing end of the microfibrils. In this way, the microfibrils would be propagated from one granule to the next by the

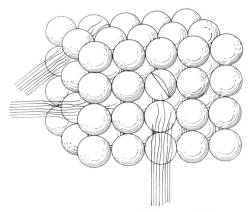

FIG. 17.—Diagrammatic Representation of the Model Proposed by Preston for Synthesis and Orientation of the Microfibrils.[34]

process of endwise synthesis. The microfibrils advance mainly in two directions at right angles, along the array of granules. Orientation along the diagonal directions may also be possible, but would be less favored.

The existence of an organized system (template) for production of microfibrils has also been suggested by Marx-Figini and Schultz.[115,173] These authors investigated the molecular weight and the amount of cellulose formed during the development of cotton fiber. Their data indicated two distinct kinetic stages, corresponding to the formation of primary and secondary walls. The first stage proceeds very slowly and yields a small amount of "primary" cellulose having a non-uniform degree of polymerization $(\overline{d.p.}_w)$ ranging from 2,000 to 6,000. The second stage proceeds much faster and provides a large amount of "secondary cellulose" having a $\overline{d.p.}_w$ of ~ 13,000 (see Fig. 18). During the second stage, the degree of polymerization is independent of variations in the kinetics or the rate of synthesis of cellulose.[189,190] This observation has led to the hypothesis that the formation of cellulose macromolecules in the secondary wall is controlled by a template or structural mechanism that accounts for the uniformity in molecular weight. The enzyme responsible for synthesis of cellulose at this stage may react only in the presence of the template within the cell. A different enzyme-system may be involved in the formation in the primary wall of cellulose that is produced without a template. By this hypothesis, the *in vitro* synthesis of cellulose (see p. 323) therefore corresponds only to the formation of this material in the primary wall.

(189) M. Marx-Figini, *Nature*, **210**, 754 (1966).
(190) M. Marx-Figini and G. V. Schulz, *Naturwissenschaften*, **53**, 466 (1966).

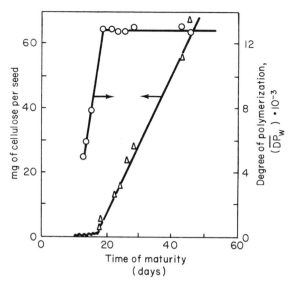

FIG. 18.—Degree of Polymerization and Amount of Cellulose Formed During the Development of Cotton Bolls, According to Marx-Figini.[189]

A similar study has been made with *Valonia* cellulose;[191,192] this showed a constant $\overline{d.p.}_w$ of $\sim 16{,}500$, regardless of the conditions of growth. Furthermore, the distribution curve indicated that almost 80% of the cellulose was monodisperse, with a $\overline{d.p.}$ of $\sim 18{,}500$.

The template originally proposed by Marx-Figini and Schultz[115] was based on microtubules, 23.0–27.0 nm in diameter, that were observed by Ledbetter and Porter[193] beneath the protoplast of cells engaged in formation of the wall (see p. 338). It was assumed that the templates that were ~ 7 μm long (about the length of the cellulose molecule) were situated inside the protein or lipoprotein microtubules. Enzymic interlocking of activated D-glucose residues supposedly takes place along the microtubule and is accompanied by a continuous folding of the resulting cellulose chain. Thus, the cellulose molecule reaches the other end of the microtubule as a folded package. There, it is connected and hydrogen-bonded to the end of the growing microfibril, forming the folded structure shown in Fig. 13b (see p. 317).

These workers have since reported that the cellulose I structure can be formed by recrystallization[173,192,194] (see p. 328). As this discovery

(191) M. Marx-Figini, *Biochim. Biophys. Acta*, **177**, 27 (1969).
(192) M. Marx-Figini, *J. Polym. Sci., Part C*, **28**, 57 (1969).
(193) M. C. Ledbetter and K. R. Porter, *J. Cell Biol.*, **19**, 239 (1963).
(194) E. Macchi, M. Marx-Figini, and E. W. Fischer, *Makromol. Chem.*, **120**, 235 (1968).

removes a major objection to the possibility of the crystallization of native cellulose from preformed molecules, they contended that the degree of polymerization is genetically controlled, but that the morphology of the microfibrils is determined only by the molecular properties of cellulose.

The possible direction and orientation of the cell-wall hemicelluloses by microtubules (see p. 337) or by a polypeptide side-chain will be discussed later (see p. 346). It should, however, be pointed out here that orientation of the matrix substance and its cross-linking may also contribute to the organization of the microfibrils.[9]

VI. Formation of the Cell Wall

The preceding discussions indicate a variety of data on and diverse views concerning the structure and orientation of microfibrils. The types of information that have been obtained, and the resulting conclusions, are obviously influenced by the systems and techniques used for investigating various aspects of the subject. During the past few years, refinements in electron microscopy and related techniques have led to a better understanding and appreciation of the role played by cell organelles in the synthesis of biological substances as an interrelated sequence of events; this emphasizes the complex, organizational system of the whole cell, and tends to lessen the significance of isolated physical experiments.

The cell organelles that have been closely related to the production of structural polysaccharides are: Golgi bodies, plasmalemma, endoplasmic reticulum, and microtubules. An excellent description of these organelles can be found in a monograph by Frey-Wyssling and Mühlethaler.[31] The role of these organelles in the formation of the different layers of cell wall,[195] and the related polysaccharides, are briefly discussed in this Section.

1. Origin of the Cell Wall

The first indication of a new wall is the appearance of stainable nodules in the equatorial plane of the cell. It was originally thought that these were thickened spindle-fibers. However, Becker,[196] through investigations with a light microscope, concluded that these nodules are small droplets that can be stained with vacuole dyes. Subsequent investigation by electron microscopy showed that the droplets are

(195) D. H. Northcote, in "Plant Cell Organelles," J. B. Pridham, ed., Academic Press, New York, 1968, p. 179.
(196) W. A. Becker, *Acta Soc. Botan. Polon.*, **11**, 139 (1934).

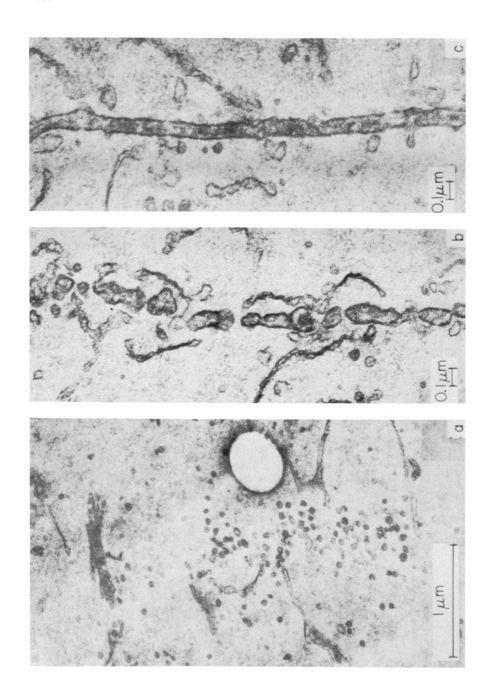

vesicles[197,198] that, according to Whaley and associates[199–201] and Frey-Wyssling and coworkers,[202] are derived from Golgi bodies. These vesicles fuse to form the cell plate or middle lamella. The cell plate appears as a free-floating layer at the last stage of mitosis, and grows from the center of the equatorial plane of the mitotic spindle outward until it joins the wall of the mother cell.[31] Through staining techniques, Becker[196] also found that the cell plate contains acidic polysaccharides, or pectins. As discussed later (see p. 340), the Golgi bodies form these and other polysaccharides found in the matrix of the wall. During this process, flat discs of Golgi cisternae swell in different places along the periphery, forming vesicles that become detached and bubble towards the cell wall. As the vesicles fuse to form the plate, microtubules are found at the growing edge of the plate, where the vesicles are still being aligned. The microtubules seem to form channels that direct the movement of the vesicles toward the area of the cell plate.[195,203–206] The cell membrane (plasmalemma) becomes visible after coalescence of the Golgi vesicles. It was therefore concluded that plasmalemma is formed from the vesicle membrane.[31,199–202]

Before joining the longitudinal wall of the mother cell, the cell plate becomes weakly birefringent, and two narrow, bright layers can be observed on both sides of the central lamella. The growing cell-plate thus consists of three layers already, namely, the middle lamella in the center and the two primary walls of future daughter cells that are being developed as the outer layers. The formation of cell plate[202] is shown in Fig. 19.

(197) R. Buvat and A. Puissant, Compt. Rend., 247, 233 (1958).
(198) K. R. Porter and J. B. Caulfield, Intern. Kongr. Elektronenmikroskopie, 4th, Berlin, 1958, Verhandl., 2, 503 (1960).
(199) W. G. Whaley, H. H. Mollenhauer, and J. H. Leech, Amer. J. Bot., 47, 401 (1960).
(200) W. G. Whaley and H. H. Mollenhauer, J. Cell Biol., 17, 216 (1963).
(201) W. G. Whaley, M. Dauwalder, and J. E. Kephart, J. Ultrastructure Res., 15, 169 (1966).
(202) A. Frey-Wyssling, J. F. López-Sáez, and K. Mühlethaler, J. Ultrastructure Res., 10, 422 (1964).
(203) K. Esau and R. H. Gill, Planta, 67, 168 (1965).
(204) A. Bajer and R. D. Allen, J. Cell Sci., 1, 445 (1966).
(205) M. C. Ledbetter, in "Formation and Fate of Cell Organelles," K. B. Warren, ed., Academic Press, New York, 1967, p. 55.
(206) J. D. Pickett-Heaps and D. H. Northcote, J. Cell Sci., 1, 109 (1966).

FIG. 19.—The Initial Stages of Cell-wall Development Shown in Electron Micrographs by Frey-Wyssling and Coworkers.[202] [(a) Accumulation of Golgi vesicles; (b) formation of cell plate by fusion of the vesicles; and (c) incorporation of matrix materials into the cell wall.]

Cell division, however, as in *Spirogyra* and other algae, is not always accompanied by cell-plate formation.[31] After the formation of plasma-lemma, the polysaccharide contents of the Golgi vesicles are transferred to the cell wall across this membrane by a reverse, pinocytosis process.[195,207] During pinocytosis, the membrane of the vesicle fuses with the plasmalemma, and the contents of the vesicle are discharged into the cell wall without any breakage of the membrane. Study of the cell membrane by the freeze-etching technique shows the scars formed by penetration of the vacuoles.[207,208] The freeze-etching technique is especially suited for observing surface layers, because it maintains the structural details of the frozen sample.

Development of the primary wall, which begins before the extension of the cell plate is completed, continues with further deposition of less-acidic polysaccharides (see p. 348). At this stage, the microtubules are randomly scattered along the cell wall,[195,206] and deposition is not confined to the cell plate. Consequently, the whole cell becomes lined with new wall-materials, but the process of cell growth and considerable longitudinal extension of the existing wall[31] prevent the old walls of the mother cell from becoming any thicker. The mechanism involved in the extension of the cell wall and the further enzymic changes of the matrix material will be discussed later (see p. 345).

The formation of the secondary wall involves (a) some changes in the type of polysaccharides laid down in the matrix,[195] and (b) apposition of the organized microfibrils (see p. 342). This process is accompanied by further changes in distribution of the cell organelles. Microtubules that were randomly distributed during the period of primary growth become organized in parallel lines close to the plasmalemma, so that their arrangement mirrors the orientation of the microfibrils on the other side of the cell membrane.[193,195,209] As already noted (see p. 334), this has led to the suggestion that microtubules may be connected with the synthesis and orientation of the cellulosic microfibrils (see p. 331). It is difficult to envisage, however, how the microfibrils could move through plasmalemma.[188] Another possibility is that microtubules may serve to direct the deposition of matrix material into the cell wall, and this, in turn, may affect orientation of the developing microfibrils[195] (see p. 337). A supporting role is played by endoplasmic reticulum, which apparently is responsible for the trans-

(207) D. H. Northcote and D. R. Lewis, *J. Cell Sci.*, **3**, 199 (1968).
(208) K. Mühlethaler, *Ann. Rev. Plant Physiol.*, **18**, 1 (1967).
(209) P. K. Hepler and E. H. Newcomb, *J. Cell Biol.*, **20**, 529 (1964).

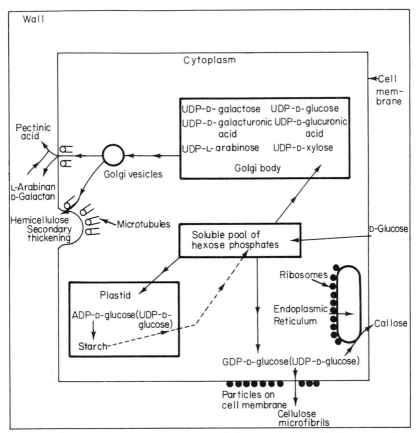

FIG. 20.—Diagrammatic Representation of the Synthesis and Transportation of Various Polysaccharides of a Growing Plant-cell.[195]

portation of photosynthetic products to plastids, where they are stored or modified before being redistributed[9] (see Fig. 20). The endoplasmic reticulum also seems to be closely related to the organization and aggregation of microtubules and to differentiation of the cell wall.

2. Biogenesis of Matrix Polysaccharides by Golgi Bodies

Golgi bodies[31,210,211] were first described by an Italian microscopist, Camillo Golgi, in 1898. However, as they are very variable and hard to detect, especially in plant cells, there has been much doubt and dis-

(210) M. Neutra and C. P. Leblond, *Sci. American*, **220**, 100 (1969).
(211) D. H. Northcote and J. D. Pickett-Heaps, *Biochem. J.*, **98**, 159 (1966).

cussion about them, and they have even been considered to be a microscopic artifact. During the past 20 years, however, all doubts have been removed by numerous studies with the electron microscope, and Golgi bodies have been shown to be present in *all* of the animal and plant cells investigated.

The Golgi body in plant cells forms a series of concentrically bent, double membranes having inflated extensions around the periphery. During the development phase, the cup-shaped vesicles open and flatten, to form a stack of about six cisternae, some 0.6–1 μm across, having the appearance of a pile of saucers.

In 1914, Ramón y Cajal observed tiny droplets of mucus in the region of the Golgi apparatus in goblet cells of intestine, and suggested that this organelle may produce the glycoprotein mucus that spreads over the lining of the intestine as a protective material. It took many years before this assumption was confirmed by observations with the electron microscope, and related techniques, and the Golgi body became recognized as a major chemical factory. Numerous studies during the past ten years have indicated that the Golgi apparatus is involved in the production and transportation, or secretion, of polysaccharides, proteins, and glycosaminoglycans (mucopolysaccharides) from the cells.[210–219] Almost all of the proteins that are secreted by cells are attached to carbohydrates, and the attachment is apparently effected by the Golgi apparatus.[210] As already noted, Golgi bodies play a significant role in cell division and the construction of cell walls. Vesicles derived from these bodies contribute polysaccharides to the cell wall, and membrane material to the plasmalemma.[199–202] Furthermore, it has been shown that the scale fragments found in the cell wall of some algae are produced by the Golgi apparatus.[220–222]

(212) L. G. Caro and G. E. Palade, *J. Cell Biol.*, **20**, 473 (1964).

(213) K. R. Porter, *Biophys. J.*, **4**, 167 (1964).

(214) M. A. Bonneville and B. R. Voeller, *J. Cell Biol.*, **18**, 703 (1963).

(215) L. G. Caro and R. P. van Tubergen, *J. Cell Biol.*, **15**, 173 (1962).

(216) G. C. Godman and N. Lane, *J. Cell Biol.*, **21**, 353 (1964).

(217) D. Fewer, J. Threadgold, and H. Sheldon, *J. Ultrastructure Res.*, **11**, 166 (1964).

(218) N. Lane, L. Caro, L. R. Otero-Vilardebó, and G. C. Godman, *J. Cell Biol.*, **21**, 339 (1964).

(219) S. R. Wellings and K. B. Deome, *J. Biophys. Biochem. Cytol.*, **9**, 479 (1961).

(220) I. Manton, *J. Cell Sci.*, **1**, 375 (1966); **2**, 265, 411 (1967).

(221) M. Parke and I. Manton, *J. Mar. Biol. Ass. U. K.*, **45**, 525 (1965); I. Manton and M. Parke, *ibid.*, **45**, 743 (1965); I. Manton, M. Parke, and H. Ettl, *J. Linnean Soc.* (London), **59**, 378 (1965).

(222) I. Manton and L. S. Peterfi, *Proc. Roy. Soc., Ser. B*, **172**, 1 (1969).

Following the initial observations by microscopy, direct evidence for the production of the cell-wall polysaccharides by the Golgi apparatus was obtained by use of autoradiography, isotopic tracing, and morphological identification methods. Northcote and Pickett-Heaps[211] incubated the root tips of wheat with tritium-labeled D-glucose. Autoradiography of the tissues showed that labeled materials appeared in the Golgi apparatus of the cells within 5 to 10 minutes. Further incubation of the tissues, for 30 to 60 minutes, with unlabeled D-glucose, chased the radioactive materials through the cytoplasm into the cell wall and the slime layer. Analysis of the radioactive materials by electrophoresis showed the presence of pectins and indicated conversion of D-glucose into D-galacturonic acid, L-arabinose, and D-galactose.[9]

It has also been shown that, in pollen tubes, both the cell wall and the contents of the Golgi vesicles can be removed by pectinase, or stained with specific, pectin reagents.[223] When these tubes were treated with tritium-labeled *myo*-inositol (which serves as a precursor of D-galacturonic acid in pectin[224,225]) and tritiated methyl-methionine (which is a methyl-group donor[226]), the labeled materials were found first in the Golgi vacuoles.[208,223]

Similar autoradiographic studies of developing xylem and phloem tissues have shown that the xylan hemicellulose[15] component of the cell wall arises in the Golgi bodies.[227–230]

In a different type of experiment, by gently breaking the cells of pea epicotyls with a razor blade and fractionating the particles by isopycnic centrifugation, a polysaccharide synthetase was produced that appeared to consist of segments of Golgi membrane, bearing vesicles and membrane aggregates.[161]

Brown and coworkers[121,231] have provided further morphological evidence for the role of the Golgi apparatus in the production of cell-wall polysaccharides. The cell wall of a *Chrysophyta* marine alga, *Pleurochrysis scherffelii*, is composed of distinct, scale-like fragments embedded in a gelatinous mass. Manton and coworkers have shown

(223) W. V. Dashek and W. G. Rosen, *Protoplasma*, **61,** 192 (1966).
(224) F. A. Loewus and S. Kelly, *Arch. Biochem. Biophys.*, **102,** 96 (1963).
(225) F. A. Loewus, *Fed. Proc.*, **24,** 855 (1965).
(226) R. Cleland, *Plant Physiol.*, **38,** 738 (1963).
(227) F. B. P. Wooding, *J. Cell Sci.*, 3, 71 (1968).
(228) D. H. Northcote and F. B. P. Wooding, *Sci. Progr.*, **56,** 35 (1968).
(229) D. H. Northcote and F. B. P. Wooding, *Proc. Roy. Soc., Ser. B*, **163,** 524 (1966).
(230) J. D. Pickett-Heaps, *Planta*, **71,** 1 (1966).
(231) R. M. Brown, Jr., W. W. Franke, H. Kleinig, H. Falk, and P. Sitte, *Science*, **166,** 894 (1969).

that these scales are produced by the Golgi apparatus.[220-222] Isolation and investigation of the scales by Brown and coworkers indicated that they consist of a network of concentric microfibrils covered with some amorphous materials, assumed to be pectic substances, which could be extracted with boiling water. The microfibrils were insoluble in alkali. The alkali-extracted material was identified as cellulose by (a) its solubility in Cuoxam, (b) hydrolysis to D-glucose, and (c) the n.m.r. spectrum of its perbenzoylated derivattive. The scale microfibrils reportedly had a cross-section of 1.0-2.5 × 2.5-4.0 nm, as compared with the 3.5 × 3.5 nm reported for elementary fibrils from other sources (see p. 315). The same distinct, fibrillar structures were also observed within the cisternae of the Golgi body, and the formation and secretion of the scales could be followed, starting from a pronounced, dilated (polymerization) center within the Golgi cisternae and proceeding to the extrusion of the scales.

These data confirm the role of Golgi bodies in the synthesis of cell-wall polysaccharides, especially the pectins and hemicelluloses that form the main bulk of the middle lamella and the primary wall. Cellulose, which is the major component of the secondary wall, seems, however, to be generally produced by a different system.

3. Biogenesis of Microfibrils by Plasmalemma Particles

After many years of investigation, the mechanism for the biogenesis of cellulose microfibrils is being gradually unraveled.[9-11,34,181,188,208] As already noted (see p. 337), the matrix substances are produced by the Golgi bodies within the cell, and are then transported through the plasmalemma and deposited within the cell wall. The microfibrils, however, have now been shown to form on the surface of the cell by apposition. The experiments of Dennis and Colvin on the synthesis of cellulose by cytoplasmic membrane of A. xylinum[181] (see p. 330), and the electron-microscope investigation by Preston and coworkers[11,34] which indicated the apposition and organization of the microfibrils by granules on the cell surface (see p. 332), have already been discussed. The cell wall in yeast contains some D-glucan and D-mannans that form fibrillar strands[26a,232] (see p. 302). Investigation of yeast cells by the freeze-etching technique (see p. 338) provided the first evidence for the formation of microfibrils by plasmalemma.[233,234] During this

(232) A. L. Houwink and D. R. Kreger, Antonie van Leeuwenhoek J. Microbiol. Serol., 19, 1 (1953).
(233) H. Moor, K. Mühlethaler, H. Waldner, and A. Frey-Wyssling, J. Biophys. Biochem. Cytol., 10, 1 (1961).
(234) H. Moor and K. Mühlethaler, J. Cell Biol., 17, 609 (1963).

study, Moor, Mühlethaler, and coworkers[233,234] found that the plasma membrane of the yeast is covered by a large number of particles having[10,188,208] a diameter of about 15.0 nm. In some areas, these particles formed a hexagonal arrangement containing 20 to 50 units. At higher resolutions, microfibrils of about 5.0 nm diameter, similar to those previously observed in yeasts, were seen to connect the hexagonal arrangements to the cell wall.[235] The yeast cell-wall can be removed with snail digestive-juice. Under proper conditions, the surviving protoplast forms a new cell-wall; this process, observed under the electron microscope, has shown a striking correlation between cell-wall formation and the occurrence of these particles in yeast. A soil amoeba, *Acanthamoeba*, also forms a cellulose membrane when it is transferred from a rich medium to a poor one containing only salts.[236] Prior to the appearance of cellulose microfibrils, the plasma membrane becomes covered with an increasing number of particles. These particles pile up in a dense layer on the plasmalemma, where the microfibrils can be seen.[188]

Staehelin[237] has shown that, in the developing cell-wall of the green alga *Chlorella*, microfibrils are laid down in the outer region of the cell wall after deposition of the matrix substances. During this process, the surface of the plasmalemma likewise becomes covered with an increasing number of particles. However, unlike the yeast particles, which appear to be permanently attached to the membrane, the particles in *Chlorella* are released, to move through the matrix layer to the outer regions of the cell wall, where they seem to form the microfibrils at a distance from the plasmalemma (see Fig. 21). Study of

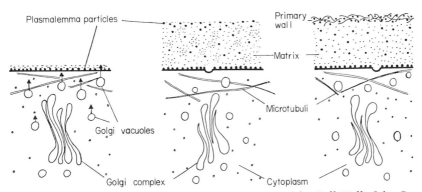

FIG. 21.—Schematic Representation of the Biogenesis of the Cell Wall of the Green Alga *Chlorella*, Showing the Role of Cell Organelles at Different Stages of Development.[208,237]

(235) E. Streiblová, *J. Bacteriol.*, **95**, 700 (1968).
(236) G. Tomlinson and E. A. Jones, *Biochim. Biophys. Acta*, **63**, 194 (1962).
(237) L. A. Staehelin, *Z. Zellforsch.*, **74**, 325 (1966).

Cyanidium caldarum, a unicellular alga, also shows the appearance of plasmalemma particles that undergo a pattern of development, rearrangement, and degradation within the life cycle of the alga.[208,238] The arrangement of the plasmalemma particles closely coincides with the growth pattern of the wall; this provides further evidence for a close relationship between the particles and the synthesis of cellulosic fibrils.[239] The arrangement also suggests that the protein particle plays a significant role in orientating the fibrils. Similar particles have been observed on the outer surface of the plasmalemma cells from higher plants, such as onions[240] and pea roots.[207]

The yeast particles were further studied by isolation and analysis of the plasmalemma. The latter was found to contain protein, polysaccharides, lipids, and ATPase activity. Treatment of the isolated plasmalemma with detergents liberated the globular particles; these were separated by density-gradient centrifugation, and analyzed. The analysis showed that the particles are composed of a mannoprotein.[241] These particles may originate from the Golgi body, which is involved in the production and transportation of glycoproteins (see p. 340) and the formation of the plasmalemma (see p. 340). This possibility has been enhanced by the observation that the membrane of the Golgi vesicles contains particles similar to those found on the plasmalemma.[188]

Further evidence for the formation of cellulose at the surface of the cell was obtained by autoradiography and chemical studies of sycamore-seedling stems that were incorporating materials mainly into the phloem and xylem walls. Wooding[227] treated these stems with radioactive D-glucose for 30 minutes, and then incubated them with unlabeled D-glucose for the same period. With these periods of incubation, only the cellulose of the developing xylem and phloem became labeled. The labeled material was found all along the boundary of the cell wall and cytoplasm, but none of it was close to the Golgi bodies or other organelles of the cytoplasm, indicating that cellulose, unlike the matrix material, is synthesized *outside* the cytoplasm.

As it is known that microfibrils radiate from the particles on the outer layer of the plasmalemma, Mühlethaler has suggested the mechanism shown in Fig. 22 for the apposition of these materials.[10] This mechanism, which closely resembles Preston's model (see p. 332),

(238) L. A. Staehelin, *Proc. Roy. Soc.*, Ser. B, **171**, 249 (1968).
(239) D. H. Northcote, *Brit. Med. Bull.*, **24**, 107 (1968).
(240) D. Branton and H. Moor, *J. Ultrastructure Res.*, **11**, 401 (1964).
(241) P. Matile, H. Moor, and K. Mühlethaler, *Arch. Mikrobiol.*, **58**, 201 (1967).

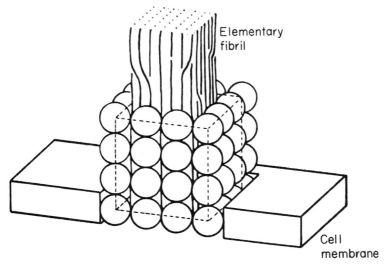

FIG. 22.—The Hypothetical Model Proposed by Mühlethaler for Biogenesis of Cellulosic Elementary Fibrils.[10]

answers the basic questions about the synthesis and orientation of the microfibrils. Other aspects of cell-wall formation, particularly the expansion mechanism and further transformation of the cell-wall material, will be discussed in the following Section.

4. Development and Transformations of the Cell Wall

Deposition of the matrix substances and formation of the microfibrils are accompanied by a sequence of related processes that lead to the development and differentiation of the cell wall; this sequence includes expansion of the wall, changes in the composition of the polysaccharides, organization and orientation of the different layers, deposition of callose for formation of pores in phloem, lignification (see p. 299), and other processes. Considerable information has been obtained about the mechanism of some of these processes and the factors that affect them; this information has been reviewed by leading molecular biologists,[9,208,242] and will very briefly be mentioned here because of its relevance to cell-wall formation and to the constitution of cell-wall polysaccharides of interest to carbohydrate chemists. According to the new concepts, the transformations of the cell wall are effected, or are assisted, by the presence of a variety of enzymes, proteins, and, perhaps, even ribonucleic acid to the extent that primary

(242) D. T. A. Lamport, Ann. Rev. Plant Physiol., 21, 235 (1970).

wall may be considered an organelle or cytological extension of the cell, rather than merely a biochemical cover.[208,242-244]

In 1888, Weissner[245] suggested that cell walls contain protein. However, this claim was later attributed to contamination, until subsequent studies showed the general distribution of protein in the primary wall of higher plants,[243,246] algae,[247-249] and fungi.[250] From the higher plants, Lamport[242,251] has isolated and purified a glycoprotein having a molecular weight of ~230,000; this material contained 95% of carbohydrate and 5% of protein. The protein fraction was rich in 4-hydroxy-L-proline, and the polysaccharide fraction was composed of L-arabinose and D-galactose residues. These investigations have led to the hypothesis that plant cell-walls contain a protein–polysaccharide network[242] (shown in Fig. 23) analogous to the peptidoglycan network of bacterial cell-walls.[252] In the proposed network, protein molecules may cross-link through disulfide bonds to increase the coherence and tensile strength of the wall.[243,253,254] Oligosaccharides of L-arabinose, glycosidically linked to 4-hydroxy-L-proline, form the polysaccharide side-chain of the network. Cleavage of the network provides acid-soluble pectic substances and alkali-soluble hemicelluloses. As the L-arabinofuranosyl bonds are readily cleaved under mildly acidic conditions, extraction with hot, dilute acids yields pectin materials. Treatment with alkali, however, breaks the link between the L-arabinose residues and other polysaccharides, to release the hemicelluloses. The glycoprotein network plays a significant role in the extension of the cell wall, and may even be responsible for the orientation of the polysaccharides.[208,242] According to Lamport's extension hypothesis,[243] during cell growth, some bonds in the glycoprotein network break, to increase the plasticity of the wall so that it can be extended (see p. 338). Therefore, auxin, which stimulates plant growth by loosening the cell wall, affects the glycoprotein

(243) D. T. A. Lamport, *Advan. Botan. Res.*, **2**, 151 (1965).
(244) D. T. A. Lamport, *Exp. Cell Res.*, **33**, 195 (1964).
(245) J. Weissner, *Ber. Deut. Botan. Ges.*, **6**, 187 (1888).
(246) D. T. A. Lamport and D. H. Northcote, *Nature*, **188**, 665 (1960).
(247) T. Punnett and E. C. Derrenbacker, *J. Gen. Microbiol.*, **44**, 105 (1966).
(248) E. W. Thompson and R. D. Preston, *Nature*, **213**, 684 (1967).
(249) I. B. Gotelli and R. Cleland, *Amer. J. Bot.*, **55**, 907 (1968).
(250) M. Novaes-Ledieu, A. Jiménez-Martinez, and J. R. Villanueva, *J. Gen. Microbiol.*, **47**, 237 (1967).
(251) D. T. A. Lamport, *Abstr. Papers. Amer. Chem. Soc. Meeting*, **158**, 47C (1969).
(252) M. R. J. Salton, "The Bacterial Cell Wall," Elsevier, New York, 1964.
(253) N. J. King and S. T. Bayley, *J. Exp. Botany*, **16**, 294 (1965).
(254) E. W. Thompson and R. D. Preston, *J. Exp. Botany*, **19**, 690 (1968).

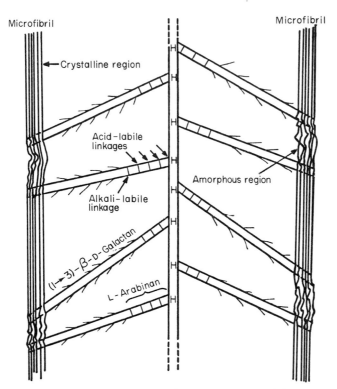

FIG. 23.—Hypothetical Structure of the Glycoprotein Network (Extension Complex) in Plant Cell-walls, as Proposed by Lamport.[242]

network rather than the pectins. Previously, it had been believed that carboxylic acid groups of pectin molecules are partly methylated and partly cross-linked with calcium and magnesium ions, and that plasticization of the cell wall takes place through the breaking of the "calcium bridge" by auxin (indoleacetic acid).[208,255]

There is considerable evidence indicating that some of the wall components are unstable and, after they are formed, change by turnover, autolysis, or degradation.[9,242] It has been reported that isolated cell-walls from oat and maize coleoptiles contain polysaccharide-degrading enzymes that are firmly bound.[256,257]

Changes in the composition of pectic substances have been inves-

(255) T. A. Bennet-Clark, in "The Chemistry and Mode of Action of Plant Growth Substances," R. L. Wain and F. Wightman, eds., Butterworths, London, 1956, p. 312; H. W. Hilton, *Advan. Carbohyd. Chem.*, **21**, 377 (1966).
(256) M. Katz and L. Ordin, *Biochim. Biophys. Acta*, **141**, 126 (1967).
(257) S. Lee, A. Kivilaan, and R. S. Bandurski, *Plant Physiol.*, **42**, 968 (1967).

tigated by Stoddart and Northcote;[258-260] they incubated sycamore-cell suspensions at different stages of growth with labeled sugars, and analyzed the polysaccharides. The results showed that the actively dividing, cambial cells contain a high proportion of a strongly acidic pectin (a D-galacturonan), which apparently occurs in very young cells and may be a part of the cell plate. As the primary cell-wall develops, there is a progressive diminution in the acidity of the pectic materials because of methylation of the carboxyl groups and incorporation of neutral blocks of L-arabino-D-galactans. It is considered that the incorporation of these blocks into the acidic polysaccharides takes place by transglycosylation within the cell wall.

The primary wall also contains relatively small proportions of randomly oriented, cellulosic microfibrils that have a low and nonuniform degree of polymerization (see p. 333). Mühlethaler[208] suggested (see p. 343) that these microfibrils are synthesized by particles loosely scattered over the plasmalemma, at the beginning of formation of the cell wall. As development of the wall proceeds, the number of these particles increases until they aggregate into regular arrays; this results in the production of densely packed and parallel-oriented microfibrils of the secondary wall, instead of the loose, microfibrillar framework of the primary wall.

Formation of the secondary wall is also accompanied by some changes in the matrix polysaccharides, and pectins are no longer deposited (see p. 338). These changes are attributed to a loss of the epimerases that lead to UDP-sugar precursors of the pectic materials.[9,261] Because the matrix polysaccharides are produced by the Golgi bodies, it is assumed that the enzymes that are involved in interconversion of the sugars required for successive stages in the development of the cell wall also reside in this organelle.

A notable feature of the development and differentiation of the cell walls in phloem tissues is the formation of pores connecting the adjacent sieve-tubes to each other and to the companion cells. At an early stage of development, walls of the sieve tubes are marked by parts of endoplasmic reticulum on both sides where the pore is to be formed.[229,262] As the sieve plate develops and the wall thickens, normal materials are deposited on the cell wall, except in the areas below the endoplasmic reticulum; these areas grow, instead, by the

(258) R. W. Stoddart and D. H. Northcote, *Biochem. J.*, **105**, 45 (1967).
(259) R. W. Stoddart and D. H. Northcote, *Biochem. J.*, **105**, 61 (1967).
(260) R. W. Stoddart, A. J. Barrett, and D. H. Northcote, *Biochem. J.*, **102**, 194 (1967).
(261) K. Zetsche, *Biochim. Biophys. Acta*, **124**, 332 (1966).
(262) K. Esau, V. I. Cheadle, and E. B. Risley, *Botan. Gaz.*, **123**, 233 (1962).

deposition of callose. The pads of callose are then eroded, to form the large pores that connect the cells of mature sieve-tubes.[9,228,229] The endoplasmic reticulum apparently serves to create the pore by supplying the enzymes required for the breakdown of callose or the removal of the breakdown products. The pores between the sieve tubes and companion cells are also developed in a similar way.[9,263]

The epimerization reactions possible, and the synthesis and transportation of the various polysaccharides in a growing plant cell-wall, are summarized in a diagram by Northcote,[9,195] shown in Fig. 20 (see p. 339).

(263) F. B. P. Wooding and D. H. Northcote, *J. Cell Biol.*, **24**, 117 (1965).

BIOSYNTHESIS OF SACCHARIDES FROM GLYCOPYRANOSYL ESTERS OF NUCLEOSIDE PYROPHOSPHATES ("SUGAR NUCLEOTIDES")*

By H. Nikaido and W. Z. Hassid†

Department of Bacteriology and Immunology and Department of Biochemistry, University of California, Berkeley, California

*The following abbreviations are used: AMP, CMP, GMP, and UMP for the 5′-phosphates of adenosine, cytidine, guanosine, and uridine, respectively; dTMP for the 5′-phosphate of thymidine, that is, 1-(2-deoxy-β-D-*erythro*-pentofuranosyl)thymine; ADP, etc., for the 5′-pyrophosphates of the foregoing nucleosides; ATP, etc., for the 5′-triphosphates of the foregoing nucleosides; XDP–sugar, where X can be the base of any nucleoside, for a nucleoside 5′-(glycosyl pyrophosphate), except for "ADP-D-ribose," which is adenosine 5′-(D-ribose 5-pyrophosphate); CDP-D-glycerol for cytidine 5′-(1-deoxy-D-glycerol-1-yl pyrophosphate); CDP-L-ribitol for cytidine 5′-(1-deoxy-L-ribitol-1-yl pyrophosphate); NAD⊕ or NADH for nicotinamide adenine dinucleotide and its reduced form; NADP⊕ or NADPH for nicotinamide adenine dinucleotide phosphate and its reduced form; CoA for coenzyme A; LPS for cell-wall lipopolysaccharide; and ACL for antigen-carrier lipid.

† H. N. is indebted to Taiji Nakae for his help in the survey of the literature, and to Dr. E. F. Neufeld who kindly read sections on the glycolipids, glycoproteins, and blood-group substances, and provided valuable criticisms. The authors are, however, entirely responsible for any omissions, errors, or prejudicial opinions that may be found in this article.

I. Introduction

The purpose of this Chapter is to summarize the progress that has been made in the area of "sugar nucleotide" metabolism since 1963, when a Chapter on that subject was written for Vol. 18 of this series of publications.[1] The discovery, in 1951, by Leloir[2,3] of the first "sugar nucleotide," namely, uridine 5'-(α-D-glucopyranosyl pyrophosphate), referred to as "uridine diphosphate D-glucose," proved to be a most important contribution to the subject of carbohydrate biochemistry and led to his being awarded a Nobel Prize in 1970. It was pointed out in the previous Chapter[1] that the discovery of the role of glycosyl esters of nucleoside pyrophosphates as glycosyl donors was primarily responsible for the rapid progress made in the last decade and a half in elucidating the mechanisms of biosynthesis

(1) E. F. Neufeld and W. Z. Hassid, *Advan. Carbohyd. Chem.*, **18**, 309 (1963).
(2) L. F. Leloir, *Arch. Biochem. Biophys.*, **33**, 186 (1951).
(3) L. F. Leloir, in "Phosphorus Metabolism," W. D. McElroy and B. Glass, eds., Johns Hopkins Press, Baltimore, Md., 1951, Vol. 1, p. 67.

of numerous glycosides, including oligosaccharides and polysaccharides in animal, micro-organism, and plant cells.

II. Occurrence of Glycosyl Esters of Nucleoside Pyrophosphates

New "sugar nucleotides" continue to be isolated, and Table I lists those that have been isolated since the previous article[1] was written. References are also given for those compounds of this category that have been enzymically synthesized. However, Table I does not generally include those compounds that have been isolated (or synthesized by enzymes) from sources *additional* to those mentioned in the previous review.[1] Compounds found under obviously non-physiological conditions have also been excluded.

The failure to detect certain sugar nucleotides in any tissue should not be interpreted as signifying that they are of little metabolic importance. The biosynthesis of most sugar nucleotides is under efficient feed-back control (see Section III, p. 363), so that, in some cases, little accumulation of these compounds takes place, even when their utilization is blocked.[31,41]

The occurrence of a sugar nucleotide in a tissue obviously suggests that it has a metabolic function in that tissue. However, it must be emphasized that some sugar nucleotides might have been produced by "cross reactions" of biosynthetic enzymes having a broad substrate-specificity. Such "nonphysiological" products may be utilized only with difficulty, or may be poor feed-back inhibitors of their own synthesis, and, consequently, they could accumulate in significant amounts. The enzymes active in the synthesis of blood-group substances can transfer sugars also to oligosaccharide acceptors; furthermore, these enzymes are found in milk (see Section VII, 4, p. 472). UDP-oligosaccharides from milk frequently contain oligosaccharide sequences found in blood-group substances, and yet they have not been shown to be active in the synthesis of any polysaccharide (including blood-group substances). These results tend to suggest that UDP-oligosaccharides are not formed by intact cells, possibly because of the intracellular compartmentalization, but that, once the cells are disrupted and the cell contents move into the milk, the UDP-oligosaccharides are produced through the interaction of solubilized glycosyl transferases with sugar nucleotides, some of which serve as acceptors of glycosyl groups. These considerations

TABLE I

Occurrence and Enzymic Synthesis of Glycosyl Esters of Nucleoside 5'-Pyrophosphates

Glycosyl nucleoside 5'-pyrophosphate			Enzymic synthesis	
Nucleoside moiety	Glycosyl moiety	Occurrence	Type of enzyme	Source of enzyme
Uridine	D-galactofuranose		isomerase(?)	fungus[4]
	D-fructose	higher plants[5,6]		
	L-rhamnose	Salmonella[7] alga[8] higher plants[9,10]		
	D-apiose [3-C-(hydroxymethyl)-D-erythrose]	higher plants[10]	decarboxylation, isomerization, and reduction from UDP-D-glucuronic acid	higher plants[11]
	2-acetamido-4-amino-2,4,6-trideoxyhexose		isomerization and transamination: from UDP-2-acetamido-2-deoxy-D-glucose	bacteria[12]
	2-acetamido-2-deoxy-D-glucuronic acid	bacteria[13,14]		
	2-amino-2-deoxy-D-glucuronic acid		UDP-D-glucose dehydrogenase	animal[15]
	2-acetamido-2-deoxy-D-galactose 4,6-disulfate	hen oviduct[16]	sulfation at C-6	hen oviduct[16]
	N-acetylmuramic acid		condensation with enolpyruvate phosphate, and then reduction	bacteria[17–19]
	O-β-D-glucopyranosyl-(1 → 4)-D-glucose			higher plants[20]

Nucleoside	Saccharide	Occurrence	Reaction	Organism
	(1 → 4)-2-acetamido-2-deoxy-D-glucose			
	O-β-D-galactopyranosyl-(1 → 4)-2-acetamido-2-deoxy-D-glucose	pig milk[22]		
	O-α-L-fucopyranosyl-(1 → 2)-O-β-D-galactopyranosyl-(1 → 4)-2-acetamido-2-deoxy-D-glucose	milk[23]		
Guanosine	D-galactose	higher plants[20,24]		
	L-galactose	Helix pomatia[25]	epimerization of GDP-D-mannose	Helix pomatia[26]
	D-xylose	higher plants[24]		
	D-rhamnose		isomerization of GDP-D-mannose and reduction	higher plants[27]
	D-mannuronic acid	brown alga[28]		
	L-guluronic acid	brown alga[28]		
Cytidine	3,6-dideoxy-D-ribo-hexose (paratose)	Salmonella paratyphi A[29]		
	3,6-dideoxy-L-arabino-hexose (ascarylose)		isomerization of CDP-D-glucose and reduction	Pasteurella pseudotuberculosis[30]
	3,6-dideoxy-D-xylo-hexose (abequose)		the same	Pasteurella[30]
	3,6-dideoxy-D-arabino-hexose (tyvelose)		the same	Salmonella,[31] Pasteurella[30]
			the same	Salmonella,[31,32] Pasteurella[30]

(continued)

TABLE I (*continued*)

Glycosyl nucleoside 5'-pyrophosphate			Enzymic synthesis	
Nucleoside moiety	Glycosyl moiety	Occurrence	Type of enzyme	Source of enzyme
	6-deoxy-3-C-methyl-2-O-methyl-L-aldohexose (vinelose)	*Azotobacter vinelandii*[33]		
	4-O-(O-methylglycolyl)-vinelose	*Azotobacter vinelandii*[34]		
Adenosine	D-fructose	higher plants[20]		
	D-mannose	corn[35]		
	D-galactose	higher plants[20,35]		
	D-mannitol	*Salmonella*[36]		
	2-acetamido-2-deoxy-D-glucose	corn[35]		
Thymidine	D-galacturonic acid	sugar beet[37]	epimerization of dTDP-D-glucose and dehydrogenation	sugar beet[38]
	D-mannuronic acid	sugar beet[37]		
	4-acetamido-4,6-dideoxy-D-glucose		reduction of dTDP-D-glucose and transamination	*Escherichia coli*[39]
	4-acetamido-4,6-dideoxy-D-galactose		the same	*Escherichia coli*[39]
	3-acetamido-3,6-dideoxyhexose		the same	*Xanthomonas campestris*[40]

(4) A. G. Trejo, G. J. F. Chittenden, J. G. Buchanan, and J. Baddiley, *Biochem. J.*, **117**, 637 (1970).

(5) Y. Umemura, M. Nakamura, and S. Funahashi, *Arch. Biochem. Biophys.*, **119**, 240 (1967).

(6) E. G. Brown and B. S. Mangat, *Biochim. Biophys. Acta*, **148**, 350 (1967).

(7) V. Ginsburg, *J. Biol. Chem.*, **241**, 3750 (1966).

(8) H. Kauss, *Biochem. Biophys. Res. Commun.*, **18**, 170 (1965).

(9) M. M. V. Hampe and N. S. Gonzalez, *Biochim. Biophys. Acta*, **148**, 566 (1967).

(10) H. Sandermann, Jr., and H. Grisebach, *Biochim. Biophys. Acta*, **156**, 435 (1968).

(11) H. Sandermann, Jr., G. T. Tisue, and H. Grisebach, *Biochim. Biophys. Acta*, **165**, 550 (1968).

(12) J. Distler, B. Kaufman, and S. Roseman, *Arch. Biochem. Biophys.*, **116**, 466 (1966).

(13) E. J. Smith, *Biochim. Biophys. Acta*, **158**, 470 (1968).

(14) P. Biely and R. W. Jeanloz, *J. Biol. Chem.*, **244**, 4929 (1969).

(15) J. E. Silbert and E. F. X. Hughes, *Biochim. Biophys. Acta*, **83**, 355 (1964).

(16) T. Harada, S. Shimizu, Y. Nakanishi, and S. Suzuki, *J. Biol. Chem.*, **242**, 2288 (1967).

(17) J. L. Strominger, *Biochim. Biophys. Acta*, **30**, 645 (1958).

(18) K. G. Gunetileke and R. A. Anwar, *J. Biol. Chem.*, **241**, 5740 (1966).

(19) K. G. Gunetileke and R. A. Anwar, *J. Biol. Chem.*, **243**, 5770 (1968).

(20) D. F. Cumming, *Biochem. J.*, **116**, 189 (1970).

(21) Y. Nakanishi, S. Shimizu, N. Takahashi, M. Sugiyama, and S. Suzuki, *J. Biol. Chem.*, **242**, 967 (1967).

(22) A. Kobata and Z. Suzuoki, *Biochim. Biophys. Acta*, **107**, 405 (1965).

(23) A. Kobata, *J. Biochem.* (Tokyo), **59**, 63 (1966).

(24) R. R. Selvendran and F. A. Isherwood, *Biochem. J.*, **105**, 723 (1967).

(25) E. M. Goudsmit and E. F. Neufeld, *Biochim. Biophys. Acta*, **121**, 192 (1966).

(26) E. M. Goudsmit and E. F. Neufeld, *Biochem. Biophys. Res. Commun.*, **26**, 730 (1967).

(27) G. A. Barber, *Biochim. Biophys. Acta*, **165**, 68 (1968).

(28) T.-Y. Lin and W. Z. Hassid, *J. Biol. Chem.*, **241**, 3283 (1966).

(29) R. M. Mayer and V. Ginsburg, *Biochem. Biophys. Res. Commun.*, **15**, 334 (1964).

(30) S. Matsuhashi, M. Matsuhashi, and J. L. Strominger, *J. Biol. Chem.*, **241**, 4267 (1966).

(31) H. Nikaido and K. Nikaido, *J. Biol. Chem.*, **241**, 1376 (1966).

(32) A. D. Elbein, *Proc. Nat. Acad. Sci. U. S.*, **53**, 803 (1965).

(33) S. Okuda, N. Suzuki, and S. Suzuki, *J. Biol. Chem.*, **242**, 958 (1967).

(34) S. Okuda, N. Suzuki, and S. Suzuki, *J. Biol. Chem.*, **243**, 6353 (1968).

(35) M. Dankert, S. Passeron, E. Recondo, and L. F. Leloir, *Biochem. Biophys. Res. Commun.*, **14**, 358 (1964).

(36) B. M. Scher and V. Ginsburg, *J. Biol. Chem.*, **243**, 2385 (1968).

(37) R. Katan and G. Avigad, *Israel J. Chem.*, **3**, 110P (1966).

(38) R. Katan and G. Avigad, *Biochem. Biophys. Res. Commun.*, **24**, 18 (1966).

(39) M. Matsuhashi and J. L. Strominger, *J. Biol. Chem.*, **239**, 2454, 4738 (1964).

(40) W. A. Volk and G. Ashwell, *Biochem. Biophys. Res. Commun.*, **12**, 116 (1963).

(41) T. Okazaki, J. L. Strominger, and R. Okazaki, *J. Bacteriol.*, **86**, 118 (1963).

indicate that extreme caution is needed in assessing the physiological function of sugar nucleotides isolated.

III. ENZYMIC SYNTHESIS OF GLYCOSYL ESTERS OF NUCLEOSIDE 5'-PYROPHOSPHATES

The several known enzymic mechanisms for the synthesis of "sugar nucleotides" have been previously reviewed.[1] It was pointed out that the mechanism of "epimerization," whereby the glycosyl moiety of the glycosyl ester of a nucleotide is transformed into one having a different group-configuration, was obscure. This process may be exemplified by one of the first such enzymic interconversions to be discovered:

$$\text{UDP-D-glucose} \overset{\text{4-epimerase}}{\rightleftharpoons} \text{UDP-D-galactose}$$

Although "epimerizations" at C-4 are the most common, examples of epimerizations at C-2 and C-5 are known.[1] Because it had been found that the reaction at C-4 requires the presence of catalytic amounts of $NAD^{\oplus}$, it was assumed that the epimerization proceeds through a 4-ketose intermediate.[42] However, the formation of such an intermediate had not been demonstrated conclusively.

Wilson and Hogness[43] showed that extensively purified UDP-D-glucose 4-epimerase from *Escherichia coli* contains 1 mole of tightly bound $NAD^{\oplus}$ per mole of enzyme, and, as an absorption peak almost identical in shape to that of NADH could be observed on addition of the substrate to the enzyme, this evidence is in accord with an oxidation-reduction mechanism involving C-4 of the hexosyl group of the UDP-hexoses.

Nelsestuen and Kirkwood[44] obtained additional evidence in support of the oxidation-reduction mechanism by study of the reaction catalyzed by a highly purified epimerase[43] containing $NAD^{\oplus}$. When treated with sodium borohydride in the presence of UDP-D-glucose, the enzyme-bound $NAD^{\oplus}$ was reduced, with a concomitant loss of enzymic activity towards UDP-D-glucose. The reduced enzyme regained practically all of its activity when it was incubated with dTDP-6-deoxy-D-*xylo*-hexos-4-ulose   ("4-keto-6-deoxy-D-glucose"), and NADH was simultaneously oxidized with a loss of tritium,

(42) E. S. Maxwell and H. de Robichon-Szulmajster, *J. Biol. Chem.*, **235**, 308 (1960).
(43) D. B. Wilson and D. S. Hogness, *J. Biol. Chem.*, **239**, 2469 (1964).
(44) G. Nelsestuen and S. Kirkwood, *Fed. Proc.*, **29**, 337 (1970).

which was incorporated into the dTDP-sugar. On hydrolysis, the tritium of this sugar nucleotide was found exclusively in D-fucose (6-deoxy-D-galactose) and 6-deoxy-D-glucose. Tritiated NADH is released when the reduced enzyme is boiled, and its reactions with specific α- or β-dehydrogenases show that sodium borohydride reduces the "β" side of the nicotinamide moiety, with a minimum of 90% stereospecificity. The results of these experiments add support to the conclusion that, during epimerization, the substrate is specifically oxidized by $NAD^{\oplus}$ to produce a 4-keto intermediate, which is then reduced to either of the two isomers possible.

UDP-D-glucose 4-epimerase from yeast has been purified to homogeneity by Darrow and Rodstrom.[44a] A molecule of the monomer (molecular weight 125,000) contains one $NAD^{\oplus}$ or NADH residue and consists of two subunits of molecular weight $\sim$60,000. Kalckar and coworkers[44b] found that the enzyme can be reduced with UMP in the presence of a sugar, which may be D-glucose, D-galactose, or the pentoses configurationally related to these hexoses, namely, D-xylose or L-arabinose. The reduced enzyme is very strongly fluorescent, and is catalytically inactive. The "pyridine nucleotide" released from the reduced enzyme seems[44b] to be identical with, or at least very similar to, NADH. The reduction was also found[44c] to alter the conformation of the enzyme protein extensively.

The only example of epimerization at C-2 known at the time of the previous review[1] involved an unusual reaction wherein the sugar was released as soon as the epimerization had occurred:

UDP-2-acetamido-2-deoxy-D-glucose $\longrightarrow$
$\qquad$ UDP + 2-acetamido-2-deoxy-D-mannose.

This observation led to the suspicion that epimerization at C-2 occurred with difficulty because of the proximity to C-1, and was impossible without the splitting of the glycosidic bond. However, Matsuhashi and Strominger[45] have since found an example of 2-epimerization in which the glycosyl group remains attached to the nucleoside 5'-pyrophosphate:

(44a) R. A. Darrow and R. Rodstrom, *Proc. Nat. Acad. Sci. U. S.*, **55**, 205 (1965); *Biochemistry*, **7**, 1645 (1968).

(44b) A. Bhaduri, A. Christensen, and H. M. Kalckar, *Biochem. Biophys. Res. Commun.*, **21**, 631 (1965); A. U. Bertland, B. Bugge, and H. M. Kalckar, *Arch. Biochem. Biophys.*, **116**, 280 (1966).

(44c) A. U. Bertland and H. M. Kalckar, *Proc. Nat. Acad. Sci. U. S.*, **61**, 629 (1968).

(45) S. Matsuhashi and J. L. Strominger, *Biochem. Biophys. Res. Commun.*, **20**, 169 (1965); S. Matsuhashi, *J. Biol. Chem.*, **241**, 4275 (1966).

CDP-paratose (-3,6-dideoxy-D-*ribo*-hexose) $\rightleftharpoons$
CDP-tyvelose (-3,6-dideoxy-D-*arabino*-hexose).

The enzyme catalyzing this reaction requires NAD⊕, and it thus resembles UDP-D-glucose 4-epimerase in this respect.

Further progress has been made in the more complex conversions, involving the change in configuration at several carbon atoms, that occur in the process of formation of 6-deoxyhexoses, namely, the conversion of GDP-D-mannose into GDP-L-fucose, and dTDP-D-glucose into dTDP-L-rhamnose.

The first conversion, involving that of GDP-D-mannose into GDP-L-fucose by enzymes from *Aerobacter aerogenes*,[46,47] represents an oxidation-reduction reaction, requiring NAD⊕, in which the hydroxyl group on C-4 is converted into a keto group, and the hydroxymethyl group into a methyl group. Subsequently, inversions take place at C-3 and C-5, to produce a second intermediate; then, reduction with NADPH at C-4 results in the formation of GDP-L-fucose. Similarly, dTDP-D-glucose is transformed into dTDP-L-rhamnose by certain micro-organisms.[48-50] This transformation involves inversion at C-3, C-4, and C-5, and reduction at C-6. In plants, a corresponding process occurs that involves the same steps, but with UDP-D-glucose instead of dTDP-D-glucose.[51]

A reaction catalyzing the inversions at C-3 and C-5 is postulated to occur in the formation of GDP-L-galactose from GDP-D-mannose;[52] also, the synthesis of 6-deoxy-D-talose from GDP-D-mannose[53] is assumed to involve a reduction at C-6 and inversion at C-4.

Further progress regarding the mechanism of conversion of dTDP-D-glucose into dTDP-6-deoxy-D-*xylo*-hexos-4-ulose ("dTDP-4-keto-6-deoxy-D-glucose") has been provided by Glaser and coworkers.[54] They studied the mechanism of this reaction by using a purified dTDP-D-glucose oxidoreductase from *Escherichia coli*. When dTDP-D-glucose-*4-d* was used as the substrate, deuterium was transferred quantitatively to C-6 of the 6-deoxy-D-*xylo*-hexos-4-ulose. The trans-

(46) V. Ginsburg, *J. Biol. Chem.*, **235**, 2196 (1960).
(47) V. Ginsburg, *J. Biol. Chem.*, **236**, 2389 (1961).
(48) L. Glaser and S. Kornfeld, *J. Biol. Chem.*, **236**, 1795 (1961).
(49) R. Okazaki, T. Okazaki, J. L. Strominger, and A. M. Michelson, *J. Biol. Chem.*, **237**, 3014 (1962).
(50) J. H. Pazur and E. W. Shuey, *J. Biol. Chem.*, **236**, 1780 (1961).
(51) G. A. Barber, *Arch. Biochem. Biophys.*, **103**, 276 (1963).
(52) J. C. Su and W. Z. Hassid, *Biochemistry*, **1**, 468 (1962).
(53) A. Markovitz, *J. Biol. Chem.*, **239**, 2091 (1964).
(54) A. Melo, W. H. Elliot, and L. Glaser, *J. Biol. Chem.*, **243**, 1467 (1968).

fer of hydrogen was shown to be intramolecular. A similar result was also obtained by Gabriel and Lindquist,[55] who used dTDP-D-glucose-4-*t*. When dTDP-D-glucose was converted into dTDP-6-deoxy-D-*xylo*-hexos-4-ulose in deuterium oxide, one atom of deuterium was incorporated at C-5 of the 6-deoxy-D-*xylo*-hexos-4-ulose moiety.[54] Similar results were obtained in the conversion of CDP-D-glucose into CDP-6-deoxy-D-*xylo*-hexos-4-ulose by CDP-D-glucose oxidoreductase from *Salmonella typhimurium*.[56] These data are in accordance with an intramolecular, oxidation-reduction mechanism for dTDP-D-glucose oxidoreductase, as shown by the following reactions.[54,55]

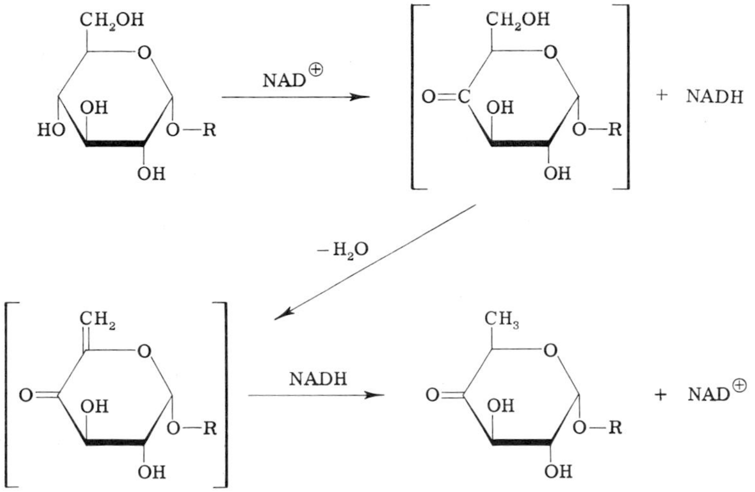

Proposed mechanism for the dTDP-D-glucose
oxidoreductase reaction

This mechanism is also consistent with earlier results obtained by the use of nucleotide derivatives of D-glucose labeled at[57] C-5 or[58] C-3.

Highly purified[59] and crystalline[60] dTDP-D-glucose oxidoreductase has now become available, and the presence of one mole of NAD$^\oplus$ per mole of enzyme has been established. That this NAD$^\oplus$ is reduced

(55) O. Gabriel and L. Lindquist, *J. Biol. Chem.*, **243**, 1479 (1968).
(56) R. D. Bevill, *Biochem. Biophys. Res. Commun.*, **30**, 595 (1968).
(57) K. Herrman and J. Lehmann, *Eur. J. Biochem.*, **3**, 369 (1968).
(58) O. Gabriel and G. Ashwell, *J. Biol. Chem.*, **240**, 4128 (1965).
(59) H. Zarkowsky and L. Glaser, *J. Biol. Chem.*, **244**, 4750 (1969).
(60) S. F. Wang and O. Gabriel, *J. Biol. Chem.*, **244**, 3430 (1969).

to NADH on addition of substrate was shown either by the change in absorption at[59] 340 nm, or by the use of a substrate analog, namely, dTDP-6-deoxy-D-glucose, which cannot undergo dehydration and reduction after oxidation to the 6-deoxy-D-*xylo*-hexos-4-ulose derivative.[61] It has been suggested that the NADH-containing enzyme has a different conformation, so that it cannot release the substrate into the medium.[62]

By the reduction of the reaction mixture with NaBT$_4$, a small proportion of nucleotide-bound D-galactose-*t* was obtained, in addition to large proportions of tritium-labeled 6-deoxyhexoses.[62a] This experiment thus directly demonstrated the presence of the 4-keto-hexose intermediate in the reaction sequence already postulated. The observation that D-glucose-*t* was not obtained presumably indicates the close association of this intermediate with the enzyme molecule, so that the reducing agent can approach the substrate from only one direction.[62a]

Melo and Glaser[63] have also shown that an enzyme system from *Pseudomonas aeruginosa* catalyzes the conversion of dTDP-6-deoxy-D-*xylo*-hexos-4-ulose into dTDP-L-rhamnose by the following reaction.

$$dTDP\text{-}6\text{-}deoxy\text{-}D\text{-}xylo\text{-}hexos\text{-}4\text{-}ulose + NADPH + H^{\oplus} \longrightarrow$$
$$NADP^{\oplus} + dTDP\text{-}L\text{-}rhamnose.$$

This conversion requires inversion of configuration at C-3 and C-5, and stereospecific reduction at C-4. Evidence has been presented that this reaction is catalyzed by two protein fractions, one of which catalyzes the exchange of the hydrogen atoms at C-3 and C-5 of dTDP-6-deoxy-D-*xylo*-hexos-4-ulose with the hydrogen atoms of water.

Progress has also been made on the biosynthesis of nucleotide-bound 3,6-dideoxyhexoses. Thus, GDP-colitose (-3,6-dideoxy-L-*xylo*-hexose) was shown to be enzymically synthesized from GDP-D-mannose by way of GDP-6-deoxy-D-*lyxo*-hexos-4-ulose;[64] CDP-paratose, CDP-adequose (-3,6-dideoxy-D-*xylo*-hexose), CDP-tyvelose, and CDP-ascarylose (-3,6-dideoxy-L-*arabino*-hexose) from CDP-D-glu-

(61) S. F. Wang and O. Gabriel, *J. Biol. Chem.*, **245**, 8 (1970).
(62) H. Zarkowsky, E. Lipkin, and L. Glaser, *Biochem. Biophys. Res. Commun.*, **38**, 787 (1970).
(62a) J. Lehmann and E. Pfeiffer, *FEBS Lett.*, **7**, 314 (1970).
(63) A. Melo and L. Glaser, *J. Biol. Chem.*, **243**, 1475 (1968).
(64) E. C. Heath and A. D. Elbein, *Proc. Nat. Acad. Sci. U. S.*, **48**, 1209 (1962); A. D. Elbein and E. C. Heath, *J. Biol. Chem.*, **240**, 1926 (1965).

cose by way of [30-32] CDP-6-deoxy-D-*xulo*-hexos-4-ulose. The mechanism of conversion of nucleoside 5'-(6-deoxyhexosyl-4-ulose pyrophosphate) ("nucleoside diphosphate 4-keto-6-deoxyhexose") into 3,6-dideoxyhexose has been studied in detail by Strominger and co-workers.[65,66] The synthesis of CDP-3,6-dideoxyhexoses from CDP-6-deoxy-D-*xylo*-hexos-4-ulose requires at least three protein fractions (E1, E2, and E3). The following reaction sequence has been found.

$$\text{CDP-6-deoxy-D-}xylo\text{-hexos-4-ulose} \xrightarrow[\text{NADPH}]{\text{E1,E3}}$$

$$\text{CDP-3,6-dideoxy-D-}erythro\text{-hexos-4-ulose} \xrightarrow[\text{NADPH}]{\text{E2}}$$

$$\text{CDP-3,6-dideoxyhexose}$$

E3 is a protein of low molecular weight, but is not identical with thioredoxin.[66] Depending on the source of protein E2, abequose, paratose, or ascarylose linked to CDP is produced. It has been mentioned earlier that CDP-tyvelose is formed from CDP-paratose by epimerization at C-2.[45]

An important discovery is the finding that metabolic regulation of saccharide synthesis can occur through allosteric control of enzymes for synthesis of "nucleotide sugars." Two types of control can be distinguished: one is a simple, feed-back inhibition, and the other, control by the energy-charge level in the cell. In the first group, the activity of the enzyme that catalyzes the first step of the pathway *specific* to the synthesis of a particular "sugar nucleotide" is feed-back inhibited by the end product, namely, the "sugar nucleotide." Thus, dTDP-L-rhamnose inhibits the first enzyme of its biosynthetic pathway, namely, dTDP-D-glucose pyrophosphorylase;[67,68] CDP-paratose[69] or CDP-abequose[31] inhibits CDP-D-glucose pyrophosphorylase; and UDP-D-xylose inhibits UDP-D-glucose dehydrogenase.[70]

An especially beautiful example of feed-back inhibition was shown by R. H. Kornfeld and Ginsburg in a pathway leading, through GDP-

(65) S. Matsuhashi and J. L. Strominger, *J. Biol. Chem.*, **242**, 3494 (1967).
(66) H. Pape and J. L. Strominger, *J. Biol. Chem.*, **244**, 3598 (1969).
(67) R. L. Bernstein and P. W. Robbins, *J. Biol. Chem.*, **240**, 391 (1965).
(68) A. Melo and L. Glaser, *J. Biol. Chem.*, **240**, 398 (1965).
(69) R. M. Mayer and V. Ginsburg, *J. Biol. Chem.*, **240**, 1900 (1965).
(70) E. F. Neufeld and C. W. Hall, *Biochem. Biophys. Res. Commun.*, **19**, 456 (1965);
    A. Bdolah and D. S. Feingold, *Biochim. Biophys. Acta*, **159**, 176 (1968).

D-mannose, to GDP-L-fucose.[71] They examined three types of bacteria, the first synthesizing polysaccharides containing D-mannose but not L-fucose, the second synthesizing those containing L-fucose but not D-mannose, and the third synthesizing those containing both. Their findings are summarized in the next Scheme, where broken lines indicate the action of feed-back inhibitors.

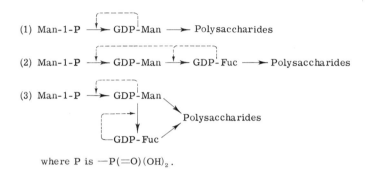

where P is —P(=O)(OH)$_2$.

Thus, in the first group of bacteria, GDP-D-mannose inhibits its own synthesis by a simple feed-back inhibition. A similar mechanism is also operative in the second group, where GDP-L-fucose inhibits both GDP-D-mannose pyrophosphorylase and GDP-D-mannose oxidoreductase. In the third group, where both "sugar nucleotides" are needed for polysaccharide synthesis, their biosynthesis is individually controlled, so that imbalance in the supply of these two sugars will not occur.

A similar example is the biosynthesis of UDP-2-acetamido-2-deoxy-D-glucose and CMP-N-acetylneuraminic acid in rat liver.[72] Here again, both sugars are needed for synthesis of complex saccharides, and each "sugar nucleotide" inhibits the first enzyme of the pathway unique to its synthesis, as shown by the dotted arrows in the following Scheme.

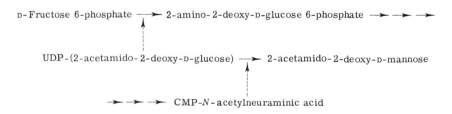

(71) R. H. Kornfeld and V. Ginsburg, *Biochim. Biophys. Acta,* **117,** 79 (1966).
(72) S. Kornfeld, R. Kornfeld, E. F. Neufeld, and P. J. O'Brien, *Proc. Nat. Acad. Sci. U. S.,* **52,** 371 (1964).

The situation is different for "sugar nucleotides" that are used for the synthesis of reserve carbohydrate. Because of the nature of the material synthesized, the enzymes are controlled mostly by the energy-charge levels in the cell. Preiss and coworkers[73] have shown this for ADP-D-glucose pyrophosphorylase, which is necessary for synthesis of glycogen and starch in bacteria and plants (see also Sections V,1 and 2; pp. 379 and 382). The results of their impressive studies, which have been summarized,[73] show that different substances activate this enzyme, according to the major route of carbon metabolism in different organisms. In higher plants where the Calvin cycle predominates,[74] glycerate 3-phosphate ("3-phosphoglycerate") is the major activator; in those bacteria that use mostly the Entner–Doudoroff pathway, D-fructose 6-phosphate; in *Rhodospirillum rubrum*, which cannot grow on D-glucose, pyruvate activates; and in *Escherichia coli*, where the Embden–Meyerhof pathway predominates, D-fructose 1,6-diphosphate is the chief activator.

UDP-D-glucose pyrophosphorylase, which is involved in the synthesis of glycogen in animals, does not seem to be under such effective control as ADP-D-glucose pyrophosphorylase of plants and bacteria, presumably because UDP-D-glucose, unlike ADP-D-glucose, has to serve also as the precursor of D-glucose, D-galactose, D-glucuronic acid, L-iduronic acid, D-xylose, and so on, in the synthesis of complex saccharides. S. Kornfeld, however, found[75] that UDP-D-glucose pyrophosphorylase from liver is inhibited by AMP, as well as by UDP-D-glucose, a result suggesting a control by energy-charge levels. Interestingly, the same enzyme from bacteria, which do not use UDP-D-glucose for the synthesis of glycogen, is not significantly inhibited by AMP.[76]

Several new types of enzyme that degrade "sugar nucleotides" have been reported. Cabib and coworkers[77] and Dankert and coworkers[78] have found an enzyme that catalyzes the following reaction in yeast and wheat germ.

$$\text{XDP-hexose} + \text{Pi} \rightleftharpoons \text{XDP} + \text{hexosyl phosphate}$$

where X is the nucleoside residue and Pi is inorganic phosphate.

(73) C. E. Furlong and J. Preiss, *J. Biol. Chem.*, **244**, 2539 (1969).
(74) M. Calvin and A. A. Benson, "The Photosynthesis of Carbon Compounds," Benjamin, Inc., New York, N. Y., 1962.
(75) S. Kornfeld, *Fed. Proc.*, **24**, 536 (1965).
(76) T. Nakae and H. Nikaido, *J. Biol. Chem.*, **246**, 4386 (1971).
(77) H. Carminatti and E. Cabib, *J. Biol. Chem.*, **240**, 2110 (1965); E. Cabib, H. Carminatti, and N. M. Woyskovsky, *ibid.* **240**, 2114 (1965).
(78) M. Dankert, I. R. J. Goncalves, and E. Recondo, *Biochim. Biophys. Acta*, **81**, 78 (1964).

Various "sugar nucleotides" were found to be utilized in this reaction. For the yeast enzyme, compounds containing D-mannose were the best substrates, and, among "D-mannose nucleotides," the activity increased[77] in the order: ADP- < GDP- < dTDP- < UDP-.

A purified enzyme, from yeast, was found to catalyze the following reaction.[79] This enzyme is highly specific, and does not act on other

$$\text{GDP-D-glucose} + H_2O \longrightarrow \text{GDP} + \text{D-glucose.}$$

"nucleotide sugars" or on other phosphoric esters.

5'-Nucleotidase from *Escherichia coli* was found to catalyze the following reaction.[80] Although other "nucleotide D-glucoses" are

$$\text{UDP-D-glucose} + H_2O \longrightarrow$$
$$\text{uridine} + \text{Pi} + \alpha\text{-D-glucopyranosyl phosphate}$$

attacked, the rates of reaction are much lower. The results of studies using intact cells indicate that this enzyme is physiologically active in the degradation of intracellular UDP-D-glucose.[81] It is interesting that *E. coli* also contains a specific protein that inhibits this enzyme.[80]

IV. SYNTHESIS OF OLIGOSACCHARIDES

1. Sucrose

The synthesis of sucrose by Leloir and collaborators has been discussed previously.[1] These workers found that UDP-D-glucose is the D-glucose donor for formation of sucrose. The synthesis takes place by means of two separate enzymes, one utilizing D-fructose as the acceptor, and the other, D-fructose 6-phosphate. The sucrose phosphate formed in the second reaction is hydrolyzed by a phosphatase, resulting in the formation of free sucrose. As the equilibrium of the reaction for the formation of sucrose phosphate lies to the right, and inasmuch as the large accumulation of sucrose in some plants could be better accounted for by hydrolysis of the sucrose phosphate with phosphatase (which is a practically irreversible reaction), it was suggested that sucrose in plants is most likely synthesized by way of the sucrose phosphate intermediate.

Later work seemed to add credence to this concept. Although small amounts of sucrose phosphate have been found among labeled,

(79) S. Sonnino, H. Carminatti, and E. Cabib, *J. Biol. Chem.*, **241**, 1009 (1966).
(80) L. Glaser, A. Melo, and R. Paul, *J. Biol. Chem.*, **242**, 1944 (1967).
(81) J. B. Ward and L. Glaser, *Arch. Biochem. Biophys.*, **134**, 612 (1969).

photosynthetic products in plants, this sugar phosphate is not readily obtainable in plant tissues, as it is apparently hydrolyzed to sucrose by a phosphatase as soon as it is produced. However, Bird and co-workers[82] found that, when tobacco-leaf chloroplasts extracted with nonaqueous solvents were used as an enzyme source, with UDP-D-glucose and D-fructose 6-phosphate as substrates, sucrose 6-phosphate was formed.[83] Acetone-extracted chloroplasts from sugar-cane leaves, in addition to forming sucrose from UDP-D-glucose and D-fructose, were also found to utilize D-fructose 6-phosphate as an acceptor for D-glucose, to produce small proportions of sucrose phosphate.[84] Moreover, the presence of enzymes that catalyze the synthesis and breakdown of sucrose phosphate in the stem and leaf tissue of sugar cane was demonstrated by Hatch.[85] These results support the view that sucrose 6-phosphate is synthesized first, and then hydrolyzed to free sucrose.

The sucrose synthetase (UDP-D-glucose:D-fructose transglucosylase) that utilizes D-fructose as the D-glucosyl acceptor for sucrose formation may serve an important function in the degradation of this disaccharide. Sucrose may be completely degraded if the UDP-D-glucose is used up in various other metabolic reactions. This assumption is supported by an observation of Milner and Avigad;[86] they found that highly purified, sugar-beet, sucrose synthetase, practically freed from traces of invertase and phosphatase, is effective in the degradation of sucrose. An investigation of the specificity of this enzyme with regard to various nucleoside pyrophosphates showed them to be effective D-glucosyl acceptors.[87] Analysis of reaction systems containing sucrose in the presence of UDP, dTDP, ADP, CDP, or GDP indicated the formation of new nucleotide components that appeared to be the D-glucosyl derivatives of the corresponding nucleoside pyrophosphates. If the value of UDP was taken as 100, the relative effectiveness (as D-glucose acceptors) of dTDP, ADP, CDP, and GDP was 52, 16, 12, and 6, respectively.

Grimes, Jones, and Albersheim[88] purified sucrose synthetase from mung beans, and determined that its molecular weight was $\sim 10^6$.

(82) I. F. Bird, H. K. Porter, and C. R. Stocking, *Biochim. Biophys. Acta*, **100**, 366 (1965).
(83) R. B. Frydman and W. Z. Hassid, *Nature*, **199**, 382 (1963).
(84) S. Haq and W. Z. Hassid, *Plant Physiol.*, **40**, 591 (1965).
(85) M. D. Hatch, *Biochem. J.*, **93**, 521 (1964).
(86) Y. Milner and G. Avigad, *Israel J. Chem.*, **2**, 316 (1964).
(87) Y. Milner and G. Avigad, *Nature*, **206**, 825 (1965).
(88) W. J. Grimes, B. L. Jones, and P. Albersheim, *J. Biol. Chem.*, **245**, 188 (1970).

They showed that the enzyme is capable of catalyzing the synthesis of sucrose from each of the "nucleoside diphosphate D-glucoses" containing, respectively, uridine, adenosine, thymidine, cytidine, and guanosine residues, and that this ability remains associated with the purified particle. UDP-D-glucose inhibits synthesis of sucrose from various other "nucleoside diphosphate D-glucoses" much more effectively than would be expected on the basis of a simple competition mechanism. The authors suggested that UDP-D-glucose controls the synthesis of the other "nucleoside diphosphate D-glucoses" from sucrose by causing a modification of the sucrose synthetase.

Results obtained by Leloir and Cardini[89] indicated that two separate enzymes are involved in the biosynthesis, in plants, of sucrose and sucrose phosphate from D-fructose and D-fructose 6-phosphate, respectively in the presence of UDP-D-glucose; these enzymes have been partially separated. Slabnik and coworkers[90] succeeded in isolating sucrose synthetase and sucrose 6-phosphate synthetase from potato tubers, and determined some of the properties of the partially purified preparations. The sucrose synthetase showed an optimum activity at 45° and was inhibited completely by ADP and some phenolic D-glucosides, whereas these had no effect on sucrose 6-phosphate synthetase.

Another important difference in the properties of the two enzymes is their specificity toward the D-glucosyl donor. Whereas sucrose 6-phosphate synthetase acts only on UDP-D-glucose, sucrose synthetase is capable of effectively utilizing the D-glucose from other "sugar nucleotides," especially ADP-D-glucose. According to Slabnik and coworkers,[90] the fact that ADP competitively inhibits the formation of sucrose, whereas it has no effect on the formation of sucrose 6-phosphate, indicates that sucrose synthetase may serve as a link between sucrose and formation of starch, whereas the sucrose 6-phosphate synthetase is probably concerned only with the synthesis of sucrose. It is suggested that, as a number of phenolic glycosides that inhibit the synthesis of starch also inhibit the formation of sucrose from UDP-D-glucose or ADP-D-glucose, there is an interconnection between the synthesis of both sucrose and starch. Sucrose 6-phosphate synthetase is excluded from this interconversion;[87,91] it acts only as a sucrose-synthesizing enzyme.

(89) L. F. Leloir and C. E. Cardini, *J. Biol. Chem.*, **214**, 157 (1955).
(90) E. Slabnik, R. B. Frydman, and C. E. Cardini, *Plant Physiol.*, **43**, 1063 (1968).
(91) M. A. R. de Fekete and C. E. Cardini, *Arch. Biochem. Biophys.*, **104**, 173 (1964).

2. α,α-Trehalose

This nonreducing disaccharide, α-D-glucopyranosyl α-D-glucopy-ranoside, was enzymically synthesized by Cabib and Leloir[1,92] (with yeast-enzyme preparation) from UDP-D-glucose plus D-glucose 6-phosphate; it was obtained as a phosphorylated derivative, namely, α,α-trehalose 6-phosphate. In a subsequent step, the phosphate group was hydrolyzed off by a phosphatase, affording the free disaccharide. The formation of α,α-trehalose was similarly shown to ensue from the same substrates by use of enzymic preparations from insects.[1]

Elbein[93,94] had found that an enzyme system from *Streptomyces hygroscopicus* and a number of other *Streptomyces* species catalyzes the transfer of D-glucose-^{14}C from GDP-D-glucose-^{14}C to D-glucose 6-phosphate, affording α,α-trehalose-^{14}C 6-phosphate. The reaction appears to be specific for GDP-D-glucose.

However, for other actinomycetes, principally *Mycobacteria* and *Nocardia* species, crude extracts were found to utilize D-glucose 6-phosphate and either UDP-D-glucose or GDP-D-glucose for the formation of α,α-trehalose 6-phosphate according to the following two reactions.[95]

UDP-D-glucose + D-glucose 6-phosphate $\longrightarrow$
$$\alpha,\alpha\text{-trehalose 6-phosphate} + \text{UDP} \quad (1)$$

GDP-D-glucose + D-glucose 6-phosphate $\longrightarrow$
$$\alpha,\alpha\text{-trehalose 6-phosphate} + \text{GDP} \quad (2)$$

It was shown[95] that extracts of *Mycobacterium smegmatis* and of several other *Mycobacteria* species can be separated into two frac-tions; both are required in order to catalyze reaction (1), whereas only one of these fractions is needed for reaction (2).

A cell-free extract of *Mycobacterium smegmatis* was separated into the two fractions by chromatography on O-(2-diethylaminoethyl)-cellulose. Fraction A catalyzed synthesis of α,α-trehalose from GDP-D-glucose, but was relatively inactive with UDP-D-glucose. However, when Fraction B was added to Fraction A, UDP-D-glucose was able to serve as an effective D-glucopyranosyl donor for synthesis of α,α-trehalose. Under these conditions, GDP-D-glucose was still

(92) E. Cabib and L. F. Leloir, *J. Biol. Chem.*, **231**, 259 (1958).
(93) A. D. Elbein, *J. Biol. Chem.*, **242**, 403 (1967).
(94) A. D. Elbein, *J. Bacteriol.*, **96**, 1623 (1968).
(95) C. Liu, B. W. Patterson, D. Lapp, and A. D. Elbein, *J. Biol. Chem.*, **244**, 3728 (1969).

active, but less so than with Fraction A alone. Fraction B could be replaced by crude α-lactalbumin, but not with purified lactalbumin or a number of other proteins. The active component of crude lactalbumin proved to be ribonucleic acid (RNA). Although poly(uridylic acid) and the RNA fraction of *M. smegmatis* are also active, they are less so than fraction B. The activity of each of these could be destroyed by treatment with ribonuclease (RNAase).

The question as to whether Fraction A contains two different enzymes for synthesis of α,α-trehalose, or whether both activities reside in one protein, has not yet been answered. This enzymic system may be somewhat analogous to the lactose system, described by Brodbeck and coworkers[96] and Brew and coworkers,[97] in which a second protein (α-lactalbumin) apparently alters the substrate specificity of the enzyme so that it can utilize D-glucose instead of 2-acetamido 2-deoxy-D-glucose as the substrate.

Roth and Sussman[98] demonstrated the presence of α,α-trehalose synthetase activity in extracts of the micro-organism *Dictyostelium discoideum*. Their data indicated that the level of α,α-trehalose synthetase activity changes considerably during the morphogenetic sequence. The presence of α,α-trehalose in this micro-organism has also been reported.

3. Lactose

Lactose was first synthesized enzymically with particulate preparations from lactating guinea-pig or mammary glands[99] by a process involving the following reaction.

$$\text{UDP-D-galactose} + \text{D-glucose} \xrightarrow[\text{transferase}]{\text{D-galactosyl}} \text{lactose} + \text{UDP}$$

Later, from cows milk, a soluble preparation was obtained containing a UDP-D-galactose:D-glucose 4-β-D-galactosyl transferase capable of synthesizing lactose from UDP-D-galactose plus D-glucose by the same reaction.[100,101] The enzyme appears to be specific for UDP-D-galactose, and none of the "D-galactosyl nucleotides" con-

(96) U. Brodbeck, W. L. Denton, N. Tanahashi, and K. E. Ebner, *J. Biol. Chem.*, **242**, 1391 (1967).

(97) K. Brew, T. C. Vanaman, and R. L. Hill, *Proc. Nat. Acad. Sci. U. S.*, **59**, 491 (1968).

(98) R. Roth and M. Sussman, *Biochim. Biophys. Acta*, **122**, 225 (1966).

(99) W. M. Watkins and W. Z. Hassid, *J. Biol. Chem.*, **237**, 1432 (1962).

(100) H. Babad and W. Z. Hassid, *J. Biol. Chem.*, **239**, PC 946 (1964).

(101) H. Babad and W. Z. Hassid, *J. Biol. Chem.*, **241**, 2672 (1966).

taining bases other than uracil (that is, guanine, adenine, cytosine, or thymine) can serve as substrate for the formation of lactose. Also, with the exception of 2-acetamido-2-deoxy-D-glucose, no other sugar can be substituted for D-glucose as the D-galactose acceptor for disaccharide formation. However, this amino sugar is only effective as a D-galactosyl acceptor to the extent of 25% relative to D-glucose. The disaccharide formed appears to be O-β-D-galactopyranosyl-$(1 \rightarrow 4)$-2-acetamido-2-deoxy-D-glucose.[99]

An attempt to incorporate a number of isomers of D-glucose labeled with ^{14}C or tritium into slices of mammary glands of lactating rats showed that D-glucose (which was incorporated intact) was the only D-galactosyl acceptor for the formation of lactose.[102]

Brodbeck and Ebner[103] found that the soluble lactose synthetase from milk can be separated into two protein components, A and B, which individually do not exhibit any catalytic activity; however, their recombination restores full lactose synthetase activity. The B fraction has been crystallized from bovine skim milk and bovine mammary tissue, and was identified as α-lactalbumin.[96] It was thus found that α-lactalbumin can be substituted for the B protein of lactose synthetase. Lactose synthetases from the milk of sheep, goats, pigs, and humans were also resolved into A and B proteins, and the fractions from these species were shown to be qualitatively inter-changeable in the rate assay of lactose synthesis.[96] Determination of the amino acid sequence of α-lactalbumin (B fraction) has shown a distinct homology in the sequence of amino acids of α-lactalbumin and hen's egg-white lysozyme,[104] suggesting that lysozyme and α-lactalbumin have evolved from a common ancestral gene.

Considerable evidence has been obtained by Brew and co-workers[97,104] indicating that protein A acts as a general D-galactosyl transferase that catalyzes the following reaction.

UDP-D-galactose + 2-acetamido-2-deoxy-D-glucose $\longrightarrow$
N-acetyl-lactosamine + UDP.

A similar reaction had previously been observed by McGuire and coworkers[105] in an investigation of D-galactosyl transferase in bovine colostrum and in a particulate fraction of rat tissues; this enzyme is believed to play an important role in the biosynthesis of carbohy-

(102) J. C. Bartley, S. Abraham, and I. L. Chaikoff, *J. Biol. Chem.*, **241**, 1132 (1966).
(103) U. Brodbeck and K. E. Ebner, *J. Biol. Chem.*, **241**, 762 (1966).
(104) K. Brew, T. C. Vanaman, and R. L. Hill, *J. Biol. Chem.*, **242**, 3747 (1967).
(105) E. J. McGuire, G. W. Jourdian, D. M. Carlson, and S. Roseman, *J. Biol. Chem.*, **240**, PC 4112 (1965).

drates found in glycoproteins and blood-group substances. The α-lactalbumin therefore appears to modify the acceptor specificity of the general D-galactosyl transferase involved in glycoprotein synthesis, so that the D-glucose becomes the better acceptor of the D-galactosyl group from UDP-D-galactose.

Fitzgerald and coworkers[106] found that the K_m for UDP-D-galactose is not influenced by the concentration of D-glucose or α-lactalbumin, but that these compounds alter the maximum velocity of the reaction. There is a reciprocal relationship between the concentration of D-glucose and α-lactalbumin in the lactose synthetase assay, and α-lactalbumin lcwers the apparent K_m of D-glucose. It has been suggested that the physiological function of α-lactalbumin is to lower the K_m of D-glucose so that it may be used maximally for the synthesis of lactose. Lactose may be synthesized at maximum rates by the A protein in the absence of α-lactalbumin but in the presence of high concentrations of D-glucose.

4. Raffinose

Raffinose is a trisaccharide, O-α-D-galactopyranosyl-(1 → 6)-α-D-glucopyranosyl β-D-fructofuranoside, found in comparatively low concentration, together with sucrose, in many higher plants. In this trisaccharide, the D-galactopyranosyl group is attached through an

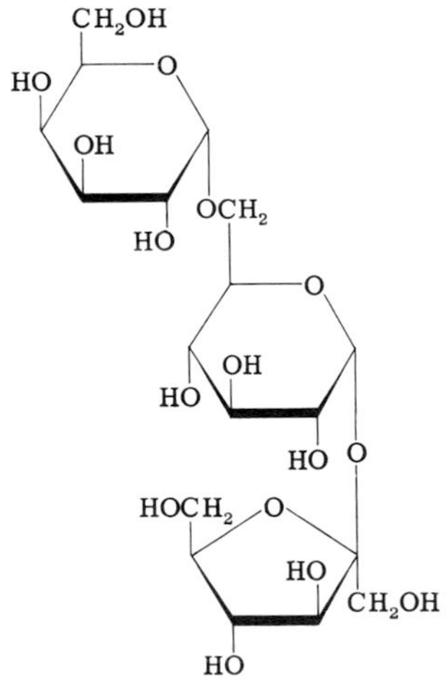

Raffinose

(106) D. K. Fitzgerald, U. Brodbeck, I. Kiyosawa, R. Mawal, B. Colvin, and K. E. Ebner, J. Biol. Chem., **245**, 2103 (1970).

α-D linkage to O-6 of the α-D-glucopyranosyl group of sucrose. However, attempts to effect a transfer of the α-D-galactopyranosyl group
of UDP-D-galactose to sucrose with enzymic preparations from
germinated, mung-bean seedlings and other plants (for the formation
of raffinose) were not successful. It appears that, if raffinose was
formed, the germinated seedlings must most probably have contained glycosidases that hydrolyzed this trisaccharide.

Bourne and coworkers[107] found that raffinose could be synthesized,
from a mixture of sucrose, α-D-galactosyl phosphate, and UTP as
substrates, by an enzyme preparation from dormant broad beans
(*Vicia faba*). The trisaccharide was subsequently synthesized by a
direct transfer of the D-galactosyl-^{14}C group from UDP-D-galactose-
^{14}C to sucrose with an enzyme preparation from mature broad beans.[108]
Apparently, the ungerminated seeds contain fewer hydrolytic enzymes than germinated ones, which accounts for the successful
enzymic synthesis of raffinose. The proportion of ^{14}C-labeled
D-galactose transferred to sucrose by the enzymic preparation from
dormant broad beans to form raffinose was 39% within 2.5 hours.
Raffinose has also been synthesized from UDP-D-galactose and
sucrose with an enzymic preparation from immature soybeans.[109]
Moreno and Cardini[110] found in extracts of wheat germ a specific
transgalactosylase that was capable of catalyzing the following
reaction.

$$\text{Raffinose} + \text{sucrose-}^{14}C \rightleftharpoons \text{raffinose-}^{14}C + \text{sucrose}$$

This result indicates that this enzyme transfers the α-D-galactosyl
group from raffinose to sucrose, thus regenerating raffinose.

Courtois and coworkers[111,112] showed that oligosaccharides and other
glycosides can be synthesized by hydrolases in plant extracts by
transfer of the sugar moiety from low-energy donors to sugars or
aglycons. Thus, synthesis of raffinose and of planteose [O-α-D-gal-
actopyranosyl-$(1 \rightarrow 6)$-O-β-D-fructofuranosyl-$(2 \rightarrow 1)$ α-D-glucopyrano-
side], which has been found in the weed *Plantago* and in seeds of
P. ovado, has been accomplished by using D-galactose as the donor in
the presence of α-D-galactosidase preparations. It was also shown[113]

(107) E. J. Bourne, J. B. Pridham, and M. W. Walter, *Biochem. J.*, **82**, 44P (1962).
(108) J. B. Pridham and W. Z. Hassid, *Plant Physiol.*, **40**, 984 (1965).
(109) T. Gomyo and M. Nakamura, *Agr. Biol. Chem.* (Tokyo), **30**, 425 (1966).
(110) A. Moreno and C. E. Cardini, *Plant Physiol.*, **41**, 909 (1966).
(111) J. E. Courtois, F. Petek, and T. Dong, *Bull. Soc. Chim. Biol.*, **43**, 1189 (1961).
(112) J. E. Courtois, *Bull. Soc. Bot. Fr.*, **115**, 309 (1968).
(113) F. Petek and J. E. Courtois, *Bull. Soc. Chim. Biol.*, **46**, 1093 (1964).

that an α-D-galactosidase from coffee beans transfers D-galactose from phenyl α-D-galactoside to cellobiose, giving rise to three isomeric trisaccharides having different structures.

Although the equilibrium for glycosidase-catalyzed reactions normally favors hydrolysis, such *in vivo* factors as rapid utilization of products, or localized high concentrations of substrate, may promote synthesis of oligosaccharides.

5. Stachyose

The tetrasaccharide stachyose consists of raffinose substituted with an additional D-galactosyl residue attached to O-6 of the D-galactosyl group of the trisaccharide. It would, therefore, be expected that,

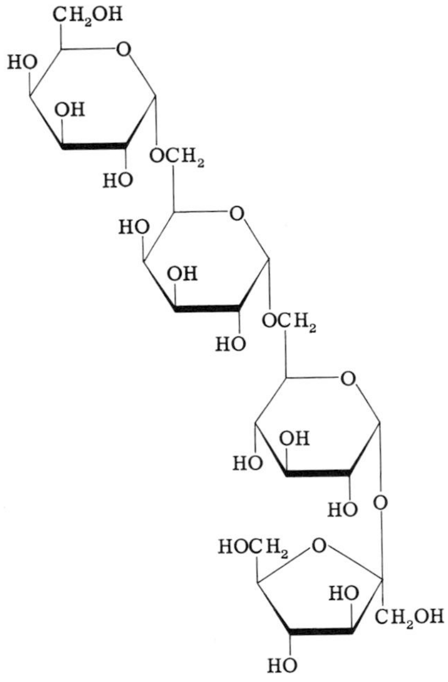

Stachyose

analogously to the process of formation of raffinose from sucrose, stachyose should be formed from raffinose and UDP-D-galactose by a transfer reaction. However, attempts to effect a transfer of D-galactose to raffinose for the formation of stachyose have been unsuccessful. Surprisingly, the donor of D-galactose for the formation of this tetrasaccharide appears to be galactinol (1-O-α-D-galactopyranosyl-*myo*-inositol).[114] Tanner and Kandler[115] found that a soluble enzyme from ripening seeds of dwarf beans (*Phaseolus vulgaris*) transfers a D-galactosyl group from galactinol to raffinose with high yield, giving rise to stachyose and *myo*-inositol.

Galactinol is found in considerable proportion in dwarf beans during a certain maturation period, and its synthesis precedes that of stachyose. The enzyme involved appears to be fairly specific with regard to the acceptor molecule. D-Galactose-^{14}C was not transferred from galactinol-^{14}C to glycerol, D-fructose, and a number of oligosaccharides. Slight acceptor specificity was observed with D-glucose, D-galactose, and lactose; but these sugars were less efficient than raffinose by a factor of 1/20 to 1/30. The most efficient acceptor of the sugars tested (besides raffinose) was melibiose, which had one-quarter the efficiency of raffinose.

In the crude extract of the ripening bean seeds, an enzyme is also present that transfers the D-galactosyl-^{14}C group from UDP-D-galactose-^{14}C, but not from ADP-D-galactose-^{14}C, to *myo*-inositol. This enzyme had previously been found in extracts of pea seeds.[116] The biosynthesis of stachyose is, therefore, assumed to occur by the following reactions.

$$\text{UDP-D-galactose} + myo\text{-inositol} \longrightarrow \text{galactinol} + \text{UDP}$$

$$\text{Galactinol} + \text{raffinose} \longrightarrow \text{stachyose} + myo\text{-inositol}$$

6. Verbascose

Verbascose is found in the roots of mullein, *Verbascum thapsus*. Tanner and coworkers[117] showed that the enzyme from seeds of *Vicia faba* that is capable of catalyzing D-galactosyl transfer from galactinol to raffinose, forming stachyose, also has the ability to effect an α-D-(1 → 6) linkage with the D-galactosyl group of stachyose, giving rise to verbascose, a member of the raffinose family.

(114) E. A. Kabat, D. L. MacDonald, C. E. Ballou, and H. O. L. Fischer, *J. Amer. Chem. Soc.*, **75**, 4507 (1953).
(115) W. Tanner and O. Kandler, *Plant Physiol.*, **41**, 1540 (1966).
(116) R. B. Frydman and E. F. Neufeld, *Biochem. Biophys. Res. Commun.*, **12**, 121 (1963).
(117) W. Tanner, L. Lehle, and O. Kandler, *Biochem. Biophys. Res. Commun.*, **29**, 166 (1967).

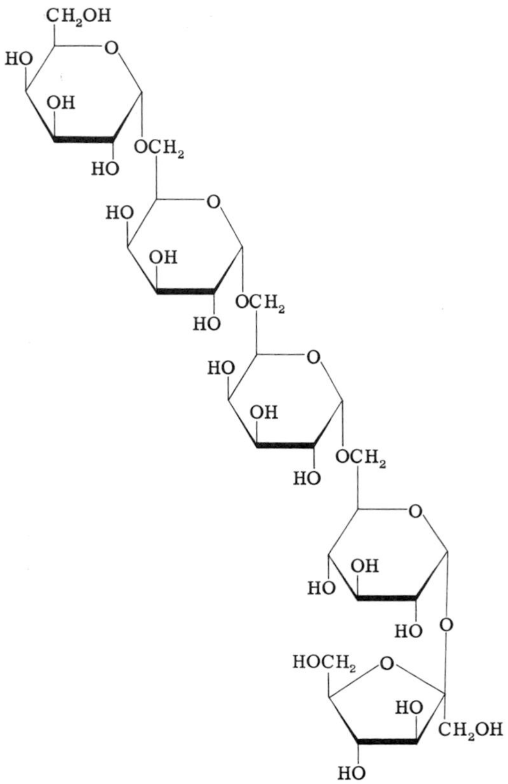

Verbascose

V. Synthesis of Homopolysaccharides

1. Glycogen

It is now generally accepted[1,118] that synthesis of glycogen occurs in mammalian tissues and yeast by the action of UDP-D-glucose:glycogen 4-α-D-glucosyltransferase, an enzyme that catalyzes the transfer of the α-D-glucopyranosyl group from UDP-D-glucose to an acceptor, forming glycogen.

The α-D-glucopyranosyl residues are united by the same α-D-(1 → 4) linkages, forming linear chains; a high-molecular-weight acceptor of the same type of linkage is required for the formation of the polysaccharide. When the linear chains become about ten resi-

(118) L. F. Leloir, 6th Int. Congr. Biochem., New York, Plenary Sessions, 33, 15 (1964).

dues long, they are intercombined by a branching enzyme, namely, $(1 \rightarrow 4)$-α-D-glucan:$(1 \rightarrow 4)$-α-D-glucan 6-glucosyltransferase, into ramified molecules having α-D-$(1 \rightarrow 6)$ links at the branch points.

The observation by Leloir and coworkers[1,119] that formation of glycogen by synthetase of animal or yeast origin is stimulated to a variable degree by the addition of D-glucose 6-phosphate has been further probed by a number of investigators. It has been found[1] that, in animal tissue, the synthetase exists in two interconvertible forms, one requiring D-glucose 6-phosphate (dependent), and the other, only slightly stimulated by this ester (independent). A similar situation has also been found in *Neurospora*[119a] and in the yeast *Saccharomyces cerevisiae*.[120]

The activities both of animal and yeast synthetases are under complicated, allosteric control by various metabolites. Although the "dependent" form is activated *in vitro* by D-glucose 6-phosphate, the intracellular concentration of this compound found in animal cells (50 to 250 μM in liver) is much lower than the concentrations needed for activation (2 mM for 50% activation of the liver enzyme). The "dependent" form of the liver enzyme can, therefore, be considered to be a completely inactive form under physiological conditions.[121] The "dependent" form of the muscle enzyme has a stronger affinity for D-glucose 6-phosphate; but this enzyme appears to be completely inhibited by physiological concentrations of ATP, ADP, and inorganic phosphate, and thus should also be considered inactive *in vivo*.[122] It is, therefore, likely that, both in muscle and liver, the regulation of glycogen synthetase activity is primarily accomplished through the interconversion of inactive (or "dependent") and active (or "independent") forms, as first suggested by Mersman and Segal.[121]

It is not yet completely clear whether the active, "independent" form is under a finer control through allosteric regulation by metabolites. If saturating concentrations of UDP-D-glucose are used for assay, the activity of the enzyme is only slightly affected by D-glucose 6-phosphate, and so it was designated the "independent" form. However, it was found that the activity of the liver enzyme is strongly affected by concentrations of D-glucose 6-phosphate in the phys-

(119) L. F. Leloir, J. M. Olavarría, S. H. Goldemberg, and H. Carminatti, *Arch. Biochem. Biophys.*, **81**, 508 (1959).

(119a) M. T. Téllez-Iñon, H. Terenzi, and H. N. Torres, *Biochim. Biophys. Acta*, **191**, 765 (1969).

(120) L. B. Rothman-Denes and E. Cabib, *Proc. Nat. Acad. Sci. U. S.*, **66**, 967 (1970).

(121) H. J. Mersmann and H. J. Segal, *Proc. Nat. Acad. Sci. U. S.*, **58**, 1688 (1967).

(122) R. Piras, L. B. Rothman, and E. Cabib, *Biochemistry*, **7**, 56 (1968).

iological range, provided that the concentrations of UDP-D-glucose are kept at a low, physiological level.[121] At first, the significance of this effect was not recognized, because inorganic phosphate was found to activate the liver enzyme to 60% of the maximal activity, at a concentration corresponding to its normal, intracellular level.[121] However, ATP was later found to counteract inorganic phosphate,[122,123] and it now appears possible that the levels of these compounds, as well as that of D-glucose 6-phosphate, regulate the activity of the active liver-enzyme.[123,124]

Krebs[125] and Atkinson[126] suggested that the course of energy metabolism may be mainly controlled by the intracellular concentrations of AMP, ADP, and ATP. According to this concept, at low levels of energy charge, only a small portion of the adenine nucleotides will be in the form of ATP, and the levels of AMP and inorganic phosphate (Pi) will be high; this will also be the state in which degradation, rather than biosynthesis, of reserve polysaccharide should be favored. Contrary to this expectation, low levels of ATP and high levels of Pi strongly activate the "independent" form of the liver synthetase. It has been hypothesized that, when the liver takes in a large amount of D-glucose, much ATP is consumed for phosphorylating this sugar, thus producing conditions that activate the glycogen synthetase.[124]

Studies by Larner and coworkers[127] showed that 3':5'-cyclic AMP is an important factor that regulates the interconversion of the two forms of glycogen synthetase. The conversion of the independent into the dependent form is catalyzed by a kinase requiring ATP and Mg^{2+}, the activity of which is increased by cyclic AMP. Huijing and Larner[128] and others[129] showed that muscle kinase appears to be similar to, but not identical with, phosphorylase b kinase,[130,131] and that the two enzymes seem to be equally sensitive to stimulation by cyclic AMP. It has been suggested[128,129] that cyclic AMP may in this instance act by increasing the affinity of an allosteric site for magnesium on the kinases. As epinephrine and other hormones increase the level of

(123) H. DeWulf, W. Stalmans, and H. G. Hers, Eur. J. Biochem., 6, 545 (1968).
(124) A. H. Gold, Biochemistry, 9, 946 (1970).
(125) H. A. Krebs, Proc. Roy. Soc. (London), Ser. B., 159, 545 (1964).
(126) D. E. Atkinson, Science, 150, 851 (1965).
(127) J. Larner, Trans. N. Y. Acad. Sci., 29, 192 (1966).
(128) F. Huijing and J. Larner, Proc. Nat. Acad. Sci. U. S., 56, 647 (1966).
(129) M. M. Appleman, L. Birnbaumer, and H. N. Torres, Arch. Biochem. Biophys., 116, 39 (1966).
(130) D. L. Friedman and J. Larner, Biochemistry, 4, 2261 (1965).
(131) E. Belocopitow, M. D. C. G. Fernandez, L. Birnbaumer, and H. N. Torres, J. Biol. Chem., 242, 1227 (1967).

cyclic AMP,[132] these observations explain the decrease in glycogen synthetase activity following the administration of these hormones.

Another hormone, namely, insulin, acts in the opposite direction, and produces a rapid rise in glycogen synthetase activity.[133,134] It has been suggested that insulin acts by changing the glycogen synthetase kinase into a form having a low affinity for cyclic AMP, so that the conversion of the independent into the dependent form (that is, inactivation of the glycogen synthetase) is retarded.[134]

The dependent form can be reconverted into the independent form through dephosphorylation by a specific phosphatase[135] that is inhibited by glycogen;[136] this can provide a mechanism for the feedback regulation of glycogen levels in tissues, through the inhibition of reactivation of glycogen synthetase by its own product.

The independent form can be converted into the dependent form by a different mechanism. It has been found[131,136] that addition of calcium ions to certain glycogen synthetase preparations produces a conversion of the independent into the dependent form. This conversion does not involve ATP, and is not affected by adenosine 3':5'-cyclic phosphate; it requires a protein[131] similar to that involved in the activation of inactive phosphorylase b kinase by Ca^{2+}. Thus, calcium appears to exert an effect on the regulation of glycogen synthetase, and it is hypothesized that the inactivation of muscle glycogen synthetase after muscle contraction may be caused by this mechanism, mediated[137] by an alteration of intracellular levels of Ca^{2+}.

In bacteria, ADP-D-glucose, not UDP-D-glucose, is the donor of D-glucosyl residues in glycogen synthesis.[138–140] ADP-D-glucose: glycogen D-glucosyl transferases from several bacterial species were found not to be affected by D-glucose 6-phosphate and other glycolytic intermediates.[140] In extracts of *Arthrobacter* and *Escherichia coli*, the

(132) G. A. Robison, R. W. Butcher, and E. W. Sutherland, *Ann. Rev. Biochem.*, **37**, 149 (1968).

(133) J. S. Bishop and J. Larner, *J. Biol. Chem.*, **242**, 1354 (1967).

(134) C. Villar-Palasi and J. I. Wenger, *Fed. Proc.*, **26**, 563 (1967).

(135) D. L. Friedman and J. Larner, *Biochim. Biophys. Acta*, **64**, 185 (1962); *Biochemistry*, **2**, 669 (1963).

(136) E. Belocopitow, M. M. Appleman, and H. N. Torres, *J. Biol. Chem.*, **240**, 3473 (1965).

(137) E. G. Krebs and E. H. Fischer, *Advan. Enzymol.*, **24**, 263 (1962).

(138) E. Greenberg and J. Preiss, *J. Biol. Chem.*, **239**, PC4314 (1964).

(139) L. Shen, H. P. Ghosh, E. Greenberg, and J. Preiss, *Biochim. Biophys. Acta*, **89**, 370 (1964).

(140) E. Greenberg and J. Preiss, *J. Biol. Chem.*, **240**, 2341 (1965).

levels of ADP-D-glucose pyrophosphorylase activity were found to be lower than that of the ADP-D-glucose:glycogen transglycolase activity.[141] Thus, the rate-limiting step in bacterial glycogen synthesis appears to correspond to the step catalyzed by ADP-D-glucose pyrophosphorylase, and this is also the step that is under efficient metabolic regulation (see also Section III; p. 365). The ADP-D-glucose pyrophosphorylase from *E. coli* is activated by a number of glycolytic intermediates, of which D-fructose 1,6-diphosphate appears to be the most effective.[141,142] D-Fructose 1,6-diphosphate activates the enzyme by increasing the maximal velocity of synthesis of ADP-D-glucose (and pyrophosphorolysis) and by increasing the affinity of the enzyme for its substrate.[142] It has been found that NADPH and pyridoxal 5′-phosphate are powerful activators of the *E. coli* enzyme.[143] This enzyme is also inhibited[142] by AMP, ADP, or Pi. As ATP stimulates the reaction, the enzyme activity is regulated by the ATP/AMP ratio within the cell, or, more precisely, by the energy-charge level of the cell,[126] high energy-charge levels favoring the production of glycogen.

It has been shown that both of the cations Mn^{2+} and Mg^{2+} can fulfil the divalent requirement for ADP-D-glucose synthesis,[142] and that both are equally effective, regardless of whether synthesis of ADP-D-glucose occurs in the presence or absence of D-fructose 1,6-diphosphate. However, the kinetics of ADP-D-glucose synthesis in the presence of Mg^{2+} were shown to differ from the kinetics[144] of the reaction in the presence of Mn^{2+}.

ADP-D-glucose pyrophosphorylase from *Rhodospirillum rubrum* has been purified to apparent homogeneity, and its molecular weight determined[73] to be 195,000. This enzyme is specifically activated by pyruvate, and is relatively insensitive to inhibition by AMP, ADP, or Pi.

Concerning the problem of the availability of an acceptor that would be needed to initiate the polymerization reaction, it has been suggested[118] that, under normal physiological conditions, some glycogen molecules or molecular fragments are probably always present, as it may be assumed that each animal cell retains some glycogen during cell division. However, it was assumed that were glycogen

(141) J. Preiss, L. Shen, and M. Partridge, *Biochem. Biophys. Res. Commun.*, **18**, 180 (1965).

(142) J. Preiss, L. Shen, E. Greenberg, and N. Gentner, *Biochemistry*, **5**, 1833 (1966).

(143) N. Gentner, E. Greenberg, and J. Preiss, *Biochem. Biophys. Res. Commun.*, **36**, 373 (1969).

(144) N. Gentner and J. Preiss, *J. Biol. Chem.*, **243**, 5882 (1968).

to disappear completely, maltose (the smallest product of degradation next to D-glucose) might remain to serve as a primer, but one having a very low rate.[145]

Gahan and Conrad[146] have published a study that throws new light on this problem. They isolated from *Aerobacter aerogenes* cells, and purified, a glycogen synthetase fraction and an activator protein that, together, catalyze *de novo* glycogen synthesis from ADP-D-glucose. Both the fraction and the activator have been shown to be devoid of glycogen. In the presence of ADP-D-glucose and Mg^{2+}, the purified glycogen synthetase catalyzes a low rate of glycogen synthesis that is stimulated several hundred-fold by supernatant activator protein. It should be noted that the glycogen synthetase has to be prepared with the utmost care in order that this stimulatory effect on *de novo* synthesis can be shown; sonic disintegration of the cells, or simple aging of the enzyme, converts the synthetase into a form having the ordinary properties, including an absolute requirement for "primer" glycogen.

Pulse-chase experiments[146] showed that, of the D-glucose residues incorporated into glycogen in the early stages of the reaction, approximately 50% remain in the external chains of the product recovered after 4- to 6-fold increase in the amount of glycogen synthesized. This result is not that to be expected from the classical mechanism of glycogen synthesis, namely, successive addition of D-glucopyranosyl groups at the nonreducing ends. In analogy with the mechanism of biosynthesis of O-antigen[147] (see also, Section VI, 6; p. 420), Gahan and Conrad proposed that short oligosaccharide chains could be produced on "carrier" molecules, and that these oligosaccharides could be joined together on the carrier so that the chain growth would occur mainly at the region of the reducing end. It was implied that this is the physiological way of synthesizing glycogen, and that, when damaged, the system degenerates into the state of ordinary "glycogen synthetases," where exogenous, glycogen "primer" becomes more active as the acceptor of glycosyl groups. Although this is an attractive hypothesis, it must be emphasized that other interpretations of the pulse-chase data are possible.[146]

No information had been available on the nature of the postulated "carrier" of D-glucosyl groups. However, Behrens and Leloir[148] have

(145) S. H. Goldemberg, *Biochim. Biophys. Acta*, **56**, 357 (1962).
(146) L. C. Gahan and H. E. Conrad, *Biochemistry*, **7**, 3979 (1968).
(147) P. W. Robbins, D. Bray, M. Dankert, and A. Wright, *Science*, **158**, 1536 (1967).
(148) N. H. Behrens and L. F. Leloir, *Proc. Nat. Acad. Sci. U. S.*, **66**, 153 (1970).

shown that a fraction from liver catalyzes transfer of the D-glucosyl group from UDP-D-glucose to a lipid acceptor that appears to be identical with the compound obtained by chemical phosphorylation of dolichol (a polyprenol, present in animal tissues, that contains 16 to 21 isoprene units, the first of which is saturated).[149] The product of the reaction appears to be dolichol 1-(β-D-glucopyranosyl phosphate). It is not yet known whether this compound is involved in the synthesis of glycogen. The liver enzyme is known to catalyze the further transfer of a D-glucosyl group from this lipid intermediate to a protein, and the hydrolysis of the glucoprotein. The following reactions thus take place in this system.

UDP-D-glucose + acceptor lipid $\longrightarrow$

D-glucose–acceptor lipid + UDP

D-Glucose–acceptor lipid + protein $\longrightarrow$

acceptor lipid + D-glucose–protein

D-Glucose-protein $\longrightarrow$ D-glucose + protein

2. Starch

The discovery of an enzyme that synthesizes glycogen from UDP-D-glucose or ADP-D-glucose stimulated search for a similar enzyme in plant tissues that would be expected to synthesize starch (see Section V,1, p. 376). Leloir and coworkers[149a] found such an enzyme(s) associated with starch grains from beans, potatoes, and corn seedlings. Freshly isolated starch granules from these sources catalyze the incorporation of the D-glucosyl-^{14}C group from UDP-D-glucose-^{14}C into a radioactive polysaccharide consisting of α-D-$(1 \rightarrow 4)$ glucosidically linked D-glucose residues. The effective synthesizing enzyme is closely bound to the starch granule, which could not be dissociated from it.

From other plants, enzymically active starch granules capable of transferring D-glucosyl groups from ADP-D-glucose and UDP-D-glucose to an acceptor, forming starch, have also been reported. Frydman[150] obtained from potato tubers a preparation of starch grains capable of catalyzing the incorporation of D-glucosyl groups from

(149) J. Burgos, F. W. Hemming, J. F. Pennock, and R. A. Morton, *Biochem. J.*, **88**, 470 (1963).

(149a) L. F. Leloir, M. A. R. de Fekete, and C. E. Cardini, *J. Biol. Chem.*, **236**, 636 (1961); E. Recondo, and L. F. Leloir, *Biochem. Biophys. Res. Commun.*, **6**, 85 (1961).

(150) R. B. Frydman, *Arch. Biochem. Biophys.*, **102**, 242 (1963).

ADP-D-glucose or UDP-D-glucose into starch. Oligosaccharides of the maltose series also served as D-glucosyl acceptors. The D-glucosyl group was transferred to either acceptor at a greater rate from ADP-D-glucose than from UDP-D-glucose. Starch grains from potato sprouts and from seeds of wrinkled peas, red and white corn, and waxy maize were also shown to catalyze the incorporation of D-glucosyl groups into starch from the same substrates. Akazawa and coworkers[151] also isolated, from ripening rice grains, starch granules as an enzyme source that synthesized ^{14}C-labeled starch from UDP-D-glucose-^{14}C. D-Glucose-^{14}C was incorporated into both the amylose and the amylopectin fractions. Similar starch granules from leaves of *Phaseolus aureus* utilized ADP-D-glucose.[152]

Later, Frydman and Cardini[153,154] obtained from sweet corn a soluble, enzyme preparation that catalyzes the transfer of the D-glucosyl group from ADP-D-glucose to phytoglycogen, which has a branched chemical structure similar to that of glycogen. The soluble, glucan synthetase is also capable of utilizing dADP-D-glucose, although to a lesser extent.[154] Soluble enzyme preparations were also obtained by the same investigators[155] from tobacco leaves and potato tubers. Furthermore, soluble glucan synthetases have been found in waxy, starchy varieties of sweet corn and in a number of other plants.[156] Ghosh and Preiss[157] also found a soluble enzyme in spinach chloroplasts.

The soluble (1 → 4)-α-D-glucan transferases obtained from different plant sources differ in specificity with regard to their primer requirements.[150] The soluble-enzyme preparation from tobacco leaves is capable of using either amylopectin, glycogen, or heated starch granules, whereas that from potato tubers can use either starch granules or soluble glucan as the acceptor.[155] The enzyme that synthesizes starch appears to be very similar to that involved in glycogen formation, but it differs in specificity regarding primer requirement.[158]

(151) T. Akazawa, T. Minamikawa, and T. Murata, *Plant Physiol.*, **39**, 371 (1964).

(152) T. Murata and T. Akazawa, *Biochem. Biophys. Res. Commun.*, **16**, 6 (1964).

(153) R. B. Frydman and C. E. Cardini, *Biochem. Biophys. Res. Commun.*, **14**, 353 (1964).

(154) R. B. Frydman and C. E. Cardini, *Biochim. Biophys. Acta*, **96**, 294 (1965).

(155) R. B. Frydman and C. E. Cardini, *Biochem. Biophys. Res. Commun.*, **17**, 407 (1964).

(156) R. B. Frydman, B. C. De Souza, and C. E. Cardini, *Biochim. Biophys. Acta*, **113**, 620 (1966).

(157) H. P. Ghosh and J. Preiss, *Biochemistry*, **4**, 1354 (1965).

(158) R. B. Frydman and C. E. Cardini, *J. Biol. Chem.*, **242**, 312 (1967).

The properties of the particulate starch synthetase have been found to change when the structure of the granules is modified by mechanical disruption.[158] If the granules are disrupted, UDP-D-glucose is no longer capable of serving as the substrate, but the activity of ADP-D-glucose is enhanced. The specificity of this enzyme is similar to that of the soluble, $(1 \rightarrow 4)$-α-D-glucan synthetase. The differences in specificity between the synthetase bound to the granule and the soluble synthetase might indicate that they are different enzymes, or that both activities are due to the same enzyme in different conformations.

From *Solanum tuberosum*, Slabnik and Frydman[159] isolated a unique phosphorylase that has no requirement for the primer addition for formation of an amylopectin-like polysaccharide in a cell-free system. The properties of this enzyme were found to differ from those of the usual potato phosphorylase; this new enzyme is assumed to be a glycoprotein, the glycosidic component of which acts as the primer. The activity of this phosphorylase disappeared at 55°, in contrast to the usual phosphorylase activity (which withstands this temperature). There was also good correlation between formation of polysaccharide and appearance of inorganic phosphate in the absence of primer. The polysaccharide formed "*de novo*" was shown to be an efficient primer for starch synthesis.[160]

This enzymic system behaves like the usual phosphorylases, synthesizing amylose at a good rate from α-D-glucopyranosyl phosphate, and synthesizing amylopectin when combined with branching enzyme. The entire system of phosphorylase–Q-enzyme is sensitive to ATP and magnesium chloride, the latter inhibiting the Q-enzyme.

The formation of a branched polysaccharide, similar to glycogen, which is present in certain plants, especially waxy maize, has not yet been clarified; however, it could be explained by interaction with starch phosphorylase or starch synthetase, causing branching.[159] Neither the mechanisms for the separate formation of the linear and branched amylopectin components, nor the proportions of each in various plants, are known at present. These problems are undoubtedly a matter of genetic determinants that are awaiting future elucidation.

Recondo and Leloir[161] found that, when ADP-D-glucose was used as the substrate and starch granules as the source of enzyme, the

(159) E. Slabnik and R. B. Frydman, *Biochem. Biophys. Res. Commun.*, **38**, 709 (1970).
(160) R. B. Frydman and C. E. Cardini, *Plant Physiol.*, **42**, 628 (1967).
(161) E. Recondo and L. F. Leloir, *Biochem. Biophys. Res. Commun.*, **6**, 85 (1961).

transfer of D-glucopyranosyl groups for formation of starch was ten times as fast as when UDP-D-glucose was used. Because a specific enzyme that catalyzes the pyrophosphorylase reaction, namely,

$$ATP + \alpha\text{-D-glucopyranosyl phosphate} \longrightarrow ADP\text{-D-glucose} + PPi$$

has been found in wheat,[162] and ADP-D-glucose has been detected in *Chlorella*,[163] and in corn[164] and rice grains,[165] it seems likely that ADP-D-glucose is a precursor of starch *in vivo*.

A rapid interconversion of sucrose and starch has for a long time been known to occur in plants *in vivo*. De Fekete and Cardini[91] demonstrated a transfer of ^{14}C-labeled D-glucose from sucrose to starch by an enzyme preparation from corn endosperm when the appropriate nucleotides were added to the reaction mixture.

They postulated[91] that UDP-D-glucose may have an important role in the initial step of transformation of sucrose into starch, whereas synthesis of starch is primarily catalyzed by ADP-D-glucose–starch transglycosylase. This concept has been based on the smaller K_m constant of the sucrose synthetase toward UDP as compared with that for ADP, and by the fact that the ADP:D-sucrose transglycosylase is specifically inhibited[166] by UDP. From these results, de Fekete and Cardini[91] postulated the following two sequences of reaction for the incorporation of D-glucose into the granule.

(a) Sucrose + ADP $\rightleftharpoons$ ADP-D-glucose
$$\text{(or UDP-D-glucose)} \longrightarrow \text{starch}$$
(b) Sucrose + UDP $\rightleftharpoons$ UDP-D-glucose $\rightleftharpoons$
$$\alpha\text{-D-glucopyranosyl phosphate} \rightleftharpoons ADP\text{-D-glucose} \longrightarrow \text{starch}$$

The enzymes that catalyze all of the above steps could be demonstrated in the corn endosperm.

Murata and coworkers,[167] working with a coupling system of sucrose synthetase and starch synthetase, both isolated from ripening rice grains, found that a more efficient transfer of D-glucose-^{14}C from sucrose-^{14}C to starch occurs in the presence of ADP as compared with UDP, indicating that ADP-D-glucose plays a prominent role in the

(162) J. Espada, *J. Biol. Chem.*, **237**, 3577 (1962).
(163) H. Kauss and O. Kandler, *Z. Naturforsch., B*, **17**, 858 (1962).
(164) E. Recondo, M. Dankert, and L. F. Leloir, *Biochem. Biophys. Res. Commun.*, **12**, 204 (1963).
(165) T. Murata, T. Minamikawa, and T. Akazawa, *Biochem. Biophys. Res. Commun.*, **13**, 439 (1963).
(166) C. E. Cardini and E. Recondo, *Plant Cell Physiol.* (Tokyo), **3**, 313 (1962).
(167) T. Murata, T. Sugiyama, and T. Akazawa, *Arch. Biochem. Biophys.*, **107**, 92 (1964); *Biochem. Biophys. Res. Commun.*, **18**, 371 (1965).

process. However, UDP-D-glucose was found to inhibit the ADP-D-glucose:sucrose transglycosylation reaction.[166] Also, the K_m value of the sucrose synthetase for ADP was higher than that for UDP and UDP-D-glucose.[91] These findings may indicate that the breakdown of sucrose proceeds through the reversal of UDP: sucrose transglucosylation, instead of directly, by way of the ADP:sucrose transglucosylation. Similar results were obtained with ripening rice grains.[151,167]

Experiments with starch granule-bound ADP-D-glucose:starch glucosyltransferase isolated from the embryos of non-waxy corn (maize) seeds indicated[168] that it is different from that of similar preparations isolated from the endosperms. This is compatible with the observation that, in the waxy mutants of corn, the activity is lowered to a very low level, whereas the activity of the embryo preparation remains the same.

In contrast to mammalian or yeast transglycosylases (see Section V,1; p. 377), the soluble, plant transglucosylases are not activated by D-glucose 6-phosphate. However, Ghosh and Preiss[169] showed that the ADP-D-glucose pyrophosphorylase activity in spinach chloroplasts is stimulated about 50-fold by glyceric acid 3-phosphate; this effect may be significant in the control of starch formation during photosynthesis. It is possible that, during fixation of carbon dioxide, the accumulation of glyceric acid 3-phosphate causes an increase of ADP-D-glucose synthesis by stimulating the ADP-D-glucose pyrophosphorylase, causing an increase in ADP-D-glucose substrate, which would then enhance the rate of starch synthesis. It therefore appears that the activation of starch synthesis occurs at the pyrophosphorylase level,[169] rather than at the trans-D-glucosylase level as in mammalian glycogen synthesis.

Partially purified ADP-D-glucose pyrophosphorylase from *Chlorella pyrenoidosa* has been shown to be activated about 20-fold by glyceric acid 3-phosphate. This enzyme appears to be distinct from UDP-D-glucose pyrophosphorylase, which is not activated by glycerate 3-phosphate.[170]

3. Cellulose

Barber and Hassid[171] have shown that peas and various tissues of other species contain an enzyme capable of forming ^{14}C-labeled

(168) T. Akatsuka and O. E. Nelson, *J. Biol. Chem.*, **241**, 2280 (1966).
(169) H. P. Ghosh and J. Preiss, *J. Biol. Chem.*, **240**, PC 960 (1965).
(170) G. G. Sanwal and J. Preiss, *Arch. Biochem. Biophys.*, **119**, 454 (1967).
(171) G. A. Barber and W. Z. Hassid, *Biochim. Biophys. Acta*, **86**, 397 (1964).

GDP-D-glucose from guanosine triphosphate (GTP) and α-D-glucopy-ranosyl phosphate. Then, Elbein and coworkers[172] and Barber and coworkers[173] found another enzyme capable of utilizing this radio-active hexosyl nucleotide as substrate for the formation of ^{14}C-labeled polysaccharide. This enzyme was found present in root tissues of mung-bean seedlings (*Phaseolus aureus*), peas, corn, squash, and string beans,[173] and in immature seed-hair of cotton.[174] This enzyme present in these plants was capable of transferring the radioactive D-glucosyl group from GDP-D-glucose-^{14}C to an unknown acceptor, to form a polysaccharide chain having chemical and enzymic prop-erties indistinguishable from those of cellulose.[172,173]

The enzyme system that polymerized the D-glucose to cellulose appeared to have a high degree of specificity for GDP-D-glucose. None of the ^{14}C-labeled-D-glucosyl nucleotides containing bases other than guanine (that is, adenine, cytosine, uracil, or thymine) could serve as substrate for the formation of cellulose. Attempts to determine the nature of the acceptor to which the enzyme transfers the first D-glucosyl group, initiating chain growth, were unsuccessful. No stimulation of incorporation of D-glucose was observed on addition of D-glucose, 2-amino-2-deoxy-D-glucose, D-mannose, D-galactose, sucrose, cellobiose, soluble or insoluble cellodextrins, or swollen cellulose. Addition of Co^{2+}, Mn^{2+}, Zn^{2+}, or Ca^{2+} enhances the incor-poration of the D-glucose into the polymer.

Based on the data obtained with plant preparations, the following mechanism for cellulose synthesis was proposed.[175,176]

$$\text{GTP} + \text{α-D-glucopyranosyl phosphate} \xrightleftharpoons{\text{pyrophosphorylase}}$$
$$\text{GDP-D-glucose} + \text{PPi}$$

The polysaccharide is then formed by another enzyme [GDP-D-glu-cose:(1 → 4)-β-D-glucan β-D-4-glucosyltransferase], which catalyzes repetitive transfers of D-glucosyl groups as shown by the following reaction.

$$n(\text{GDP-D-glucose}) + \text{acceptor} \xrightarrow{\text{glucotransferase}}$$
$$\text{acceptor-}[(1 → 4)\text{-β-D-glucosyl}]_n + n(\text{GDP})$$
$$\text{(cellulose)}$$

(172) A. D. Elbein, G. A. Barber, and W. Z. Hassid, *J. Amer. Chem. Soc.*, **86**, 309 (1964).
(173) G. A. Barber, A. D. Elbein, and W. Z. Hassid, *J. Biol. Chem.*, **239**, 4056 (1964).
(174) G. A. Barber and W. Z. Hassid, *Nature*, **207**, 295 (1965).
(175) W. Z. Hassid, *Ann. Rev. Plant Physiol.*, **18**, 253 (1967).
(176) W. Z. Hassid, *Science*, **165**, 137 (1969).

Subsequently, other workers presented data that suggested that UDP-D-glucose may also be an effective donor for synthesis of cellulose. Thus, Brummond and Gibbons,[177,178] using an incubation mixture that was a modification of that used by Barber and coworkers,[173] reported that an enzyme preparation from the bean of *Lupinus albus* is capable of incorporating ^{14}C-labeled D-glucose from UDP-D-glucose-^{14}C into a mixture of polysaccharides ranging from water-soluble to alkali-insoluble. They reported that the alkali-insoluble fraction (about 7%) is synthesized more readily from UDP-D-glucose than from GDP-D-glucose. However, repetition of this work with *Lupinus albus*[179] yielded a polysaccharide, similar to the one previously obtained by Feingold and coworkers[180] from mung beans, that contained only β-D-(1 → 3)-glucosyl linkages.

Ordin and Hall[181] found that particulate preparations from oat (*Avena sativa*) coleoptiles could utilize UDP-D-glucose as substrate for polysaccharide formation. Upon degradation of the polysaccharide derived from UDP-D-glucose with an impure cellulase, cellobiose and, to a lesser extent, a substance identified as a trisaccharide containing β-D-(1 → 4), β-D-(1 → 3)-glucosyl linkages were obtained. However, they found that, when GDP-D-glucose was used as substrate, only cellobiose and cellotriose were obtained, indicating the production of a polysaccharide containing only β-D-(1 → 4) linkages (cellulose).

Villemez and coworkers[182] worked with an incubation mixture containing sucrose and albumin, and reported the production of an insoluble polysaccharide by action of a particulate enzyme preparation of mung beans on radioactive UDP-D-glucose as substrate; hydrolysis of this polysaccharide with acid produced, in addition to cellobiose, laminarabiose [in which the D-glucose residues are joined by β-D-(1 → 3) glucosyl linkages]. Franz and Meier[183] also reported that UDP-D-glucose-^{14}C, as well as GDP-D-glucose-^{14}C, are readily utilized as precursors of cellulose by cotton hairs at different stages of growth.

(177) D. O. Brummond and A. P. Gibbons, *Biochem. Biophys. Res. Commun.*, **17**, 156 (1964).
(178) D. O. Brummond and A. P. Gibbons, *Biochem. Z.*, **342**, 308 (1965).
(179) H. M. Flowers, K. K. Batra, J. Kemp, and W. Z. Hassid, *Plant Physiol.*, **43**, 1703 (1968).
(180) D. S. Feingold, E. F. Neufeld, and W. Z. Hassid, *J. Biol. Chem.*, **233**, 783 (1958).
(181) L. Ordin and M. A. Hall, *Plant Physiol.*, **42**, 205 (1967); **43**, 473 (1968).
(182) C. L. Villemez, Jr., G. Franz, and W. Z. Hassid, *Plant Physiol.*, **42**, 1219 (1967).
(183) G. Franz and H. Meier, *Phytochemistry*, **8**, 579 (1969).

Whereas the results of Ordin and Hall[181] with enzyme preparations from oat coleoptiles and UDP-D-glucose as substrate could be substantiated, a later investigation by Flowers and coworkers[184] who used chemical methods showed that particulate enzyme preparations from mung beans behaved differently from those previously reported.[182]

Examination of the polymers synthesized by particulate enzyme preparations from mung beans (*Phaseolus aureus*) and those of the *Lupinus albus* variety, by use of GDP-D-glucose and UDP-D-glucose, showed that a $(1 \rightarrow 4)$-β-D-glucan is formed only from GDP-D-glucose,[184] and a $(1 \rightarrow 3)$-β-D-glucan when UDP-D-glucose[179] is the substrate.

Two chemical methods were used for distinguishing between the $(1 \rightarrow 3)$- and $(1 \rightarrow 4)$-linkages: one was the lead tetraacetate method,[185] in which oxidation of a disaccharide such as laminarabiose, obtained from degradation of $(1 \rightarrow 3)$-glucan, produces D-arabinose from the reducing end, whereas oxidation of cellobiose, obtained from degradation of cellulose, which contains $(1 \rightarrow 4)$-linkages, affords erythrose. Similarly, the Smith method,[185] namely, periodate oxidation of cellobiose and subsequent reduction with borohydride followed by acid hydrolysis, yields erythritol from the reducing end, whereas laminarabiose forms arabinitol.

Both the particulate and the digitonin-solubilized enzyme systems from mung beans and *Lupinus albus* catalyze the biosynthesis of alkali-insoluble D-glucans from UDP-D-glucose. Some 90 to 95% of the polymer produced from this substrate is soluble in hot, dilute alkali. The inter-D-glucosidic linkages of both the alkali-soluble and alkali-insoluble polymers are identical, being β-D-$(1 \rightarrow 3)$. Examination of the products of degradation of the polymers produced by enzymic preparations showed that only $(1 \rightarrow 3)$-β-D-glucans are formed.[179] No evidence was obtained for the formation of any measurable proportions of β-D-$(1 \rightarrow 4)$ glucosidic linkages.

On the other hand, when GDP-D-glucose was used as the substrate, both the particulate and the digitonin-solubilized enzyme systems from mung beans and from *Lupinus albus* yielded only $(1 \rightarrow 4)$-β-D-glucan (cellulose), practically all insoluble.

A small proportion of D-mannose was shown to be present in the

(184) H. M. Flowers, K. K. Batra, J. Kemp, and W. Z. Hassid, *J. Biol. Chem.*, **244**, 4969 (1969).
(185) A. J. Charlson and A. S. Perlin, *Can. J. Chem.*, **34**, 1200 (1956); M. L. Wolfrom and A. Thompson, *Methods Carbohyd. Chem.*, **3**, 143 (1963).

polymer; it was isolated as a mixed D-glucose–D-mannose disaccharide from the hydrolysis products. This result correlates with the fact that pure cellulose is not found in the growing roots or shoots of higher plants; the common minor component is usually D-mannose.[186] The incorporation of radioactive D-glucose to form the $(1 \rightarrow 4)$-β-D-glucan was found to be an extremely rapid reaction that reaches half-maximal incorporation in about half a minute, although a considerable proportion of the substrate still remains.

Radioactive D-glucan synthesized by the particulate enzyme system from mung beans with ^{14}C-labeled UDP-D-glucose as the substrate was also hydrolyzed by a highly purified exo-β-D-$(1 \rightarrow 3)$-glucanase.[187] Both the soluble and the insoluble glucan were degraded to the extent of 91%. The alkali-insoluble glucan formed from GDP-D-glucose by the action of the particulate system from the same plant, a polymer which, from chemical data, is known to be a β-D-$(1 \rightarrow 4)$-glucan (cellulose), was not acted upon at all by this glucanase.[188] The fact that the alkali-insoluble polysaccharide synthesized from GDP-D-glucose does not release any D-glucose on treatment with the highly purified glucanase, which only degrades oligosaccharides and polysaccharides containing β-D-$(1 \rightarrow 3)$-linkages, confirmed the results obtained from chemical data, namely, that the enzyme system of mung beans produces a β-D-$(1 \rightarrow 4)$-linked glucan (cellulose) from GDP-D-glucose.

A soluble, cellulose synthetase was prepared from a particulate enzyme of mung-bean (*Phaseolus aureus*) seedlings by extraction with digitonin, precipitation of the protein with ammonium sulfate, and high-speed centrifugation. The soluble enzyme was found incapable of incorporating the D-glucosyl group from GDP-D-glucose into $(1 \rightarrow 4)$-β-D-glucan; however, addition of Mg^{2+} to the solution rendered the enzyme active. The addition to the solution of such compounds as D-glucose, cellobiose, maltose, or soluble cellodextrins did not enhance the activity of the soluble enzyme. The fact that only magnesium chloride could render the soluble enzyme active indicated that the activator or primer is combined with the enzyme, and therefore does not require an additional primer, but only Mg^{2+}.

As mentioned earlier (this page), the particulate enzyme reacted

(186) D. T. Dennis and R. D. Preston, *Nature*, **191**, 667 (1961).
(187) F. I. Huotari, T. E. Nelson, F. Smith, and S. Kirkwood, *J. Biol. Chem.*, **243**, 952 (1968).
(188) K. K. Batra and W. Z. Hassid, *Plant Physiol.*, **44**, 755 (1969).

very rapidly; the D-glucosyl group from GDP-D-glucose was not incorporated into cellulose after 2 minutes of incubation. For the soluble enzyme, the incorporation of radioactivity into the polymer was proportional to enzyme concentration and the elapsed time up to a period of 5 minutes. Addition of small proportions of (ethylene-dinitrilo)tetraacetate and magnesium chloride caused the reaction to proceed linearly with time, up to 25 minutes. The value of K_m for the GDP-D-glucose was found to be 80 μM.

Despite some unexplained results concerning cellulose synthesis with a cell-free system in plants, the data seem, in general, to be consistent with the assumption that GDP-D-glucose is the "sugar nucleotide" from which the β-D-(1 → 4)-linked chain of cellulose originates. However, there is evidence for the presence of another enzyme system that, in some plants, produces from UDP-D-glucose a polymer that appears to be a mixed, β-D-(1 → 4), β-D-(1 → 3)-linked polysaccharide.[181,184]

Experiments *in vitro* with mung beans[173,184] indicate that GDP-D-glucose is the direct donor of D-glucose for the formation of cellulose in plants.[188a] Extraction of the water-insoluble material with organic solvents resulted in negligible proportions of radioactive material in the solution. However, the work of Colvin[189] suggested the presence of a D-glucose–lipid precursor that passes the cytoplastic membrane and is extracellularly converted into cellulose. Villemez and Clark[190] claimed to have detected, in polysaccharide synthesis by crude, mung-bean particulate preparations, a lipid intermediate similar to that known for bacterial and yeast systems (see Sections V,6, p. 396; and VI,6 and 7, pp. 418 and 428). Kauss[191] found that particulate, mung-bean enzyme preparations incorporate the D-glucosyl group from ^{14}C-labeled UDP-D-glucose into lipophilic products identified as a mixture of β-sitosteryl D-glucoside and stig-masteryl D-glucoside; the same particulate preparation is also capable of synthesizing glucan. However, he found that these steroid D-glucosides do not appear to be intermediates in the synthesis of glucan. Nevertheless, in view of the available data, further exploration of the possibility that glycosyl-lipids serve as intermediates in the synthesis of plant polysaccharides seems worth while.

(188a) For a detailed treatment of the biogenesis of cellulose microfibrils, see F. Shafizadeh and G. D. McGinnis, This Volume, p. 297.

(189) J. R. Colvin, *Can. J. Biochem. Physiol.*, **39**, 1921 (1961).

(190) C. L. Villemez, Jr., and A. F. Clark, *Biochem. Biophys. Res. Commun.*, **36**, 57 (1969).

(191) H. Kauss, *Z. Naturforsch., B*, **23**, 1522 (1968).

4. $(1 \rightarrow 3)$-β-D-Glucan

In the Section on cellulose biosynthesis (see p. 386), considerable discussion was devoted to the biosynthesis of this glucan concerning its nucleoside (D-glucopyranosyl pyrophosphate) precursor. It appears that UDP-D-glucose is the precursor from which a glucan containing only β-D-$(1 \rightarrow 3)$-glucosyl linkages is produced, although in plants there are enzyme systems that produce glucans containing mixed β-D-$(1 \rightarrow 3)$, β-D-$(1 \rightarrow 4)$ glucosidic linkages.[179,181]

As $(1 \rightarrow 3)$-β-D-glucan is widely distributed in higher as well as in lower plants,[192] it is of interest to include additional available information regarding its biosynthesis.

Experiments with particulate enzyme preparations from mung beans (*Phaseolus aureus*) and *Lupinus albus* beans showed that incorporation of radioactive D-glucose from UDP-D-glucose-^{14}C into the glucan increased with increase of concentration of UDP-D-glucose-^{14}C substrate in the incubation mixture, and reached a maximum of 50% of the substrate radioactivity incorporated.[179]

It was found that both the particulate and the digitonin-solubilized enzyme system from mung beans (*Phaseolus aureus*) and *Lupinus albus* beans catalyze the biosynthesis of alkali-soluble and alkali-insoluble glucans from UDP-D-glucose. From 90 to 95% of the $(1 \rightarrow 3)$-β-D-glucan produced from this substrate is soluble in hot, dilute alkali.[188] Treatment with alkali presumably causes solubilization of the fractions having lower degrees of polymerization; the rest (a relatively small proportion) of the polymer is of higher molecular weight. Examination by chemical methods of the structure of both the alkali-soluble and alkali-insoluble fraction showed them to have the same inter-D-glucosidic linkages. Analysis of the oligosaccharides obtained from the partially degraded polymers showed them to contain only β-D-$(1 \rightarrow 3)$-linkages. There was no evidence for the presence of measurable proportions of any other D-glucosidic linkage, thus confirming the results of Feingold and coworkers[180] on the synthesis of callose.

Variation in the concentration of UDP-D-glucose (over a 1000-fold range) does not cause any change in the structure of the radioactively labeled polymer produced. The reaction rate at high concentrations of substrate is considerable, and this makes possible the production of milligram quantities of solid polymer. The polymer is, however, contaminated with some protein and D-glucose.[179]

The biosynthesis of a β-D-$(1 \rightarrow 3)$-linked glucan, paramylon, which

(192) D. J. Manners, *Ann. Repts. Progr. Chem.*, **63**, 590 (1966).

occurs as an insoluble polymer in *Euglena gracilis*, has been investigated by Goldemberg and Marechal.[193] As with callose,[180] the donor is UDP-D-glucose and no added acceptor is required. The reaction (measured by the release of UDP) proceeded normally with *Euglena* particulate enzyme in the presence of a potent $(1 \rightarrow 3)$-β-D-glucan glucanohydrolase from snail-gut juice, which caused hydrolysis of polysaccharide to D-glucose as rapidly as it was formed.[194] Therefore, there was no indication that the endogenous acceptor is split off from the particulate enzyme under these conditions or by preincubation with the $(1 \rightarrow 3)$-β-D-glucan glucanohydrolase. It was, therefore, assumed that either the endogenous acceptor has different properties, perhaps similar to those of the untreated paramylon grains (which are resistant to the glucanohydrolase), or is not attacked when combined with the synthetase.

In a subsequent study,[194] the same authors showed that extraction of the particulate enzyme with sodium deoxycholate results in smaller particles, which are precipitated after dialysis and can be sedimented by centrifugation at 100,000 g; this enzyme does not seem to be bound to the polysaccharide fraction.

Addition of laminarabiose to the incubation mixture increases the incorporation of D-glucosyl group from the UDP-D-glucose substrate into paramylon by about 20 to 30%. There is no indication that the laminarabiose serves as an acceptor that is incorporated into the polysaccharide.

5. $(1 \rightarrow 2)$-β-D-Glucan

A glucan has been synthesized from UDP-D-glucose-^{14}C by the action of a particulate enzymic preparation obtained from two strains of *Rhizobium japonicum*. The enzymic system requires Mg^{2+} or Mn^{2+} in order to be active.[195]

This polysaccharide appears to be identical with a glucan (isolated from cultures of *Agrobacterium tumefaciens*) in which the main linkage between the D-glucose residues is[196,197] β-D-$(1 \rightarrow 2)$. The synthetic glucan obtained with active preparations from *Rhizobium japonicum* contains, in addition to this linkage, some β-D-$(1 \rightarrow 3)$-

(193) S. H. Goldemberg and L. R. Marechal, *Biochim. Biophys. Acta,* **71,** 743 (1963).
(194) L. R. Marechal and S. H. Goldemberg, *J. Biol. Chem.,* **239,** 3163 (1964).
(195) R. A. Dedonder and W. Z. Hassid, *Biochim. Biophys. Acta,* **90,** 239 (1964).
(196) R. Hodgson, A. J. Riker, and W. H. Peterson, *J. Biol. Chem.,* **158,** 89 (1945).
(197) E. W. Putman, A. L. Potter, R. Hodgson, and W. Z. Hassid, *J. Amer. Chem. Soc.,* **72,** 5024 (1950).

and β-D-(1 → 6)-linkages. Nuclear magnetic resonance data show that the β-D-(1 → 2)-linkages account for more than 80% of the total inter-D-glucosidic bonds of the synthetic glucan.

6. Mannan

Mannan was discovered in bakers' yeast (*Saccharomyces cerevisiae*) by Salkowski[198] in 1894; at that time, it was termed "yeast gum." The first detailed studies of its structure were made by the methylation procedure by Haworth and coworkers.[199,200] The results showed that the polysaccharide consists entirely of D-mannose residues and is highly branched, the D-mannose residues being combined by (1 → 2)-, (1 → 3)-, and (1 → 6)-linkages. Its molecular weight, as determined by measurements of osmotic pressure, is that calculated for a polymer containing 200–400 hexose residues. The high dextro-rotation of this mannan indicated that the D-mannose residues are combined primarily by α-D linkages, although the presence of some β-D linkages could not be excluded.

Ballou and coworkers[201,202] studied in greater detail the structure of yeast mannan produced by *S. cerevisiae* grown in a culture medium containing D-glucose. They showed that controlled acetolysis by the method of Gorin and Perlin[203] selectively cleaves (1 → 6)-linkages of this polysaccharide, and mainly produces (in excellent yield) di-, tri-, and tetra-saccharides having relatively stable (1 → 2)- and (1 → 3)-linkages. By using shorter periods of acetolysis, they could obtain penta-, hexa-, and hepta-saccharides in which the (1 → 6)-linkages from the backbone were not cleaved. These results suggested that the smaller units are connected by (1 → 6)-linkages to give the larger fragments.

Direct proof for the (1 → 6) backbone was then obtained by G. H. Jones and Ballou,[204] who found that a bacterial α-D-mannosidase removed the (1 → 2)- and (1 → 3)-linked side-chains from the mannan of *S. cerevisiae*, leaving the backbone structure of (1 → 6)-linked D-mannose residues intact. The structure of yeast mannan can be represented as follows.

(198) E. Salkowski, *Ber.*, **27**, 497 (1894).
(199) W. N. Haworth, E. L. Hirst, and F. A. Isherwood, *J. Chem. Soc.*, 784 (1937).
(200) W. N. Haworth, R. L. Heath, and S. Peat, *J. Chem. Soc.*, 833 (1941).
(201) Y. C. Lee and C. E. Ballou, *Biochemistry*, **4**, 257 (1965).
(202) T. S. Stewart, P. B. Mendershausen, and C. E. Ballou, *Biochemistry*, **7**, 1843 (1968).
(203) P. A. J. Gorin and A. S. Perlin, *Can. J. Chem.*, **34**, 1796 (1965).
(204) G. H. Jones and C. E. Ballou, *J. Biol. Chem.*, **243**, 2442 (1968).

.... α-D-Man-(1 → 6)-α-D-Man-(1 → 6)-α-D-Man-(1 → 6) ....
      R   R   R

where R = α-D-Man-(1 → 2)-,

          α-D-Man-(1 → 2)-α-D-Man-(1 → 2)-,

          α-D-Man-(1 → 3)-α-D-Man-(1 → 2)-,

    or α-D-Man-(1 → 3)-α-D-Man-(1 → 2)-α-D-Man-(1 → 2)-.

O-Phosphono-D-mannans have been isolated from several strains of yeast.[205–207] Stewart and Ballou[208] have observed a great variation in the phosphorus content of different mannans, the highest being in *Kloeckera brevis* mannan, which has one phosphorus atom for every nine D-mannose residues. These phosphonomannans appeared to have a structure similar to that of the mannan already described, except for the presence of phosphate groups.[208]

Some information has become available pertaining to the problem of the biosynthesis of D-mannan in yeasts. Algranati and coworkers[209] obtained evidence, for the first time, that GDP-D-mannose, which is present in *Saccharomyces carlsbergensis*, is a precursor to this polysaccharide. Behrens and Cabib[210] studied in greater detail the problem of its biosynthesis by action of a particulate preparation from the same yeast, with GDP-D-mannose as the substrate. The particles were found to catalyze the incorporation of D-mannose-^{14}C (from GDP-D-mannose-^{14}C) into a branched, D-mannose polysaccharide, without the need for an added primer. By comparing the oligosaccharides obtained by controlled acetolysis of the enzymically synthesized, radioactive mannan with those derived from *S. cerevisiae* mannan, Behrens and Cabib[210] showed that these oligosaccharides, also, contained (1 → 2)-, (1 → 3)-, and (1 → 6)-linkages. They therefore concluded that the enzymically synthesized mannan has a structure similar to that of that isolated from *S. cerevisiae* cells.

As the particulate enzyme was produced by centrifugation of lysed protoplasts of the yeast, an endogenous primer could be expected to be present in the particles. The optimal pH for the activity lay between 5.5 and 7.2. For activity, there was an almost-absolute requirement for Mn^{2+} which could not be replaced by Mg^{2+}. The reaction product remained attached to the particles, from which it

(205) A. R. Jeanes and P. R. Watson, *Can. J. Chem.*, **40**, 1318 (1962).
(206) P. J. Mill, *J. Gen. Microbiol.*, **44**, 329 (1966).
(207) M. E. Slodki, *Biochim. Biophys. Acta*, **57**, 525 (1962).
(208) T. S. Stewart and C. E. Ballou, *Biochemistry*, **7**, 1855 (1968).
(209) I. D. Algranati, H. Carminatti, and E. Cabib, *Biochem. Biophys. Res. Commun.*, **12**, 504 (1963).
(210) N. H. Behrens and E. Cabib, *J. Biol. Chem.*, **243**, 502 (1968).

could be partially released by such mild treatment as incubation at 37° or heating at 100° at pH 7.5.

Scher and coworkers[211] showed that an alkali-stable, acid-labile lipoid material was synthesized from GDP-D-mannose and an endogenous lipid in crude, cell-free preparations of *Micrococcus lysodeikticus*. This compound has been shown to be a polyisoprenyl (D-mannosyl phosphate); it can serve as a D-mannosyl donor in the enzymic synthesis of a homopolysaccharide of D-mannose. The polyisoprenol was found to be a C_{55} compound containing two internal, *trans* double bonds. It was shown that only the D-mannosyl group of GDP-D-mannose is transferred to the polyisoprenyl phosphate ("undecaprenyl 1-phosphate"), and that the reaction is readily reversible.

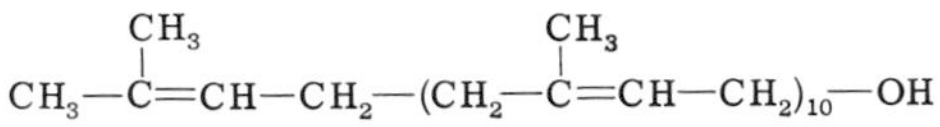

$$CH_3-\underset{\underset{CH_3}{|}}{C}=CH-CH_2-(CH_2-\underset{\underset{CH_3}{|}}{C}=CH-CH_2)_{10}-OH$$

Undecaprenol

$$\text{GDP-D-mannose + undecaprenyl 1-phosphate} \underset{}{\overset{Mg^{2+}}{\rightleftharpoons}}$$
$$\text{undecaprenyl (D-mannosyl phosphate) + GDP} \qquad (3)$$

$$\text{Undecaprenyl (D-mannosyl phosphate) + (D-mannose)}_n \longrightarrow$$
$$\text{(D-mannose)}_{n+1} \text{ + undecaprenyl phosphate} \qquad (4)$$

It was later shown[212] that purified undecaprenyl 1-phosphate can serve as the acceptor in reaction 3, provided that Mg^{2+} and a surfactant such as phosphatidylglycerol is added. An isomer of undecaprenyl 1-phosphate, namely, ficaprenyl phosphate, which has three

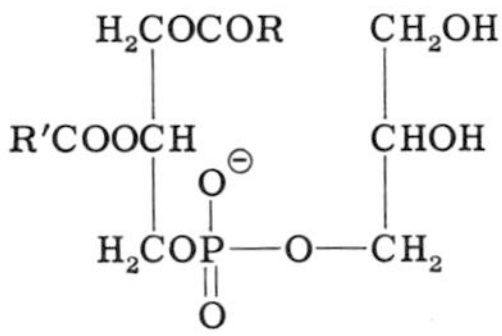

$$\begin{array}{cc}
H_2COCOR & CH_2OH \\
| & | \\
R'COOCH & CHOH \\
| \quad\; O^{\ominus} & | \\
H_2COP\!-\!\!O\!-\!\!CH_2 \\
\;\;\;\| \\
\;\;\;O
\end{array}$$

"Phosphatidylglycerol"

where RCO and R'CO are fatty acid radicals.

(211) M. Scher, W. J. Lennarz, and C. C. Sweeley, *Proc. Nat. Acad. Sci. U. S.*, **59**, 1313 (1968).

(212) M. Lahav, T. H. Chiu, and W. J. Lennarz, *J. Biol. Chem.*, **244**, 5890 (1969).

internal, *trans* double bonds, was found active as the acceptor. It was also found[213] that the second reaction does not require a divalent cation, but is inhibited by Triton X-100, to which the enzyme catalyzing reaction 3 is insensitive.

The polysaccharide product has been respectively studied by acetolysis, exhaustive methylation, and treatment with α-D-mannosidase.[213] Like yeast mannan, it contains D-mannose residues linked through $(1 \rightarrow 2)$-, $(1 \rightarrow 3)$-, and $(1 \rightarrow 6)$-linkages, but the degree of branching is very low. About 80 percent of the D-mannose residues incorporated were found at nonreducing, terminal positions.

Tanner's results[214] indicate that, in yeast (*S. cerevisiae*), as well as in *M. lysodeikticus*, a liphophilic D-mannosyl intermediate is the immediate precursor for biosynthesis of mannan. He showed that D-mannosyl residues were transferred to the lipid intermediate, and some evidence was provided indicating that the linkage to the lipid might be through a phosphoric diester. The identity of the lipid is, however, not yet known. Incorporation of D-mannose-^{14}C from GDP-D-mannose-^{14}C into the lipid requires metal ions, but, in contrast to the Mn^{2+} requirement for biosynthesis of D-mannan, Mg^{2+} appears to be able to replace Mn^{2+} completely. A particulate enzyme system, optimal in Mg^{2+} but deficient in Mn^{2+} level, produced a higher than normal, steady-state level of D-mannosyl-^{14}C-lipid and a decreased rate of synthesis of D-mannan, thus indicating an intermediate role of D-mannosyl-lipid for its biosynthesis. Possibly, the lipid intermediate provides a mechanism for passage of the complex through the cytoplasmic membrane, after which the next step in the biosynthesis of D-mannan can occur.

Kauss[215] has similarly observed the enzymic synthesis of D-mannosyl-lipid by use of a particulate enzyme from mung-bean shoots. The results suggested that only the D-mannosyl group of GDP-D-mannose is transferred, because GDP, but not GMP, was incorporated into GDP-D-mannose by an exhange reaction. There is some evidence suggesting that the lipid is a polyisoprenoid compound, because prior incubation of the cells with mevalonic-5-*t* acid resulted in the labeling of the acceptor lipid.

Sentandreu and Northcote[216] studied the cell-wall synthesis of *S. cerevisiae* by following the incorporation of ^{14}C-labeled D-glucose

(213) M. Scher and W. J. Lennarz, *J. Biol. Chem.*, **244**, 2777 (1969).
(214) W. Tanner, *Biochem. Biophys. Res. Commun.*, **35**, 144 (1969).
(215) H. Kauss, *FEBS Lett.*, **5**, 81 (1969).
(216) R. Sentandreu and D. H. Northcote, *Biochem. J.*, **115**, 231 (1969).

and L-threonine-^{14}C into the cytoplasm and wall. They found that both radioactive D-glucose and radioactive L-threonine were incorporated into D-mannan glycopeptides, which contained D-mannan of high molecular weight and low-molecular-weight mono- and oligo-saccharide units composed of D-mannose, both types of carbohydrate being attached to the peptide. Results of 60-sec., pulse-chase experiments with D-glucose-^{14}C and the whole cells, followed by incubation in nonradioactive media, showed that the radioactivity of the D-mannose and D-mannose oligosaccharides, attached to the peptides by O-D-mannosyl links to hydroxyamino acid, remained constant during the chase. Were transglycosylation operative, the radioactivity would be expected to move into the large mannan molecule also attached to the peptide. This suggested that the D-mannose and the D-mannose oligosaccharides, once formed and incorporated into the glycopeptide, do not undergo transglycosylation into the large mannan molecule (as was previously considered possible).[217]

Bretthauer and coworkers[218] isolated from *Hansenula holstii* a particulate enzyme capable of catalyzing the transfer of D-mannose-^{14}C from GDP-D-mannose-^{14}C to endogenous, mannan-acceptor molecules. The same particulate enzyme fraction was found to catalyze the transfer of phosphate-^{32}P from GDP-^{32}P-D-mannose to endogenous mannan. Divalent-metal ions proved to be required, Mn^{2+} being the most effective for transfer of phosphate-^{32}P and D-mannose-^{14}C. The incorporation of both phosphate-^{32}P and D-mannose-^{14}C was inhibited to the extent of 70% by either 0.5 mM GMP or GDP.

The acceptor molecules have at the nonreducing end D-mannosyl groups to which the entire D-mannosyl phosphate moiety is assumed to be transferred from GDP-D-mannose. Hydrolysis with a strong acid liberates this acceptor D-mannosyl group having the attached incorporated phosphate-^{32}P group, as D-mannose 6-phosphate-^{32}P. That this phosphate is present in the polymer in the phosphoric diester linkage is indicated by the fact that it becomes susceptible to hydrolysis by phosphate monoesterase after the removal of the incorporated D-mannose residues by mild hydrolysis with acid. On this basis, the synthesis of a $(1 \to 6)$-phosphoric diester linkage between two D-mannose residues has been proposed.

(217) R. Sentandreu and D. H. Northcote, *Biochem. J.*, **109**, 419 (1968).
(218) R. K. Bretthauer, L. P. Kozak, and W. E. Irwin, *Biochem. Biophys. Res. Commun.*, **37**, 820 (1969).

7. Chitin

The observation by Glaser and Brown[219] that chitin is synthesized from UDP-2-acetamido-2-deoxy-D-glucose has been confirmed in several systems. Incorporation of labeled 2-acetamido-2-deoxy-D-glucose from its UDP derivative has been found by the use of a particulate fraction from an insect;[220] the product was assumed to be chitin, but it was not actually characterized. Similar incorporation has also been observed with particulate fractions from fungi;[221,222] here, the product was characterized by its hydrolysis with chitinase. In one system, the dependence of incorporation on chitodextrin and the increase of V_{max} in the presence of (non-radioactive) 2-acetamido-2-deoxy-D-glucose were observed,[222] confirming the results of Glaser and Brown.[219]

8. Pectins (D-Galacturonans)

Pectins are complex, macromolecular compounds that occur in all higher plants. They are found in the cell walls, particularly in the intercellular layers, and in some ripe fruits and plant juices.

The basic building unit of pectin is D-galactopyranosyluronic acid, and the linear skeleton of the uronic acid residues is connected by α-D-(1 → 4)-glycosidic linkages. These D-galacturonans are associated with other polysaccharides, mainly, D-galactan and L-arabinan.[223] Some of the carboxyl groups of the D-galacturonic acid residues in the chain are methylated.

A nucleoside pyrophosphoric ester of this uronic acid, UDP-D-galacturonic acid, was shown to be present in mung beans.[224] In this plant were also found enzymes that led to the formation of this uronic acid derivative of a nucleoside pyrophosphate by the following pathway:[225,226]

UDP-D-glucose $\xrightarrow{\text{dehydrogenase}}$

UDP-D-glucuronic acid $\underset{\text{epimerase}}{\rightleftharpoons}$

UDP-D-galacturonic acid

(219) L. Glaser and D. H. Brown, J. Biol. Chem., 228, 729 (1957).

(220) E. G. Jaworski, L. C. Wang, and G. Marco, Nature, 198, 790 (1963).

(221) E. G. Jaworski, L. C. Wang, and W. D. Carpenter, Phytopathology, 55, 1309 (1965).

(222) C. A. Porter and E. G. Jaworski, Biochemistry, 5, 1149 (1966).

(223) G. H. Beavan and J. K. N. Jones, J. Chem. Soc., 1218 (1947).

(224) E. F. Neufeld and D. S. Feingold, Biochim. Biophys. Acta, 53, 589 (1961).

(225) J. L. Strominger and L. W. Mapson, Biochem. J., 66, 567 (1957).

(226) D. S. Feingold, E. F. Neufeld, and W. Z. Hassid, J. Biol. Chem., 235, 910 (1960).

The sequence of these enzymic reactions led to the postulate that UDP-D-glucose is a precursor of UDP-D-galacturonic acid, from which the pectin (D-galacturonan) is synthesized. The results of experiments performed *in vitro* proved to be in accord with this hypothesis.

It has been found[227] that particulate preparations from mung bean (*Phaseolus aureus*) seedlings catalyze the formation of a D-galacturonic-^{14}C acid chain from UDP-D-galacturonic-^{14}C acid. *Penicilium chrysogenum* D-galacturonanase completely hydrolyzed this synthetic, radioactive product to D-galacturonic-^{14}C acid. Partial degradation of the product with an exo-D-galacturonan transeliminase yielded[228,229] radioactive 4-O-(4-deoxy-β-L-*threo*-hex-4-enosyluronic acid)-D-galacturonic acid. The action of the latter enzyme is known to be specific for hydrolysis of the D-galacturonic acid chain.

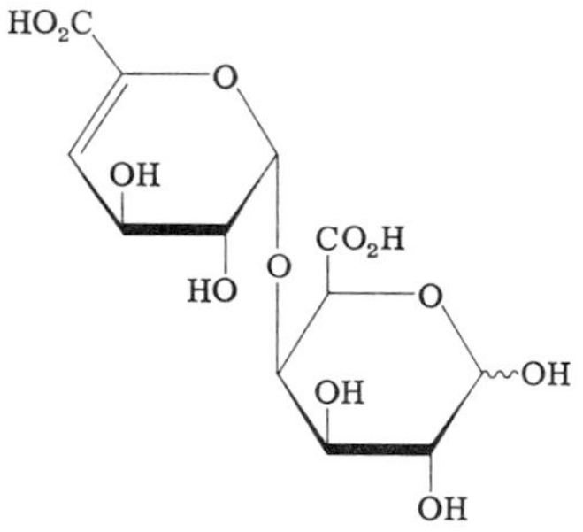

4- O -(4-Deoxy-β-L- *threo* -hex-4-enosyluronic
acid)-D-galacturonic acid

Experiments with a number of plants showed that the enzyme that utilizes UDP-D-galacturonic acid as the donor of the D-galactosyluronic acid group for the formation of D-galacturonan is fairly specific; however, some incorporation of this uronic acid into D-galacturonan was also observed with a particulate enzyme preparation from tomatoes when dTDP-D-galacturonic acid and, to a lesser extent, CDP-D-galacturonic acid were used as donors.[230]

The D-galacturonic acid transferase has been found to have an apparent Michaelis constant of 1.7 μM for UDP-D-galacturonic acid;

(227) C. L. Villemez, Jr., T. Y. Lin, and W. Z. Hassid, *Proc. Nat. Acad. Sci. U. S.*, **54**, 1626 (1965).
(228) J. D. Macmillan and R. H. Vaughn, *Biochemistry*, **3**, 564 (1964).
(229) J. D. Macmillan, H. J. Phaff, and R. H. Vaughn, *Biochemistry*, **3**, 572 (1964).
(230) T. Y. Lin, A. D. Elbein, and J. C. Su, *Biochem. Biophys. Res. Commun.*, **22**, 650 (1966).

at a substrate concentration of 37 μM, the particulate enzyme catalyzes the polymerization of D-galacturonic acid residues at the rate[231] of 4.7 μmoles/mg of protein/min.

As some of the carboxyl groups of the D-galacturonic acid residues of pectin are esterified with methyl groups, it has been assumed[232] that the immediate precursor of pectin is not UDP-D-galacturonic acid but a derivative of methyl D-galactosyluronate, namely, UDP-(methyl D-galactosyluronate). However, experimental results failed to substantiate this assumption. When UDP-methyl D-galacturonate was used as the substrate, the methyl D-galactosyluronate group from this "methyl uronate nucleotide" could not be incorporated into the D-galacturonan chain by the mung-bean enzyme-system.[227,231] In this system, the esterification of the D-galacturonic acid occurs[233] by a transfer of the methyl group from "S-adenosyl-L-methionine," analogous to the case in which the 4-methyl ether groups are transferred to D-glucuronic acid of hemicellulose.[234] The data indicate that methylation occurs at a later stage in the synthesis of the polymer.[233,235] The formation of the D-galacturonan chains from UDP-D-galacturonate and from UDP-D-glucuronate appears to be independent of the methylation process.

Kauss and coworkers[236,237] found that D-galacturonan methyl transferase does not methylate exogenous D-galacturonan. In contrast, the formation of pectin methyl ester is greatly increased if additional D-galacturonan is preformed in the enzyme preparation by incubation with UDP-D-galacturonate followed by digestion with phosphate diesterase; this indicates that the enzymes responsible for polymerization and methylation, as well as the pectic substances, are localized together in structural compartments. There is evidence that the complex is established by lipid membranes which keep the newly formed pectin and the esterase of the cells apart, as long as the integrity of the structure is not disturbed.

(231) C. L. Villemez, Jr., A. L. Swanson, and W. Z. Hassid, *Arch. Biochem. Biophys.*, **116**, 446 (1966).
(232) P. Albersheim and J. Bonner, *J. Biol. Chem.*, **234**, 3105 (1959); P. Albersheim, *ibid.*, **238**, 1608 (1963).
(233) H. Kauss and W. Z. Hassid, *J. Biol. Chem.*, **242**, 3449 (1967).
(234) H. Kauss and W. Z. Hassid, *J. Biol. Chem.*, **242**, 1680 (1967).
(235) H. Kauss, A. L. Swanson, and W. Z. Hassid, *Biochem. Biophys. Res. Commun.*, **26**, 234 (1967).
(236) H. Kauss, A. L. Swanson, R. Arnold, and W. Odzuck, *Biochim. Biophys. Acta*, **192**, 55 (1969).
(237) H. Kauss and A. L. Swanson, *Z. Naturforsch.*, **24B**, 28 (1969).

myo-Inositol is known to occur in higher plants, both free and as its hexaphosphate, phytic acid.[238] Loewus[239] showed that, in plants, *myo*-inositol is capable of undergoing enzymic scission between C-1 and C-6 to produce D-glucuronic acid. This reaction is similar to the oxidative cleavage demonstrated by Charalampous[240] in the rat kidney. Loewus and coworkers[241] found that D-glucuronic acid and its lactone are directly converted by detached plant tissues into the D-galactosyluronic groups of pectin. When *myo*-inositol-2-*t* or 2-[14]*C*) was supplied to strawberry fruit or parsley leaves and allowed to be metabolized, the D-galactosyluronic compounds obtained from the degraded pectin contained most of the radioactivity in C-5. Other experiments[242] with parsley leaves produced similar results. It appears that, in the plant, D-glucuronic acid formed from the ruptured *myo*-inositol ring undergoes a number of known enzymic reactions, resulting in the formation[243] of UDP-D-galacturonic acid, which serves as donor of the D-galactosyluronic group for pectin formation. *myo*-Inositol thus appears to be a potential source of the pectin in plants.

9. Xylan

Xylan is a polysaccharide consisting mainly of D-xylopyranose residues and a small proportion of L-arabinofuranose residues. The polysaccharide exists, mainly in association with cellulose, in the secondary cell-walls of practically all higher plants.

As previously pointed out,[1] considerable data have been accumulated from experiments *in vivo* indicating that D-xylose in plants originates from D-glucose by a series of known enzymic reactions. It was assumed that, after this pentose is formed, it is enzymically polymerized to D-xylan.

The results of later work with plant enzymic preparations *in vitro* agreed with the postulated mechanism of synthesis of xylan *in vivo*. Particulate enzyme preparations from corn shoots[244] and immature corn-cobs[245] were capable of transferring D-xylosyl-[14]*C* groups from

(238) S. J. Angyal and L. Anderson, *Advan. Carbohyd. Chem.*, **14**, 135 (1959).
(239) F. A. Loewus, *Fed. Proc.*, **24**, 855 (1965).
(240) F. C. Charalampous, *J. Biol. Chem.*, **235**, 1286 (1960).
(241) F. A. Loewus, R. Jang, and C. G. Seegmiller, *J. Biol. Chem.*, **232**, 533 (1958).
(242) F. A. Loewus and S. Kelly, *Biochem. Biophys. Res. Commun.*, **7**, 204 (1962); F. A. Loewus, S. Kelly, and E. F. Neufeld, *Proc. Nat. Acad. Sci. U. S.*, **48**, 421 (1962).
(243) W. Z. Hassid, E. F. Neufeld, and D. S. Feingold, *Proc. Nat. Acad. Sci. U. S.*, **45**, 905 (1959).
(244) J. B. Pridham and W. Z. Hassid, *Biochem. J.*, **100**, 21P (1966).
(245) R. W. Bailey and W. Z. Hassid, *Proc. Nat. Acad. Sci. U. S.*, **56**, 1586 (1966).

UDP-D-xylose-^{14}C to some unknown acceptor, to form a polysaccharide that appears to be $(1 \rightarrow 4)$-linked xylan. Total hydrolysis of the polysaccharide produced D-xylose-^{14}C and L-arabinose-^{14}C, indicating that the particulate preparation contains an epimerase that converts UDP-D-xylose into UDP-L-arabinose. The labeled polymer closely resembles the D-xylan of plants in practically all its physical and chemical properties. Partial hydrolysis with fuming hydrochloric acid afforded a homologous series of oligosaccharides of low molecular weight that were shown to be D-xylodextrins and which were chromatographically identical with the authentic plant $(1 \rightarrow 4)$-β-D-xylodextrins. Complete hydrolysis of the polymer with acid resulted in the production of both ^{14}C-labeled D-xylose and ^{14}C-labeled L-arabinose in the ratio of 4:1. Treatment with weak acid (0.01 N), which is known to hydrolyze furanoses, liberated L-arabinose-^{14}C, but practically no D-xylose-^{14}C, indicating that some of the L-arabinose residues have the furanoid ring.

VI. Synthesis of Heteropolysaccharides

1. D-Gluco-D-mannan

D-Gluco-D-mannans are plant cell-wall constituents which are considered to belong to the hemicellulose groups that are closely associated with cellulose.

It has been shown[173] that extracts from mung-bean seedlings incorporate the D-glucosyl-^{14}C group of GDP-D-glucose-^{14}C into cellulose, and that the addition of GDP-D-mannose to the incubation mixture increases the incorporation of D-glucose-^{14}C into an alkali-insoluble polysaccharide that is not cellulose. Elbein and Hassid,[246] and Elbein[247] subsequently found that this polysaccharide is D-gluco-D-mannan. When ^{14}C-labeled GDP-D-mannose is used as the sole substrate, there is produced a radioactive D-gluco-D-mannan that requires Mg^{2+} for its synthesis.

The insoluble polysaccharide was characterized by examination of the soluble, ^{14}C-labeled oligosaccharides isolated by a variety of treatments, including enzymic hydrolysis, partial hydrolysis with acid, or acetolysis. From the polymer containing D-mannose-^{14}C, several oligosaccharides were isolated having various ratios of D-mannose to D-glucose, from 1:1 to 3 or 4:1. One of these oligosaccharides was a disaccharide tentatively identified as an O-β-D-man-

(246) A. D. Elbein and W. Z. Hassid, *Biochem. Biophys. Res. Commun.*, **23**, 311 (1966).
(247) A. D. Elbein, *J. Biol. Chem.*, **244**, 1608 (1969).

nopyranosyl-D-glucopyranose. In this disaccharide, the label was almost exclusively on the D-mannosyl group. Also, the identical, ^{14}C-labeled disaccharide was obtained from the polymer synthesized in the presence of GDP-D-glucose-^{14}C and GDP-D-mannose, except that the radioactivity was in the D-glucose residue. The identity of these disaccharides indicated that both D-mannose-^{14}C and D-glucose-^{14}C are incorporated into the same polymer. The polymer, as well as the oligosaccharide, was susceptible to hydrolysis by enzymes specific for β-D linkages. Methylation experiments on the polymer (as well as on the oligosaccharides), and periodate oxidation of the oligosaccharides, indicated that the D-mannose residues are joined in the polysaccharide by $(1 \rightarrow 4)$-linkages.

Based on the ratios of D-mannose to D-glucose in the larger oligosaccharides, the polymer formed from GDP-D-mannose-^{14}C or GDP-D-glucose-[14]C (in the presence of D-mannose) appears to be a β-D-$(1 \rightarrow 4)$-linked D-gluco-D-mannan containing 3 or 4 D-mannose residues per D-glucose residue.

2. DL-Galactan

An investigation by Su and Hassid[248] of the carbohydrates of a marine red alga, *Porphyra perforata*, showed that the ethanol-insoluble fraction of this plant contains a galactan that, on hydrolysis, gives approximately equal proportions of D- and L- galactose. The polysaccharide consists of residues of D-galactose, 6-O-methyl-D-galactose, 3,6-anhydro-L-galactose, and of sulfate (ester) group in the molar ratios of ~1:1:2:1.

An exploration of the "sugar nucleotides" in this alga[52] disclosed the presence of UDP-D-galactose, GDP-L-galactose, UDP-D-glucose, GDP-D-mannose, and UDP-D-glucuronic acid. Moreover, AMP, UMP, GMP, IMP, ADP, UDP, IDP, NAD$^\oplus$, NADP$^\oplus$, and a new nucleotide, adenosine 3',5'-pyrophosphate, were isolated.

As *Porphyra perforata* contains complex saccharides consisting mainly of both D- and L-galactose residues, together with "sugar nucleotides" containing the two enantiomorphs of galactose, it may be assumed that these "sugar nucleotides" are the precursors of the complex galactose-containing carbohydrates.

Inasmuch as GDP-L-galactose is found together with GDP-D-mannose in this red alga, it is suggested that the formation of "L-galactose nucleotide" probably occurs by a mechanism similar to

(248) J. C. Su and W. Z. Hassid, *Biochemistry*, 1, 474 (1962).

the epimerization reactions in the conversion of GDP-D-mannose into GDP-L-fucose,[46,47] in which one of the glycosyl groups has the L-*galacto* configuration.

On the basis of this consideration, it is likely that the pathway for the formation of DL-galactan involves the following sequence of reactions.

D-Mannopyranosyl phosphate $\longrightarrow$ GDP-D-mannose $\longrightarrow$

GDP-L-galactose $\longrightarrow$ DL-galactan

The origin of the sulfuric acid group of the galactan is not known. However, in this connection, it is of interest to compare the structure of "active sulfate" (adenylyl sulfate 3'-phosphate), which is known to be a sulfate donor,[249] with that of the adenosine 3',5'-pyrophosphate discovered in *Porphyra perforata* The adenosine 3',5'-pyrophosphate in this alga may possibly activate the inorganic sulfate, causing the formation of "active sulfate," which then effects the enzymic sulfation of the polysaccharide.

3. Hemicellulose

Hemicellulose refers to a chemically ill-defined group of hetero-polysaccharides of plant cell-walls that are closely associated with cellulose. They are composed of different hexoses and pentoses, and uronic acids. The hemicelluloses may be separated[250] into two fractions, A and B; however, there is no clear demarcation of structure between the two fractions. Hemicellulose B usually contains a higher proportion of uronic acid, chiefly 4-O-methyl-D-glucuronic acid, than the A fraction. The methyl ethers of D-aldobiouronic acids containing D-glucuronic acid, D-galacturonic acid, or D-mannuronic acid are isolated chiefly as the 4-O-methylaldobiouronic acids, and are hydrolyzed with difficulty by acid.

As previously mentioned, the methyl donor[234] for the formation of 4-O-methyl-D-glucuronic acid of hemicellulose B proved to be S-adenosyl-L-methionine, as in the formation of the methyl ester of pectin.[233]

From immature corn-cobs, a particulate preparation containing hemicellulose B was found capable of transferring the ^{14}C-labeled methyl group from S-adenosyl-L-methionine-[14]*C* to a macromolecular acceptor present in the particles.[234] The radioactive product was

(249) F. D'Abramo and F. Lipmann, *Biochim. Biophys. Acta*, **25**, 211 (1957).
(250) T. E. Timell, *Advan. Carbohyd. Chem.*, **19**, 247 (1964); **20**, 409 (1965).

shown to be hemicellulose B labeled in the 4-*O*-methyl-D-glucuronic acid residues, and it was isolated mainly as 4-*O*-methyl-D-glucosyl-uronic acid-(1 → 2)-D-xylose after partial hydrolysis with acid. In another experiment with the particulate enzyme from the same plant, Kauss[251] obtained a (4-*O*-methyl-D-glucosyluronic acid)-D-galactose together with (4-*O*-methyl-D-glucosyluronic acid)-D-xylose. He also found that the particulate enzyme preparations from immature corn-cobs contain, in addition to the methyl transferase, an enzyme that introduces the D-glucosyluronic group from UDP-D-glucuronic acid into hemicellulose B.

Particulate enzyme preparations from mung-bean shoots are also capable of transferring the methyl groups from S-adenosyl-L-methi-onine to the D-glucuronic acid residue of polysaccharides present in the particles.[252] On partial hydrolysis of the polysaccharides, a mixture was produced that consisted mainly of 4-*O*-methyl-β-D-glucosyluronic acid-(1 → 6)-D-galactose and minor proportions of 4-*O*-methyl-α-D-glucosyluronic acid-(1 → 2)-D-xylose and 4-*O*-methyl-D-glucuronic acid; this indicates that the products of the transfer reaction consist mainly of polysaccharides in which 4-*O*-methyl-D-glucuronic acid is linked to D-galactose, presumably, 4-*O*-methyl-D-glucurono-D-galactan.

The introduction of the methyl ether groups of plant heteropoly-saccharides at a macromolecular level is similar to the finding that C- and N-methyl groups of RNA and DNA are also introduced into preformed macromolecules.[253]

4. Alginic Acid

Alginic acid is a glycuronan known to be a major structural con-stituent of marine brown algae. It is composed of a mixture of β-D-(1 → 4)-linked polymers of D-mannuronic acid and its 5-epimer, namely, L-guluronic acid. The D-mannuronic acid is always the preponderant component (about 80%) of this polymer.[254,255]

As this glycuronan is a mixture of two polymers, one consisting of D-mannuronic acid residues and the other of L-guluronic acid resi-dues, it was suspected that this brown alga contains the "nucleoside diphosphate uronic acid" precursors from which this mixed glycuro-

(251) H. Kauss, *Biochim. Biophys. Acta*, **148**, 572 (1967).
(252) H. Kauss, *Phytochemistry*, **8**, 985 (1969).
(253) E. Borek and P. R. Srinivasan, *Ann. Rev. Biochem.*, **35**, 275 (1966).
(254) F. G. Fischer and H. Dörfel, *Z. Physiol. Chem.*, **302**, 186 (1955).
(255) S. K. Chanda, E. L. Hirst, E. G. V. Percival, and A. G. Ross,*J. Chem. Soc.*, 1833 (1952).

nan is formed. A search[28] for nucleotides in the brown alga *Fucus gardneri* Silva revealed the presence of UDP-D-mannuronic acid and UDP-L-guluronic acid; these "nucleoside diphosphate uronic acids" were isolated, and their structures were thoroughly studied.

It was also found that enzyme preparations from the *Fucus gardneri* alga contained all the enzyme activities required for the pathway leading to the synthesis of the glycuronan. The following enzymes were shown to be present: hexokinase, phosphomannomutase, GDP-D-mannose pyrophosphorylase, GDP-D-mannose dehydrogenase, and D-mannuronic acid transferase.[256] Active enzymic preparations could be obtained when the brown algal homogenate was treated with poly(vinylpyrrolidinone). Starting with D-mannose, the enzyme systems effect a number of consecutive reactions, resulting in the formation of the D-mannuronic acid chain:

$$\alpha\text{-D-Mannose} \xrightarrow[\text{ATP}]{\text{kinase}} \alpha\text{-D-mannose 6-phosphate} \xrightarrow{\substack{\text{mannose phosphate} \\ \text{mutase}}}$$
$$\alpha\text{-D-mannosyl phosphate}$$

$$\alpha\text{-D-mannosyl phosphate} + \text{GTP} \xrightarrow{\text{pyrophosphorylase}} \text{GDP-D-mannose} + \text{PPi}$$

$$\text{GDP-D-mannose} + 2\text{NAD(P)}^+ + \text{H}_2\text{O} \xrightarrow{\text{dehydrogenase}}$$
$$\text{GDP-D-mannuronic acid} + 2\text{NAD(P)H} + 2\text{H}^\oplus$$

$$\text{GDP-D-mannuronic acid} \xrightarrow{\text{transferase}} \text{D-mannuronan} + \text{GDP}$$

L-Guluronic acid, which is the minor constituent of the alginic acid polymer and which is required for the complete synthesis of this polymer, is probably synthesized through interconversion of GDP-D-mannuronic acid and GDP-L-guluronic acid by a 5-epimerase. Such an enzymic reaction has, however, not yet been demonstrated.

The complete synthesis of alginic acid probably involves a succession of transfers of uronic acid residues from the two GDP-uronic acids to an acceptor molecule, forming the (1 → 4)-linked glycuronan chain:

$$\begin{matrix} \text{GDP-D-mannuronic acid} \\ + \\ \text{GDP-L-guluronic acid} \end{matrix} \xrightarrow[\text{acceptor}]{\text{glycosyl transferase(s)}} \text{alginic acid} + \text{GDP}$$

(256) T. Y. Lin and W. Z. Hassid, *J. Biol. Chem.*, **241**, 5284 (1966).

5. Capsular Polysaccharides of Micro-organisms

Studies by Smith and coworkers on the biosynthesis of Types I and III pneumococcal polysaccharides have been discussed in Vol. 18 of this Series.[1] Since then, the biosynthesis of Type XIV pneumococcal polysaccharide has been investigated.[257,258] This polysaccharide contains 2-acetamido-2-deoxy-D-glucose, D-glucose, and D-galactose residues in the ratios of 2:1:3. When a particulate fraction from Type XIV pneumococcus was incubated with UDP derivatives of these sugars, each labeled with ^{14}C, the sugars were incorporated into a polysaccharide,[257,258] as well as into glycolipids.[257-259] The polysaccharide resembles the Type XIV capsular polysaccharide, and reacts with anti-XIV sera. However, it is slightly different, in that (a) 2-acetamido-2-deoxy-D-glucose, D-glucose, and D-galactose are incorporated in the ratios of 1:1:2, not 2:1:3, and (b) the product reacts more slowly with antibodies, and, on double-diffusion plates, forms a band that suggests a higher rate of migration.[257] It is not yet clear whether the in vitro incorporation represents the synthesis of new chains of polysaccharides, elongation of already existing chains, or addition of short branches to incomplete chains. The possible role of lipid intermediates also remains unknown.

The capsular polysaccharide of Aerobacter aerogenes was found to contain a repeating unit having the sequence D-galactosyl-(D-glucosyl-uronic acid)-D-mannosyl-D-galactose.[260] Incubation of the cell-envelope fraction from this organism with UDP-D-galactose-^{14}C results in the formation of a compound that appears to be a D-galactosyl phosphate–lipid.[260] Presumably, the D-galactosyl phosphate portion is transferred as a unit from UDP-D-galactose, because uridine 5'-phosphate inhibits this reaction. In the second step, it appears that the D-mannosyl group of GDP-D-mannose is transferred onto the lipid intermediate, presumably forming D-mannosyl–D-galactosyl phosphate–lipid.[260] This system thus resembles that operative in the biosynthesis of peptidoglycan and of the O side-chain portion of cell-wall lipopolysaccharide, in that repeating units are synthesized on a lipid carrier. The identity of the lipid in this system is not yet known, but it has been found that undecaprenyl phosphate from Micrococcus lysodeikticus functions efficiently in this system.[212]

(257) J. Distler and S. Roseman, Proc. Nat. Acad. Sci. U. S., **51**, 897 (1964).
(258) J. Distler, B. Kaufman, and S. Roseman, Methods Enzymol.. **8**, 450 (1966).
(259) B. Kaufman, F. D. Kundig, J. Distler, and S. Roseman, Biochem. Biophys. Res. Commun., **18**, 312 (1965).
(260) F. A. Troy and E. C. Heath, Fed. Proc., **27**, 345 (1968).

Escherichia coli K-235 produces an extracellular homopolymer of N-acetylneuraminic acid. This substance, named colominic acid, has been synthesized[260a] *in vitro* from CMP-N-acetylneuraminic acid in the presence of a purified particulate fraction from this strain of *E. coli*. The activity of the enzyme is stimulated by ammonium sulfate (1.2–1.5*M*), and by the addition of exogenous colominic acid (especially of aged, slightly degraded preparations).

A nonfermentative yeast, *Cryptococcus laurentii*, produces a capsular polysaccharide that contains[261] D-mannose, D-xylose, and D-glucuronic acid residues in the approximate molar ratios of 5:2:1. Structural studies[262] indicate that the polysaccharide has a backbone of D-mannose residues, with D-xylose and D-glucuronic acid residues as end groups. A particulate fraction from this organism was shown to catalyze the transfer of the D-xylosyl group from its UDP derivative to acceptor polysaccharides.[263] The transferase can be solubilized with digitonin. Partially de-D-xylosylated capsular polysaccharide from the parent strain is most active as the acceptor, and similarly treated polysaccharides from related species are active to a lesser degree. Neither untreated capsular polysaccharides from these strains nor *S. cerevisiae* mannan showed any activity as an acceptor.

It has since been found that oligosaccharides serve as acceptors of D-glycosyl transfer in this system.[264] Thus, D-mannose is transferred from its GDP derivative to α-D-(1 $\rightarrow$ 2)-linked D-manno-biose and -triose, and a D-xylosyl group is transferred to α-D-(1 $\rightarrow$ 3)-linked D-mannobiose. In each instance, the glycosyl group seems to be transferred to the nonreducing end of the oligosaccharide acceptor.

6. Cell-wall Lipopolysaccharide of Gram-negative Bacteria

Gram-negative bacteria contain large proportions of lipopolysaccharide (LPS) in their cell walls. In most cases, the innermost portion consists of a lipid (called lipid A) to which an acidic polysaccharide (core polysaccharide) is linked.[265] The latter further carries a neutral

(260a) D. Aminoff, F. Dodyk, and S. Roseman, *J. Biol. Chem.*, **238**, PC1177 (1963); D. Aminoff and F. D. Kundig, *Methods Enzymol.*, **8**, 419 (1966).
(261) M. J. Abercrombie, J. K. N. Jones, M. V. Lock, M. B. Perry, and R. J. Stoodley, *Can. J. Chem.*, **38**, 1617 (1960).
(262) A. R. Jeanes, J. E. Pittsley, and P. R. Watson, *J. Appl. Polym. Sci.*, **8**, 2005 (1964).
(263) A. Cohen and D. S. Feingold, *Biochemistry*, **6**, 2933 (1967).
(264) J. S. Schutzbach and H. Ankel, *FEBS Lett.*, **5**, 145 (1969).
(265) O. Lüderitz, A.-M. Staub, and O. Westphal, *Bacteriol. Rev.*, **30**, 192 (1966).

polysaccharide, designated as O side-chains (see Fig. 1).[265] Considerable progress has been made on the biosynthesis of this very complex macromolecule, especially in *Salmonella* and *Escherichia coli*.[265-270]

a. **Biosynthesis of the Core Polysaccharide.**—Little is known about the biosynthesis of lipid A. Lipid A from *Salmonella* has a skeleton comprised of 2-amino-2-deoxy-β-D-glucosyl-(1 → 6)-2-amino-2-deoxy-D-glucosyl phosphate; it is assumed that the phosphate group can form a phosphoric diester bridge to C-4 of the nonreducing terminal 2-amino-2-deoxy-D-glucosyl residue of the neighboring chain, thus serving as a cross-link.[271] All the available hydroxyl and amino groups of these sugar residues are acylated, and 3-hydroxymyristic acid (3-hydroxytetradecanoic acid) comprises a large proportion of these acyl groups.[272] This fatty acid is found exclusively in lipid A, but its transfer to partially deacylated lipid A in a cell-free system has not yet been achieved. However, 3-hydroxymyristoyl acyl carrier protein has been shown to transfer its acyl moiety to a lysophosphatidyl-2-aminoethanol in the presence of an enzyme system from *E. coli*;[273] it is not known whether this reaction is related to synthesis of lipid A.

The biosynthesis of the core presumably begins with the transfer of a 3-deoxy-D-*manno*-octulosonic acid (KDO) group from CMP–KDO to lipid A. This reaction was demonstrated[274] in a cell-free system from *E. coli* O111. A soluble enzyme catalyzed the transfer of "KDO" onto added lipid A. However, treatment of lipid A with 0.2 *M* sodium hydroxide, which presumably splits off O-acyl groups, dramatically increased the amount of "KDO" transferred; this could mean that "KDO" is normally transferred to a precursor (of lipid A) that is not yet acylated, but it is equally possible that the effect merely reflects

(266) B. L. Horecker, *Ann. Rev. Microbiol.*, **20**, 253 (1966).

(267) O. Lüderitz, K. Jann, and R. Wheat, in "Comprehensive Biochemistry," M. Florkin and E. H. Stotz, eds., Elsevier, Amsterdam, 1968, Vol. 26A.

(268) H. Nikaido, *Advan. Enzymol.*, **31**, 77 (1968).

(269) M. J. Osborn, *Ann. Rev. Biochem.*, **38**, 501 (1969).

(270) S. J. Ajl, G. Weinbaum, and S. Kadis, eds., "Microbial Toxins," Academic Press, Inc., New York, 1971, Vol. 4.

(270a) C. G. Hellerqvist, B. Lindberg, S. Svensson, T. Holme, and A. A. Lindberg, *Carbohyd. Res.*, **8**, 43 (1968); **9**, 237 (1969).

(270b) H. Nikaido, *J. Biol. Chem.*, **244**, 2835 (1969).

(271) J. Gmeiner, O. Lüderitz, and O. Westphal, *Eur. J. Biochem.*, **7**, 370 (1969).

(272) O. Westphal and O. Lüderitz, *Angew. Chem.*, **66**, 407 (1954); A. J. Burton and H. E. Carter, *Biochemistry*, **3**, 411 (1964).

(273) S. S. Taylor and E. C. Heath, *J. Biol. Chem.*, **244**, 6605 (1969).

(274) E. C. Heath, R. M. Mayer, R. D. Edstrom, and C. A. Beaudreau, *Ann. N. Y. Acad. Sci.*, **133**, 315 (1966).

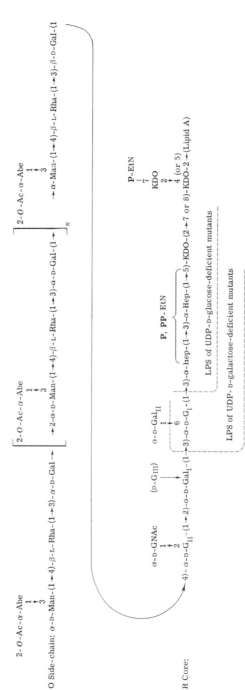

Fig. 1.—Structure of a Subunit of Cell-Wall Lipopolysaccharide from *Salmonella typhimurium* (O antigen: 4, 5, 12₁, 12₃). (Several subunits are presumably cross-linked through phosphoric diester linkages. This structure was elucidated as the result of contributions from many laboratories. For earlier results, see the review by Lüderitz and coworkers.[265] The more recent contributions are found in Refs. 270, 270a, 270b, 275, and 299. Abe, abequose; Man, D-mannose; Rha, L-rhamnose; Gal, D-galactose; GNAc, 2-acetamido-2-deoxy-D-glucose; G, D-glucose; Hep, L-*glycero*-D-*manno*-heptose; EtN, 2-aminoethanol; KDO, 3-deoxy-D-*manno*-octulosonic acid; and Ac, acetyl.)

the increased solubility of the alkali-treated preparation. When the alkali-treated lipid A was further treated with 0.5 M hydrochloric acid for one hour at 100°, a phosphorylated oligosaccharide containing 2-deoxy-2-(3-hydroxymyristamido)-D-glucose residues was obtained; this oligosaccharide was also a good acceptor of "KDO." The enzymically transferred "KDO" appears to be glycosidically linked, as the potential carbonyl group involving C-2 cannot be reduced by sodium borohydride. The site of attachment of "KDO" in the acceptor molecule is not yet known; presumably, it is one of the hydroxyl groups of the sugar or that of the 3-hydroxymyristoyl group. Little is known regarding the biosynthesis of the remaining portion of the inner region of the "core" polysaccharide. It has been suggested that a branched trisaccharide of "KDO" is present at least in *Salmonella* LPS,[275] but the enzymic addition of the second and third "KDO" residues has not yet been looked for. The origin of the phosphate of 2-aminoethanol[275,276] and the pyrophosphate of 2-aminoethanol[269] is unknown. The nature of the "activated" (that is, nucleotide-linked) form of L-*glycero*-D-*manno*-heptose is not known; information about the enzymic transfer of these residues is thus still lacking.

The transfer of the first D-glucose residue (G_I of Fig. 1) was demonstrated by the use of *Salmonella* mutants defective in phosphoglucoisomerase.[277] In the early stages of the study, it was difficult to show enzymic transfer of glycosyl residues in the synthesis of LPS, mainly because the amount of endogenous acceptors was extremely limited. Although *Salmonella* or *E. coli* cells contain fairly large proportions of LPS, representing 2 to 3% of the dry weight of the bacteria, most of the LPS molecules are already in a completed form and, therefore, cannot function as acceptors of glycosyl groups. This difficulty was circumvented for the first time in the D-galactosyl transferase system by the use of genetic mutants unable to synthesize UDP-D-galactose.[278–280] These mutants lack the enzyme UDP-D-glucose 4-epimerase, and, consequently, they synthesize LPS only up to the step where UDP-D-galactose becomes requisite.[281] Thus,

(275) W. Dröge, V. Lehmann, O. Lüderitz, and O. Westphal, *Eur. J. Biochem.*, **14**, 175 (1970).
(276) A. P. Grollman and M. J. Osborn, *Biochemistry*, **3**, 1571 (1964).
(277) L. Rothfield, M. J. Osborn, and B. L. Horecker, *J. Biol. Chem.*, **239**, 2788 (1964).
(278) H. Nikaido, *Proc. Nat. Acad. Sci. U. S.*, **48**, 1337 (1962).
(279) M. J. Osborn, S. M. Rosen, L. Rothfield, and B. L. Horecker, *Proc. Nat. Acad. Sci. U. S.*, **48**, 1831 (1962).
(280) R. D. Edstrom and E. C. Heath, *Biochem. Biophys. Res. Commun.*, **16**, 576 (1964); *J. Biol. Chem.*, **242**, 3581 (1967).
(281) T. Fukasawa and H. Nikaido, *Biochim. Biophys. Acta*, **48**, 470 (1961).

the "cell envelope" fraction from these mutants contains only this kind of "defective" LPS (see Fig. 1), and the envelope-bound D-galactosyl transferase readily transfers the D-galactosyl group of UDP-D-galactose to these endogenous, incomplete LPS acceptors, once UDP-D-galactose has been added to the system[278-280] A similar principle was utilized in the study of D-glucosyl transferase. Mutants lacking phosphoglucoisomerase[277] (if grown with D-fructose as the only source of carbon) or UDP-D-glucose pyrophosphorylase[282] synthesize LPS that contains L-*glycero*-D-*manno*-heptose residues but not D-glucose or D-galactose residues. Thus, addition of UDP-D-glucose-^{14}C to the cell-envelope fraction of these mutants resulted in the rapid incorporation of the D-glucosyl group (G_I of Fig. 1) into the endogenous LPS.

As described earlier, the enzymic transfer of the next sugar residue, that of D-galactose (Gal_I of Fig. 1) was accomplished by using the cell-envelope preparation from mutants defective in UDP-D-galactose synthesis. The site of attachment of the transferred D-galactosyl group was difficult to determine, because the particular D-galactosyl linkage was extremely acid-labile. This problem was finally solved by oxidizing the primary alcohol group at C-6 of the transferred D-galactosyl group into an aldehyde group by the action of D-galactose oxidase, and then oxidizing this to a carboxyl group with hypobromite. As is well known, the glycosidic linkages of hexosiduronic acids are resistant to acid hydrolysis, and thus, mild treatment of the oxidized LPS with acid produced a disaccharide identified as 3-O-α-D-galactosyluronic acid-D-glucose.[283] This result obviously showed that the D-galactosyl group is linked to the D-glucose residue of the acceptor LPS through an α-D-(1 → 3)-linkage.

In the early stages of the study on core synthesis, the endogenous LPS present in the cell-envelope preparation served as the acceptor of transferred glycosyl groups; for this reason, detailed work on the properties of glycosyl transferase was difficult. The situation was, however, greatly improved when LPS in heated, cell-envelope preparations was found to serve as an exogenous acceptor.[277] By the use of this acceptor, enzyme activities were also detected in the soluble fraction of sonic extracts of *Salmonella typhimurium*,[277] and it was established that D-glucosyl and D-galactosyl transferases have

(282) T. Fukasawa, K. Jokura, and K. Kurahashi, *Biochem. Biophys. Res. Commun.*, **7**, 121 (1962); *Biochim. Biophys. Acta*, **74**, 608 (1963); H. J. Risse, O. Lüderitz, and O. Westphal, *Eur. J. Biochem.*, **1**, 233 (1967).

(283) S. M. Rosen, M. J. Osborn, and B. L. Horecker, *J. Biol. Chem.*, **239**, 3196 (1964).

strict acceptor specificities, so that each of them only transfers glycosyl groups to the D-glucose-deficient and D-galactose-deficient LPS (in the heated cell-envelope preparations), respectively.[277] This important finding established that the sequence of various sugars in the core is determined by the donor and acceptor specificities of glycosyl transferases that act in succession.

The soluble D-galactosyl transferase has now been highly purified, and found to contain an unidentified lipid.[284] Extraction of this lipid prevented the aggregation of enzyme molecules, but the presence of this lipid was not necessary for enzyme activity.

Although the heated, cell-envelope fraction was a good acceptor of glycosyl groups, purified LPS was completely inactive. This paradox was solved when it was found that the removal of phospholipids from cell-envelope preparations led to the loss of their acceptor activity.[285] Phospholipids could be added back to the extracted, cell-envelope fraction or even to the purified LPS, and heating and slow cooling of the mixture was found to produce phospholipid–LPS or phospholipid –cell-envelope complex, each of which was fully active as an acceptor of glycosyl groups.[285] Electron-microscope studies showed that phospholipid molecules were inserted between LPS molecules, and that these two kinds of molecules together formed mixed micelles in the aqueous environment.[286] Among the phospholipids tested, those containing two unsaturated fatty acids had the highest activity, and those containing two saturated fatty acids were inactive.[287] Fatty acids containing cyclopropane rings also produced active phospholipids.[287] Thus, the degree of close stacking of apolar side-chains of phospholipids in micellar aggregates seemed to have a crucial influence on their activity. The polar portion of the phospholipids also had some influence; although most polar groups occurring in bacterial phospholipids produced active lipids, phosphatidylcholine, which rarely occurs in bacteria, was completely inactive.[287]

This system is thus an interesting example in which phospholipids are required in the glycosyl transferase reaction. That the phospholipids react by complexing with the LPS acceptor was suggested by the results just described,[285–287] but decisive experiments on the role of phospholipids were conducted as follows. When the complexes of phospholipids and the proper LPS acceptor were mixed

(284) A. Endo and L. Rothfield, *Biochemistry*, **8**, 3508 (1969).
(285) L. Rothfield and B. L. Horecker, *Proc. Nat. Acad. Sci. U. S.*, **52**, 939 (1964).
(286) L. Rothfield and R. W. Horne, *J. Bacteriol.*, **93**, 1705 (1967).
(287) L. Rothfield and M. Pearlman, *J. Biol. Chem.*, **241**, 1386 (1966).

with the soluble fraction of crude extracts of S. *typhimurium*, specific glycosyl transferases were bound to the complex, and thus disappeared from the nonsedimentable fraction.[288,289] The LPS–phospholipid complexes appear, therefore, to be the true substrates of the transferases. LPS–phospholipid and LPS–phospholipid–enzyme complexes were also isolated, by centrifugation through sucrose density-gradients.[290]

The binding of the transferases in this system is very specific. LPS alone does not bind enzyme; LPS preparations of different structure, when complexed with phospholipids, bind only the enzyme involved in the transfer of the next glycosyl group. Thus, the complex of D-glucose-deficient LPS and phospholipids binds only UDP-D-glucose:LPS D-glucosyl transferase; the complex containing galactose-deficient LPS binds only UDP-D-galactose: LPS D-galactosyl transferase. These results again indicate the strict specificity with which the glycosyl transferases interact with acceptors.

It has now been found that short treatment of a UDP-D-glucose-4-epimerase-less mutant of E. *coli* with dilute (ethylenedinitrilo)tetraacetic acid specifically releases a ternary complex of LPS, phosphatidyl-2-aminoethanol, and UDP-D-galactose:LPS D-galactosyl transferase,[291] without causing significant lysis or killing of the cells.[292]

The detailed mechanism by which phospholipids produce active substrates out of LPS is not completely clear. The observation that phospholipid molecules become inserted between LPS molecules suggests that phospholipids might function simply by "supporting" and holding LPS molecules in a particular conformation, so that LPS could interact with the active site of the enzyme.[293] If this is so, phospholipids should decrease the apparent K_m of glycosyl transferase toward LPS, but should not affect the V_{max} of the reaction. Later studies have, however, shown[294] that there was a 13-fold increase in V_{max}. Phospholipids, by themselves, were also found able to combine with the purified D-galactosyl transferase.[294] Furthermore, phosphatidyl-2-aminoethanol containing saturated but relatively short fatty acids, for example, 1,2-di-O-decanoyl-sn-glycerol 3-(2-

(288) L. Rothfield and M. Takeshita, *Biochem. Biophys. Res. Commun.*, **20**, 521 (1965).
(289) L. Rothfield and M. Takeshita, *Ann. N. Y. Acad. Sci.*, **133**, 384 (1966).
(290) M. M. Weiser and L. Rothfield, *J. Biol. Chem.*, **243**, 1320 (1968).
(291) S. B. Levy and L. Leive, *J. Biol. Chem.*, **245**, 585 (1970).
(292) L. Leive, *Biochem. Biophys. Res. Commun.*, **18**, 13 (1965).
(293) L. Rothfield, M. Takeshita, M. Pearlman, and R. W. Horne, *Fed. Proc.*, **25**, 1495 (1966).
(294) A. Endo and L. Rothfield, *Biochemistry*, **8**, 3508 (1969).

aminoethyl phosphate)[294a] was found to be inactive in stimulating the D-galactosyl transferase reaction, and was unable to combine with the enzyme despite the fact that it could form stable com-

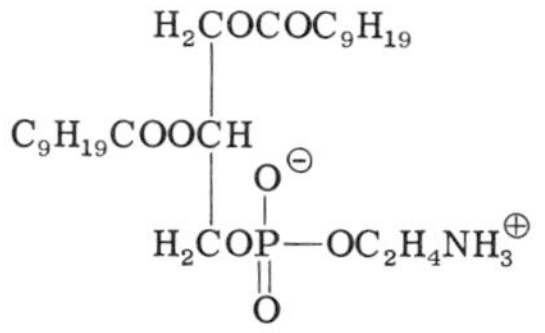

$$H_2COCOC_9H_{19}$$
$$C_9H_{19}COOCH$$
$$\overset{|}{\underset{O}{\overset{O^\ominus}{|}}}$$
$$H_2COP{-}OC_2H_4NH_3^\oplus$$

1, 2-Di- O -decanoyl- sn -glycerol
3-(2-aminoethyl phosphate)

plexes with LPS. From these results, it is now postulated that the D-galactosyl transferase has two binding sites, one specific for the acceptor LPS, and the other for phospholipid, and that the function of phospholipids is to combine with the enzyme *and activate it*, as well as to form a complex with LPS.[294]

D-Glucosyl (G_{II} of Fig. 1) and 2-acetamido-2-deoxy-D-glucosyl (GNAc of Fig. 1) groups are successively transferred from their UDP derivatives to the LPS containing Gal_I and G_I (see Fig. 1) in the presence of a cell-envelope fraction from *Salmonella*.[295,296] Simultaneous presence of various "sugar nucleotides" is unnecessary, and, after the transfer of one glycosyl group, the cell-envelope fraction can be recovered by centrifugation, washed free of the "sugar nucleotide" containing that sugar, and then be used as the acceptor for the transfer of the next glycosyl group.[295] The successive transfer of glycosyl groups in the sequence has been demonstrated in *E. coli*.[280] These experiments thus clearly demonstrate that glycosyl groups are sequentially added in the biosynthesis of "core" polysaccharide. Although transferases responsible for the transfer of G_{II} and GNAc groups (see Fig. 1; p. 411) have not yet been extensively studied, it is expected that they have strict substrate and acceptor specificities and work in succession, by analogy to the G_I and Gal_I transferases.

In vitro transfer of the second D-galactosyl group (Gal_{II} of Fig. 1) is usually inefficient. Although this reaction has been shown to occur with cell-envelope fractions of *Salmonella*,[297,298] the details are as

(294a) For definition of the term sn-glycerol, see *J. Biol. Chem.*, **242**, 4845 (1967).
(295) M. J. Osborn, S. M. Rosen, L. Rothfield, L. D. Zeleznick, and B. L. Horecker, *Science*, **145**, 783 (1964).
(296) M. J. Osborn and L. D'Ari, *Biochem. Biophys. Res. Commun.*, **16**, 568 (1964).
(297) P. Mühlradt, H. J. Risse, O. Lüderitz, and O. Westphal, *Eur. J. Biochem.*, **4**, 139 (1968).
(298) P. Mühlradt, *Eur. J. Biochem.*, **18**, 20 (1971).

yet unrevealed. In some strains, a third D-glucose residue seems to exist in the core (G_{III} of Fig. 1),[299] but its enzymic addition has not been demonstrated.

One, or both, of the L-*glycero*-D-*manno*-heptose residues in LPS are phosphorylated, and different polysaccharide chains may be cross-linked by phosphoric diester linkages.[300] Some mutants of *Salmonella* produce LPS lacking these phosphate groups,[297] and the LPS in the cell-envelope fraction from such mutants acts as an acceptor of the terminal phosphate group of ATP in a cell-free system.[298,301] It is of interest that LPS lacking these phosphate groups can accept the first D-glucosyl group (G_I of Fig. 1), but not the next glycosyl group, namely, D-galactosyl (Gal_I), either in intact cells or in a cell-free system.[297] The phosphorylating enzyme seems to prefer LPS that carries the G_I group to that devoid of this group, and it has been proposed that phosphorylation normally takes place after the addition of G_I but before the addition of Gal_I and Gal_{II} groups.[298]

b. Biosynthesis of O Side-Chain.—Biosynthesis of O side-chains has been studied in *Salmonella anatum* and in *Salmonella typhimurium*. In these organisms (and in other *Salmonella* serotypes thus far investigated) the O side-chain is made up of oligosaccharide "repeating units." For *S. anatum*, the repeating unit consists of D-mannosyl-L-rhamnosyl-D-galactose, and for *S. typhimurium*, of abequosyl-D-mannosyl-L-rhamnosyl-D-galactose. In the early stages of investigation on the biosynthesis of O side-chains, "sugar nucleo-tide" precursors containing these component sugars were added to the cell-envelope fraction of the organisms just mentioned, and the formation of incorporation products containing the repeating-unit structure expected was shown mostly by partial hydrolysis with acid.[302-305] As the source of cell envelopes, mutants that could not synthesize O side-chains (owing to the defective synthesis of dTDP-L-rhamnose[304,305] or of GDP-D-mannose[303]) were used, although the cell envelopes from the wild type of organism were also found to be quite active.[302]

When only dTDP-L-rhamnose and UDP-D-galactose were added

(299) G. Hämmerling, O. Lüderitz, and O. Westphal, *Eur. J. Biochem.*, **15**, 48 (1970).
(300) R. Cherniak and M. J. Osborn, *Fed. Proc.*, **25**, 410 (1966).
(301) P. Mühlradt, *Eur. J. Biochem.*, **11**, 241 (1969).
(302) P. W. Robbins, A. Wright, and J. L. Bellows, *Proc. Nat. Acad. Sci. U. S.*, **52**, 1302 (1964).
(303) L. D. Zeleznick, S. M. Rosen, M. Saltmarsh-Andrew, M. J. Osborn, and B. L. Horecker, *Proc. Nat. Acad. Sci. U. S.*, **53**, 207 (1965).
(304) H. Nikaido, *Biochemistry*, **4**, 1550 (1965).
(305) H. Nikaido and K. Nikaido, *Biochem. Biophys. Res. Commun.*, **19**, 322 (1965).

to *S. anatum* cell-envelope fraction[306] or to the cell-envelope fraction from a *S. typhimurium* mutant having defective core-synthesis,[307] L-rhamnose and D-galactose were found to be incorporated into a fraction that was readily extracted by such organic solvents as butyl alcohol or chloroform–methanol (LPS is not extracted into these solvents). Furthermore, when, after preincubation with dTDP-L-rhamnose and UDP-D-galactose, GDP-D-mannose was added, the sugars in the fraction soluble in organic solvent were converted into a polysaccharide (O side-chain) that could not be extracted into organic solvents.[306,307] In this way, it was found that lipid intermediates are involved in the biosynthesis of a polysaccharide (see Scheme 1).

Most of the studies on the nature of lipid intermediates were conducted with the "disaccharide–lipid" produced when only dTDP-L-rhamnose and UDP-D-galactose are added to the cell-envelope fraction. Alkaline hydrolysis of this lipid intermediate afforded L-rhamnosyl-D-galactosyl phosphate,[306,307] whereas treatment with 45% aqueous phenol at 68° or mild hydrolysis with acid gave an L-rhamnosyl-D-galactose.[307] By the use of UDP-D-galactose labeled with ^{32}P, the intact D-galactosyl phosphate moiety of this compound was found to be transferred to the acceptor lipid.[308,309] It was suspected that the phosphate group of the disaccharide phosphate is linked, in a pyrophosphate linkage, to a phosphate group of the acceptor lipid, because the rate of hydrolytic release of inorganic phosphate from the disaccharide–lipid in acid was exactly the same as the rate of release of inorganic phosphate from UDP-D-glucose under similar conditions.[308] The purified disaccharide–lipid contained two molar proportions of phosphate group per mole of disaccharide,[308] a result consistent with the hypothesis just mentioned. The structure of the lipid moiety was difficult to elucidate, because acid and alkaline hydrolysis produced no fragment that could be readily indentified.[308] However, application of mass spectrometry solved the problem, indicating clearly that the lipid was a phosphate of a C_{55} isoprenoid alcohol;[310] this lipid has been called[308] antigen-carrier lipid phosphate

(306) I. M. Weiner, T. Higuchi, L. Rothfield, M. Saltmarsh-Andrew, M. J. Osborn, and B. L. Horecker, *Proc. Nat. Acad. Sci. U. S.*, **54**, 228 (1965).
(307) A. Wright, M. Dankert, and P. W. Robbins, *Proc. Nat. Acad. Sci. U. S.*, **54**, 235 (1965).
(308) M. Dankert, A. Wright, W. S. Kelley, and P. W. Robbins, *Arch. Biochem. Biophys.*, **116**, 425 (1966).
(309) I. M. Weiner, T. Higuchi, M. J. Osborn, and B. L. Horecker, *Ann. N. Y. Acad. Sci.*, **133**, 391 (1966).
(310) A. Wright, M. Dankert, P. Fennessey, and P. W. Robbins, *Proc. Nat. Acad. Sci. U. S.*, **57**, 1798 (1967).

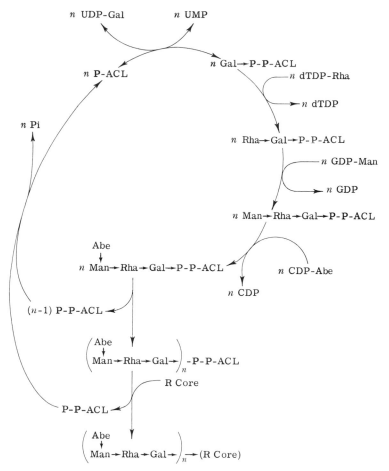

Scheme 1. Lipid Cycle in the Biosynthesis of S. *typhimurium* O Side-chains. (P-ACL, antigen-carrier-lipid phosphate. For other abbreviations, see Fig. 1., p. 411.)

(P-ACL), or undecaprenyl phosphate (see p. 396). (The enzymic synthesis of this lipid has been reported.[311]) The disaccharide-lipid, or L-rhamnosyl-D-galactosyl-PP-ACL, therefore corresponds to undecaprenyl (L-rhamnosyl-D-galactosyl pyrophosphate).

The first reaction, the transfer of D-galactosyl phosphate from UDP-D-galactose to P-ACL, has been studied in detail.[312] The equilibrium constant appears to lie somewhere near 0.5, indicating the "high energy" nature of the bond formed. Purified P-ACL has been shown

(311) J. G. Christenson, S. K. Gross, and P. W. Robbins, *J. Biol. Chem.*, **244**, 5436 (1969).
(312) M. J. Osborn and R. Yuan Tze-Yuen, *J. Biol. Chem.*, **243**, 5145 (1968).

to be effective as the acceptor when added to cell-envelope fraction in the presence of nonionic detergents.[269] Cell-envelope fraction containing D-galactosyl-PP-ACL, but not UDP-D-galactose, can readily transfer L-rhamnose from dTDP-L-rhamnose to this intermediate, thus producing the disaccharide–lipid.[312] Furthermore, cell-envelope fractions containing L-rhamnosyl-D-galactosyl–PP-ACL can incorporate D-mannose from GDP-D-mannose in the absence of UDP-D-galactose and dTDP-L-rhamnose.[306,307] These results clearly establish the sequential nature of the reactions leading to the biosynthesis of the repeating oligosaccharide unit (see Scheme 1).

When D-mannosyl-L-rhamnosyl-D-galactosyl-PP-ACL is produced, the trisaccharide units rapidly polymerize to form O side-chain polysaccharide.[306,307] If cell envelopes from strains unable to synthesize the core are utilized, the polysaccharide is not transferred to the core, and presumably stays attached to P-ACL. This conclusion was based on the observation that the polysaccharide could be extracted only after mild, acid hydrolysis and that a phosphate group was present at the reducing end of the polysaccharide thus obtained.[313] In mutants defective in core synthesis, O side-chains were, indeed, found to accumulate, presumably in a form still linked to P-ACL.[314,315]

That the repeating units are normally polymerized while still attached to P-ACL, and *then* transferred to the core, was demonstrated in two ways. (1) Pulse-chase experiments that used a phosphomannoisomerase mutant of *S. typhimurium*, where D-mannose-^{14}C is incorporated only into O side-chains, indicated a precursor–product type of relationship between "free" O side-chains (presumably linked to P-ACL before extraction) and LPS-linked, O side-chains.[316] (2) O Side-chains were shown to be elongated at the reducing end, rather than at the nonreducing end.[317,318] This mechanism, which is later discussed in detail, tends to rule out the possibility that the O side-chain grows by the sequential addition of repeating units on the core LPS.

When an O side-chain is elongated, a monomeric repeating-unit has to interact with an oligomer of repeating units. Both are linked to carrier molecules, most probably to P-ACL. Here, two possible mechanisms, shown in Fig. 2, may be distinguished. In one mechanism, the monomeric repeating-unit is added to the nonreducing

(313) J. L. Kent and M. J. Osborn, *Biochemistry*, **7**, 4409 (1968).
(314) I. Beckmann, T. V. Subbaiah, and B. A. D. Stocker, *Nature*, **201**, 1299 (1964).
(315) J. L. Kent and M. J. Osborn, *Biochemistry*, **7**, 4396 (1968).
(316) J. L. Kent and M. J. Osborn, *Biochemistry*, **7**, 4419 (1968).
(317) D. Bray and P. W. Robbins, *Biochem. Biophys. Res. Commun.*, **28**, 334 (1967).
(318) M. J. Osborn and I. M. Weiner, *J. Biol. Chem.*, **243**, 2631 (1968).

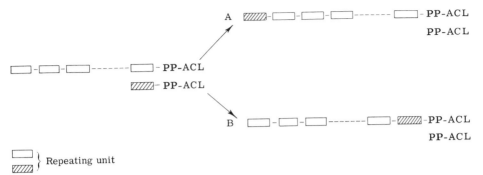

FIG. 2.—Two Theoretically Possible Methods of Chain Elongation in O Side-chain Synthesis, Considered by Bray and Robbins.[317] (In method A, a repeating unit monomer is added to the nonreducing end of the growing O side-chain, which is linked to PP-ACL. In method B, a polymer of the repeating units is transferred onto the nonreducing, terminal sugar of a repeating unit monomer, linked to PP-ACL.)

end of the polymeric repeating-unit; in the other, the polymeric repeating-unit is added to the nonreducing end of the monomeric repeating-unit. Chemically, both mechanisms proceed through the transfer of a substituted D-galactosyl residue onto a nonreducing terminal D-mannosyl residue of another repeating unit in *Salmonella anatum*. But, in the former mechanism, O side-chain "grows" at the nonreducing end, whereas, in the latter, the addition of new repeating units takes place at the reducing end of the polysaccharide. Pulse-labeling experiments using a mutant deficient in core synthesis clearly showed[147,317] that chain elongation is accomplished by the latter mechanism, that is, by growth at the "reducing end." This situation is rather unusual, because most polysaccharides are known to be synthesized by the transfer of glycosyl groups to the nonreducing end.[147] It must, however, be emphasized that these examples of "growth at the nonreducing end" all involve direct transfer of a glycosyl group from a "sugar nucleotide," unlike the elongation of O side-chains.

Abequose residues in *S. typhimurium* occur as branches of O side-chain, and are not found in its "main chain" (see Fig. 1). Indeed, the enzymic polymerization readily occurs in the absence of CDP-abequose,[305,306] hence, it was thought possible that abequose residues are added after the synthesis of the main chain of O side-chain polysaccharide is completed. However, this hypothesis appears unlikely in view of the following results. (1) Under certain experimental conditions, the incorporation of D-mannose, L-rhamnose, and D-galactose is strongly stimulated by the simultaneous presence of CDP-abequose.[305] (2) By skilful manipulation of the conditions of reaction, especially the temperature, D-mannosyl-L-rhamnosyl-D-galactosyl-PP-ACL was shown to be a good acceptor of abequosyl transfer from CDP-abequose, whereas a cell-envelope fraction containing the P-ACL-linked polymer of this trisaccharide was inactive.[318] Also, below 20°, the polymerization of repeating units was shown to be considerably faster in the presence than in the absence of CDP-abequose.[318] (3) A mutant that was defective in the biosynthesis of CDP-abequose was found unable to polymerize the trisaccharide repeating units in intact cells.[319] These results all favor the postulated sequence shown in Scheme 1 (see p. 419), where abequose is added to the trisaccharide-PP-ACL, and the resultant tetrasaccharide-PP–ACL participates in the polymerization reaction.

According to Scheme 1, the last reaction in the synthesis of LPS is the joining of completed O side-chain to the core. This unusual reaction, involving the joining of two macromolecules, was found to occur in a cell-free system.[320] O Side-chains linked to P-ACL were generated *in vitro* in a cell-envelope preparation from a core-deficient mutant, and this particulate preparation was incubated with an exogenous, core LPS; this resulted in efficient transfer of the O side-chain to the core, a transfer catalyzed by an enzyme or enzymes present in the cell-envelope fraction.[320] Unfortunately, a satisfactory method is not yet known for the extraction of O side-chain polysaccharide linked to PP-ACL, and further resolution of this system has not yet been achieved.

Synthesis of LPS can be modified in several ways. Bacteriophages[321] or a chromosomal gene[322] can determine the production of a system responsible for the D-glucosylation of LPS. The LPS synthesized in these strains contains one D-glucose residue for every repeating unit of the O side-chain. Cell envelopes from one of these strains were shown to catalyze the transfer of D-glucosyl groups from UDP-D-glucose to endogenous LPS in the absence of net synthesis of LPS;[323] this result clearly indicated that the D-glucosylation is an example of the "modification reaction." It has been found that this D-glucosylation proceeds in two steps, as follows.[324,325]

(319) R. Yuasa, M. Levinthal, and H. Nikaido, *J. Bacteriol.*, **100**, 433 (1969).

(320) M. A. Cynkin and M. J. Osborn, *Fed. Proc.*, **27**, 293 (1968).

(321) P. W. Robbins and T. Uchida, *Fed. Proc.*, **21**, 702 (1962).

(322) P. H. Mäkelä and O. Mäkelä, *Ann. Med. Exp. Fenn.* (Helsinki), **44**, 310 (1966).

(323) T. Uchida, T. Makino, K. Kurahashi, and H. Uetake, *Biochem. Biophys. Res. Commun.*, **21**, 354 (1965).

(324) A. Wright, *Fed. Proc.*, **28**, 658 (1969); *J. Bacteriol.*, **105**, 927 (1971).

(325) H. Nikaido, K. Nikaido, T. Nakae, and P. H. Mäkelä, *J. Biol. Chem.*, **246**, 3902 (1971); K. Nikaido and H. Nikaido, *ibid.*, **246**, 3912 (1971); M. Takeshita and P. H. Mäkelä, *ibid.*, **246**, 3920 (1971).

$$\text{UDP-D-glucose} + \text{P-ACL} \rightleftharpoons \text{D-glucosyl-P-ACL} + \text{UDP} \qquad (5)$$

$$\text{D-Glucosyl-P-ACL} + \text{acceptor} \longrightarrow \text{D-glucosyl-acceptor} + \text{P-ACL} \qquad (6)$$

In reaction 5, only the D-glucosyl groups are transferred to P-ACL (in contrast to the first reaction in O side-chain synthesis, where the entire D-galactosyl phosphate portion is transferred); this was revealed by the inhibition of the "forward" reaction by UDP, but not by UMP, and by the use of UDP-D-glucose labeled with ^{32}P. Despite the fact that this is a straightforward glycosyl transferase reaction, it is freely reversible, and the equilibrium constant appears to be about 0.5. Catalytic hydrogenation of D-glucosyl-P-ACL releases a D-glucosyl phosphate, that, in paper chromatography, behaves slightly differently from authentic α-D-glucopyranosyl phosphate;[325] it is not a cyclic phosphate, as the phosphate group is completely split off by the action of the alkaline phosphate monoesterase of *E. coli*.[325] As this compound is not a substrate of phosphoglucomutase, and as the anomeric hydroxyl group of the D-glucose is engaged by a phosphate group, the compound is probably β-D-glucopyranosyl phosphate.[324] Because a D-glucose residue is linked to a D-galactose residue through an α-D-$(1 \rightarrow 4)$-linkage in the product of the second reaction,[324-326] the reaction sequence involves double inversion of the anomeric configuration of the D-glucosyl groups: from UDP-α-D-glucose to β-D-glucosyl-P-ACL, and then to α-D-glucosyl-LPS.

In the cell-free systems that use cell envelopes, the major "acceptor" of the D-glucosyl groups is LPS. In intact cells, however, an O side-chain linked to P-ACL can serve as an acceptor, and the evidence so far accumulated indicates that D-glucosyl groups are transferred to the growing O side-chains in such a way that the addition of a new repeating-unit oligosaccharide is immediately followed by the D-glucosylation of the penultimate repeating-unit.[325]

LPS can also be acetylated. Cell-envelope fractions from *Salmonella anatum* were found to transfer acetyl groups from acetyl-CoA to endogenous LPS, or to exogenously added oligosaccharides.[327,328] The D-galactose residues in these acceptor molecules become acetylated, probably at O-6. The stage of synthesis of LPS at which the acetylation occurs in intact cells is not yet known.

c. Biosynthesis of T_1 Side-Chains.—Some strains of *Salmonella* can produce a completely different side-chain, called T_1, which also

(326) R. Tinelli and A.-M. Staub, *Bull. Soc. Chim. Biol.*, **42**, 583 (1960).

(327) A. Wright, J. M. Keller, and P. W. Robbins, *Fed. Proc.*, **23**, 271 (1964).

(328) P. W. Robbins, J. M. Keller, A. Wright, and R. L. Bernstein, *J. Biol. Chem.*, **240**, 384 (1965).

becomes linked to the R core. This side-chain contains about equal proportions of D-ribofuranose residues and D-galactofuranose residues.[329] Experiments with intact cells of various mutants that are defective in D-galactose metabolism have shown that the D-galactofuranose residues in T_1 side-chains are derived from the D-galactopyranosyl group of UDP-D-galactopyranose.[330,331] When cell-envelope fractions from the T_1 strains were incubated with UDP-D-galactopyranose-^{14}C, ^{14}C-labeled T_1 side-chains were synthesized in which both the D-galactofuranose and D-ribofuranose residues were labeled[330] with ^{14}C. The mechanism of conversion of D-galactopyranosyl into D-galactofuranosyl groups in this system remains obscure, but the enzymic synthesis of UDP-D-galactofuranose with a mold enzyme has been described.[4] The efficient conversion of the D-galactosyl-^{14}C groups into D-ribofuranosyl-^{14}C groups initially suggested the possible presence of a specific pathway; however, the results of later experiments that used hexoses labeled specifically at C-1 or C-6, and of experiments that employed a mutant defective in phosphoglucoisomerase, suggested that the D-ribofuranosyl groups are produced through such conventional pathways as the transketolase–transaldolase pathway or the pentose phosphate shunt.[331]

7. Cell-wall Peptidoglycan

All bacteria (except *Mycoplasma* and extreme halophiles) and blue-green algae are known to contain a complex, cross-linked, peptidoglycan in their cell walls. The glycan portion has a disaccharide repeating-unit, namely, 2-acetamido-2-deoxy-β-D-glucopyranosyl-(1 → 4)-2-acetamido-3-O-(D-1-carboxyethyl)-2-deoxy-β-D-glucopyranose; these are linked through (1 → 4)-linkages. The glycose residue at the reducing end of this disaccharide is that of N-acetylmuramic acid; it carries a short peptide chain, linked to it through an amide bond between the carboxyl group of the N-acetylmuramic acid and the amino group of the N-terminal amino acid. The structure of the peptidoglycan from *Staphylococcus aureus* is shown in Fig. 3. The D-glutamic acid residue and the D-alanine residue seem to be present in peptidoglycan preparations from most bacteria, but L-alanine and L-lysine residues are frequently replaced by other amino acids in

(329) M. Berst, C. G. Hellerqvist, B. Lindberg, O. Lüderitz, S. Svensson, and O. Westphal, *Eur. J. Biochem.*, **11**, 353 (1969).

(330) M. Sarvas and H. Nikaido, *Bacteriol. Proc.*, 59, (1969); H. Nikaido and M. Sarvas, *J. Bacteriol.*, **105**, 1073 (1971).

(331) M. Sarvas and H. Nikaido, *J. Bacteriol.*, **105**, 1063 (1971).

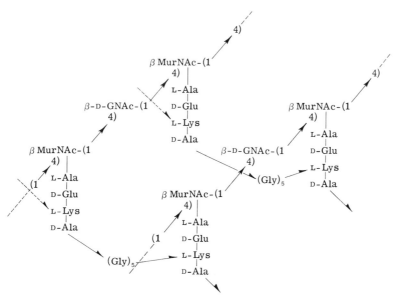

FIG. 3.—Schematic Structure of the Peptidoglycan from *Staphylococcus aureus.*
[MurNAc, *N*-acetylmuramic acid; GNAc, 2-acetamido-2-deoxy-D-glucose; Ala, alanine;
Glu, glutamic acid; Lys, lysine; Gly, glycine. Arrows (→ and →) show glycosidic and
peptide linkages, respectively. In the latter, arrows are drawn from the amino acid
supplying the carboxyl group to the amino acid supplying the amino group.]

other species of bacteria. The structure of the peptidoglycans has
been reviewed in detail.[332]

The biosynthesis of peptidoglycan has been elucidated mainly
through the efforts of Park, Strominger, and their coworkers. Only
a brief account will be given, as this subject is too large for detailed coverage here; moreover, the subject has been reviewed repeatedly,[269,332–338] and many reactions in this pathway are involved
in the transfer of amino acids, not of sugars. Because the synthesis
of peptidoglycans has been studied in great detail for *Staphylococcus
aureus*, the present discussion will be confined primarily to this

(332) J.-M. Ghuysen, *Bacteriol. Rev.*, **32**, 425 (1968).
(333) J. L. Strominger, *Fed. Proc.*, **21**, 134 (1962).
(334) W. Weidel and H. Pelzer, *Advan. Enzymol.*, **26**, 193 (1964).
(335) H. J. Rogers, *Symp. Soc. Gen. Microbiol.*, **15**, 186 (1965).
(336) H. H. Martin, *Ann. Rev. Biochem.*, **35**, 457 (1966).
(337) J. L. Strominger, K. Izaki, M. Matsuhashi, and D. J. Tipper, *Fed. Proc.*, **26**, 9
      (1967).
(338) J.-M. Ghuysen, J. L. Strominger, and D. J. Tipper, in "Comprehensive Biochemis-
      try," M. Florkin and E. Stotz, eds., Elsevier, Amsterdam, 1968, Vol. 26A, p. 53.

organism. For other organisms, some differences in the pathway are found, depending on the variations in the structure of the resulting peptidoglycan; for these results, the reader is referred to recent reviews.[269,337,338]

In the first stage of peptidoglycan synthesis, UDP-N-acetylmuramic acid is synthesized, and then a pentapeptide chain becomes attached to the carboxyl group of N-acetylmuramic acid (see Scheme 2). Strominger[17] found that the reaction of enolpyruvate phosphate with UDP-2-acetamido-2-deoxy-D-glucose results in the formation of a compound that appeared to be a UDP-2-acetamido-2-deoxy-D-glucose–pyruvate enol ether, and he predicted that its reduction would afford UDP-N-acetylmuramic acid. The occurrence of these reactions has now been confirmed, and the intermediate product has been characterized in greater detail.[18,19]

Amino acids are then sequentially attached to the molecule of

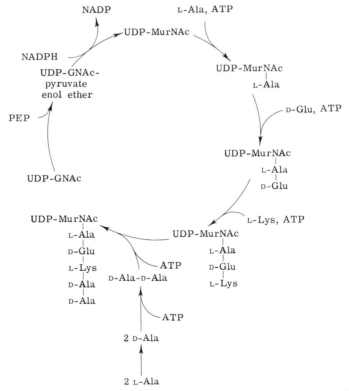

Scheme 2. Biosynthesis of UDP-N-acetylmuramyl-pentapeptide. (MurNAc = N-acetylmuramic acid, PEP = enolpyruvate phosphate.)

UDP-*N*-acetylmuramic acid.[339-341] The enzymes involved are highly specific, and the sequence of the amino acids is determined solely by the substrate and acceptor specificities of these enzymes, which require a divalent cation and ATP. One exception to the rule of sequential addition is the terminal dipeptide sequence, D-alanyl-D-alanine; this dipeptide is first assembled and then transferred to the UDP-*N*-acetylmuramyl-tripeptide[342] (see Scheme 2).

It will be recalled that some of these UDP-*N*-acetylmuramyl compounds were isolated by Park and Johnson,[343] from penicillin-inhibited cells of *Staphylococcus aureus*, at about the same time that Leloir's group isolated the first "sugar nucleotide," UDP-D-glucose, from yeast.[344] Other conditions also produce accumulation of some of these compounds. Thus, D-cycloserine is known to induce an accumulation of UDP-*N*-acetylmuramy-tripeptide in *S. aureus*; this is easily explained by the observation that D-cycloserine is a powerful, competitive inhibitor both of alanine racemase and D-alanyl-D-alanine synthetase.[345]

When UDP-2-acetamido-2-deoxy-D-glucose and UDP-*N*-acetyl-muramyl-pentapeptide are together added to broken bacterial-cell preparations, a new product is formed that contains the sugar and sugar–peptide portions of the precursors.[346] If cells broken by means of a sonic oscillator are used, it is necessary to carry out the incubations on a filter-paper support;[346] the precise function played by the filter paper is not yet known. In contrast, if enzymes are prepared by grinding cells with alumina, filter-paper support is unnecessary and the product contains a linear glycan backbone in which the two sugar residues occur in the alternate sequence expected.[347]

Two unexpected findings in this system led Strominger and coworkers to the discovery of the lipid intermediates and of the complex sequence of reactions involved in the synthesis of "linear" peptidoglycan. One was the observation that UMP and inorganic

(339) E. Ito and J. L. Strominger, *J. Biol. Chem.*, **235**, 5 (1960).
(340) E. Ito and J. L. Strominger, *J. Biol. Chem.*, **237**, 2689 (1962).
(341) S. G. Nathenson, J. L. Strominger, and E. Ito, *J. Biol. Chem.*, **239**, 1773 (1964).
(342) E. Ito and J. L. Strominger, *J. Biol. Chem.*, **237**, 2696 (1962).
(343) J. T. Park and M. Johnson, *J. Biol. Chem.*, **179**, 585 (1949).
(344) R. Caputto, L. F. Leloir, C. E. Cardini, and A. C. Paladini, *J. Biol. Chem.*, **184**, 333 (1950).
(345) U. Roze and J. L. Strominger, *Mol. Pharmacol.*, **2**, 92 (1966).
(346) P. M. Meadow, J. S. Anderson, and J. L. Strominger, *Biochem. Biophys. Res. Commun.*, **14**, 382 (1964).
(347) J. S. Anderson, M. Matsuhashi, M. A. Haskin, and J. L. Strominger, *Proc. Nat. Acad. Sci. U.S.*, **53**, 881 (1965); *J. Biol. Chem.*, **242**, 3180 (1967).

phosphate, not UDP, are formed as reaction products from UDP-N-acetylmuramyl-pentapeptide.[347] The second was discovery of the presence of a fast-moving, radioactive material when the incorporation products were chromatographed on paper.[347]

It was found that the first reaction in synthesis of linear peptidoglycan involves the transfer of pentapeptidyl-N-acetylmuramyl 1-phosphate to an endogenous, lipid acceptor from UDP-N-acetylmuramyl-pentapeptide, with concomitant production of UMP (see Reaction I of Scheme 3).[347] This reaction is fully reversible,[348] and its kinetics have been studied in detail.[349] Evidence was presented that the glycosyl phosphate is attached to the lipid through a pyrophosphate bridge.[350] Strominger and coworkers[351] found that the endogenous, lipid acceptor could be extracted from the lyophilized,

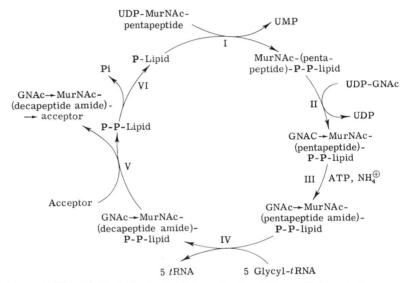

Scheme 3. The Lipid Cycle in the Biosynthesis of "Linear" Peptidoglycan in *S. aureus*. [Although the amidation reaction (III) is shown here as though it precedes the addition of pentaglycine peptide (IV), there is no evidence as to the sequence in which these reactions occur. The amidation reaction proceeds equally well before the addition of 2-acetamido-2-deoxy-D-glucose residues (reaction II), at least in a cell-free system.[352]]

(348) W. G. Struve and F. C. Neuhaus, *Biochem. Biophys. Res. Commun.*, **18**, 6 (1965).
(349) M. G. Heydanek, Jr., W. G. Struve, and F. C. Neuhaus, *Biochemistry*, **8**, 1214 (1969).
(350) C. P. Dietrich, M. Matsuhashi, and J. L. Strominger, *Biochem. Biophys. Res. Commun.*, **21**, 619 (1965).
(351) C. P. Dietrich, A. V. Colucci, and J. L. Strominger, *J. Biol. Chem.*, **242**, 3218 (1967).
(352) G. Siewert and J. L. Strominger, *J. Biol. Chem.*, **243**, 783 (1968).

particulate enzyme with chloroform–methanol at −17°. The enzyme activity survives this treatment, and the extracted enzyme shows an absolute requirement for exogenous, lipid acceptor.[351]

In the next step, a 2-acetamido-2-deoxy-D-glucosyl group is transferred from its UDP derivative onto the "monosaccharide–lipid" intermediate, to form the "disaccharide–lipid" intermediate (see Reaction II of Scheme 3).[347] This is a glycosyl-transfer reaction of the more usual type, and UDP is released as a product of the reaction.[347]

Normally, the disaccharide–pentapeptide unit is further modified while it is still attached to the carrier lipid. In S. aureus, this process involves the amidation of the 1-carboxyl group of D-glutamic acid in the presence of $NH_4^{\oplus}$ and ATP.[352] (This D-glutamic acid residue is linked through its 5-carboxyl group to L-lysine;[353] its 1-carboxyl group is therefore free before amidation). Another modification is the addition of the pentaglycine chain to the 6-amino group of L-lysine in the intermediate. Here glycyl t-RNA is the direct donor of the glycine residues.[354,355] For the three species of glycine-specific t-RNA obtained by reverse-phase, column chromatography, all were active in peptidoglycan synthesis, but only two were found to "recognize" the known codons for glycine.[356] The results of time-course experiments indicated that the five glycine residues are added sequentially[357,358] and prior assembly of the pentaglycine unit on t-RNA was not observed.

At one time, the lipid carrier was believed to be a glycerophosphatide,[359] but further purification showed conclusively that it is a C_{55}-isoprenoid alcohol phosphate.[360] Its nuclear magnetic resonance spectrum showed that two of its internal double bonds have the trans configuration, and that this isoprenoid alcohol differs from ficaprenol, a C_{55}-isoprenoid alcohol from fig leaves, which has three internal, trans, double bonds.[361]

(353) E. Muñoz, J.-M. Ghuysen, M. Leyh-Bouille, J. F. Petit, H. Heymann, E. Bricas, and J. Lefrancier, Biochemistry, 5, 3748 (1966).

(354) A. N. Chatterjee and J. T. Park. Proc. Nat. Acad. Sci. U.S., 51, 9 (1964).

(355) M. Matsuhashi, C. P. Dietrich, and J. L. Strominger, Proc. Nat. Acad. Sci. U.S., 54, 587 (1965); J. Biol. Chem., 242, 3191 (1967).

(356) R. M. Bumstead, J. L. Dahl, D. Söll, and J. L. Strominger, J. Biol. Chem., 243, 779 (1968).

(357) J. Thorndike and J. T. Park, Biochem. Biophys. Res. Commun., 35, 642 (1969).

(358) T. Kamiryo and M. Matsuhashi, Biochem. Biophys. Res. Commun., 36, 215 (1969).

(359) J. S. Anderson and J. L. Strominger, Biochem. Biophys. Res. Commun., 21, 516 (1966).

(360) Y. Higashi, J. L. Strominger, and C. C. Sweeley, Proc. Nat. Acad. Sci. U.S., 57, 1878 (1967).

(361) Y. Higashi, J. L. Strominger, and C. C. Sweeley, J. Biol. Chem., 245, 3697 (1970).

This lipid carrier is, therefore, identical with, or at least very similar to, the C_{55} lipid involved in the synthesis of D-mannan in *Micrococcus lysodeikticus*; the latter lipid, also, has two internal, *trans*, double bonds.[211] The former lipid behaves very similarly to the carrier lipid involved in O side-chain synthesis in Gram-negative bacteria,[269] and can substitute for the latter lipid in the O antigen system.[269] This result is not, however, a rigorous proof of identity, as the *M. lysodeikticus* system is known to be able to use ficaprenol phosphate as efficiently as undecaprenol phosphate.[212]

At any rate, the polyisoprenoid carrier-lipids thus far isolated from bacteria are either C_{50} or C_{55} compounds. In contrast, a polyisoprenyl (D-glucosyl phosphate) containing 16 to 21 isoprene residues has been found in animals,[148] as already discussed in Section V,1 (see p. 381). Because many free polyprenols, differing in chain length, in degree of saturation, and in the number of *cis* and *trans* groups, have been isolated from different organisms,[362] it is likely, as with "sugar nucleotides," that a large group of polyprenol sugar phosphates containing various sugar and polyprenol moieties will be found.

It has been proposed that lipid intermediates of peptidoglycan synthesis are formed in order that the hydrophilic precursors of peptidoglycan can be transported across the hydrophobic interior of the cell membrane;[347] such transport is necessary, because peptidoglycan is located on the outer side of the cell membrane, whereas its UDP-linked precursors are found in the cytoplasm. Lipid intermediates have been shown, or at least suspected, to be involved also in the biosynthesis of the O antigen portion of the bacterial lipopolysaccharides,[306,307,324,325] bacterial D-mannan,[211] plant polysaccharides,[214,215] bacterial capsular polysaccharide,[260] glycoprotein,[148] and teichoic acid (see Section VII, 6; p. 481). It seems likely that these intermediates also function as donors of glycosyl groups in a hydrophobic environment, either within or around the membrane, an area into which "sugar nucleotides" would be unable to penetrate.

The repeating unit synthesized on the lipid carrier is then transferred, presumably to the growing end of the linear peptidoglycan. This reaction is known to be inhibited by low concentrations of two antibiotic substances, ristocetin and vancomycin.[347] It is not yet known whether the chain elongation occurs at the reducing or nonreducing end of the glycan chain (see also, Ref. 147).

Another product of this reaction, undecaprenyl pyrophosphate, accumulates both in cell-free incubation mixtures[363] and in intact

(362) F. W. Hemming, *Biochem. J.*, **113**, 23P (1969).
(363) G. Siewert and J. L. Strominger, *Proc. Nat. Acad. Sci. U.S.*, **57**, 767 (1967).

cells,[364] when its conversion into undecaprenyl phosphate by a specific phosphatase is inhibited by the antibiotic bacitracin.[363]

The last reaction in peptidoglycan synthesis is the formation of cross-links between peptide moieties; that this reaction is inhibited by penicillin was first indicated by Martin,[365] and later by other investigators.[366,367] Of particular importance was the observation that the peptidoglycan formed in the presence of penicillin has more than one D-alanine residue per chain, whereas the peptidoglycan found in normal cells contains only one D-alanine residue per chain. These results, together with the observation that the enzymically synthesized "linear" (not cross-linked) peptidoglycan fully retained the two D-alanine residues originally present in the precursor UDP-N-acetyl-muramyl-pentapeptide,[355] suggested that the cross-linking is achieved by a penicillin-sensitive, transpeptidation reaction. In this reaction, the carboxyl group of the subterminal D-alanine residue first becomes linked to the enzyme, at the same time releasing the terminal D-alanine residue into the medium, and then the carboxyl group of the subterminal D-alanine residue becomes linked to a free amino group of the neighboring chain (see Scheme 4). In S. aureus, the amino group of the N-terminal glycine residue of the pentaglycine structure becomes linked to the carboxyl group of D-alanine. The cross-linking reaction has to occur in a place that is outside the cell membrane and is, presumably, deficient in ATP, but this difficulty is overcome by an ingenious mechanism wherein the bond energy between two D-alanine residues is used for the formation of cross-linking bonds.

This cross-linking reaction has not yet been demonstrated with cell-free systems from S. aureus. A similar cross-linking reaction has, however, been shown to occur with a particulate preparation from Escherichia coli.[368] In this system, the carboxyl group of the sub-terminal, D-alanine residue becomes linked to the free amino group of the meso-diaminopimelic acid (meso-2,6-diaminoheptanedioic acid) residue in another peptide chain, and the reaction is strongly inhibited by low concentrations of penicillin.[368] It has been proposed that penicillin acts as a structural analog of the D-alanyl-D-alanine portion of the peptide,[367] and that inactive, penicilloyl-enzyme is formed, with the opening of the β-lactam ring of penicillin.[337,367] The

(364) K. J. Stone and J. L. Strominger, Fed. Proc., 29, 933 Abstr. (1970).

(365) H. H. Martin, J. Gen. Microbiol., 36, 441 (1964).

(366) E. M. Wise, Jr., and J. T. Park, Proc. Nat. Acad. Sci. U.S., 54, 75 (1965).

(367) D. J. Tipper and J. L. Strominger, Proc. Nat. Acad. Sci. U.S., 54, 1133 (1965); J. Biol. Chem., 243, 3169 (1968).

(368) K. Izaki, M. Matsuhashi, and J. L. Strominger, Proc. Nat. Acad. Sci. U.S., 55, 656 (1966); J. Biol. Chem., 243, 3180 (1968).

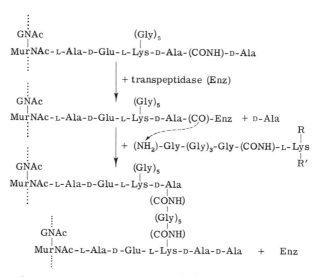

Scheme 4. Formation of Cross-links by Transpeptidation.

assumed penicilloyl transpeptidase has not yet been isolated, but penicilloyl D-alanine-carboxypeptidase has been demonstrated in a particulate preparation from *Bacillus subtilis*.[369]

8. Glycosaminoglycans ("Mucopolysaccharides")

The major portion of animal glycosaminoglycans ("mucopolysaccharides") consists of a polymer of disaccharide repeating-units:[370] O-(2-acetamido-2-deoxy-β-D-glucopyranosyl)-(1 → 4)-D-glucopyranosyluronic acid in hyaluronic acid; O-(2-acetamido-2-deoxy-β-D-galactopyranosyl)-(1 → 4)-D-glucopyranosyluronic acid sulfated at C-4 or C-6 of the hexosamine in chondroitin 4-sulfate or 6-sulfate, respectively; highly sulfated O-(2-amino-2-deoxy-α-D-glucopyranosyl)-(1 → 4)-D-gluco(or L-ido)pyranosyluronic acid in heparin; and O-(2-acetamido-2-deoxy-β-D-galactopyranosyl)-(1 → 4)-L-idopyranosyluronic acid sulfated at C-4 of the hexosamine in dermatan sulfate. However, it has become clear that most, if not all, mucopolysaccharides occur in Nature in a form covalently attached to a protein "backbone," although definitive evidence is still lacking in the case of hyaluronic acid. The region where polysaccharide is attached to

(369) P. J. Lawrence and J. L. Strominger, *J. Biol. Chem.*, **245**, 3660 (1970).
(370) R. W. Jeanloz, *Advan. Enzymol.*, **25**, 433 (1963); R. W. Jeanloz, in "The Chemical Physiology of Mucopolysaccharides," C. Quintarelli, ed., Little, Brown and Co., Boston, 1968.

protein ("linkage region") was found to consist of the same sequence in heparin,[371] chondroitin 4-sulfate,[372,373] dermatan sulfate,[374] heparitin sulfate,[375] and chondroitin 6-sulfate:[376]

$$\beta\text{-D-GU-}(1 \rightarrow 3)\text{-}\beta\text{-D-Gal-}(1 \rightarrow 3)\text{-}\beta\text{-D-Gal-}(1 \rightarrow 4)\text{-}\beta\text{-D-Xyl} \rightarrow \text{Ser.}$$

Thus, the polysaccharide is attached to polypeptide by an O-β-D-xylopyranosyl-L-serine linkage.[372]

In many instances, N-acetylhexosamine was shown to be linked to the (nonreducing) terminal D-glucosyluronic acid group of the sequence, and another D-glucuronic acid residue was, in turn, found to be linked to the N-acetylhexosamine residue mentioned. Thus, the alternate sequence of hexosamine and hexuronic acid is already beginning at this point, but the sugar residues can be different from those found in the repeating units comprising the major portion of the polysaccharides. For example, in heparin, 2-amino-2-deoxy-D-glucose residues in the linkage area are N-acetylated,[377] whereas, in the main chain, they are N-sulfated. In dermatan sulfate, D-glucuronic acid is present in the linkage region,[374] whereas the main chain contains L-iduronic acid residues, although a few D-glucuronic acid residues are also found in this region.[378]

Keratan sulfate is unusual among mucopolysaccharides in that it contains variable proportions of 6-deoxyhexoses and sialic acid, but no uronic acid residues. It is also linked to a protein backbone, but the linkages are unusual and are of the type found in "glycoproteins." Keratan sulfate from cartilage is attached to the protein by O-(2-acetamido-2-deoxy-D-galactosyl)-L-serine linkages;[379] corneal keratan sulfate is suspected to have N-(2-acetamido-2-deoxy-β-D-glucosyl)-L-asparagine linkages.[380] In bovine cartilage, keratan sulfate and chondroitin sulfate are attached to the same protein backbone.[381]

Hyaluronic acid is also suspected of having a protein "backbone." An arabinose (instead of D-xylose) has been reported to be present

(371) U. Lindahl and L. Rodén, *J. Biol. Chem.*, **240**, 2821 (1965).
(372) U. Lindahl and L. Rodén, *J. Biol. Chem.*, **241**, 2113 (1966).
(373) L. Rodén and R. Smith, *J. Biol. Chem.*, **241**, 5949 (1966).
(374) L.-A. Fransson, *Biochim. Biophys. Acta*, **156**, 311 (1968).
(375) J. Knecht, J. A. Cifonelli, and A. Dorfman, *J. Biol. Chem.*, **242**, 4652 (1967).
(376) T. Helting and L. Rodén, *Biochim. Biophys. Acta*, **170**, 301 (1968).
(377) U. Lindahl, *Biochim. Biophys. Acta*, **130**, 368 (1966).
(378) L.-A. Fransson and L. Rodén, *J. Biol. Chem.*, **242**, 4170 (1967).
(379) B. A. Bray, R. Lieberman, and K. Meyer, *J. Biol. Chem.*, **242**, 3373 (1967).
(380) N. Seno, K. Meyer, B. Anderson, and P. Hoffman, *J. Biol. Chem.*, **240**, 1005 (1965).
(381) P. Hoffman, T. A. Mashburn, Jr., and K. Meyer, *J. Biol. Chem.*, **242**, 3805 (1967).

in a hyaluronidase-digestible fraction from brain[382] and in a hyaluronic acid preparation from bovine, vitreous humor.[383] However, oligosaccharide–serine compounds containing an arabinose have not yet been isolated. The results of degradation of hyaluronic acid–protein complex from umbilical cord suggested that several hyaluronic acid chains were bound to a common polypeptide, but, again, no oligosaccharides from the linkage region could be isolated.[384]

a. **Biosynthesis of the Linkage Region.**—The transfer of the first D-xylosyl group was first demonstrated by the use of a particulate fraction from hen oviduct.[385] When UDP-D-xylose-^{14}C and Mn^{2+} were added to this preparation, D-xylosyl groups were transferred to an endogenous acceptor. At least part of the acceptor appeared to be proteinaceous, as about half of the material having radioactivity incorporated became soluble in trichloroacetic acid after pronase digestion.[385] Also, some of the D-xylose transferred is presumably attached to L-serine (or L-threonine) residues, because a fraction of it (39% in one experiment) was released on treatment with alkali, presumably through a β-elimination reaction.

The D-xylosyl transfer-reaction has subsequently been demonstrated in better characterized systems. A "soluble" fraction from mast-cell tumor, a tissue very active in biosynthesis of heparin, was found to catalyze the enzymic transfer of D-xylose to an endogenous acceptor;[386] there were two systems, and these could be separated by fractionation with ammonium sulfate, followed by dialysis. The fraction precipitated during dialysis does not require Mn^{2+}, and the D-xylose residues in the incorporation product are released as xylitol after reduction with alkaline sodium borohydride. D-Xylosyl-L-serine was identified among the peptides produced by pronase digestion of the product. Thus, this fraction appears to catalyze the first reaction in the biosynthesis of the carbohydrate portion of heparin. In contrast, the fraction not precipitated during dialysis required Mn^{2+} for activity, and the D-xylosyl groups were incorporated into an unidentified compound containing alkali-stable linkages.

(382) A. H. Wardi, W. S. Allen, D. L. Turner, and Z. Stary, *Arch. Biochem. Biophys.*, **117**, 44 (1966).

(383) A. H. Wardi, W. S. Allen, D. L. Turner, and Z. Stary, *Biochim. Biophys. Acta*, **192**, 151 (1969).

(384) L. Rodén and M. B. Mathews, *Fed. Proc.*, **27**, 529 (1968).

(385) E. E. Grebner, C. W. Hall, and E. F. Neufeld, *Biochem. Biophys. Res. Commun.*, **22**, 672 (1966).

(386) E. E. Grebner, C. W. Hall, and E. F. Neufeld, *Arch. Biochem. Biophys.*, **116**, 391 (1966).

A very similar reaction was observed with enzyme preparations from chick-embryo cartilage.[387] Most of the reaction product contained D-xylose in an alkali-labile linkage, and a xylosyl-serine was isolated from the pronase-digested incorporation product. As the enzyme had been prepared from a tissue very active in chondroitin 4-sulfate synthesis, this reaction is assumed to be the first step in the synthesis of the polysaccharide moiety of this compound. Here, again, after centrifuging at 105,000 g, the supernatant fraction was at least as active as the particulate fraction.

The transfer of the second glycosyl group, D-galactosyl, from UDP-D-galactose to an endogenous acceptor has also been demonstrated in the same system.[387] This reaction is stimulated by $Mn^{2\oplus}$, but the addition of UDP-D-xylose has no effect. Again, the supernatant fraction obtained after centrifuging at 105,000 g was quite active. Degradation of the product with alkaline sodium borohydride yielded an oligosaccharide that behaved like a D-galactosyl-D-xylitol.

More-detailed studies have been performed with this system.[388] When UDP-D-galactose-^{14}C and UDP-D-xylose were added to homogenates of chick-embryo cartilage, the ^{14}C-labeled sugar was incorporated into endogenous acceptors, and partial hydrolysis with acid produced, *inter alia*, 4-O-β-D-galactosyl-D-xylose and O-β-D-galactosyl-$(1 \rightarrow 3)$-O-β-D-galactosyl-$(1 \rightarrow 4)$-D-xylose. These results indicated that the incorporation reaction constitutes the synthesis of the linkage region of chondroitin sulfate–protein complex. It was also found that compounds of low molecular weight act as acceptors in these reactions. Thus, D-xylose and D-xylosyl-L-serine are acceptors for the transfer of the "first" D-galactosyl group, and 4-O-β-D-galactosyl-D-xylose and 4-O-β-D-galactosyl-β-D-xylosyl-L-serine (but not free D-galactose) serve as acceptors for the transfer of the "second" D-galactosyl group. The enzymes catalyzing these reactions appear to be separate, because no competition was observed between these two groups of acceptors. However, both enzymes showed similar pH maxima and similar K_m for UDP-D-galactose, required rather high concentrations of Mn^{2+}, and were localized in the pellet fraction obtained by centrifuging at 300,000 g.

The next sugar to be transferred in the biosynthesis of the linkage region is D-glucuronic acid. The transfer of this sugar from UDP-D-glu-

(387) H. C. Robinson, A. Telser, and A. Dorfman, *Proc. Nat. Acad. Sci. U.S.*, **56**, 1859 (1966).
(388) T. Helting and L. Rodén, *Biochem. Biophys. Res. Commun.*, **31**, 786 (1968); *J. Biol. Chem.*, **244**, 2790 (1969).

copyranosyluronic acid to exogenous, oligosaccharide acceptors has
been demonstrated by the use of a pellet fraction obtained from chick-
embryo cartilage[389] by centrifugation at 300,000 g. O-β-D-Galactosyl
$(1 \rightarrow 3)$-O-β-D-galactosyl-$(1 \rightarrow 4)$-O-β-D-xylosyl-L-serine was shown to
be the best acceptor, although 3-O-β-D-galactosyl-D-galactose was also
active. This enzyme is apparently different from the D-glucosyluronic
transferase involved in the biosynthesis of the main polysaccharide
portion of the chondroitin sulfate, as there were differences in the
pH–activity and Mn^{2+} concentration–activity curves, and the transfer
of the D-glucosyluronic group to 3-O-β-D-galactosyl-D-galactose was
not inhibited by the addition of a pentasaccharide (having a terminal
2-acetamido-2-deoxy-D-galactosyl group) from the main polysaccharide portion of chondroitin 6-sulfate.[389]

The mode of chain initiation in the synthesis of hyaluronic acid
is not yet clear. In group A *Streptococcus*, inhibitors of protein
synthesis do not interfere with the synthesis of hyaluronic acid,[390] a
result suggesting the absence of polypeptide "backbone" from
streptococcal hyaluronic acid. Although the endogenous hyaluronic
acid is bound to membrane(s), it can be separated from the membrane
under mild conditions.[390]

b. Biosynthesis of the Main Polysaccharide Portion.—The main
polysaccharidic portion of mucopolysaccharides usually contains one
hexuronic acid component and one 2-amino-2-deoxyhexose component in alternating sequence. It is synthesized in cell-free systems
when uridine pyrophosphate derivatives of these components are
added to particulate fractions from cells synthesizing mucopolysaccharides. Synthesis of hyaluronic acid from UDP-D-glucuronic acid
and UDP-2-acetamido-2-deoxy-D-glucose by preparations from group
A streptococci and from chicken sarcoma has already been discussed.[1]
Similarly, incubation of "microsomal" fraction from chick-embryo
cartilage with UDP-D-glucuronic acid and UDP-2-acetamido-2-deoxy-
D-galactose resulted in the incorporation of labeled sugars into a
polysaccharide.[391,392] This product, which, in ion-exchange chromatography, behaved like partially sulfated chondroitin sulfate, was
degraded by testicular and streptococcal hyaluronidases in the same
way as desulfated chondroitin sulfate. Also, it afforded the disaccharide chondrosin on partial hydrolysis with acid. Thus, the *in*

(389) T. Helting and L. Rodén, *J. Biol. Chem.*, **244**, 2799 (1969).
(390) A. C. Stoolmiller and A. Dorfman, *Fed. Proc.*, **28**, 900 (1969).
(391) J. E. Silbert, *J. Biol. Chem.*, **239**, 1310 (1964).
(392) R. L. Perlman, A. Telser, and A. Dorfman, *J. Biol. Chem.*, **239**, 3623 (1964).

vitro product appears to be a precursor, which is converted into chondroitin sulfate by sulfation.

Similar results have been obtained for the biosynthesis of heparin. "Microsomal" fractions from mouse mast-cell tumors, which actively synthesize heparin, were found to incorporate equimolar amounts of D-glucuronic acid and 2-acetamido-2-deoxy-D-glucose from their UDP derivatives.[393] As 2-amino-2-deoxy-D-glucose residues in heparin are *N*-sulfated but not *N*-acetylated, it was theorized that UDP-2-amino-2-deoxy-D-glucose might possibly be the precursor; but, in the cell-free system, this "sugar nucleotide" was completely inactive. In ion-exchange chromatography, the product synthesized in the presence of UDP-D-glucuronic acid and UDP-2-acetamido-2-deoxy-D-glucose behaves like hyaluronic acid, but it is resistant to testicular hyaluronidase, and is degraded by heparinase. These results indicated that the product was a biosynthetic precursor of heparin; its subsequent *N*- and *O*-sulfation reactions have since been demonstrated.[394]

The sulfation reaction in mucopolysaccharide synthesis has been studied in several laboratories by using both endogenous and exogenous, desulfated mucopolysaccharide as acceptors. It had been believed that sulfate transferases involved in mucopolysaccharide synthesis exist only in the supernatant fraction (see Ref. 395, for example), but reports on membrane-bound sulfate transferases are rapidly increasing.[394,396−400] It is possible that the higher transferase activity in the soluble fraction is an artifact caused by the presence of larger proportions of endogenous acceptors in the supernatant fluid (see also, Ref. 395). It seems reasonable to assume that the sulfation takes place within the same membrane-bound, enzyme complex that is involved in the synthesis of the polysaccharide chain. In this connection, it must be emphasized that Silbert's system, which is localized in "microsomes," can transfer more than one sulfate group for one disaccharide repeating–unit in the acceptor;[394]

(393) J. E. Silbert, *J. Biol. Chem.*, **238**, 3542 (1963).
(394) J. E. Silbert, *J. Biol. Chem.*, **242**, 5146, 5153 (1967).
(395) E. Meezan and E. A. Davidson, *J. Biol. Chem.*, **242**, 1685, 4956 (1967).
(396) J. E. Silbert, *J. Biol. Chem.*, **242**, 2301 (1967).
(397) S. DeLuca and J. E. Silbert, *J. Biol. Chem.*, **243**, 2725 (1968).
(398) J. E. Silbert and S. DeLuca, *Biochem. Biophys. Res. Commun.*, **31**, 990 (1968).
(399) L. I. Rice, L. Spolter, Z. Tokes, R. Eisenman, and W. Marx, *Arch. Biochem. Biophys.*, **118**, 374 (1967).
(400) A. S. Balasubramanian, N. S. Joun, and W. Marx, *Arch. Biochem. Biophys.*, **128**, 623 (1969).

this behavior is in striking contrast to that of the earlier, "supernatant" systems, which transfer only such small proportions of sulfate group that the chemical behavior of the acceptor mucopolysaccharides is not measurably modified.

The hypothesis that the repeating units are sulfated as soon as they are added to the growing polysaccharide chain would be attractive; in the cell-free system, however, sulfation seems to occur on preformed polysaccharides as readily as on polysaccharide chains in the process of elongation.[394] In any event, sulfation of such polysaccharides as heparin must be a complex process, as sulfate groups are transferred to several different positions on the disaccharide repeating unit. The enzymic sulfation of the amino group was found to be accompanied by simultaneous release of free acetate groups from the heparin precursor.[394]

As the main polysaccharide portion of mucopolysaccharides consists of disaccharide repeating units, it was thought that its biosynthesis might possibly proceed through the formation of disaccharide intermediates. This possibility was tested with a system from chick-embryo cartilage[401] synthesizing chondroitin sulfate. It was found that desulfated oligosaccharides from chondroitin sulfate served as acceptors of glycosyl groups, and that those having a terminal D-glucosyluronic acid group accepted only 2-acetamido-2-deoxy-D-galactose from UDP-2-acetamido-2-deoxy-D-galactose, whereas oligosaccharides having a terminal 2-acetamido-2-deoxy-D-galactosyl group accepted only the D-glucosyluronic group from UDP-D-glucuronic acid. These results suggested that the biosynthesis proceeds through sequential transfer of alternate monosaccharide residues. However, they do not rigorously rule out the involvement of disaccharide intermediates. Let us assume that A and B are the two monosaccharide residues comprising a repeating unit, and that the biosynthesis proceeds as follows.

$$\text{UDP–A} + \text{acceptor} \longrightarrow \text{A–acceptor} + \text{UDP} \tag{7}$$

$$\text{UDP–B} + \text{A–acceptor} \longrightarrow \text{B–A–acceptor} + \text{UDP} \tag{8}$$

$$\begin{aligned}\text{B–A–acceptor} + \text{mucopolysaccharide} \longrightarrow \\ \text{B–A–mucopolysaccharide} + \text{acceptor} \tag{9}\end{aligned}$$

It is conceivable that oligosaccharides having a nonreducing, terminal A residue will act as an analog of A–acceptor and will accept the mono-

(401) A. Telser, H. C. Robinson, and A. Dorfman, Arch. Biochem. Biophys., 116, 458 (1966).

saccharide B. If the enzyme that catalyzes reaction (9) does not have a strict substrate-specificity *in vitro*, it might transfer the monosaccharide A, as well as the disaccharide B–A, from the saccharide–acceptor intermediate, and exactly the type of results reported by Telser and coworkers[401] will be obtained. It must be mentioned that the transfer of incomplete repeating units from the oligosaccharide–lipid intermediate occurs in a cell-free system synthesizing the O-antigen portion of bacterial lipopolysaccharide,[304] and that this finding was once erroneously taken to indicate the absence of oligosaccharide intermediates in this system.

The involvement of lipid intermediates has also been examined in a system synthesizing hyaluronic acid in group A *Streptococcus*.[402,403] An incorporation product soluble in organic solvents was not isolated,[402] and UDP was found to be one of the products of the reaction.[402,403] Although these results do not support the idea that lipid intermediates are involved, they also do not rigorously rule out this possibility.

VII. Synthesis of Other Compounds

1. Miscellaneous Glycosides

Collins and coworkers[404] showed that the steroid 2-acetamido-2-deoxy-D-glucosyl transferase of rabbit liver, which has also been found in rabbit kidney and intestine, is situated in the microsomes and that it requires UDP-2-acetamido-2-deoxy-D-glucose and steroid D-glucosiduronic acids as substrates. The transferase has a pH optimum of 8.0 in phosphate buffer, and a K_m of 68 μM for 17α-estradiol-3-yl β-D-glucosiduronic acid. It will transfer 2-acetamido-2-deoxy-D-glucose to 17-epiesteriol-3-yl and 16,17-epiestriol-3-yl D-glucosiduronic acids, but not to corresponding derivatives of estriol or 16-epiestriol. Himaya and coworkers[405] found that these estrogenic compounds seemed to have stimulated 2-acetamido-2-deoxy-D-glucosyl transferase synthesis in rabbit-kidney tissue, although they may also have produced an activation of the enzyme.

UDP-D-glucosyluronic transferase has been found in a number of organs, including liver, kidney, stomach, and intestine, in several

(402) A. C. Stoolmiller and A. Dorfman, *J. Biol. Chem.*, **244**, 236 (1969).
(403) N. Ishimoto and J. L. Strominger, *Biochim. Biophys. Acta*, **148**, 296 (1967).
(404) D. C. Collins, H. Jirku, and D. S. Layne, *J. Biol. Chem.*, **243**, 2928 (1968).
(405) A. Himaya, D. C. Collins, D. G. Williamson, and D. S. Layne, *Biochem. J.*, **113**, 445 (1969).

species.[406] Dahm and Breuer[407] reported the partial purification of a soluble UDP-D-glucosyluronic transferase that is localized in the ground plasma of human intestine and that catalyzes the formation of estrogen D-glucosiduronic acids.

Halac and Bonevardi[408] studied the UDP-D-glucosyluronic transferase activity of rabbit liver by the use of (ethylenedinitrilo)tetraacetic acid (EDTA), with phenolphthalein as the acceptor. Treatment of the enzyme with EDTA resulted in 10- to 20-fold activation of transferase activity, but this activation was largely undetectable unless albumin and Mg^{2+} were included in the assay mixture.

Halac and Reff[409] found that liver fractions from rats are capable of conjugating bilirubin and p-nitrophenol with D-glucuronic acid. The D-glucosyluronic transferase has been solubilized by treating EDTA-activated microsomes with deoxycholate. The results were compatible with the existence of a UDP-D-glucosyluronic transferase for bilirubin that differs from the enzyme that catalyzes the formation of p-nitrophenyl β-D-glucosiduronic acid.

Lippel and Olson[410] reported that, after the intraportal injection of retinol-6,7-$^{14}C_2$ to rats, retinyl β-D-glucosiduronic acid appeared in the bile. Both 1-O-retinoyl-β-D-glucuronic acid and retinyl β-D-glucosiduronic acid are also synthesized in vitro when washed, rat-liver microsomes are incubated with UDP-D-glucuronic acid and either retinoic acid or retinol, respectively. The synthesis of 1-O-retinoyl-β-D-glucuronic acid was also demonstrated in microsomes of the kidney and in particulate fractions of the intestinal mucosa.

Formation of a D-glucoside of o-aminophenol and, probably, also of those of p-nitrophenol and phenolphthalein has been demonstrated in the mollusks Arion ater and Helix pomatia by Dutton.[411] The UDP-D-glucosyltransferase appears specific to the nucleotide and sugar moieties of the donor substrate. It was pointed out[411] that vertebrates use D-glucosyluronic transfer as a mechanism for "detoxifying" foreign phenols, but this mechanism cannot be used by invertebrates, although they possess D-glucosyluronic transferases for the synthesis of endogenous compounds; instead, the invertebrates appear to use β-D-glucosyl transfer for the modification of foreign phenolic compounds.

(406) G. J. Dutton and I. D. E. Storey, Methods Enzymol., 5, 159 (1962).
(407) K. Dahm and H. Breuer, Biochim. Biophys. Acta, 113, 404 (1966).
(408) E. Halac and E. Bonevardi, Biochim. Biophys. Acta, 67, 498 (1963).
(409) E. Halac and A. Reff, Biochim. Biophys. Acta, 139, 328 (1967).
(410) K. Lippel and J. A. Olson, J. Lipid Res., 9, 168 (1968).
(411) G. J. Dutton, Arch. Biochem. Biophys., 116, 399 (1966).

Hahlbrock and Conn[412] have isolated an enzyme capable of catalyzing the transfer of D-glucose from UDP-D-glucose to 2-hydroxyisobutyronitrile (2-hydroxy-2-methylpropionitrile; acetone cyanohydrin) from flax seedlings (*Linum usitatissium* L.). The product of this reaction was shown to be identical with linamarin, one of the two cyanogenic D-glucosides produced by the flax plant. The partially purified UDP-D-glucose:ketone cyanohydrin β-D-glucosyltransferase exhibited a high degree of specificity for the two aliphatic side-chains of the cyanohydrins of acetone and butanone, as well as for UDP-D-glucose as the D-glucose donor.

1-Thio-2-phenylacetohydroximate was found by Matsuo and Underhill[413] to occur naturally in *Tropaeolum majus*, and to be derived from 2-phenylacetaldehyde oxime. An enzyme preparation from leaves of *T. majus* has been shown to catalyze the D-glucosylation of sodium 1-thio-2-phenylacetohydroximate by UDP-D-glucose, to produce 1-thio-β-D-glucopyranosyl 2-phenylacetohydroximate.

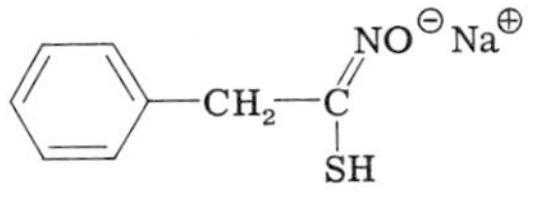

Sodium 1-thio-2-phenyl-
acetohydroximate

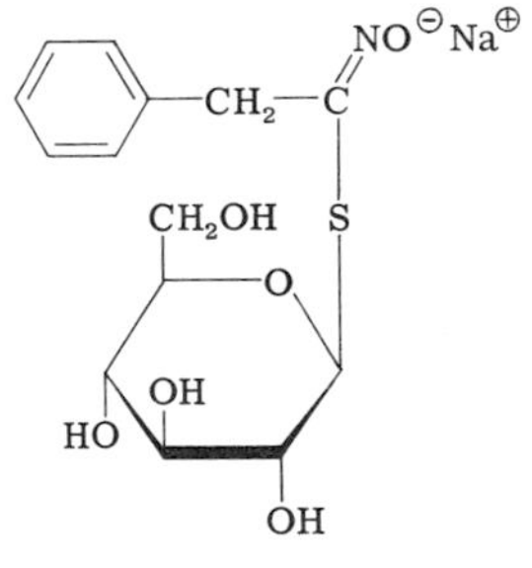

1-Thio-β-D-glucopyranosyl
2-phenylacetohydroximate
("Sodium desulfobenzylglucosinolate")

Ortman and coworkers[414] have found that enzymes from parsley catalyze the sequential transfer of D-glucosyl and D-apiosyl groups from their UDP derivatives to apigenin (4′,5,7-trihydroxyflavone).

(412) K. Hahlbrock and E. E. Conn, *J. Biol. Chem.*, **245**, 917 (1970).
(413) M. Matsuo and E. W. Underhill, *Biochem. Biophys. Res. Commun.*, **36**, 18 (1969).
(414) R. Ortmann, H. Sandermann, Jr., and H. Grisebach, *FEBS Lett.*, **7**, 164 (1970).

The final product is apiin, 7-O-[β-D-apiofuranosyl-(1 → 2)-β-D-glucosyl]-4′,5,7-trihydroxyflavone. Similar reactions were found to occur with the 3′-methoxy derivative of apigenin, namely, chrysoeriol, as the substrate.

2. Glycolipids

a. **Sphingolipids.**—Glycolipids of this group have been found to be synthesized by successive glycosyl-transfer reactions from "sugar nucleotides." Contributions by Roseman and coworkers[415–421] have been important in this area. The presumed pathway of synthesis in animals is shown in Scheme 5. For the chemistry of glycosphingolipids, the reader is referred to a recent review.[422] Reaction 1, in which the D-glucosyl group of UDP-D-glucose is transferred to ceramide (N-acylsphingosine), was demonstrated by the use of extracts from embryonic-chick brain[415] and rat brain.[423] The enzymic activity appears to be bound to a membrane fraction, and the reaction is enhanced by the addition of detergents.[415] This enzyme is unique among enzymes of sphingoglycolipid synthesis in that it does not require divalent cations.[415] Sphingosine, dihydrosphingosine, and D-galactosylsphingosine are inactive as acceptors.[415]

The second reaction (2) on the pathway to the gangliosides involves the transfer of the β-D-galactosyl group from UDP-D-galactose onto O-4 of the D-glucosyl group of β-D-glucosylceramide; this reaction was demonstrated for preparations derived from rat spleen,[424,425] embryonic-chick brain,[415] and frog brain.[426] In embryonic-chick brain, the enzymic activity is found in a particulate fraction, which also contains another β-D-galactosyltransferase, catalyzing[415] reaction 8. The physical separation of these two activities has not yet been

(415) S. Basu, Fed. Proc., **27**, 346 (1968); S. Basu, B. Kaufman, and S. Roseman, J. Biol. Chem., **243**, 5802 (1968).
(416) S. Basu and B. Kaufman, Fed. Proc., **24**, 479 (1965).
(417) B. Kaufman and S. Basu, Methods Enzymol., **8**, 365 (1966).
(418) B. Kaufman, S. Basu, and S. Roseman, J. Biol. Chem., **243**, 5804 (1968).
(419) S. Basu, B. Kaufman, and S. Roseman, J. Biol. Chem., **240**, 4115 (1965).
(420) J. C. Steigerwald, B. Kaufman, S. Basu, and S. Roseman, Fed. Proc., **25**, 587 (1966).
(421) S. Basu, M. Basu, H. Den, and S. Roseman, Fed. Proc., **29**, 410 Abstr. (1970).
(422) J. Kiss, Advan. Carbohyd. Chem. Biochem., **24**, 381 (1969).
(423) J. A. Curtino, R. O. Caledron, and R. Caputto, Fed. Proc., **27**, 346 (1968).
(424) G. Hauser, Biochem. Biophys. Res. Commun., **28**, 502 (1967).
(425) J. Hildebrand and G. Hauser, J. Biol. Chem., **244**, 5170 (1969).
(426) M. C. M. Yip, G. Yip, and J. A. Dain, Abstr. Papers Amer. Chem. Soc. Meeting, **156**, Biol. 70 (1968).

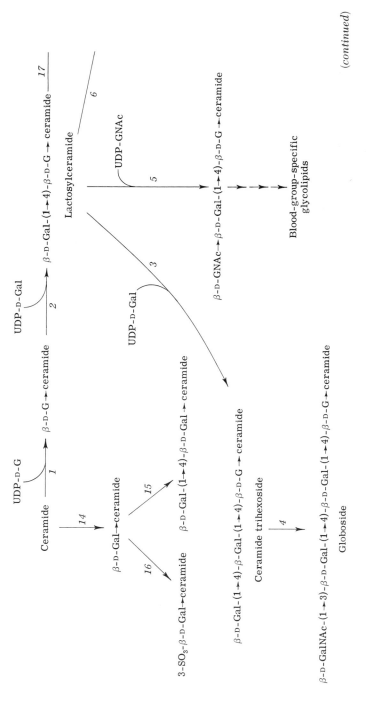

(continued)

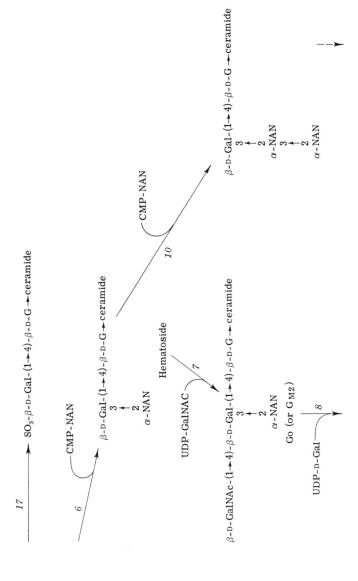

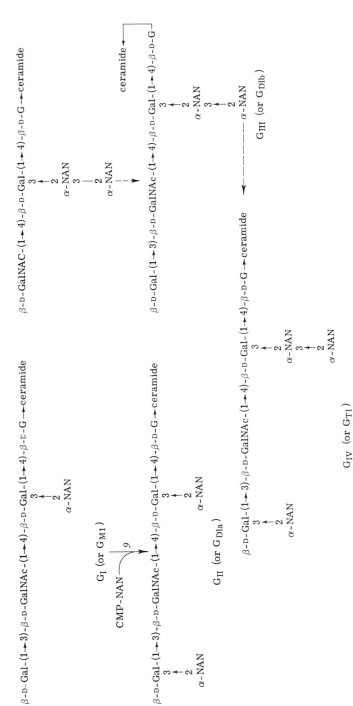

Scheme 5. **Biosynthesis of Glycosphingolipids. (For the structure of various sphingolipids, see the review by Kiss.[422] Structure of ceramide trihexoside and globoside is based on Refs. 428 and 429, respectively.)**

achieved, but the following evidence indicated the presence of two distinct enzymes.[415] (a) There was no competition between reaction 2 and reaction 8. (b) Reaction 8 was observed only in the presence of Mn^{2+}, but Mg^{2+} was as active as Mn^{2+} in stimulating reaction 2. The strict substrate-specificity of this D-galactosyl transferase is seen from the observation that sphingosine, ceramide, D-galactosylceramide, lactosylceramide, or hematoside did not serve as acceptors. In rat spleen, the D-galactosyl transferase catalyzing reaction 2 was shown[425] to be distinct from the D-galactosyl transferase catalyzing reaction 3. This conclusion is based on (a) a very different susceptibility to heat inactivation, (b) lack of competition between substrates for these two reactions, (c) different degrees of inhibition by various sphingolipids, and (d) different requirements for divalent cations. These two enzymes were also shown to be distinct from the D-galactosyl transferase catalyzing reaction 8. Kidney homogenates of female C3H/He mice were found to be much less active in catalyzing reaction 2 than the corresponding homogenates from the males.[427]

Lactosylceramide is apparently an important branch point in glycosphingolipid synthesis (see Scheme 5); it can be D-galactosylated, eventually to afford globoside, a major glycolipid of erythrocytes in certain species (reactions 3 and 4); it can accept 2-acetamido-2-deoxy-D-glucose (reaction 5) and, presumably, lead eventually to blood-group-specific glycolipids; or it can accept an N-acetylneuraminosyl group (reaction 6), and enter the pathway leading to a variety of gangliosides.

In the first pathway, lactosylceramide was shown to be converted into a ceramide trihexoside by a D-galactosyl transferase present in the "microsomal" fractions of rat spleen (reaction 3).[425] The difference between the properties of this D-galactosyl transferase and those of the enzyme catalyzing reaction 2 has already been discussed. The ceramide trihexoside is known to occur in a variety of organs (see Ref. 425) including spleen.[430] It is possible that, in some tissues, this compound is further converted into globoside by the addition of 2-acetamido-2-deoxy-D-galactosyl groups; this hypothetical reaction 4 has not yet been demonstrated in vitro.

In the second pathway, a 2-acetamido-2-deoxy-D-glucosyl group is transferred from its UDP ester to lactosylceramide (reaction 5). This reaction was demonstrated with an enzyme from rabbit bone-marrow;

(427) J. B. Hay and G. M. Gray, Biochem. Biophys. Res. Commun., 38, 527 (1970).
(428) A. Makita and T. Yamakawa, J. Biochem. (Tokyo), 55, 365 (1964).
(429) A. Makita, M. Iwanaga, and T. Yamakawa, J. Biochem. (Tokyo), 55, 202 (1964).
(430) A. Makita and T. Yamakawa, J. Biochem. (Tokyo), 51, 124 (1962); A. Wagner, Clin. Chim. Acta, 10, 175 (1964).

it requires Mn^{2+}, and, specifically, lactosylceramide as the acceptor.[421] Glycolipids (of erythrocytes) having blood-group specificity are known to contain a 2-acetamido-2-deoxy-D-glucose residue linked to the D-galactose residue of the lactosylceramide.[431,432] It may, therefore, be assumed that reaction 5 is the first specific step in the biosynthesis of this type of glycolipid.

In the third pathway, lactosylceramide accepts an N-acetylneuraminyl group from its CMP derivative, thus entering a pathway specific to ganglioside synthesis. Roseman and coworkers[416–418] have demonstrated the presence of a sialyl transferase catalyzing this reaction (reaction 6 of Scheme 5) in embryonic-chicken brain; at the same time, they found in the same tissue two other sialyl transferases, catalyzing reaction 9 and 10, respectively. That these enzymes were different from each other was indicated by the following observations.[418] (a) The enzyme that catalyzes reaction 9 can be completely heat-inactivated under conditions where the bulk of the other two activities are preserved. (b) Substrate-competition experiments indicated that reaction 6 and reaction 10 are catalyzed by two independent enzymes. (c) The enzyme catalyzing reaction 10 is strongly stimulated by the presence of histone or poly(L-lysine), and also requires phosphatidyl-2-aminoethanol; these effects are not observed with the other enzymes. These results emphasize the fact that each biosynthetic step is catalyzed by a specific glycosyl transferase having strict substrate and acceptor specificities. All the enzymes thus far described appear to be situated in the synaptosome (nerve ending) fraction in embryonic-chick brain.[418,433]

The enzyme catalyzing reaction 6 was also found in the brain of young rats.[434] In addition to lactosylceramide, lactose was also claimed to serve as a substrate. Care is, however, needed in interpreting these results, because chick brain is known to contain a fourth sialyl transferase that utilizes various β-D-galactosides as acceptors, and which is, presumably, involved in glycoprotein synthesis.[417]

Enzyme activity catalyzing reaction 7 was demonstrated in a particulate fraction from chick-embryo brain.[419,420] This enzyme converts hematoside into ganglioside G_0 (according to the terminology of Kuhn and Wiegandt[435]) (G_{M2} according to Svennerholm,[436] or "Tay–Sachs ganglioside") by transferring the 2-acetamido-2-deoxy-

(431) S. Hakomori and R. W. Jeanloz, Fed. Proc., 24, 231 (1965).
(432) S. Hakomori and G. D. Strycharz, Biochemistry, 7, 1279 (1968).
(433) H. Den and B. Kaufman, Fed. Proc., 27, 346 (1968).
(434) A. Arce, H. F. Maccioni, and R. Caputto, Arch. Biochem. Biophys., 116, 52 (1966).
(435) R. Kuhn and H. Wiegandt, Chem. Ber., 96, 866 (1963).
(436) L. Svennerholm, J. Neurochem., 10, 613 (1963).

β-D-galactosyl group from UDP-2-acetamido-2-deoxy-D-galactose.

Theoretically, there is an alternative pathway for the synthesis of ganglioside G_0. Instead of going through hematoside, the 2-acetamido-2-deoxy-D-galactosyl group might first be added to lactosylceramide, to form O-(2-acetamido-2-deoxy-β-D-galactopyranosyl)-(1 → 4)-O-β-D-galactopyranosyl-(1 → 4)-O-β-D-glucopyranosylceramide, and then sialic acid might be transferred to this lipid in order to complete the synthesis of ganglioside G_0. The former reaction was, indeed, demonstrated with rat-brain homogenate,[437] and the latter reaction was assumed to be involved in the *in vitro* conversion of O-(2-acetamido-2-deoxy-β-D-galactopyranosyl)-(1 → 4)-O-β-D-galactopyranosyl-(1 → 4)-O-β-D-glucopyranosylceramide into ganglioside G_1 in the presence of homogenate of rat kidney.[438] However, in neither of these studies has the possible presence of the enzymes catalyzing reactions 6 and 7 been examined. Furthermore, in embryonic-chick brain, at least, this alternative pathway is unlikely to be predominant, because the rate of the enzymic transfer of 2-acetamido-2-deoxy-D-galactose to hematoside is ten times that of its transfer to lactosylceramide.[420]

The conversion of ganglioside G_0 into G_1 ("monosialoganglioside") is accomplished by the transfer of the β-D-galactosyl group from UDP-D-galactose (reaction 8); this was shown with particulate enzymes from embryonic-chick brain,[419] adult-frog brain,[439] and rat brain.[440] In the first system, various other glycolipids, including O-(2-acetamido-2-deoxy-β-D-galactopyranosyl)-(1 → 4)-O-β-D-galactopyranosyl-(1 → 4)-O-β-D-glucopyranosylceramide, were shown to be inactive as the acceptor.[419] This result clearly indicates that a sialic acid residue must be present before the transfer of this (nonreducing) terminal D-galactosyl group occurs. As already mentioned, this D-galactosyl transferase is different from the enzyme that catalyzes[415] reaction 2.

The "di-(N-acetylneuraminosyl)-lactosylceramide" [O-(N-acetyl-α-neuraminosyl)-(2 → 3)-O-(N-acetyl-α-neuraminosyl)-(2 → 3)-O-β-D-galactosyl-(1 → 4)-O-β-D-glucosylceramide], produced through reaction 10, can be assumed to undergo a sequence of reactions (11 to 13) similar to those reactions of the pathway leading to G_{II}. However, information is not yet available on these hypothetical reactions.

(437) S. Handa and R. M. Burton, *Fed. Proc.*, **25**, 587 (1966).

(438) J. N. Kanfer, R. S. Blacklow, L. Warren, and R. O. Brady, *Biochem. Biophys. Res. Commun.*, **14**, 287 (1964).

(439) J. A. Yiamouyiannis and J. A. Dain, *Lipids*, **3**, 378 (1968).

(440) G. B. Yip and J. A. Dain, *Biochim. Biophys. Acta*, **206**, 252 (1970).

At one time, it was believed that cerebroside (β-D-galactosylcer-amide) is synthesized as follows.

$$\text{Sphingosine} + \text{UDP-D-galactose} \longrightarrow \text{psychosine} + \text{UDP} \quad (18)$$

$$\text{Psychosine} + \text{fatty acyl-CoA} \longrightarrow \text{cerebroside} + \text{CoA} \quad (19)$$

Reaction 18 was shown to occur in the presence of guinea-pig or rat-brain microsomes.[441] Ceramides were stated to be inactive as the acceptor of D-galactose residues, but it appears that only N-acetyl-sphingosine was used. Also, inactivity of any of the substrates in these reactions could be due to the lack of proper emulsification.[441] Reaction 19 was postulated by R. O. Brady,[442] when he reported on the first enzymic synthesis of cerebrosides with rat-brain microsomes and stearoyl-^{14}C-CoA. In a reaction mixture containing psychosine as an acceptor, twice as much stearic acid-^{14}C was incorporated as in the mixture containing a combination of sphingosine and UDP-D-glucose instead of psychosine.[442]

These results are at variance with the pathway of biosynthesis of D-glucosylceramide, where acylation of sphingosine precedes D-glu-cosylation.[415] A re-examination of the problem suggested that the re-sults just described probably do not correspond to the physiological pathway, and that, normally, sphingosine is first acylated to form a ceramide, and that this is D-galactosylated (reaction 14). Thus, it has been shown that the D-galactosyl group is transferred from UDP-D-galactose both to ceramides containing hydroxy fatty acids[443] and to ceramides containing nonhydroxy fatty acids,[444] when a crude, microsomal fraction from mouse brain was used. Acylation of ex-ogenous psychosine was not found under these conditions. Similar results were also obtained with preparations from rat brain.[445] The biosynthesis of ceramides by the transfer of fatty acids from their CoA derivatives to sphingosine has been demonstrated.[446-448]

D-Galactosylceramide was shown to be sulfated at C-3 of the D-galactosyl group in the presence of adenylyl 3'-phosphate sulfate

(441) W. W. Cleland and E. P. Kennedy, *J. Biol. Chem.*, **235**, 45 (1960).

(442) R. O. Brady, *J. Biol. Chem.*, **237**, PC2416 (1962).

(443) P. Morell and N. S. Radin, *Biochemistry*, **8**, 506 (1969).

(444) P. Morell and N. S. Radin, *Fed. Proc.*, **29**, 409 Abstr. (1970).

(445) Y. Fujino and M. Nakano, *Biochem. J.*, **113**, 573 (1969).

(446) M. Sribney, *Biochim. Biophys. Acta*, **125**, 542 (1966).

(447) P. Morell and N. S. Radin, *J. Biol. Chem.*, **245**, 342 (1970).

(448) P. E. Braun, P. Morell, and N. S. Radin, *J. Biol. Chem.*, **245**, 335 (1970).

("PAPS"), to yield sulfatide (reaction *16*).[449] In contrast to the enzymes catalyzing the addition of sugar residues in glycolipid synthesis, the enzyme catalyzing this reaction was found in the soluble
fraction from brains of young rats.[449] When this reaction was examined
in greater detail,[450] particulate fractions from brains of young rats
were found to contain about 75% of the total activity. Two sulfuryl
transferases were observed: one, specific for glycolipids, and the
other, for soluble D-galactosides. The former enzyme transfers sulfate
to C-3 of the D-galactosyl group in D-galactosylceramide (reaction *15*,
Scheme 5; p. 443) and in lactosylceramide (reaction *17*). The latter
compound is the best acceptor in this reaction; this is significant,
in view of the finding of a large amount of lactosylceramide sulfate
in kidney.[451]

Di-D-galactosylceramide, which is a major dihexosylceramide in
the kidney of certain strains of mice, was shown[452] to be synthesized
in a cell-free system by way of reactions *14* and *15*. It is interesting
that kidney homogenates from female C57BL mice are much less
active in catalyzing reaction *14*, and that the concentration of di-D-galactosylceramide in female kidney is also much lower than that
found in male.[452]

b. Glycerolipids.—Chloroplasts contain 1,2-diglycerides having
either one or two D-galactosyl groups attached to the third hydroxyl
group of glycerol. Spinach-leaf chloroplasts were shown to
incorporate the D-galactosyl group from UDP-D-galactose into an
endogenous acceptor; the products were, apparently, 1,2-di-*O*-acyl-
3-*O*-β-D-galactosyl-*sn*-glycerol and 1,2-di-*O*-acyl-3-*O*-(6-*O*-α-D-galac-
tosyl-β-D-galactosyl)-*sn*-glycerol.[453] This pathway was examined in
more detail by using spinach-leaf, chloroplast preparations.[454,455] Products of the reaction were characterized as D-galactosyl-, di-D-galacto-
syl-, and tri-D-galactosyl-diglycerides. Because tri-D-galactosyl-diglyc-
eride is not found in spinach leaves, breakdown of the regulatory

(449) G. M. McKhann, R. Levy, and W. Ho, *Biochem. Biophys. Res. Commun.*, **20**, 109
        (1965).
(450) F. A. Cumar, H. S. Barra, H. J. Maccioni, and R. Caputto, *J. Biol. Chem.*, **243**, 3807
        (1968).
(451) A. Stoffyn, P. Stoffyn, and E. Martensson, *Biochim. Biophys. Acta*, **152**, 353
        (1968).
(452) L. Coles and G. M. Gray, *Biochem. Biophys. Res. Commun.*, **38**, 520 (1970).
(453) E. F. Neufeld and C. W. Hall, *Biochem. Biophys. Res. Commun.*, **14**, 503 (1964).
(454) A. Ongun and J. B. Mudd, *J. Biol. Chem.*, **243**, 1558 (1968).
(455) J. B. Mudd, H. H. D. M. Van Vliet, and L. L. M. Van Deenen, *J. Lipid Res.*, **10**, 623
        (1969).

process under cell-free conditions was suspected. The kinetics of the incorporation of D-galactose-^{14}C into various products, as well as pulse-chase experiments, indicated that the mono-D-galactosyl-diglyceride was the precursor of di-D-galactosyl and tri-D-galactosyl compounds. Diolein was active as the acceptor of the first D-galactosyl group, although the transfer of the second and the third D-galactosyl groups did not occur. This suggests that, to allow the transfer of these D-galactosyl groups, the fatty acids must be more highly unsaturated (as in endogenous, chloroplast lipids). Dipalmitin did not accept even the first D-galactosyl group. An overall synthesis of mono-D-galactosyl-diglyceride from acyl-"acyl carrier protein," D-glycerol 1-phosphate ("L-α-glycerophosphate"), and UDP-D-galactose has also been demonstrated by the use of crude extracts of *Euglena gracilis*.[456] It is noteworthy that acyl-"acyl carrier protein," which is completely inactive in the synthesis of phospholipid in this organism, serves as a precursor despite the fact that both the phospholipid synthesis and D-galactolipid synthesis are thought to proceed through the stage of phosphatidic acid.[456]

Very similar reactions were demonstrated by the use of a particulate fraction from young rat-brain.[457,458] Incubation with 1,2-diglyceride and UDP-D-galactose afforded β-D-galactosyldiglyceride,[457] whereas a 3-O-(α-D-galactosyl-D-galactosyl)diglyceride was produced in the presence of deoxycholate.[458] In the latter reaction, the conversion of diglyceride-^{14}C into di-D-galactosyl-diglyceride is inhibited by the addition of mono-(D-galactosyl-^{12}C)-diglyceride.[458] These results are consistent with the assumption that the biosynthesis proceeds by the stepwise addition of single D-galactosyl groups.

A Triton X-100 extract of HeLa cells was found to incorporate D-mannosyl and L-fucosyl groups from their respective GDP esters into endogenous lipid(s).[459] The products were poorly characterized, but diglyceride stimulated incorporation with "acetone-powder" enzymes from which most of the endogenous acceptors had presumably been removed. The extent of stimulation is not known, as the levels of incorporation without diglyceride were not given; however, of the diglycerides tested, only dipalmitin was reported to be active for the L-fucose system, and only diolein for the D-mannose system.

Many micro-organisms, especially Gram-positive bacteria, are now

(456) O. Renkonen and K. Bloch, *J. Biol. Chem.*, **244**, 4899 (1969).
(457) D. A. Wenger, J. W. Petitpas, and R. A. Pieringer, *Fed. Proc.*, **26**, 765 (1967); *Biochemistry*, **7**, 3700 (1968).
(458) K. Subba Rao, D. A. Wenger, and R. A. Pieringer, *Fed. Proc.*, **27**, 346 (1968).
(459) H. B. Bosmann, *Biochim. Biophys. Acta*, **187**, 122 (1969).

known to synthesize glycosyl-diglycerides. The biosynthesis of these lipids has been examined in three cell-free systems. Distler and Roseman[257,259] found that a particulate fraction from *Diplococcus pneumoniae* Type XIV first transferred an α-D-glucosyl group to an endogenous acceptor, and then transferred an α-D-galactosyl group to this glycolipid. The end product was identified as 1,2-di-*O*-acyl-3-*O*-(2-*O*-α-D-galactopyranosyl-α-D-glucopyranosyl)-*sn*-glycerol, and the stepwise nature of the reaction was established by the utilization of added 3-*O*-α-D-glucopyranosyl-diglyceride in the second step.

Lennarz[460] has found that *Micrococcus lysodeikticus* contains an enzyme system for the synthesis of 1,2-di-*O*-acyl-3-*O*-(3-*O*-α-D-mannosyl-α-D-mannosyl)-*sn*-glycerol, as follows.

GDP-D-mannose + diglyceride $\longrightarrow$

α-D-mannosyl-diglyceride + GDP

GDP-D-mannose + α-D-mannosyl-diglyceride $\longrightarrow$

α-D-mannosyl-(1 $\longrightarrow$ 3)-α-D-mannosyl-diglyceride

The first reaction was demonstrated by the use of acetone-extracted, membrane fractions from this organism. GDP-D-mannose, 1,2-diglyceride, Mg^{2+}, high ionic strength, and an anionic detergent were required. 1,2-Diglycerides containing branched-chain, fatty acids (as found in *M. lysodeikticus*) are most active, and diglyceride prepared from pig-liver lecithin, for example, shows only 10% of the activity of the former. Dipalmitin is completely inactive. As the anionic detergent, Na^{+} salts of branched-chain, fatty acids, again, are most active. The second reaction is catalyzed by an enzyme found in the soluble fraction, and requires the presence of D-mannosyl-diglyceride, GDP-D-mannose, and Mg^{2+}. No detergent requirement has been demonstrated for this reaction.

The third example, the biosynthesis of 1,2-di-*O*-acyl-3-*O*-(2-*O*-α-D-glucopyranosyl-α-D-glucopyranosyl)-*sn*-glycerol was examined with a particulate fraction from *Streptococcus faecalis*.[461] UDP-D-glucose is the donor of the D-glucosyl group, and the addition of two D-glucosyl groups is sequential, as isolated mono-D-glucosyl-diglyceride functions as a good acceptor for the second D-glucosyl group. 2,3-Di-*O*-acyl-*sn*-glycerol is inactive as an acceptor, but all 1,2-di-*O*-acyl-*sn*-glycerols thus far tested (from dicaprin to distearin) are active.

(460) W. J. Lennarz, *J. Biol. Chem.*, **239**, PC3110 (1964); W. J. Lennarz and B. Talamo, *ibid.*, **241**, 2707 (1966).
(461) R. A. Pieringer, *J. Biol. Chem.*, **243**, 4894 (1968).

All detergents tested were inhibitory, and the best results were obtained by mixing lyophilized enzyme with a benzene solution of the diglyceride, and evaporating off the solvent.

c. D-Mannophosphoinositides.—When particulate fractions from *Mycobacteria* are incubated with phosphatidyl-*myo*-inositol and GDP-D-mannose-^{14}C, various "D-mannophosphoinositides" are produced.[462–464] The reaction sequence appears to be rather complex. With particulate fractions from *M. phlei*, under certain conditions, the major product was thought to be 2-deoxy-1-O-phosphatidyl-L-*myo*-inositol-2-yl α-D-mannopyranoside.[462] However, this product is

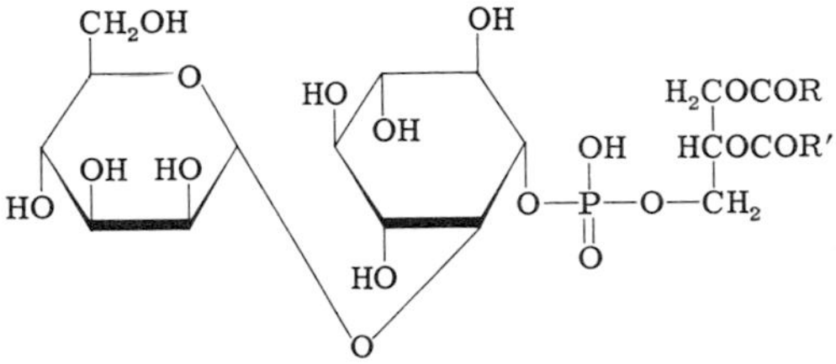

2-Deoxy-1-O-phosphatidyl-L-*myo*-
inositol-2-yl α-D-mannopyranoside

now thought to contain additional acyl group(s).[463] Results with a particulate preparation from *M. tuberculosis* suggested a somewhat similar situation.[464] Here, the D-mannosyl-^{14}C group from GDP-D-mannose-^{14}C was incorporated into endogenous lipids, and both mono- and di-D-mannosyl-"phosphoinositides" were produced. When the latter compound was degraded, however, only the D-mannosyl group linked to O-6 of *myo*-inositol was labeled; that linked to O-2 was not. Thus, mono-(D-mannosyl-^{14}C)-"phosphoinositide" produced *in vitro* cannot function as an acceptor of the second D-mannosyl group, although endogenous mono-D-mannosyl-phosphoinositide apparently can. It is assumed[464] that the endogenous compound is different from the enzymically produced compound, perhaps in bearing extra acyl groups; this acylation presumably does not occur with the cell-free preparation from *M. tuberculosis*.

With the particulate fraction of *M. phlei* (which carries out the most complex series of conversion reactions), Brennan and Ballou[463] found that the addition of a phosphatidyl-*myo*-inositol and GDP-D-mannose

(462) D. L. Hill and C. E. Ballou, *J. Biol. Chem.*, **241**, 895 (1966).
(463) P. Brennan and C. E. Ballou, *J. Biol. Chem.*, **242**, 3046 (1967).
(464) K. Takayama and D. S. Goldman, *Biochim. Biophys. Acta*, **176**, 196 (1969).

gives rise to three products. All of them are di-D-mannosyl compounds and thus have α-D-mannopyranosyl groups attached to O-2 and O-6 of *myo*-inositol. Furthermore, two of the products were found to contain one, and two, acyl groups, respectively, in addition to the two acyl groups in the phosphatidic acid portion. The donors of these acyl groups are probably present in the enzyme preparation.

d. Sterol D-glucosides.—Particulate preparations from immature, soybean seeds,[465] mung-bean shoots,[191] and starch grains from immature endosperm of sweet corn[466] catalyze the transfer of the D-glucosyl group from UDP-D-glucose onto such endogenous sterols as β-sitosterol, stigmasterol, and campesterol. A similar reaction is also obtained by the use of leaf discs from various plants.[467] In some systems, acylated sterol D-glucosides, presumably containing an acyl group linked to O-6 of D-glucose, are also formed.[191,466,467] In a case where the specificity of the donor nucleotide was tested,[465] only UDP-D-glucose and dTDP-D-glucose were active. Sterols added to the reaction mixture sometimes give a small stimulation.[465,466] In none of the systems has a requirement for divalent cations been found. Kauss[191] examined the possibility that sterol D-glucosides could serve as "lipid intermediates" in the synthesis of D-glucose-containing polysaccharides, but the results were negative (see also, Section V,3, p. 391).

e. Other Lipids.—Soluble enzymes from *Pseudomonas aeruginosa* were found to catalyze the stepwise transfer of two L-rhamnosyl groups from their dTDP ester to 3-(3-hydroxydecanoyloxy)decanoate to form the O-α-L-rhamnosyl-(1 → 2)-α-L-rhamnosyl derivative of this compound.[468] The two enzymes catalyzing these reactions have been separated by adsorption onto, and elution from, calcium phosphate gel.

The functions of lipid intermediates in the biosynthesis of a mannan, a bacterial capsular polysaccharide, a cell-wall lipopolysaccharide, and a cell-wall peptidoglycan are respectively discussed in Sections dealing with each (see Sections V,6, p. 396; VI,5,6, and 7, pp. 408, 418, 428). In all these syntheses, the glycosyl groups are

(465) C. T. Hou, Y. Umemura, M. Nakamura, and S. Funahashi, *J. Biochem.* (Tokyo), **63**, 351 (1968).
(466) N. Lavintman and C. E. Cardini, *Biochim. Biophys. Acta*, **201**, 508 (1970).
(467) W. Eichenberger and D. W. Newman, *Biochem. Biophys. Res. Commun.*, **32**, 366 (1968).
(468) M. M. Burger, L. Glaser, and R. M. Burton, *J. Biol. Chem.*, **238**, 2595 (1963); *Methods Enzymol.*, **8**, 441 (1966).

linked to the hydroxyl group of polyisoprenol, either through a pyrophosphate or through a phosphoric diester bridge. A polyisoprenyl (β-D-glucosyl phosphate) has also been synthesized by an animal enzyme,[148] but its function is not known (see Section V,1, p. 381). A few glycosyllipids of uncertain structure have been suggested as being intermediates in glycoprotein synthesis; these will be discussed in Section VII, 3 (p. 471).

3. Glycoproteins

a. **Structure of glycoproteins.**—Glycoproteins can be classified into three groups according to the mode of attachment of the carbohydrate moiety to the protein portion. In the first group, D-galactose is linked glycosidically to the hydroxyl group of 5-hydroxy-L-lysine,[469–472] as in collagen. In the second group, 2-acetamido-2-deoxy-D-galactose is linked glycosidically to the hydroxyl group of either an L-threonine or an L-serine residue of the protein, as in "soluble," blood-group substances of ovarian-cyst fluid[473] or in submaxillary mucins.[474] In the third group, 2-acetamido-2-deoxy-β-D-glucose is linked, through a glycosylamine linkage, to the amide nitrogen atom of an L-asparagine residue of the protein.[475] Thus, this linkage can be expressed as N-(2-acetamido-2-deoxy-β-D-glucopyranosyl)-L-asparagine. This kind of linkage occurs widely, both in secreted, soluble glycoproteins (such as serum proteins, digestive enzymes, egg albumin, and immunoglobulins,) and in the membrane glycoproteins. In this article, these types of carbohydrate chains will be referred to hereafter as the first, second, and third type, respectively.

(469) W. T. Butler and L. W. Cunningham, *J. Biol. Chem.*, **241**, 3882 (1966).
(470) R. G. Spiro, *J. Biol. Chem.*, **242**, 1923 (1967).
(471) L. W. Cunningham, J. D. Ford, and J. P. Segrest, *J. Biol. Chem.*, **242**, 2570 (1967).
(472) L. W. Cunningham and J. D. Ford, *J. Biol. Chem.*, **243**, 2390 (1968).
(473) E. A. Kabat, E. W. Bassett, K. Pryzwansky, K. O. Lloyd, M. E. Kaplan, and E. G. Layug, *Biochemistry*, **4**, 1632 (1965).
(474) B. Anderson, N. Seno, P. Sampson, J. G. Riley, P. Hoffman, and K. Meyer, *J. Biol. Chem.*, **239**, PC2716 (1964); K. Tanaka and W. Pigman, *ibid.*, **240**, PC1487 (1965); S. Harbon, G. Herman, B. Rossignol, P. Jollès, and H. Clauser, *Biochem. Biophys. Res. Commun.*, **17**, 57 (1964); K. Tanaka, M. Bertolini, and W. Pigman, *ibid.*, **16**, 404 (1964); V. P. Bhavanandan, F. Buddecke, R. Carubelli, and A. Gottschalk, *ibid.*, **16**, 353 (1964); S. Harbon, G. Herman, and H. Clauser, *Eur. J. Biochem.*, **4**, 265 (1968).
(475) G. S. Marks, R. D. Marshall, and A. Neuberger, *Biochem. J.*, **87**, 274 (1963); R. D. Marshall and A. Neuberger, *Biochemistry*, **3**, 1596 (1964); I. Yamashina, K. Ban-I, and M. Makino, *Biochim. Biophys. Acta*, **78**, 382 (1963); V. P. Bogdanov, E. D. Kaverzneva, and A. P. Andreyeva, *ibid.*, **83**, 69 (1964).

Most glycoproteins contain more than one carbohydrate chain, attached to different sites on the same polypeptide. In many, these carbohydrate units are not identical. In some, one protein or polypeptide can even contain carbohydrate chains attached through *different types of linkages.* For example, in H chains of rabbit IgG immunoglobulin, there is always a carbohydrate unit attached through the third type of linkage mentioned, but 35% of the chains also contain a unit attached through a linkage of the second type.[476] A similar situation was found for the H chain of a human IgA myeloma globulin.[477] Another example is the glomerular-basement membrane, which has one complex carbohydrate chain linked to L-asparagine for every ten D-glucosyl-D-galactose units linked to 5-hydroxy-L-lysine.[470] In the present Chapter, the discussion will be divided according to the type of *glycoprotein* synthesized, but it is clear from the foregoing that, in reality, the biosynthesis of each type of *carbohydrate unit* will be discussed. A review on the structure and biosynthesis of glycoproteins has appeared.[477a]

b. General Mechanism for Synthesis of Glycoproteins.—In earlier studies on the synthesis of glycoproteins, intact animals, perfused organs, tissue slices, or cell suspensions were utilized.[478-482] Radioactive precursors of the carbohydrate moiety and of the protein moiety were added, and, at various intervals of time, the extent of incorporation into various subcellular fractions was determined. It was found that, in contrast to radioactive amino acids (which are incorporated into proteins in ribosomal fractions), hexoses and hexosamines are incorporated into glycoproteins mostly at the "smooth" endoplasmic reticulum.[476-480] (The problem of biosynthetic sites will be discussed in greater detail on p. 466.) These results were interpreted as meaning that the carbohydrate portion of the glycoproteins is added, by action of membrane enzymes, after the complete polypeptide chain has been synthesized on the ribosomes. This view is supported by the following results. (a) In collagen, the carbohydrate moiety is linked to the 5-hydroxy-L-lysine residues.[469-472]

(476) D. S. Smyth and S. Utsumi, *Nature,* **216,** 332 (1967).
(477) G. Dawson and J. R. Clamp, *Biochem. J.,* **107,** 341 (1968).
(477a) R. G. Spiro, *Ann. Rev. Biochem.,* **39,** 599 (1970).
(478) E. J. Sarcione, *J. Biol. Chem.,* **239,** 1686 (1964).
(479) E. J. Sarcione, M. Bohne, and M. Leahy, *Biochemistry,* **3,** 1973 (1964).
(480) G. B. Robinson, J. Molnar, and R. J. Winzler, *J. Biol. Chem.,* **239,** 1134 (1964).
(481) G. M. W. Cook, M. T. Laico, and E. H. Eylar, *Proc. Nat. Acad. Sci. U.S.,* **54,** 247 (1965).
(482) R. G. Spiro and M. J. Spiro, *J. Biol. Chem.,* **241,** 1271 (1966).

These 5-hydroxy-L-lysine residues are known to be formed by the oxidation of L-lysine residues present in completed polypeptides.[483,484] (b) A search for aminoacyl tRNA containing 2-acetamido-2-deoxy-D-glucose produced negative results.[485] (c) Although puromycin inhibited polypeptide synthesis, it did not inhibit the attachment of carbohydrate residues to the preformed polypeptide molecules.[481,482] (d) Incorporation studies with cell-free systems also supported this conclusion, as discussed later.

c. **Biosynthesis of Carbohydrate Chains.**—In order to demonstrate an effective transfer of sugar units in cell-free systems, proper acceptor molecules have to be prepared, and then added to the reaction mixture. These acceptor molecules are glycoproteins from which certain, or all, carbohydrate units have been removed, either by treatment with glycosidases or by chemical degradation. Sometimes, biosynthetic reactions are inhibited, so that incomplete glycoproteins could be produced, and isolated for use in these studies.[486]

(i) **Carbohydrate Residues Linked to 5-Hydroxy-L-lysine.**—The group of glycoproteins in which D-galactosyl groups are linked to 5-hydroxy-L-lysine residues usually have carbohydrate chains of simple structure; either a single D-galactosyl group is linked to the 5-hydroxy-L-lysine, or 2-O-α-D-glucosyl-β-D-galactose[486a] is present. For collagen, or glomerular-basement membrane, this disaccharide unit can be degraded by the use of the "Smith degradation" technique.[486b,487] By use of this "carbohydrate-free," glomerular-basement membrane as the acceptor, an enzyme that transfers the D-galactosyl group from its UDP ester was found in rat-kidney cortex.[486b] This enzyme requires Mn^{2+} and an acceptor of high molecular weight; free 5-hydroxy-L-lysine was inactive as an acceptor. Similarly, a D-galactosyl transferase has been extracted from embryonic, guinea-pig skin in the presence of 0.1% of Triton X-100, and purified by gel filtration.[487] The enzyme has an absolute requirement for Mn^{2+} (or

(483) D. J. Prockop, E. Weinstein, and T. Mulveny, *Biochem. Biophys. Res. Commun.*, **22**, 124 (1966).
(484) K. I. Kivirikko and D. J. Prockop, *Proc. Nat. Acad. Sci. U. S.*, **57**, 782 (1967).
(485) H. Sinohara and H. H. Sky-Peck, *Biochim. Biophys. Acta*, **101**, 90 (1965).
(486) N. Blumenkrantz, D. J. Prockop, and J. Rosenbloom, *Abstr. Papers Amer. Chem. Soc. Meeting*, **156**, Biol. 286 (1968).
(486a) R. G. Spiro, *J. Biol. Chem.*, **242**, 4813 (1967).
(486b) R. G. Spiro and M. J. Spiro, *Fed. Proc.*, **27**, 345 (1968).
(487) H. B. Bosmann and E. H. Eylar, *Biochem. Biophys. Res. Commun.*, **33**, 340 (1968).

Co^{2+}), UDP-D-galactose, and the acceptor. That the D-galactosyl groups are transferred to the 5-hydroxy-L-lysine residues in the acceptor was shown by the isolation of radioactive 5-(D-galactosyloxy)-L-lysine after degradation of the reaction product with hot alkali.[487] The enzyme also transfers D-galactosyl groups to free 5-hydroxy-L-lysine, but not to L-threonine or L-serine.

The second step in the biosynthesis of this type of carbohydrate chain is the transfer of the D-glucosyl group from UDP-D-glucose to the D-galactosyl group linked to 5-hydroxy-L-lysine. This was shown by using an enzyme from rat-kidney cortex, with 5-(D-galactosyloxy)-L-lysine as the acceptor.[486b] Free D-galactose and lactose were inactive. In another study, advantage was taken of the fact that the D-glucose residues of the collagen molecule can readily be split off by brief hydrolysis with acid.[488] These "D-glucose-free" collagen molecules serve as acceptors for the D-glucosyl transferase.[488] This enzyme was also purified from embryonic, guinea-pig skin; it was activated best by Mn^{2+}, and less by Co^{2+}, Mg^{2+}, and Ca^{2+}. Although "desialyzed" porcine submaxillary mucin, fetuin, and α_1-acid glycoprotein contain carbohydrate units having (nonreducing) terminal D-galactosyl groups, they cannot serve as acceptors in this reaction. Free D-galactose is also inactive.[488] As D-galactose is present in most glycoproteins, it is, perhaps, necessary to have an enzyme that "recognizes" a broad area of its acceptor molecule (probably including the 5-hydroxy-L-lysine residue) in order to avoid indiscriminate D-glucosylation of D-galactose residues in various glycoproteins.[488] The intracellular localization of this enzyme is quite different from that of the glycosyl transferases involved in the synthesis of more common glycoproteins; although the latter enzymes are all found in the internal, smooth-membrane fraction, the D-glucosyl transferase was found only in a fraction identified[489] as "plasma membrane."

(ii) Carbohydrate Chains Linked to L-Serine or L-Threonine.—In the second group of glycoproteins, in which 2-acetamido-2-deoxy-D-galactose is linked to L-serine or L-threonine residues of a polypeptide, the carbohydrate moieties frequently have rather simple structures. In ovine submaxillary mucin, the carbohydrate chain is a disaccharide, namely, sialyl-(2 → 6)-2-acetamido-2-deoxy-D-galactose. The anomeric configuration of the 2-acetamido-2-deoxy-D-galactose residue has been found[490] to be α. Successive treatment

(488) H. B. Bosmann and E. H. Eylar, *Biochem. Biophys. Res. Commun.,* **30,** 89 (1968).

(489) A. Hagopian, H. B. Bosmann, and E. H. Eylar, *Arch. Biochem. Biophys.,* **128,** 387 (1968).

(490) B. Weissmann and D. F. Hinrichsen, *Biochemistry,* **8,** 2034 (1969).

of this glycoprotein with sialidase and a *Clostridium* hexosaminidase removed most of the carbohydrate units, and McGuire and Roseman[491] used this product as the acceptor in an *in vitro* system. A glycosyl transferase, purified from extracts with dilute phosphate buffer of sheep submaxillary glands, was shown to transfer the 2-acetamido-2-deoxy-D-galactosyl group from its UDP ester to L-serine and L-threonine residues of the acceptor already described. The enzyme was purified by such conventional techniques as fractionation with ammonium sulfate and ion-exchange chromatography, and required Mn^{2+} or Co^{2+} for activity.[491] As an acceptor, L-serine, L-threonine, and peptides obtained by pronase digestion of the "acceptor protein" were completely inactive.[491] As most proteins contain L-serine and L-threonine residues, this strict acceptor-specificity of the 2-acetamido-2-deoxy-α-D-galactosyl transferase can be beneficial for avoiding the accidental transfer of the sugar residues to proteins other than mucins.

Essentially similar results have been obtained by Eylar and co-workers[492-494] on the biosynthesis of bovine submaxillary mucin. The enzyme is firmly bound to a membrane; solubilization by Triton X-100 results in a large increase in enzymic activity.[495] The enzyme was purified by gel filtration, and its properties were studied.[494] Among various substances studied, the only one (besides carbohydrate-free mucins) that was active as the acceptor of 2-acetamido-2-deoxy-D-galactose was a basic protein, called "encephalitogen," isolated from bovine, spinal-cord myelin.[492]

To the protein containing 2-acetamido-2-deoxy-D-galactose residues, N-acetylneuraminic acid is then transferred from CMP-sialic acid.[496] The enzyme catalyzing this reaction has been purified from phosphate-buffer extracts of sheep submaxillary gland, and found not to require divalent cations.[497] Free 2-acetamido-2-deoxy-D-galactose and various oligosaccharides containing this sugar are inactive as acceptors. However, most compounds of high molecular weight containing (nonreducing) terminal 2-acetamido-2-deoxy-D-galactosyl groups appear to be active,[497] and, in this sense, the enzyme is much

(491) E. J. McGuire and S. Roseman, *J. Biol. Chem.*, **242**, 3745 (1967).
(492) A. Hagopian and E. H. Eylar, *Arch. Biochem. Biophys.*, **126**, 785 (1968).
(493) A. Hagopian and E. H. Eylar, *Arch. Biochem. Biophys.*, **128**, 422 (1968).
(494) A. Hagopian and E. H. Eylar, *Arch. Biochem. Biophys.*, **129**, 515 (1969).
(495) A. Hagopian and E. H. Eylar, *Arch. Biochem. Biophys.*, **129**, 447 (1969).
(496) D. M. Carlson, E. J. McGuire, G. W. Jourdian, and S. Roseman, *Fed. Proc.*, **23**, 380 (1966).
(497) D. M. Carlson, E. J. McGuire, and G. W. Jourdian, *Methods Enzymol.*, **8**, 361 (1966).

less strict in its acceptor specificity than the 2-acetamido-2-deoxy-D-galactose transferase.

Porcine submaxillary mucin has a more complex carbohydrate chain[498] in which sialic acid and D-galactose are both linked to the 2-acetamido-2-deoxy-D-galactose residue that comprises the reducing end of the chain. A particulate preparation from pig submaxillary-gland was shown to transfer the D-galactosyl group from its UDP ester to the sialidase-treated, ovine submaxillary-mucin.[499] This enzyme could be differentiated from another D-galactosyl transferase that exists in the same tissue and is, presumably, involved in the biosynthesis of the glycoproteins of the third type.[499]

The localization of enzymes participating in the synthesis of the carbohydrate chains of ovine submaxillary mucin was studied by incubating submaxillary-gland slices with 2-amino-2-deoxy-D-glucose-^{14}C and then fractionating various subcellular components.[500] The Golgi apparatus contained little radioactivity, and the largest degree of incorporation was found in the "rough, endoplasmic reticulum." Hydrolysis of this fraction with acid produced 2-acetamido-2-deoxy-D-galactose-^{14}C, and the radioactivity incorporated was released by treatment with alkali; these results indicated that the incorporation is a step in the biosynthesis of mucin. In this connection, it may be recalled that the addition of the first monosaccharide unit (2-acetamido-2-deoxy-D-glucose) in the biosynthesis of the third type of glycoprotein is also claimed by some workers to occur while the polypeptide is still attached to ribosomes (see later). However, the situation is not at all clear, because an *in vitro* study indicated that the 2-acetamido-2-deoxy-D-galactose transferase is localized in the "plasma membrane" fraction in bovine submaxillary gland.[495]

(iii) Carbohydrate Chains Linked to L-Asparagine.—The third group of glycoproteins have their carbohydrate units attached through 2-acetamido-2-deoxy-D-glucose residues linked to the amide nitrogen atom of L-asparagine residues. A survey of the literature indicated that the L-asparagine residue carrying the carbohydrate chain always has an L-serine or L-threonine residue on its carboxyl end.[501] This

(498) D. M. Carlson, *J. Biol. Chem.*, **243**, 616 (1968); R. L. Katzman and E. H. Eylar, *Arch. Biochem. Biophys.*, **127**, 323 (1968).

(499) H. Schachter and E. J. McGuire, *Fed. Proc.*, **27**, 345 (1968).

(500) B. Rossignol, G. Herman, and H. Clauser, *Biochem. Biophys. Res. Commun.*, **34**, 111 (1969).

(501) E. H. Eylar, *J. Theor. Biol.*, **10**, 89 (1966).

rule has been presented in a more refined form by Neuberger and Marshall,[501a] who proposed that the sequence Asn-X-Thr (or Ser) is necessary for the attachment of this type of carbohydrate chain. X can be any amino acid (hydroxy or non-hydroxy).

In this group, there seem, basically, to be two kinds of carbohydrate unit. One type contains only 2-acetamido-2-deoxy-D-glucose and D-mannose residues; the other type is more complex and contains, in addition to residues of those two sugars, those of D-galactose, N-acetyl- or N-glycolyl-neuraminic acid, and, occasionally, L-fucose. Some of these glycoproteins contain carbohydrate chains of both types. Jackson and Hirs[502] have suggested that the former type of carbohydrate side-chain occurs when the amino acid on the carboxyl end of the L-asparagine residue (which forms the attachment site for carbohydrate) has an apolar side-chain, and the latter, when it has a polar side-chain. In fact, porcine pancreatic ribonuclease contains both types of carbohydrate unit, and the amino acid sequences around these carbohydrate side-chains follow this rule.[502] It is possible that the transferases "recognize" the different amino acid sequences, or that the side chains occurring in a hydrophobic region are less accessible for certain sugar transferases.[502]

In both types of carbohydrate chain, the more proximal, "core" region seems to consist of several D-mannose and 2-acetamido-2-deoxy-D-glucose residues.[477a] In one type, additional D-mannose (and possibly 2-acetamido-2-deoxy-D-glucose) residues are present in the peripheral region, in the other, residues of such sugars as D-galactose, sialic acid, and L-fucose comprise this region.[477a] Indeed, in many cases in which detailed structural investigations have been performed, the peripheral regions of the more complex carbohydrate chains were found to have the following common sequence: N-acetyl-neuraminosyl–β-D-galactosyl–(2-acetamido-2-deoxy-β-D-glucosyl). Examples of glycoproteins having this type of carbohydrate chain include α_1-acid glycoprotein,[503] fetuin,[504] thyroglobulin,[505] and IgA immunoglobulin.[477]

At present, little is known about the biosynthesis of the D-mannose-(2-acetamido-2-deoxy-D-glucose) type of chain and of the "core"

(501a) A. Neuberger and R. D. Marshall, in "Symposium Foods: Carbohydrates and Their Roles," H. W. Schultz, R. F. Cain, and R. W. Wrolstad, eds., Avi, Westport, Conn., 1969.

(502) R. L. Jackson and C. H. W. Hirs, *J. Biol. Chem.*, **245**, 624 (1970).

(503) E. H. Eylar and R. W. Jeanloz, *J. Biol. Chem.*, **237**, 622 (1962).

(504) R. G. Spiro, *J. Biol. Chem.*, **237**, 646 (1962).

(505) M. J. Spiro and R. G. Spiro, *J. Biol. Chem.*, **243**, 6520 (1968).

portion of more complex chains, except that the incorporation of ^{14}C from GDP-D-mannose-^{14}C into an endogenous acceptor has been shown with particulate fractions from Ehrlich ascites tumor[506] and from sheep thyroid-gland.[506a] The latter preparation also has a 2-ace-tamido-2-deoxy-D-glucosyl transferase activity which, it has been suggested (on the basis of its distribution among subcellular fractions), participates in the biosynthesis of the "core" portion.[506a] (In the mold *Aspergillus niger*, enzymic transfer of the D-mannosyl group from its GDP ester to a D-mannosidase-treated glycoprotein has been demonstrated.[507] A similar D-mannosyl transfer reaction to an endogenous, glycoprotein acceptor has been described in *Aspergillus oryzae*.[508])

Our knowledge is, therefore, mostly limited to the facts concerning the biosynthesis of the peripheral trisaccharide sequence. Thus, treatment of this kind of glycoprotein with sialidase removes the sialic acid residues (which always occupy the nonreducing, terminal position), exposing the underlying D-galactose residues. These sialidase-treated glycoproteins then serve as good acceptors of sialic acid from CMP-sialic acid. The sialyl transferase catalyzing this reaction was purified from goat colostrum,[509,510] and was shown not to require any cations. Although desialylized glycoproteins are the most efficient acceptors, β-D-galactosides of low molecular weight were also active. The specificity studies showed that (1) the favored penultimate sugar was 2-acetamido-2-deoxy-D-glucose, and (2) among the linkages between the terminal D-galactose and the subterminal 2-acetamido-2-deoxy-D-glucose, β-D-(1 → 4)-linkage was markedly favored over the β-D-(1 → 3)- and β-D-(1 → 6)-linkages. The linkage synthesized was mostly N-acetylneuraminosyl–(2 → 6)-D-galactose. A very similar sialyl transferase has been purified from thyroid gland;[505] this enzyme transfers sialic acid to the terminal D-galactose residues of desialyzed glycoproteins of the third type, but not to 5-(D-galactosyloxy)-L-lysine from collagen or to β-D-galactosides of low molecular weight. A similar reaction was also demon-

(506) J. Molnar, H. Chao, and G. Markovic, *Arch. Biochem. Biophys.*, **134**, 533 (1969).
(506a) S. Bouchilloux, O. Chabaud, M. Michel-Bechet, M. Ferrand, and A. M. Athouël-Haon, *Biochem. Biophys. Res. Commun.*, **40**, 314 (1970).
(507) J. H. Pazur, D. L. Simpson, and H. R. Knull, *Biochem. Biophys. Res. Commun.*, **36**, 394 (1969).
(508) M. Richard, R. Letoublon, P. Louisot, and R. Got, *FEBS Lett.*, **6**, 80 (1970).
(509) B. Bartholomew, G. W. Jourdian, and S. Roseman, *Proc. 6th Intern. Congr. Biochem., New York*, Abstracts VI-503 (1964).
(510) B. A. Bartholomew and G. W. Jourdian, *Methods Enzymol.*, **8**, 368 (1966).

strated for a deoxycholate-insoluble fraction from rat-liver microsomes as the enzyme, with sialidase-treated ceruloplasmin as the acceptor.[511]

A particulate preparation from rat mammary-gland also transfers sialic acid to the (nonreducing) terminal β-D-galactosyl groups in various compounds.[512,513] This enzyme is quite different in its properties from the sialyl transferases already described. (a) It transfers sialic acid to O-3 of D-galactose, in contrast to the colostrum enzyme, which produces a (2 → 6)-linkage. (b) It is most active with smaller oligosaccharides, and acts only very slowly with desialyzed glycoproteins as acceptor. (c) It shows no particular "preference" toward the sugar to which β-D-galactose is attached. These results suggest that the rat, mammary-gland enzyme is probably not involved in the biosynthesis of the third type of glycoprotein carbohydrate unit, but its physiological function is not yet clear.

When the glycoproteins are successively treated with sialidase and β-D-galactosidase, there are produced acceptor proteins to which the D-galactosyl group can be transferred from UDP-D-galactose.[105] The enzyme catalyzing this reaction has also been purified from goat colostrum, and was shown[105] to require Mn^{2+}. The product of the reaction was active as the acceptor in the sialyl transferase reaction already mentioned, and partial hydrolysis of the product with acid produced O-β-D-galactosyl-(1 → 4)-2-acetamido-2-deoxy-D-glucose, indicating that the enzyme transferred β-D-galactosyl groups to O-4 of terminal 2-acetamido-2-deoxy-D-glucose residues of the acceptor. This enzyme does not appear to have a strict acceptor specificity, and even free 2-acetamido-2-deoxy-D-glucose acts as an acceptor. It is interesting that crude extracts exhibited much more activity toward desialyzed glycoprotein than toward free 2-acetamido-2-deoxy-D-glucose, whereas the purified preparation was about equally active toward these two acceptors. Although the enzyme could replace 85% of the D-galactose residues that had been removed from orosomucoid by β-D-galactosidase, this result seems to suggest the presence of still other D-galactosyl transferase(s) in the crude extract. A similar β-D-galactosyl transferase has been solubilized from membranes of Ehrlich ascites tumor-cells by the use of nonionic detergents.[514]

(511) J. Hickman, G. Ashwell, A. G. Morell, C. J. A. van den Hamer, and I. H. Scheinberg, J. Biol. Chem., 245, 759 (1970).
(512) G. W. Jourdian, D. M. Carlson, and S. Roseman, Biochem. Biophys. Res. Commun., 10, 352 (1963).
(513) D. M. Carlson and G. W. Jourdian, Methods Enzymol., 8, 358 (1966).
(514) J. F. Caccam and E. H. Eylar, Arch. Biochem. Biophys., 137, 315 (1970).

A similar D-galactosyl transferase has been purified about 1,000-fold from thyroid gland.[515] Its properties appear to be similar to those of the colostrum enzyme already described. As thyroid gland mostly synthesizes a single species of glycoprotein, namely, thyroglobulin, it seems to contain only one major transferase for each sugar residue, thus simplifying the task of identifying the transferase responsible for each step of glycoprotein synthesis.

One of the most interesting findings in this area has been made by Brew, Vanaman, and Hill,[97] who showed that the membrane-bound β-D-galactosyl transferase is probably identical with the A protein of lactose synthetase. Thus, this enzyme is usually involved in glycoprotein synthesis, but, if a large amount of α-lactalbumin is made in mammary gland in response to hormonal stimuli, α-lactalbumin alters its specificity, and the enzyme functions as lactose synthetase as long as α-lactalbumin is being produced. These results are discussed in greater detail in Section IV,3 (see p. 371).

If the desialyzed, β-D-galactosidase-treated glycoprotein is further treated with a *Clostridium perfingens* hexosaminidase, underlying D-mannose residues are exposed, and the product now accepts the 2-acetamido-2-deoxy-D-glucosyl group from its UDP ester. The transferase was purified from goat colostrum,[516] and found to require Mn^{2+}. The site of attachment of the transferred 2-acetamido-2-deoxy-D-glucosyl groups has not yet been elucidated; it is, however, assumed to be the (nonreducing) terminal D-mannosyl groups of the acceptor glycoprotein. With this enzyme, free D-mannose, the methyl D-mannopyranosides, and yeast mannan were inactive as acceptors.[516] Because ribonuclease A, to which 2-acetamido-2-deoxy-D-glucose-containing oligosaccharides can be attached *in vivo* (to form ribonuclease B), could not act as acceptor in this reaction,[516] this enzyme appears to be different from the hypothetical enzyme that transfers the 2-acetamido-2-deoxy-D-glucosyl group to the amide nitrogen atom of L-asparagine residues in the polypeptide chain.

A similar 2-acetamido-2-deoxy-D-glucosyl transferase has been demonstrated in the smooth-membrane fraction from cultured mouse-cells.[517]

Two L-fucosyl transferases presumably involved in glycoprotein

(515) M. J. Spiro and R. G. Spiro, *J. Biol. Chem.*, **243**, 6529 (1968).
(516) I. R. Johnson, E. J. McGuire, G. W. Jourdian, and S. Roseman, *J. Biol. Chem.*, **241**, 5735 (1966).
(517) H. B. Bosmann, *FEBS Lett.*, **8**, 29 (1970).

synthesis have been found in smooth, internal membranes from HeLa cells.[518] They are different in pH optima, temperature optima, and divalent-cation requirement. One of them transfers the L-fucosyl group from GDP-L-fucose to fetuin or to α_1-acid glycoprotein from which terminal sialic acid and D-galactose residues have been removed. Thus, the L-fucose is assumed to become linked to the terminal 2-acetamido-2-deoxy-β-D-glucose residue. The other enzyme requires porcine, submaxillary mucin[498] from which both the sialic acid and the L-fucose residues have been removed. Thus, this enzyme presumably transfers L-fucose to the terminal D-galactose residue, thereby restoring the original structure of the acceptor. Despite the fact that these enzymes synthesize the same linkages as do the L-fucosyl transferases found in human milk, the former enzymes cannot use milk oligosaccharides as acceptors.

If the transfer of sugars proceeds in a stepwise manner, and if the substrate and acceptor specificities of the transferases are the only factors that determine the sequence of the sugars, errors in and failures of transfer reactions may be expected, especially in the biosynthesis of complex carbohydrate chains. In fact, microheterogeneity is a pronounced characteristic of carbohydrate chains in glycoproteins. To cite just one example, IgA immunoglobulin contains six carbohydrate chains per molecule, but digestion of this protein produces 12 different glycopeptides.[477] Most of this heterogeneity seems to have been caused by the absence of some of the sugars that occupy nonreducing terminal positions, especially L-fucose and N-acetylneuraminic acid.[477]

(iv) **Glycosyl Transfer to Endogenous Acceptors.**—In some studies, endogenous acceptors present in the enzyme preparations were utilized. Thus, the transfer of sialic acid residues from CMP-sialic acid to endogenous glycoproteins has been demonstrated by the use of rat-liver microsomes;[519] incorporation of D-galactose, 2-acetamido-2-deoxy-D-glucose, and 2-acetamido-2-deoxy-D-galactose was shown in the presence of "postmicrosomal particulate" fraction from Ehrlich ascites carcinoma, ATP, and D-galactose-^{14}C or 2-amino-2-deoxy-D-glucose-^{14}C;[520] and transfer of D-galactosyl groups from UDP-D-galac-

(518) H. B. Bosmann, A. Hagopian, and E. H. Eylar, *Arch. Biochem. Biophys.*, **128**, 470 (1968).
(519) P. J. O'Brien, M. R. Canady, C. W. Hall, and E. F. Neufeld, *Biochim. Biophys. Acta,* **117**, 331 (1966).
(520) E. H. Eylar and G. M. W. Cook, *Proc. Nat. Acad. Sci. U.S.*, **54**, 1678 (1965).

tose to endogenous acceptors was shown by the use of a 1% deoxy-cholate-soluble fraction of rat-liver microsomes.[521]

Although information obtained by this type of study is limited, owing to uncertainities regarding the nature of endogenous acceptors, these studies can also provide significant results that cannot be obtained by experiments utilizing exogenous acceptors. For example, studies on the incorporation of N-acetylneuraminic acid and D-galactosyl groups from their CMP and UDP esters, with a particulate preparation from bovine retina,[522] showed that the K_m value for UDP-D-galactose (6.5 μM), for example, was much smaller than the value obtained[105] in the soluble system (250 μM). It was concluded[522] that this is, presumably, due to the efficient organization of enzyme and acceptor within the membrane. In this system, the authors[522] could also show that part of the N-acetylneuraminic acid transfer is dependent on the presence of UDP-D-galactose, and, also, that part of the D-galactosyl-^{14}C groups incorporated from UDP-galactose-^{14}C in the presence of CMP-N-acetylneuraminic acid cannot be removed by β-D-galactosidase unless the incorporation product is first treated with sialidase.[523] This result clearly establishes the sequential transfer of D-galactose, followed by N-acetylneuraminic acid, and illustrates the surprisingly large amount of information that can sometimes be gleaned by the use of endogenous acceptors.

(v) **Further Studies on the Sites of Biosynthesis.**—From studies of the enzymic transfer of monosaccharides, it is clear that the carbohydrate chains of glycoproteins are produced by the stepwise addition of sugars. Thus, especially in the glycoproteins of the third type (where carbohydrate chains can be very complex in structure), the addition of the "first" sugar residue, that of 2-acetamido-2-deoxy-β-D-glucose, might be thought to occur at sites different from those involved in the transfer of other peripheral sugar residues. This problem has been studied mostly by labeling the intact cells, followed by fractionation. In all cases examined, if a tissue or an organ is labeled by administration of 2-amino-2-deoxy-D-glucose-^{14}C less than 30 min before sacrifice, most of the labeled glycoproteins are found in association with "membranes" and only a small proportion, if any, is found associated with ribosomes.[478–482,506,524–533] However, some

(521) E. J. Sarcione and P. J. Carmody, *Biochem. Biophys. Res. Commun.*, **22,** 689 (1966).
(522) P. J. O'Brien and C. G. Muellenberg, *Biochim. Biophys. Acta,* **167,** 268 (1968).
(523) P. J. O'Brien and C. G. Muellenberg, *Biochim. Biophys. Acta,* **158,** 189 (1968).
(524) L. Helgeland, *Biochim. Biophys. Acta,* **101,** 106 (1965).

investigators assume that ribosome-bound, nascent polypeptides from liver really contain some 2-acetamido-2-deoxy-D-glucose residues, presumably corresponding to the innermost sugar residue linked to L-asparagine of the polypeptides.[524–528,532] A strong argument for this interpretation was presented by Molnar and Sy[526] and Lawford and Schaechter,[528] who have shown that treatment with puromycin *in vitro*, of the liver ribosomes labeled *in vivo* with 2-acetamido-2-deoxy-D-glucose-^{14}C, resulted in the release of ^{14}C-labeled peptides from ribosomes.

Conflicting results have, however, been reported by workers who used other tissues. Thus, in thyroid[529,530] and in Ehrlich ascites tumor-cells,[481] ribosomes are claimed to be free of ^{14}C from 2-amino-2-deoxy-D-glucose-^{14}C, whereas some incorporation was reported in intestinal mucosa.[534] The situation is also controversial in a system extensively studied in several laboratories, namely, cells secreting immunoglobulin. The presence of 2-amino-2-deoxy-D-glucose-^{14}C in nascent polypeptides has been reported in such a system,[535] but Swenson and Kern[536] pointed out that most of the incorporation product obtained by supplying 2-amino-2-deoxy-D-glucose-^{14}C to such cells was not an immunoglobulin. When immunoglobulin was specifically precipitated by antisera, these authors found that only 5% of the precipitable radioactivity was located in the microsome fraction;[536] they further argued that, since about 25% of the 2-acetamido-2-deoxy-D-glucose residues in the immunoglobulin corresponded to the residues linked to L-asparagine, 25% (not 5%) of the incorporated activity should be found in polysome-microsomes if

(525) J. Molnar, G. B. Robinson, and R. J. Winzler, *J. Biol. Chem.*, **240**, 1882 (1965).

(526) J. Molnar and D. Sy, *Biochemistry*, **6**, 1941 (1967).

(527) J. L. Simkin and J. C. Jamieson, *Biochem. J.*, **103**, 153 (1967).

(528) G. R. Lawford and H. Schachter, *J. Biol. Chem.*, **241**, 5408 (1966).

(529) S. Bouchilloux and C. Cheftel, *Biochem. Biophys. Res. Commun.*, **23**, 305 (1966).

(530) C. Cheftel and S. Bouchilloux, *Biochim. Biophys. Acta*, **170**, 15 (1968).

(531) C. Cheftel, S. Bouchilloux, and O. Chabaud, *Biochim. Biophys. Acta*, **170**, 29 (1968).

(532) T. Hallinan, C. N. Murty, and J. H. Grant, *Arch. Biochem. Biophys.*, **125**, 715 (1968).

(533) H. B. Bosmann, A. Hagopian, and E. H. Eylar, *Arch. Biochem. Biophys.*, **130**, 573 (1969).

(534) P. Louisot, J. Frot-Coutaz, G. Bertagnolio, R. Got, and L. Colobert, *Biochem. Biophys. Res. Commun.*, **28**, 385 (1967).

(535) F. Melchers and P. M. Knopf, *Cold Spring Harbor Symp. Quant. Biol.*, **32**, 255 (1967).

(536) R. M. Swenson and M. Kern, *Proc. Nat. Acad. Sci. U.S.*, **59**, 546 (1968).

these residues are added onto nascent polypeptides.[536] However, it is possible that the size of the pool of the nascent polypeptides is much smaller than that of the immature immunoglobulin already released from polysomes, and Swenson and Kern's argument does not appear to be valid. In studies from other laboratories, some incorporation of label from 2-amino-2-deoxy-D-glucose-^{14}C into rough, endoplasmic reticulum[537] and polysome-associated, nascent polypeptide[538,539] has been reported, and autoradiographic studies with immunoglobulin-secreting cells[540] showed that radioactive 2-amino-2-deoxy-D-glucose is incorporated both at rough, endoplasmic reticulum and at the Golgi apparatus. In some of these studies,[537,538] the radioactive product was identified as immunoglobulin. In any event, it is difficult to assess the quantitative significance of the incorporation observed, because it has not been possible to determine the specific radioactivity of intracellular precursors in a given compartment within the cell.

The incompletely glycosylated intermediates are assumed to travel along the internal membrane systems, at the same time acquiring more and more sugar residues.[482,528] It has, indeed, been proposed by Eylar[501] that secreted proteins are glycosylated, so that immature forms of such proteins can be recognized and segregated into a compartment in which the sugars serve as markers that facilitate the orderly transportation of these proteins toward the cell suface. Such a process is strikingly exhibited in the case of immunoglobulin. Thus, there is a time lag of about 20 minutes between the transfer of the major portion of the carbohydrate chains to the synthesized polypeptide and the final excretion of immunoglobulin into the medium,[535,536] but there is no lag between the acquisition of sialic acid residues by intracellular immunoglobulin and its secretion.[536] This clearly shows that sialic acid is added to the protein just before it is secreted. Swenson and Kern[541] also showed that, after their release from ribosomes, the immunoglobulin polypeptides seem to be located exclusively in the cisternae of the smooth, endoplasmic reticulum. In contrast, when the polypeptides have acquired residues of 2-acetamido-2-deoxy-D-glucose, D-mannose, and D-galactose, they are found in the "soluble" fraction.[542] It seems likely that these glyco-

(537) J. W. Uhr and I. Schenkein, Proc. Nat. Acad. Sci. U.S., **66**, 952 (1970).
(538) C. J. Sherr and J. W. Uhr, Proc. Nat. Acad. Sci. U.S., **64**, 381 (1969).
(539) N. J. Cowan and G. B. Robinson, FEBS Lett., **8**, 6 (1970).
(540) D. Zagury, J. W. Uhr, J. D. Jamieson, and G. E. Palade, J. Cell Biol., **46**, 52 (1970).
(541) R. M. Swenson and M. Kern, Proc. Nat. Acad. Sci., **57**, 417 (1967).
(542) R. M. Swenson and M. Kern, J. Biol. Chem., **242**, 3242 (1967).

protein molecules are actually situated in some fragile organelles, such as Golgi (see later), and are released into the soluble fraction during the fractionation of subcellular components (see also, Ref. 537). In any event, the movement of immunoglobulin precursors from ribosomes to smooth, endoplasmic reticulum to another compartment, as they acquire more and more carbohydrate, was clearly shown by these studies.

The hypothesis of the slow, orderly acquisition of carbohydrate units by glycoprotein precursors is also supported by the isolation of incompletely glycosylated, precursor proteins from membrane fractions[527,543] and by the presence[519-523,544] of membrane-associated, endogenous "acceptors."

There exists a great deal of data as to the place at which the synthesis of the peripheral portion of glycoprotein carbohydrate units takes place, and it is not possible to go into them in detail here. One idea gaining favor is the participation of the Golgi apparatus; this was first suggested by the results of autoradiographic studies[545] on the biosynthesis of carbohydrate–protein complexes of "mucopolysaccharide" type. Autoradiography could simply be indicating sites of concentration, instead of sites of synthesis, but *in vitro* studies using endogenous acceptors have shown that the transfer of the 2-acetamido-2-deoxy-D-glucosyl group from its UDP ester[546] occurs in a Golgi-rich fraction in liver.[547] Careful studies with exogenous acceptors also established that all three enzymes involved in the biosynthesis of the N-acetylneuraminosyl-β-D-galactosyl-(2-acetamido-2-deoxy-β-D-glucose) trisaccharide sequence of glycoproteins are localized in the same membrane-fraction, identified as Golgi, in liver.[548] It was found that UDP-D-galactose:2-acetamido-2-deoxy-D-glucose-β-D-galactosyl transferase, probably identical with the β-D-galactosyl transferase involved in the synthesis of the trisaccharide sequence mentioned, is a good marker for the purification of Golgi apparatus from liver;[549] its specific activity in the isolated Golgi

(543) Y.-T. Li, S.-C. Li, and M. R. Shetlar, *J. Biol. Chem.*, **243**, 656 (1968).

(544) R. P. D'Amico and M. Kern, *J. Biol. Chem.*, **243**, 3425 (1968).

(545) M. Peterson and C. P. Leblond, *J. Cell Biol.*, **21**, 143 (1964); M. Neutra and C. P. Leblond, *ibid.*, **30**, 137 (1966).

(546) R. R. Wagner and M. A. Cynkin, *Arch. Biochem. Biophys.*, **129**, 242 (1969).

(547) R. R. Wagner and M. A. Cynkin, *Biochem. Biophys. Res. Commun.*, **35**, 139 (1969).

(548) H. Schachter, I. Jabbal, R. L. Hudgin, L. Pinteric, E. J. McGuire, and S. Roseman, *J. Biol. Chem.*, **245**, 1090 (1970).

(549) B. Fleischer, S. Fleischer, and H. Ozawa, *J. Cell Biol.*, **43**, 59 (1969).

fraction was 40 times that in the crude homogenate.[549] Autoradiographic studies with immunoglobulin-secreting cells also indicated that D-galactose is incorporated into protein(s) almost exclusively in Golgi apparatus.[540] When sections of animal cells were stained by a method that shows the location of carbohydrates under the electron microscope, Golgi apparatus from both secretory and non-secretory cells were found to contain carbohydrates.[549a] Furthermore, it was found that the last saccule on one side of the Golgi stack is stained strongly, and the last saccule on the other side, not at all, and that a clear gradient of reactivity occurs between these two saccules.[549a] These results suggest that more and more carbohydrate is attached to the polypeptide chain as the glycoproteins move toward one face of the Golgi stack.

These results should, however, not be taken to mean that all glycosylation reactions in glycoprotein synthesis take place at Golgi apparatus. There are two limitations on the experiments mentioned. (a) The tissue studied was mostly involved in the biosynthesis of secreted glycoproteins. (b) The enzymes studied were involved only in the terminal stage of glycoprotein biosynthesis. As regards the first point, there are some indications that the site of glycosylation of membrane-glycoprotein components may be different from that of secreted glycoproteins, both from *in vivo* incorporation experiments[506,533] and enzyme-localization studies.[489,495] Eylar and co-workers[489,495] apparently suppose that glycosyl transferases involved in the synthesis of secreted proteins are located in "plasma membranes," in contrast to those for the synthesis of membrane components. The identification of "plasma membranes" might be debatable, but the data seem to show that different glycosyl transferases are situated in different membranes in the same cell.[489] [Nevertheless, it is still possible that the differences in distribution observed are artifacts caused by (a) the presence of enzymes degrading either the substrate or the product of the reaction, or (b) the differences in the accessibility of membrane-associated enzymes to exogenous, acceptor molecules.]

Concerning the second point, it has already been mentioned that the most proximal 2-acetamido-2-deoxy-D-glucose residue of the carbohydrate chains of the third type seems to become attached at a site different from that for other sugars. Incubation of thyroid lobes *in vitro* with radioactive D-mannose or D-galactose showed that

(549a) A. Rambourg, W. Hernandez, and C. P. Leblond, *J. Cell Biol.*, **40**, 395 (1969).

D-mannose residues are transferred before the thyroglobulin sub-units are aggregated, whereas D-galactose residues are transferred onto already aggregated 17–18 S thyroglobulin.[550] Autoradiography of sections of thyroid lobes incubated under these conditions showed that D-mannose-t is first incorporated at rough, endoplasmic reticulum and then moves into Golgi, whereas D-galactose-t is incorporated exclusively at Golgi.[550a] These results thus strongly support the hypothesis that the biosynthesis of the "core" region of the car-bohydrate units takes place before glycoproteins reach Golgi ap-paratus. It should also be noted that *in vivo* incorporation data all seem to indicate that the attachment site of N-acetylneuraminic acid is different from that of more proximal 2-acetamido-2-deoxy-D-glu-cose.[525,528] Glycosylation of submaxillary mucins, also, has been reported to take place in non-Golgi fractions.[495,531]

(vi) **Investigations on Possible Lipid Intermediates.**—Some in-vestigators have made observations that may suggest the presence of lipid intermediates in the biosynthesis of animal glycopro-teins.[148,551,551a] Smooth-membrane fraction from rabbit liver transfers the D-mannosyl group of GDP-D-mannose onto an endogenous lipid; the behavior of synthesized "D-mannose–lipid" on column chromatography or alkaline and acid hydrolysis[551] was reminiscent of the behavior of polyisoprenyl (D-mannosyl phosphate) isolated from *Micrococcus lysodeikticus*,[211] although side-by-side comparison has not been made.

Microsome fractions from mouse brain were shown to incorporate D-mannosyl groups from GDP-D-mannose into a product extractable with organic solvents.[551a] The D-mannosyl group in the product is readily cleaved by acid, but not by alkali. It was hypothesized[551] that this could be a lipid-bound intermediate involved in the biosynthesis of a complex saccharide.

The system described by Behrens and Leloir[148] in liver has been much better characterized. Here, D-glucosyl transfer from UDP-D-glucose produces dolichol (β-D-glucosyl phosphate), and then the D-glucosyl groups are transferred to an endogenous protein. How-ever, except for collagens, glycoproteins containing D-glucose res-

(550) A. Herscovics, *Biochem. J.*, **112**, 709 (1969).
(550a) P. Whur, A. Herscovics, and C. P. Leblond, *J. Cell Biol.*, **43**, 289 (1969).
(551) J. F. Caccam, J. J. Jackson, and E. H. Eylar, *Biochem. Biophys. Res. Commun.*, **35**, 505 (1969).
(551a) M. Zatz and S. H. Barondes, *Biochem. Biophys. Res. Commun.*, **36**, 511 (1969).

idues are not known, and the protein synthesized in this system does not appear to be a collagen.[148] These reactions have been discussed in Section V,1 (see p. 381).

4. Blood-group Substances

The biosynthesis of substances that determine ABO and Lewis blood-group specificities has been elucidated in detail. These substances occur as protein–polysaccharide complexes in secretions,[552] as glycosphingolipids on the surface of cells,[431,432,553] and as oligosaccharides in human milk.[554] Although the precise structure has been worked out only for glycoproteins[552] and oligosaccharides,[554] the immunological specificity suggests that all three groups of substances have the same, or at least a similar, structure at the nonreducing end of the carbohydrate chain. Blood-group glycolipids from erythrocytes have now been highly purified, and their sugar composition appears to support this assumption[432] (see also, Refs. 431 and 555). The pathways for the biosynthesis of blood-group substances are shown in Scheme 6.

A 2-acetamido-2-deoxy-D-galactosyl transferase activity was shown to be present in hog gastric-mucosa. This enzyme converts the glycoprotein having H specificity into an A substance[556,557] (see reaction IV, Scheme 6); O erythrocytes are also converted by the same enzyme into those having blood-group A activity.[558] Similarly, "microsome" fractions from type H hog gastric-mucosa were found to transfer the L-fucosyl group from GDP-L-fucose to an endogenous acceptor.[559] The incorporation product was precipitated by an eel serum having anti-H activity; therefore the incorporation presumably corresponds to the addition of a 2-O-α-L-fucosyl group (L-Fuc$_{II}$ of Scheme 6) to the terminal β-D-galactosyl group of the endogenous acceptor (reaction III, Scheme 6).

Human milk contains both the glycosyl transferases and acceptor

(552) W. M. Watkins, Science, 152, 172 (1966).
(552a) K. O. Lloyd and E. A. Kabat, Proc. Nat. Acad. Sci. U. S., 61, 1470 (1968).
(553) T. Yamakawa, R. Irie, and M. Iwanaga, J. Biochem. (Tokyo), 48, 490 (1960).
(554) R. Kuhn, H. H. Baer, and A. Gauhe, Ann., 611, 242 (1958).
(555) S. Hakomori, J. Koscielak, K. J. Bloch, and R. W. Jeanloz, J. Immunol., 98, 31 (1967).
(556) H. Tuppy and W. L. Staudenbauer, Nature, 210, 316 (1966).
(557) H. Tuppy and H. Schenkel-Brunner, Eur. J. Biochem., 10, 152 (1969).
(558) H. Schenkel-Brunner and H. Tuppy, Nature, 223, 1272 (1969).
(559) A. P. Grollman and D. M. Marcus, Biochem. Biophys. Res. Commun., 25, 542 (1966).

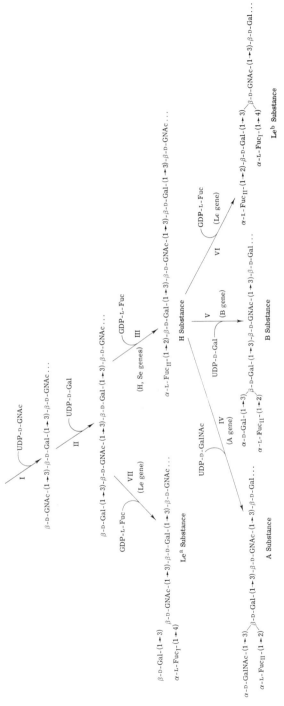

Scheme 6. Biosynthesis of Blood-group Substances. (The structure of the proximal portion is shown in a simplified form. According to Lloyd and Kabat,[552a] two chains are linked to O-3 and O-6 of the β-D-galactose residue comprising the nonreducing, terminal residue in the sequence at the top of the Scheme. Only the chains linked to O-3 are shown here.)

oligosaccharides needed for the synthesis of blood-group substance, thus representing an ideal system for *in vitro* studies. The work by Ginsburg and coworkers[560] has so far produced the following information. (1) 2-α-L-Fucosyl transferase transfers an L-fucosyl group to O-2 of a β-D-galactopyranosyl group[560] (reaction III, Scheme 6). Lactose is a good acceptor, but methyl β-D-galactopyranoside is not. Apparently, lactose acts as an analog for the "lactosamine" moiety in the ABO blood-group substances (see Scheme 6), and this enzyme catalyzes the transfer of the L-fucosyl group symbolized as L-Fuc$_{II}$ in Scheme 6. This conclusion is supported by the finding that the enzyme is present in "secretors," but not in "nonsecretors" (who fail to transfer this L-fucosyl group to blood-group glycoproteins[560]). (2) 4-α-L-fucosyl transferase transfers the L-fucosyl group to O-4 of the 2-acetamido-2-deoxy-β-D-glucosyl group.[560] Only those oligosaccharides having the structure β-D-Gal-(1 $\rightarrow$ 3)-β-D-GNAc-(1 $\rightarrow$ 3)-β-D-Gal-(1 $\rightarrow$ 4)-D-G appear to have been tested, and the transfer of the α-L-fucosyl group took place even when the (nonreducing) D-galactosyl group was already substituted by α-L-fucosyl at O-2 ("Lacto-N-fucopentaose I," see Ref. 505). That this is the enzyme responsible for the transfer of L-Fuc$_I$ in Scheme 6 (reactions VI and VII) was also shown by the fact that Le^{a-b-} women do not have this activity in their milk.[561] Very similar results on this enzyme have been obtained by other workers.[561a] (3) 2-Acetamido-2-deoxy-α-D-galactosyl transferase transfers the (nonreducing) terminal 2-acetamido-2-deoxy-α-D-galactosyl group (see Scheme 6, reaction IV).[562,563] (4) α-D-Galactosyl transferase transfers the (nonreducing) terminal D-galactosyl group (see Scheme 6, reaction V).[564] Milk from AB donors contains both of these enzymes, that from A donors contains only 2-acetamido-2-deoxy-α-D-galactosyl transferase, and that from B donors, only α-D-galactosyl transferases. Many oligosaccharides were tested as acceptors, and only those containing the 2-O-α-L-fucosyl-β-D-galactose structure were active. Furthermore, when a second L-fucosyl group was present on the 2-acetamido-2-deoxy-D-glucosyl group (L-Fuc$_I$, Scheme 6), no transfer was observed.

(560) L. Shen, E. F. Grollman, and V. Ginsburg, *Proc. Nat. Acad. Sci. U.S.*, **59**, 224 (1968).

(561) E. F. Grollman, A. Kobata, and V. Ginsburg, *Fed. Proc.*, **27**, 345 (1968).

(561a) Z. Jarkovsky, D. M. Marcus, and A. P. Grollman, *Biochemistry*, **9**, 1123 (1970).

(562) A. Kobata, E. F. Grollman, and V. Ginsburg, *Arch. Biochem. Biophys.*, **124**, 609 (1968).

(563) A. Kobata and V. Ginsburg, *J. Biol. Chem.*, **245**, 1484 (1970).

(564) A. Kobata, E. F. Grollman, and V. Ginsburg, *Biochem. Biophys. Res. Commun.*, **32**, 272 (1968).

Similar results were obtained by using oligosaccharide acceptors and particulate enzymes from submaxillary glands and stomach mucosal-linings. Activities demonstrated were 2-acetamido-2-deoxy-α-D-galactosyl transferase, determined by A gene (reaction IV);[565,566] α-D-galactosyl transferase, determined by B gene (reaction V);[565,567,568] 2-α-L-fucosyl transferase, found only in "sectors" (reaction III);[569] 4-α-L-fucosyl transferase, presumably determined by the Le gene (reactions VI and VII);[566] β-D-galactosyl transferase, which uses 2-acetamido-2-deoxy-D-glucose as acceptor (probably reaction II);[567] and 2-acetamido-2-deoxy-β-D-glucosyl transferase, which uses O-β-D-galactosyl-(1 → 4)-2-acetamido-2-deoxy-D-glucose as acceptor (probably reaction I).[567] The last two enzymes may be involved in the biosynthesis of the common region of blood-group oligosaccharides. Furthermore, when O-β-D-galactosyl-(1 → 4)-2-acetamido-2-deoxy-D-glucose was used as the acceptor, α-L-fucosyl groups were also transferred to O-3 of 2-acetamido-2-deoxy-D-glucose.[569] It is known that some carbohydrate chains of blood-group substances, which contain a β-D-galactosyl group linked (1 → 4) to 2-acetamido-2-deoxy-D-glucose, also contain an α-L-fucosyl group linked (1 → 3) to the latter sugar.[570]

5. D-Glucosyl 2'-Deoxyribonucleate

The 2'-deoxyribonucleic acids (DNAs) of T2, T4, and T6 bacteriophages are D-glucosylated at the hydroxyl groups of the 5-(hydroxymethyl)cytosine residues. The phage genomes code for different D-glucosyl transferases, all of which utilize UDP-D-glucose, and the proportion of different substituent groups (α-D-glucosyl, β-D-glucosyl, and α-gentiobiosyl) in the DNA of each phage is largely accounted for by the nature of reactions catalyzed by these D-glucosyl transferases (see Ref. 1). There remained, however, two outstanding questions regarding these D-glucosylation reactions. (1) Although the equilibrium constant of the reaction favors an almost complete D-glucosylation of all available 5-(hydroxymethyl)cytosine residues, in T2 DNA only 75% of them are D-glucosylated.[571] (2) Despite

(565) C. Race, D. Ziderman, and W. M. Watkins, *Biochem. J.*, **107**, 733 (1968).

(566) V. M. Hearn, Z. G. Smith, and W. M. Watkins, *Biochem. J.*, **109**, 315 (1968).

(567) D. Ziderman, S. Gompertz, Z. G. Smith, and W. M. Watkins, *Biochem. Biophys. Res. Commun.*, **29**, 56 (1967).

(568) C. Race and W. M. Watkins, *Biochem. J.*, **114**, 86P (1969).

(569) M. A. Chester and W. M. Watkins, *Biochem. Biophys. Res. Commun.*, **34**, 835 (1969).

(570) K. O. Lloyd, S. Beychok, and E. A. Kabat, *Biochemistry*, **6**, 1448 (1967).

(571) I. R. Lehman and E. A. Pratt, *J. Biol. Chem.*, **235**, 3254 (1960).

the fact that the β-D-glucosyl transferase of T4-infected cells can D-glucosylate almost all of the 5-(hydroxymethyl)cytosine residues in a cell-free system,[572] only 30% of the available sites are substituted[571] by β-D-glucosyl groups in T4 DNA. Investigations have disclosed that quite different mechanisms operate in these two cases; they illustrate quite well the different ways by which the D-glucosyl transfer reaction can be restricted.

For T2 DNA, the degradation studies showed that certain nucleotide sequences decrease the ability of 5-(hydroxymethyl)cytosine residues to accept α-D-glucosyl groups:[573] (a) when a 5-(hydroxymethyl)cytidine residue is linked through O-5' to another 5-(hydroxymethyl)cytidine 3'-phosphate residue, the former residue can never accept an α-D-glucosyl group;[573] and (b) and when a 5-(hydroxymethyl)cytidine residue is linked through O-3' to a purine nucleoside 5'-phosphate, the ability to accept an α-D-glucosyl group is decreased to a certain extent.[573] The first type of restriction also exists for T4 and T6 α-D-glucosyl transferases, but the second type only seems to apply with T2 α-transferase.[574] These examples, then, illustrate the restraint imposed by the acceptor specificity.

However, the kind of restriction just described cannot explain the fact that β-D-glucosyl groups occupy only 30% of the available sites in T4 DNA. Detailed analysis has shown that, although T4 α-transferase cannot transfer a D-glucosyl group to a 5-(hydroxymethyl)-cytidine residue having another 5-(hydroxymethyl)cytidine group on its "5' side", β-transferase can transfer D-glucosyl groups to all 5-(hydroxymethyl)cytosine residues, no matter what the neighboring bases may be.[575] The smaller number of β-D-glucosyl groups transferred, then, is most easily explained by a competition between α- and β-transferases for acceptor sites.[575] However, the possibility still cannot be excluded that portions of DNA are differently accessible to α- and β-transferases.[572]

6. Teichoic Acids

The teichoic acids are a group of polymers present chiefly in the cell walls of a number of Gram-positive bacteria.[576] They were dis-

(572) J. Josse and A. Kornberg, J. Biol. Chem., **237**, 1968 (1962).

(573) M. R. Lunt, J.-C. Siebke, and K. Burton, Biochem. J., **92**, 27 (1964).

(574) M. R. Lunt and E. A. Newton, Biochem. J., **95**, 717 (1965); A. de Waard, Biochim. Biophys. Acta, **92**, 286 (1964).

(575) A. de Waard, T. E. C. M. Ubbink, and W. Beckman, Eur. J. Biochem., **2**, 303 (1967).

(576) A. R. Archibald and J. Baddiley, Advan. Carbohyd. Chem., **21**, 323 (1966); A. R. Archibald, J. Baddiley, and N. L. Blumsom, Advan. Enzymol., **30**, 223 (1968).

covered by Baddiley and coworkers[577] in 1958. Two types of these acids are known, one containing chains of glycerol phosphate, and another, chains of ribitol phosphate, with a variety of sugars and amino acids attached to them. The precursors of the polymeric alcohol phosphates are considered to be CDP-D-glycerol and CDP-L-ribitol, previously isolated from *Lactobacillus arabinosus*.[578,579] Some of the cell walls of the Gram-positive bacteria contain up to 50% of teichoic acid, on the basis of dry weight.

The glycerol or ribitol residues are usually joined through phosphoric diester linkages to form a chain, and sugars and D-alanine ester residues are attached to hydroxyl groups of the alditols. Considerable variations occur in the nature and number of glycosyl residues, but only residues of D-glucose, 2-acetamido-2-deoxy-D-glucose, D-galactose, and, possibly, D-mannose have been found in these compounds.[576,580] Both α- and β-D-glycosidic linkages and mono-, di-, or tri-saccharide units may occur. The proportions of the sugar residues, as well as the ratio of α to β linkages, have been found to differ in teichoic acids isolated from different micro-organisms or from different strains of the same micro-organism.

Glycerol teichoic acids are more widely distributed than the ribitol compounds. Some of them appear to be located mostly in the region between the wall and the underlying protoplast membrane.[581] For detailed structures of the various teichoic acids, the reader should consult reviews by Baddiley and coworkers[576] and Baddiley.[580]

Glaser and Burger[582,583] studied the biosynthesis of teichoic acids with particulate preparations from *Bacillus licheniformis* (ATCC 9945) and *B. subtilis* (NCTC 3610). They found that these preparations catalyze the synthesis of poly(glycerol phosphate) according to the reaction

$$n \text{ (CDP-glycerol)} \longrightarrow \text{(glycerol phosphate)}_n + n \text{ CMP.}$$

The synthetic material, like natural teichoic acid, is a $(1 \rightarrow 3)$-linked polymer of glycerol phosphate. A similar preparation from *Lactobacillus plantarum* (ATCC 8014) catalyzes the synthesis of poly(ribitol

(577) J. J. Armstrong, J. Baddiley, J. G. Buchanan, B. Carss, and G. R. Greenberg, *J. Chem. Soc.*, 4344 (1958).

(578) J. Baddiley, J. G. Buchanan, B. Carss, and A. P. Mathias, *J. Chem. Soc.*, 4583 (1956).

(579) J. Baddiley and A. P. Mathias, *J. Chem. Soc.*, 2723 (1954).

(580) J. Baddiley, in "Current Biochemical Energetics," N. D. Kaplan and E. P. Kennedy, eds., Academic Press, New York, N. Y., 1966, p. 371.

(581) J. B. Hay, A. J. Wicken, and J. Baddiley, *Biochim. Biophys. Acta*, **71**, 188 (1963).

(582) M. M. Burger, and L. Glaser, *Biochim. Biophys. Acta*, **64**, 575 (1962).

(583) M. M. Burger and L. Glaser, *J. Biol. Chem.*, **239**, 3168 (1964); L. Glaser, *ibid*, **239**, 3178 (1964).

phosphate) from CDP-L-ribitol. Both enzymes show an absolute requirement for Ca^{2+} and Mg^{2+}.

Ishimoto and Strominger[584] also reported the presence in *Staphylococcus aureus* of a poly(ribitol phosphate) synthetase that catalyzes the transfer of the ribitol phosphate group from CDP-L-ribitol to an acceptor by the following reaction.

$$n \text{ (CDP-L-ribitol)} + \text{acceptor} \longrightarrow \text{acceptor--(P-L-ribitol)}_n + n \text{ CMP}$$

The poly(ribitol phosphate) and poly(glycerol phosphate) synthesized can further be glycosylated, as was first shown by Nathenson and Strominger[585] and Burger.[586] Both systems are particulate in nature. The former system, from *Staphylococcus aureus* (Copenhagen), transfers the 2-acetamido-2-deoxy-D-glucosyl group from its UDP ester to ribitol residues of exogenous poly(ribitol phosphate) in which some of the ribitol residues already bear 2-acetamido-2-deoxy-α-D-glucosyl substituents.[585] The parent organism contains this sugar linked through both α- and β-D-linkages, and the cell-free system also links about 70% of the 2-acetamido-2-deoxy-D-glucosyl groups through a β-D-linkage, and the rest through an α-D-linkage. When the poly(ribitol phosphate) simultaneously being generated *in situ* from CDP-ribitol was compared with the poly(ribitol phosphate) synthesized *in situ* in the absence of UDP-2-acetamido-2-deoxy-D-glucose, the transfer of the sugar to the former acceptor was found to be the more efficient. This discovery suggested that ribitol phosphate units are normally glycosylated soon after their addition to the growing chain, presumably owing to the close spatial association of enzymes within the membrane.[585] Burger's system,[586] from *Bacillus subtilis*, transfers the D-glucosyl group from UDP-D-glucose to poly(glycerol phosphate), generated either *in situ* from CDP-glycerol, or added as a polymer after purification.[586]

Many teichoic acids are serologically active and often constitute the group-specific components of bacteria.[587–589] By the use of specific antibodies, it was established that *B. subtilis* W-23 contains two teichoic acids in the cell wall; both are ribitol teichoic acids, but

(584) N. Ishimoto and J. L. Strominger, *J. Biol. Chem.*, **241**, 639 (1966).

(585) S. G. Nathenson and J. L. Strominger, *J. Biol. Chem.*, **237**, PC 3839 (1962); **238**, 3161 (1963).

(586) M. M. Burger, *Biochim. Biophys. Acta*, **71**, 495 (1963).

(587) M. McCarty, *J. Exp. Med.*, **109**, 361 (1959); W. Juergens, A. Sanderson, and J. L. Strominger, *Bull. Soc. Chim. Biol.*, **42**, 1669 (1961).

(588) J. Baddiley and A. L. Davison, *J. Gen. Microbiol.*, **24**, 295 (1961).

(589) A. J. Wicken, S. D. Elliott, and J. Baddiley, *J. Gen. Microbiol.*, **31**, 231 (1963).

one type carries no glycosyl substituents, whereas, in the other type, each D-ribitol residue bears one D-glucosyl substituent.[590] In a cell-free system, D-glucosyl groups are transferred to poly(ribitol phosphate), and the D-glucose-free teichoic acid that occurs naturally in the cell wall was found to be a good acceptor.[591] Thus, in intact cells, there must be a complete compartmentalization of the systems that synthesize the two teichoic acids, so that one kind of teichoic acid can emerge without being D-glucosylated at all. In other cases where two cell-wall teichoic acids are formed, similar metabolic segregation also seems to exist. Thus, in *Staphylococcus aureus* Copenhagen, any ribitol teichoic acid chain carries 2-acetamido-2-deoxy-D-glucose side-chains, either all as α-D anomers, or all as β-D anomers.[592] In *Bacillus licheniformis* (mentioned earlier), a single chain is fully substituted either by D-glucosyl or by D-galactosyl groups.[590] Separation of two types of "membrane," each of which is specific for the synthesis of one kind of teichoic acid, has been reported by Burger,[593] who used *B. licheniformis*.

An increasing number of polymers have been found that do not possess the structure of the "classical" teichoic acids, and yet are structurally related to them. Many of these "teichoic acids" contain reducing sugars as a part of the main chain, and utilize glycosidic linkages of these sugars in building up the long, chain-like polymers. Burger and Glaser[594] were the first to demonstrate the *in vitro* biosynthesis of this class of compound, using particulate fractions from *B. licheniformis*. The reactions found are as follows.

$$n \text{ UDP-D-Gal} + n \text{ CDP-D-glycerol} \longrightarrow (3\text{-}O\text{-}\beta\text{-D-galactopyranosyl-}$$
$$\text{D-glycerol 1-phosphate})_n + n \text{ UDP} + n \text{ CMP}$$

$$n \text{ UDP-D-G} + n \text{ CDP-D-glycerol} \longrightarrow (3\text{-}O\text{-}\alpha\text{-D-glucopyranosyl-}$$
$$\text{D-glycerol 1-phosphate})_n + n \text{ UDP} + n \text{ CMP}$$

In the polymers synthesized, the repeating units are joined together by phosphoric diester linkages between C-1 of D-glycerol and C-6 of the hexose. For these reactions, there is a question as to whether the repeating unit is preassembled and then polymerized, or whether each component of the chain is added in a strictly sequential way. Burger and Glaser[594] could find no evidence for the presence of a nucleo-

(590) M. M. Burger, *Proc. Nat. Acad. Sci. U. S.*, **56**, 910 (1966).
(591) T. Chin, M. M. Burger, and L. Glaser, *Arch. Biochem. Biophys.*, **116**, 358 (1966).
(592) M. Torii, E. A. Kabat, and A. E. Bezer, *J. Exp. Med.*, **120**, 13 (1964).
(593) M. M. Burger, *Fed. Proc.*, **27**, 293 (1968).
(594) M. M. Burger and L. Glaser, *J. Biol. Chem.*, **241**, 494 (1966).

tide-linked intermediate carrying the entire repeating unit. Further-more, pre-incubation of the particulate enzyme with CDP-D-glyc-erol-^{12}C alone increased the subsequent incorporation of hexosyl-^{14}C from UDP-D-glucose-^{14}C or UDP-D-galactose-^{14}C, even when, be-tween the two incubations, the particulate enzyme had been washed free from contaminating nucleotides. The simplest interpretation is that, during the first incubation, D-glycerol 1-phosphate units are added to the terminal position of the growing chains, thus increas-ing the number of glycerol terminal units, and that, during the second incubation, hexosyl-^{14}C groups are added to such terminal units. These results therefore favor the sequential-transfer hypothesis, but, as Burger and Glaser[594] pointed out, they do not exclude the pos-sibility of prior assembly of repeating units on a membrane-associated carrier. Such carriers have since then been found in connection with lipopolysaccharide and peptidoglycan synthesis, and "lipid inter-mediate" has also been observed as a precursor in the biosynthesis of another "teichoic acid," described next.

Another type of "teichoic acid," which contains a hexopyranosyl phosphate as a part of the main chain, has been characterized by Baddiley and coworkers.[594a] The polymer from *Staphylococcus lactis* I3 has the following structure.

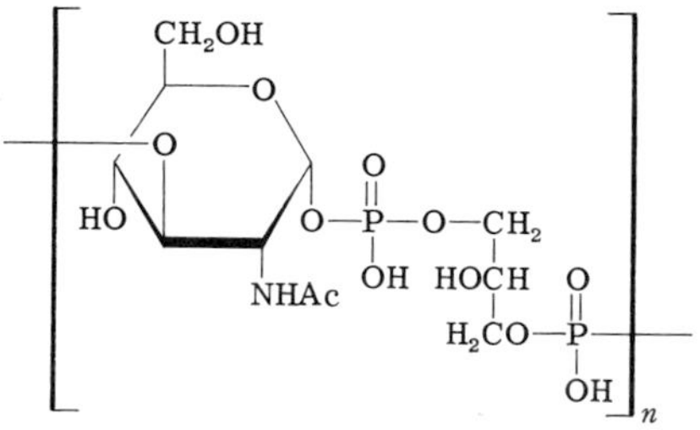

The synthesis of this polymer in a cell-free system has been achieved,[595] and it was found that UDP-2-acetamido-2-deoxy-D-glu-copyranose donates the entire 2-acetamido-2-deoxy-D-glucopyranosyl phosphate moiety intact, and that CDP-D-glycerol donates the glyc-erol phosphate moiety. The presence of "lipid intermediates" in this system has been reported.[596]

Staphylococcus lactis (NCTC 2102) contains a wall polymer that shows still further deviation from the "usual" structure of teichoic acids; in this polymer, the repeating unit is 2-acetamido-2-deoxy-

(594a) A. R. Archibald, J. Baddiley, and D. Button, *Biochem. J.*, **95**, 8C (1965).
(595) N. L. Blumsom, L. J. Douglas, and J. Baddiley, *Biochem. J.*, **100**, 26C (1966).
(596) L. J. Douglas and J. Baddiley, *FEBS Lett.*, **1**, 114 (1968).

α-D-glucopyranosyl phosphate, the units being linked together by phosphoric diester bridges to the 6-hydroxyl group of the sugar. There is no ribitol or glycerol in this polymer. Brooks and Baddiley[597] found that a particulate fraction from this organism catalyzes the synthesis of this polymer from UDP-2-acetamido-2-deoxy-D-glucose; this was the only substrate required for synthesis of the polymer, and labeled substrate was used to show that 2-acetamido-2-deoxy-D-glucosyl phosphate is transferred as an intact unit from the substrate to the polymer. High concentrations of Mg^{2+} and Mn^{2+} were required for optimum activity. End-group analysis during synthesis *in vitro* showed that newly formed chains contain up to ∼ 15 repeating units.

The same authors have also shown[598] that the 2-acetamido-2-deoxy-α-D-glucopyranosyl phosphate group of the "sugar nucleotide" is first transferred to a phospholipid acceptor, producing the hexopyranosyl phosphate–lipid, and that the hexopyranosyl phosphate moiety is then transferred to the growing polymer chain. The system is thus similar to that operating in the biosynthesis of O antigen chains and cell-wall peptidoglycan, which uses polyisoprenyl phosphate as the carrier lipid. However, the identity of the lipid involved in the biosynthesis of S. *lactis* wall polymer is not yet known. The reaction, in this case also, involves the transfer of hexopyranosyl phosphate groups, in contrast to the transfer of glycosyl groups that occurs in the synthesis of O antigen and peptidoglycan. Brooks and Baddiley[597] also studied the direction of chain growth in this system. Pulse-labeling indicated that the chain extension occurs by transfer from the nucleotide to the "sugar end" of the chain, that is, to the end that is not attached to the peptidoglycan in the wall. The results indicated that the biosynthesis of bacterial-wall polymers comprising glycosyl phosphate units proceeds by successive addition of units to the "nonreducing" end of the chain. The mechanism thus differs from that for the synthesis of O-antigenic polysaccharides in bacteria, where transfer occurs to the "reducing end" of the growing polymer.[147,599]

7. Poly[Adenosine 5'-(D-Ribose 5-Pyrophosphate)]

In 1963, Chambon and coworkers[600] discovered that the adenylic acid moiety of ATP is incorporated into a polymer in the presence

(597) D. Brooks and J. Baddiley, *Biochem. J.*, 113, 635 (1969).
(598) D. Brooks and J. Baddiley, *Biochem. J.*, 115, 307 (1969).
(599) H. Hussey, D. Brooks, and J. Baddiley, *Nature*, 221, 665 (1969).
(600) P. Chambon, J. D. Weill, and P. Mandel, *Biochem. Biophys. Res. Commun.*, 11, 39 (1963).

of "nicotinamide mononucleotide" and particulate fractions from mammalian nuclei. A few years later, these workers,[601] as well as two other groups,[602-604] found that the immediate precursor of the polymer is nicotinamide adenine dinucleotide, generated from ATP and nicotinamide mononucleotide. It was also found that the product is a polymer of the 5-O-(adenosine 5'-pyrophosphato)-D-ribose portion (commonly called "ADP-ribose") from nicotinamide adenine dinucleotide.[601-604]

Chambon and coworkers[601] also established the structure of the product as follows. Hydrolysis by snake-venom pyrophosphatase released an isomer of "ADP-ribose," as shown in the scheme. This

"ADP-ribose" unit from
nicotinamide adenine
dinucleotide

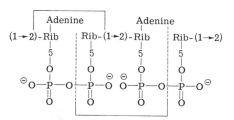

An isomer of "ADP-ribose"
obtained after digestion with
venom pyrophosphatase.

product of digestion was further treated with phosphoric monoesterase, which removed both of the phosphate groups. Hydrolysis of the "ADP-ribose" isomer with dilute acid gave adenosine 5'-monophosphate and D-ribose 5-phosphate. Methylation followed by acid hydrolysis gave both 2,3,5-tri-O-methyl-D-ribose and 3,5-di-O-methyl-D-ribose. These results clearly indicated a structure in which "ADP-ribose" units are joined together by D-ribofuranosyl-(1 → 2)-D-ribofuranose linkages, as shown in the scheme. Thus, this polymer is synthesized by a transglycosylation reaction in which the D-ribosyl group is transferred from the ring nitrogen atom of the nicotinamide

(601) P. Chambon, J. D. Weill, J. Doly, M. T. Strosser, and P. Mandel, *Biochem. Biophys. Res. Commun.*, **25**, 638 (1966).
(602) T. Sugimura, S. Fujimura, S. Hasegawa, and Y. Kawamura, *Biochim. Biophys. Acta*, **138**, 438 (1967).
(603) Y. Nishizuka, K. Ueda, K. Nakazawa, and O. Hayaishi, *J. Biol. Chem.*, **242**, 3164 (1967).
(604) R. H. Reeder, K. Ueda, T. Honjo, Y. Nishizuka, and O. Hayaishi, *J. Biol. Chem.*, **242**, 3172 (1967).

residue onto O-2 of another D-ribose unit. This reaction is, therefore, very different from the glycosyl-transfer reactions that involve "sugar nucleotides," thus far discussed.

It has now been found that the "ADP-ribose" moiety of nicotinamide adenine dinucleotide is also transferred onto some proteins.[605-607] When histone serves as an acceptor, several "ADP-ribose" units are transferred in succession, so that a short chain of oligo-(ADP-ribose), linked covalently to the protein, is formed.[605] In another reaction, "transferase II," a soluble enzyme involved in protein synthesis in mammalian cells, acts as an acceptor of a single "ADP-ribose" unit in the presence of diphtheria toxin.[606,607] Treatment of the product with venom pyrophosphatase releases adenosine 5'-monophosphate, but the D-ribose 5-phosphate portion still remains attached to the protein; it is, therefore, assumed that the linkage involves C-1 of D-ribose.[606] The transferase II that carries the "ADP-ribose" unit is completely inactive, but it can be reactivated by incubating with nicotinamide and diphtheria toxin. Under these conditions, the reaction is reversed, generating free transferase II protein and nicotinamide adenine dinucleotide. Thus, diphtheria toxin was shown to have a very specific transglycosylase activity; the mechanism of this reaction has been studied in detail.[608]

(605) Y. Nishizuka, K. Ueda, T. Honjo, and O. Hayaishi, *J. Biol. Chem.*, **243**, 3765 (1968).
(606) T. Honjo, Y. Nishizuka, O. Hayaishi, and I. Kato, *J. Biol. Chem.*, **243**, 3553 (1968).
(607) D. M. Gill, A. M. Pappenheimer, Jr., R. Brown, and J. T. Kurnick, *J. Exp. Med.*, **129**, 1 (1969).
(608) R. S. Goor and E. S. Maxwell, *J. Biol. Chem.*, **245**, 616 (1970).

AUTHOR INDEX

Numbers in parentheses are reference numbers and indicate that an author's work is referred to, although his name is not cited in the text. Numbers in italics show the page on which the complete reference is listed.

Goor, R. S., 483

Gorin, P. A. J., 171, 173, 188(63, 64), 189(64), 190(63, 64), 191(64, 68), 202, 259(145), 261(152), 273 (174), *276*, 281(10), 282(56), 287 (179), 289(179, 191, 197), 290 (179, 203), *291, 292, 295, 296*, 302, 394

Got, R., 462, 467

Gotelli, I. B., 346

Gottschalk, A., 455

Goudsmit, E. M., 355(25, 26), 357

Gralén, N., 310

Gramera, R. E., 284(98), *293*

Grant, G. A., 199, 200(25), 203(25)

Grant, J. H., 466(532), 467

Gray, G. M., 446, 450

Greathouse, G. A., 320, 322, 324(137)

Grebner, E. E., 434

Green, P. B., 331

Greenberg, G. R., 477

Greenberg, E., 379, 380

Gregersen, N., 177, 181(73), 182(73)

Grewe, R., 172

Grey, A. A., 111

Grimes, W. J., 367

Grinberg, B., 312, 313(93)

Grindley, T. B., 114

Grisebach, H., 354(10, 11), 357, 441

Grollman, A. P., 412, 472

Grollman, E. F., 474

Gromet, Z., 322

Gross, S. K., 419

Groth, P., 110

Grüssner, A., 198, 200(7), 245(7), 248, 264(7), 265(7)

Grützmacher, H. F., 229

Grundschober, F., 260(150), *276*

Grushetskiĭ, K. M., 67

Günther, I., 316

Gütter, E., 313

Guibé, L., 56

Guilloux, E. R., 117

Gump, K.-H., 260(148), 262(148), *276*

Gunetileke, K. G., 354(18, 19), 357, 426(18, 19)

Gupta, P., 221, 236(83), 256(83)

Gurr, G. E., 82(198), 83

Gut, M., 283(75a), 284(108), 286(75a), *292, 293*

Guthrie, R. D., 78, 115, 258(177), 276 (177), 277

Gutowsky, H. S., 90

H

Haas, H. J., 283(88), 289(196), *293*

Hämmerling, G., 417

Haga, M., 166, 272(173), 277, 286 (158a), 290(200a), *294, 296*

Hagopian, A., 458, 459, 460(495), 465, 466(533), 467, 470(489, 495), 471 (495)

Hahlbrock, K., 441

Haines, A. H., 287(182), *295*

Hakomori, S., 447, 472(431, 432)

Halac, E., 440

Hall, C. W., 363, 434, 450, 465, 469 (519)

Hall, L. D., 55, 56, 57, 58, 59, 60(103, 104), 74(48), 88(59, 65), 108 (225), 109(81), 111(66, 81), 112 (66), 116(103, 104), 117(77, 80), 118(294), 119(300), 120(77, 80, 300), 121(77, 80), 122, 124, 186, 218, 222(79b), 226, 287(179a), *295*

Hall, M. A., 325, 388, 389, 391(181), 392(181)

Hallinan, T., 466(532), 467

Ham, J. T., 82(202a), 83

Hamilton, W. C., 69

Hammer, H., 82(198), 83, 110

Hampe, M. M. V., 354(9), 357

Hanack, M., 53

Handa, S., 448

Hanessian, S., 56, 163, 164(48, 50, 51, 52, 53), 165(52), 166(52, 53), 285 (130, 159), 286(162), 290(130), *294, 295*

Hanle, J., 311

Hann, R. M., 234, 252, 285(135), *294*

Hansen, H. N., 82(197b), 83

Haq, S., 367

Harada, H., 303, 304, 305(29), 306(29)

Harada, N., 63, 64(126)

Harada, T., 354(16), 357

Harbon, S., 455

Hardegger, E., 281(6), *291*

Harris, M., 310, 320

Harrison, R., 285(152), *294*

Hart, H., 133

W

SUBJECT INDEX

A

Abequose, residues in lipopolysaccharides, 421

Acetalation
of D-fructose, mechanism of, 218–220
mechanism of, 211–218
of L-sorbose, mechanism of, 205

Acetals
benzylidene, reaction with N-bromo-succinimide, 163–168
cyclic
of ketoses, 197–277
hydrolysis of, 203
preparation of, 199
stability of, 202
cyclic sugar, conformations of, 118
dithio-, sugar, 13
rearrangements of, 220

Acetamido group, effect on acyloxonium rearrangements, 161

Acetic acid–sulfuric acid, cyclitol rearrangement in, 188–191

Acetic anhydride–zinc chloride, rearrangements of saccharides in, 192

Acetohydroximate, 1-thio-β-D-gluco-pyranosyl 2-phenyl-, biosynthesis of, 441

—, 1-thio-2-phenyl-, biosynthesis of, 441

Acetoxonium salts, 130

Acetoxyl group, reactivity of, in acyloxonium salt formation, 131

Acetyl group, determination of, in carbohydrates, 15

Acyloxy groups
neighboring-group reactions of, 128
reactivities of, 131

Adenine, 9-α-D-mannopyranosyl-, hydrochloride, conformation of, 108

Adenosine
3′,5′-pyrophosphate, in marine algae, 404
5′-pyrophosphate glycosyl esters, occurrence and enzymic synthesis of, 356
5′-triphosphate, in cellulose synthesis, 322

Adonose, purification of, 236

Alditols
acetals, rearrangements of, 221
acetates, conformation of, 72
conformational and crystal-structure analyses of, 69
preparation of, 15
rearrangements in hydrogen fluoride, 173–176
—, anhydro-, rearrangements in hydrogen fluoride, 173–175

Aldohexopyranoses
conformational analysis of, 67
conformational equilibria, 91
—, 1,6-anhydro-β-D-, conformations of, 116

Aldohexopyranosides, methyl 4,6-O-benzylidene-α-D-, conformations of, 115

Aldopentopyranoses
anomeric effect and conformational equilibria of, 103, 104
conformation of, 59
in solution, 85
conformational analysis of, 67
conformational equilibria, 91–98
tetraacetates, conformation in chloroform solution, 86

Aldopyranoses
conformation in solution, 84, 85
conformational analysis of, 66
infrared spectroscopy in, 54
nuclear magnetic resonance spectroscopy in, 57
—, 1-thio, conformation of, 87

Aldopyranosides, methyl
cuprammonia complexes in conformational analysis, 64
infrared spectroscopy and conformational analysis of, 54
—, methyl D-
conformation in solution, 84
optical rotation and conformation of, 61
—, methyl 2,3- and 3,4-anhydro-, conformations of, 124

CUMULATIVE AUTHOR INDEX FOR VOLS. 1–26

CUMULATIVE SUBJECT INDEX FOR VOLS. 1–26

539

ERRATA

VOLUME 24

Page 386, line 7. For "perioxide" read "peroxide."

Page 417, Reference 181a. Change 181a to 200a, and move to page 421.

Page 420, line 4. For "benzyloxcarbonyl" read "benzyloxycarbonyl."

Page 421, first line under formulas. For superscript "181a" read superscript "201a."